A HISTORY OF TECHNOLOGY
&
INVENTION

Progress Through the Ages

VOLUME I

Contributors to This Work

CHITA DE LA CALLE, assistant, Museum of Man, Part Five

E. M. CASTAGNOL, agricultural expert, Part Three, Chapter 10: Agricultural Techniques

GEORGES CONTENAU, Honorary Chief Curator, National Museums, Part Two, Chapter 6

MAURICE DAUMAS, Director, Museum of the National Conservatory of Arts and Crafts, General Preface and Part One, Chapter 5 (collaborator)

JEAN DESHAYES, lecturer, Faculty of Letters, Lyon, Part Two, Chapter 6

MAURICE DESTOMBES, corresponding member, French School of the Far East, Paris, Part Three, Chapter 10: Measuring and Determining Time and Space

MAURICE DURAND, Director of Studies, Practical School of Higher Learning, Sorbonne, Part Three, Chapter 10 (collaborator)

PAUL-MARIE DUVAL, Director of Studies, Practical School of Higher Learning, Sorbonne, Part Two, Chapter 9

DANIEL FAUCHER, Honorary Dean, Faculty of Letters, Toulouse, corresponding member of the Institute of France, Part One, Chapter 4

JEAN FILLIOZAT, Professor, College of France, Part Three, Chapter 11

BERTRAND GILLE, Professor, Faculty of Letters and Humanities, Clermont-Ferrand, Part Six

GEORGES GOYON, in charge of research, National Scientific Research Center, former participant in the excavation of Tanis, Egypt, Part Two, Chapter 7

ANDRÉ HAUDRICOURT, in charge of research, National Scientific Research Center, Part One, Chapter 5 (collaborator)

DR. PIERRE HUARD, Professor, University of Rennes, Director, Research Center of Science and Technology of the Far East at the French School of the Far East, Paris, Part Three, Chapter 10 (collaborator)

ANDRÉ LEROI-GOURHAN, Professor, Sorbonne, Part One, Chapters 1, 2, and 3

MELVIN KRANZBERG, Professor of History, Case Western Reserve University, and Editor in Chief of *Technology and Culture,* the bibliographies specially prepared for this edition

JACQUES PAYEN, archivist and palaeographer, Director of Studies, College of Applied Advanced Research, Index and tables

ROBERT SCHRIMPF, engineer, E.P.C.I., Part Three, Chapter 10, Chemistry

JEAN THÉODORIDÈS, Head of Research, National Scientific Research Center, Part Four, Chapter 13

GASTON WIET, Professor, College of France, Part Four, Chapter 12

MING WONG, assistant, University of Rennes, in charge of research, French School of the Far East, Part Three, Chapter 10 (collaborator)

With the collaboration of the Documentation Center of the History of Technology

A HISTORY OF TECHNOLOGY
&
INVENTION

Progress Through the Ages

VOLUME I

*The Origins of Technological Civilization
to 1450*

EDITED BY

MAURICE DAUMAS

TRANSLATED BY

EILEEN B. HENNESSY

JOHN MURRAY

FIRST PUBLISHED IN GREAT BRITAIN
BY JOHN MURRAY
50 ALBEMARLE STREET, LONDON W1X 4BD

ORIGINALLY PUBLISHED AS *Histoire Generale des Techniques*,
UNDER THE DIRECTION OF MAURICE DAUMAS.
© 1962 BY PRESSES UNIVERSITAIRES DE FRANCE

ENGLISH TRANSLATION © 1969 BY CROWN PUBLISHERS, INC.
PRINTED IN THE UNITED STATES OF AMERICA

0 7195 3730 4

CONTENTS

PART TWO

MEDITERRANEAN ANTIQUITY

PART FOUR

ISLAM AND BYZANTIUM

PART FIVE

PRE-COLUMBIAN AMERICA

PART SIX

THE MEDIEVAL AGE OF THE WEST (Fifth Century to 1350)
By BERTRAND GILLE

A HISTORY OF TECHNOLOGY
&
INVENTION

Progress Through the Ages

VOLUME I

PREFACE

THIS WORK attempts to trace a history with which we are still not too well acquainted: the history of the methods that man has discovered and utilized to improve the conditions of his existence. He had at his disposal the natural materials, both organic and inert, found on, or a short distance beneath, the surface of the earth. Once his essential needs had been satisfied, he could have remained indifferent to the variety of these materials and their individual characteristics, as do all the other animals. We are not aware, for example, that a continual effort over several hundreds of millennia encouraged any other animal species to perfect its hunting techniques or manner of arranging its shelter. This was not the case with man. Perhaps the most characteristic indication of his appearance on earth was the birth of this constant need for progress, which has never slackened since the beginning of the Quaternary era. While civilizations are mortal, each one has, before succumbing to its fate, prepared a heritage of which its successors were never unaware.

Technology in prehistoric times

There is no evidence to tell us whether industrial man was born in throwing a stone, breaking a branch, or producing fire. Every civilization known to us, even very early civilizations that are known to archaeologists only through sparse remains, appears to be rich in manual experience. This fact permits us to suppose that they have a very lengthy past. Each one of them was already very old at the time it was creating the products whose remains now bear witness to its existence.

At the beginning of history as such, the utilization and transformation of raw materials had already created an industry; that is, some human beings had already acquired from their predecessors a certain technique to which they devoted all their skill and a part of their time.

More precisely, evidence of industry, in our modern sense of the term, exists in all the civilizations of early antiquity, which are better known to us than those of prehistory. We suddenly find ourselves in the presence of genuine industrial workshops whose techniques remained unchanged during several millennia.

The date of the appearance of these activities is also unknown; only the fact of their presence has been determined. These are not "primitive" industries that permitted only a preliminary emancipation and that gradually made way for more refined techniques (for example, the shaping of stones into tools). The industrial workshops of the early ages of history are those of the future — for our age and future ages will continue to spin, weave natural fibers, fire clay vases, and work metals.

1

We shall never know when or how they were invented; nor do we know where they were born. While we may be correct to a certain extent in thinking that from the end of the fourth millennium until the first millennium Central Asia was probably an important center for the dissemination of technology, many other areas on the face of the earth played a similar role, both then and later. Southeastern Europe and Asia Minor transmitted early metallurgical traditions to the Mediterranean peoples; the southeastern Mediterranean area in turn exercised a decisive influence on both Asia and Europe during the second and first millennia. From the dawn of history, industrial activity was in evidence in other regions (eastern and southern Asia, South Africa, and all of western Europe), but it is not possible to establish definite relationships among them.

When we are able to begin written history, we realize that very different peoples not only possessed approximately the same technical knowledge but also that they established relations with one another in order to exchange information about methods of manufacturing, finished products, and especially raw materials. Industry broke upon the world like a mold whose spores are carried by the wind. It is impossible to claim, as has been attempted, that it is the creation of a particular area. On the contrary, the facts at our disposal seem to indicate that invention appeared simultaneously in various areas.

The circumstances of invention

Simultaneity in the invention of metallurgical techniques, for example, or in that of pottery-making techniques, may have lasted from several millennia to several centuries. Such periods are comparable to the forty or fifty years required for the establishment of a new industry between the seventeenth and nineteenth centuries, to the ten to fifteen years required at the end of the nineteenth century, and to the three or four years necessary in the second half of the twentieth century.

Another feature that has remained unchanged down to modern times is that invention is never the product of a single man, but rather that of a period and a society; it is born in definite historical circumstances. This explains its simultaneity, since the circumstances required to make a given invention effective can be realized within a very short time interval in several places between which there was no exchange of information on this subject. It is not often that alleged precursors actually were the initiators of a given invention.

Throughout history, in fact, it is possible to discover antecedents for most technical improvements, and they are generally errors. For example, it can be claimed that the Greek mechanicians of the Alexandrian age were the precursors of automation. In their books (insofar as these are known to us) occur numerous descriptions of ingenious methods for moving inanimate objects without apparent human intervention. We do not know whether these methods were put to use, but we are certain that the devices created (if they were) could not have produced satisfactory results. The artisans of that time had neither the materials nor the tools nor the professional traditions that would have permitted them to construct workable machines on these principles. Not until fifteen centuries later were some of these machines produced by manufacturers who had not read the Greek authors. The same could be said for the devices noted by Leonardo da

Vinci in his notebooks, which we now attempt to regard as inspired forerunners.

The desire to build machines, vehicles, and domestic devices that would make certain work less fatiguing or daily activities more pleasant has always produced dreams and unfinished projects. For one of these projects to become an invention, the group of techniques involved must have reached a stage of perfection sufficient not only to ensure the creation of the new device but also to enable its creation to represent a profitable acquisition for a large number of individuals.

The same is true for the perfecting of machines or their application in fields different from the ones for which they were originally conceived. Technical progress is not a child of invention alone; the continuous evolution of the methods acquired also contributed to it. This evolution was subject to the same restrictions. Thus certain traditional tools of the manual crafts scarcely changed in form for ten or more centuries, because they were perfectly suited to the requirements of these crafts. It was only when new materials, and in particular certain grades of steel, came into existence that they were modified. Beginning in the first half of the seventeenth century, clockmakers created a special set of equipment, made of iron and brass, for their own use. Some of these tools (lathes, for example) may have served as models for the perfecting of carpenters' and cabinetmakers' lathes, and for the construction of lathes for the working of large metal objects. However, for several centuries the early artisans continued making their lathes of wood, which was their customary material. As for the mechanics' lathes, they came into use only at the end of the eighteenth century; the oldest we know, that of Jacques de Vaucanson (which dates from around 1763), has an exact replica of the cage of a clockmaker's lathe, enlarged to suitable proportions. Until this period iron, which was still relatively rare, was not used in the construction of machines.

The collective experience and its transmission Whether it is due to progressive evolution or to invention, the development of technical methods is the result of a collective experience that is constantly being accumulated. Each generation continues to inherit the experience of all its predecessors; in the field of technology, progress is a sum total. In contrast to what we have seen in the history of the sciences, for example, new efforts in technology were not hesitant, or at least were much less so. In fact, technology seems to have followed a continuous ascending curve without being forced to propose uncertain solutions. It seems that it did not even experience periods of stagnation or regression. (This naturally supposes that we are considering humanity as a whole, not limited groups of peoples. Certain groups, while they enjoyed stabilized technical levels over a period of several centuries, always profited in the long run from the experience of other groups, without causing any retardation in the latters' technical progress. This lesson can be learned from the twentieth century.)

For only slightly more than two centuries has technical knowledge been transmitted by methods other than word and deed. Before acquired experience began to benefit, in modern times, from general methods of dissemination, it was transmitted only by individual instruction, from the earliest ages down to approximately the century of Louis XIV. Beginning in the sixteenth century, treatises and encyclopedias were printed in large numbers, but for a long time they were novelties rather than textbooks, for only a limited public. Moreover, one century more or less makes no difference in the proportion of time between all the preceding millennia and the very short period during which the book was a factor in technical progress.

During the ages when direct contacts were indispensable, the rapidity of progress continued to be linked with the frequency of these contacts, that is, with the demographic development of humanity. We have no idea of the density of population of the regions inhabited during prehistoric times. We have reason to believe that at first it was very sparse. Aside from the difficulty of creating the first tools from nothing, the dispersion of the human race suffices to explain the slowness of these early stages. The miracle is that they were accomplished, and then everything became possible. In the course of this work, we shall at once see that from the ancient empires to the civilizations of late antiquity the development was already evident; it is already apparent that technical activity had benefited from the much larger number of participants. While Europe remained for several centuries at the level it had reached (and then only with difficulty) as a result of the great invasions, technical progress was rapidly continuing in the Far East.

This rapid progress of China and her neighbors was to certain extent beneficial to the Occident. Ultimately, however, it was stunted. It did not progress beyond a certain level of civilization that was highly developed in all areas, each one of which formed the parts of a homogeneous ensemble. Once it had arrived at this point, for several centuries the Far East failed to realize the need for progress. Its techniques remained stationary at a level that the Occident reached and began to surpass in the course of the seventeenth century.

The adventure of Western technology is very different. The period of stagnation that followed the fall of the Roman Empire seems to have lasted no more than two or three centuries. Demographic development indisputably influenced technical progress until the middle of the following century. We shall see that the crafts were perfected and became more numerous; then the first machines appeared, increased in number, and were constantly improved upon as consumer needs grew. The medieval period was a time of continuous improvement, a period that witnessed the appearance and development of those great collective inventions whose genesis is still very obscure, ranging from water mills and windmills to the manufacturing of paper and eyeglasses, from the harness and navigation of oceans to the exploitation of mines.

The pressure of demand alone was not enough to accelerate technical progress. A steady increase in the number of technicians was also necessary, and this was made possible by the demographic expansion of Western Europe. Even during periods in which knowledge was still transmitted by traditional methods, the fact that the workers and their "teams" were becoming increasingly numerous set a new rhythm for the development of technology. While we do not have exact information on the rate of this development from about the tenth to the sixteenth centuries, we can follow it with greater precision during the three succeeding centuries.

The number of technicians and the acceleration of progress

We note that the activity and efficiency of the "technicians," like their number, were constantly increasing. This idea of the numerical influence of the protagonists on the rhythm of technological progress has always been neglected. Yet this may well be the cause of the spectacular achievements of our age, the rapid succession of which astonishes our contemporaries and gives them a feeling of unquestionable superiority over earlier generations. Actually, each modern technician, considered individually, possesses no more professional talents or qualities than even his earliest predecessors. One hundred men studying the same problem at the same time obtain much greater results than a single man devoting himself to the same work for a period one hundred times as long. Moreover, technological progress itself acted as its own stimulant; it continually created (and is still creating) improved methods favorable to its acceleration.

Naturally, many factors other than those mentioned here intervened in favor of technological progress. Commercial and industrial organization and financial activities have evolved along with technology, and all these phenomena have always been closely linked in quite complex fashion. They have been the object of numerous studies dealing with every historical period. Very often, in fact, the history of technology has been described only in terms of these activities. The political, military, and social events that for a long time constituted traditional history were without question closely linked to other events throughout the ages. Here we need only mention these circumstances; they have been frequently studied elsewhere.

The relations between technology and the sciences

For various reasons we shall investigate more thoroughly the effects of scientific progress on the development of technology. It is traditional to consider the latter as subordinate to the former. Only in the last century, however, has science exercised a strong influence on technology, whereas technology has been offering scientists subjects for research probably since the beginning of human thought.

For more than twenty centuries the relations between science and technology, in the sense that interests us here, remained fragmentary. They undoubtedly began with the early contributions of astronomy and arithmetic. The great scientific activity of the century of Pericles was not expressed in any appreciable gain in technology. While the expansion of Chinese thought was accompanied by the proliferaton of highly developed techniques, scientific thinking of that period contributed practically nothing to these techniques.

The builders of the medieval cathedrals apparently derived nothing from the mathematicians, in a period when navigation and medicine had scarcely begun to make use of the discoveries of science. Toward the end of the sixteenth century the contribution of science became more evident; the most notable example is the application by Christian Huygens of the isochronism of the oscillations of the pendulum, discovered by Galileo, to the regulation of clocks. However, this is an isolated example. Men had been constructing compasses for a long time before the first modern study on magnetism appeared, and William

Gilbert's *De Magnete* was of no use to navigators. Similarly, the problem of the calculation of longitude on the open sea resulted in the establishment of the Greenwich Observatory, but it was not completely solved until around the middle of the eighteenth century when clockmakers, utilizing only their own resources, learned how to construct satisfactory chronometers.

Matters continued in this fashion until around the middle of the nineteenth century. The steam engine had been in operation for approximately seventy years before an attempt (which succeeded only fifty years later) was made to establish a theoretical explanation of its function. Similarly, the construction of machine tools preceded the theoretical works of the nineteenth-century mechanicians; the production of mineral acids, Lavoisier's chemical system.

Only at this stage did exchanges of information between these two areas of activity begin to balance each other. The chemical industry very soon benefited by the discoveries of the organic chemists of the first half of the nineteenth century; electrochemistry and the electric telegraph appeared very shortly after the works of Alessandro Volta, Humphrey Davy, André Ampère, and Michael Faraday, but electrotechnology had to await the inventions of Antonio Pacinotti and Zénobe Gramme toward the end of the century. Metallurgy, the expansion of which after 1850 encouraged the beginning of the contemporary rhythm of industrial production, received a vital stimulus from scientific research only toward the end of the century.

The traditional forms of production, however, did not disappear easily under the pressure of modern techniques. The windmill and water mill remained in use almost until our own times; André Leroi-Gourhan recalls in this book that it is not such a long time since cutters of gun flints disappeared. Naturally, we are referring here only to the numerous examples of coexistence among the highly industrialized groups of peoples. These ancient techniques have, of course, long since ceased to be generators of progress. They themselves had achieved a stage of perfection they could not surpass, as happened in the case of Chinese technology in general four or five centuries ago. In this period, then, it is by the creation of new techniques, often inspired by science, that progress continued to accelerate.

Very often technology is still the initiating factor. The first internal combustion engines operated without the assistance of thermodynamics; the first airplanes flew without the help of aerodynamics. The science of radio electronics was born after the first radio broadcasts, and electroacoustics after the recording of sounds, while the first plactics owed nothing to the theories of chemical synthesis.

What has just been said in no way detracts from the merits of scientific research and its creative power. It was only during the first half of the nineteenth century that this creative power began to influence technology, and only since the beginning of the twentieth century has science become in turn the principal factor of innovation, but its influence had already been decisive. Even in cases where it did not play this role, it was an increasingly powerful auxiliary of technology; the field opened up by invention could not have been exploited as rapidly and efficiently, using only its own resources. By reversing the random examples mentioned above, we are able to understand the precise nature of the relations between science and technology. This aspect of contemporary technical progress is of such primary importance that we must examine it more thoroughly in another volume of this work.

The history of technology as a recent discipline

To recall, even briefly, the circumstances under which our civilization has developed, it was necessary at the outset to dispel certain illusions cultivated not only by the public at large but also by the majority of technicians. We have already mentioned how seriously most people misjudge the importance and the quality of the contributions of the past. This attitude and misjudgment are at the origin of the admiration and fear with which most of our contemporaries view their own achievements. Only a thorough knowledge of the history of technology can completely reconcile them with their era. Among the various historical disciplines, the history of technology is one of the youngest. The collection of facts it has gathered and verified is not yet very extensive. Frequently there are gaps in our knowledge.

While the history of technology has often been studied, most of these studies, as we have already noted, are the work of economists and sometimes sociologists. By determining the facts that are proved, and by providing the technical information necessary for understanding step by step the nature of various perfections and inventions, this study hopes to promote the history of technology itself. By making a method of general, basic information available to a relatively large public, such a work will be able to arouse new interest in a developing field of investigation.

The nature of this book

We have begun publication of precisely this work with the present volume, which is to be followed by three others. The subject was admittedly a vast one, and it seemed necessary above all to devote all available space to the description of techniques and their development. This aspect of the history of technology is the one most often neglected—and the one on which all complementary studies (particularly those that deal with the various factors that determined the development of this history) should be based. Thus the political, social, and economic context of technological progress is mentioned only when it is indispensable to do so.

We were equally selective in our definition of the term "techniques." Today the general connotation given to this word is the same as that of the word "arts" in the eighteenth century. In order to remain within our limitations we considered as "techniques" only those human activities whose object it is to collect, adapt, and transform raw materials in order to improve the conditions of human existence. Thus such activities as, for example, accountancy, banking, and the conduct of military operations are outside our subject. We do not deal with language, but with the methods of transmitting, recording, and writing it — paper, the proliferation of written texts, and so on.

While remaining within rather modest size limitations, considering the vast subject matter, the work will be within the reach of a larger number of readers. Had we gone beyond these limitations, this would have become a work of the encyclopedic type in which specialists verify the preliminary factors of their research or check the novelty of their results.

Chronologically the material has been divided as follows:

After a discussion of what is known of the ages that preceded the first historical civilizations, we have briefly treated of the latter, taken by geographical zone, approximately in the order as they are usually presented by history, describing first the technological levels reached by the peoples of Mediterranean antiquity (including the Romans), then those of the Byzantine Empire and Islam. Thus this first volume is devoted to "The Origins of Technological Civilization." Then follow chapters dealing with the Far East, southern Asia, and pre-Columbian America, thus grouping the civilizations that did not directly flower into the Occidental industrial era. We are beginning to realize that the medieval period was an extraordinary time of transition. Far from being unproductive, it witnessed the fruitful development of invention, bringing to a successful conclusion the sometimes divergent and always slow, tentative efforts of the early ages of technological civilization. The last portion of this volume shows how during the medieval period the rhythm of progress was established on a different footing.

The beginnings of the modern industrial era are described in the second volume under the title "The First Stages of Mechanization." A classical chronological division of general history, the Renaissance, constitutes the first part of this volume, and discusses the general technological level attained by Western Europe during this period. The remaining sections are devoted to the period extending from the end of the sixteenth to the beginning of the eighteenth centuries, after which geographical and chronological divisions are abandoned for a plan of subdivisions based on major techniques or groups of techniques. This subdivision is better suited to the character of the past two centuries, the technological history of which is traced in the following volumes. Without neglecting the interdependence of development of the various technological areas, it permits us to follow the line of evolution of the most important regions, an evolution that is becoming less and less the exclusive property of a single country or even continent.

In this volume the expansion of mechanization (the subject of the third volume) will be studied from the beginning of the eighteenth to the middle of the nineteenth centuries. This period witnessed the phenomenon that, beginning in the eighteenth century and especially after the middle of the nineteenth, various authors began to call the "Industrial Revolution." While the term has a certain significance for economic history, it has none for the historian of technology. Out of respect for a commonly held notion, we have been obliged to introduce the idea of a second industrial revolution, which certain authorities place in the first half of the nineteenth century, others in the second. It would be necessary to situate three or four additional "revolutions" in the first sixty years of our century.

Actually, the great rupture in the methods and rhythm of industrial production seems to have occurred around 1850–1860. It is this period that we have chosen as the dividing line between the last two volumes. "The Industrial Civilization," the technical foundations on which it was gradually erected, and the rhythm of the development of technology during the past century, will be the subject of the fourth volume.

This work required such a wide variety of knowledge that it could be achieved only by collaboration. We have called upon the most competent specialists for each chapter. The methods of study include those of the prehistorian, the archaeologist, the archivist, and the research worker, and in all cases these must be complemented by the studies of the technician as such. Depending on the subject and the period studied, the historian of technology is plagued either by the mediocrity or rarity of his sources or, on the contrary, by their abundance. While technical literature became so abundant, beginning in the middle of the last century, that it is often impossible for the historian to accomplish an exhaustive search, the available original sources on such ancient civilizations as those of Mesopotamia, the medieval Arabs, or ancient India have until now been hardly (and sometimes not at all) studied by the historians of technology. Thus every chapter in these volumes represents an original work rather than a synthesis of earlier works.

It would be impossible to conclude this general preface without thanking all the specialists who agreed to collaborate on the four volumes of this work. Each of them understood the interest it presents for the history of technology as well as for general culture, and almost all of them agreed to add this supplementary task to their already demanding work. To the writing of the text as such was added the never very appealing obligations of adapting this text to the requirements of a collective work, and obtaining for the editor the indispensable pictorial documentation — for a history of technology is inconceivable without illustrations. We have tried whenever possible to show the reader the object in question, for a brief description can give only a vague idea of it. Even technicians do not always understand what is being discussed if they do not have a picture before them, and we had in mind the readers of varying educational backgrounds whom we wished to interest in this work.

The editor understood this necessity perfectly, and made available to us the means necessary for an abundant illustration. Lastly, he had sufficient confidence in the authors and the director of the work to permit the doubling of the number of volumes in the course of the preparatory work. I wish to express my thanks to him for this understanding — the best possible form of encouragement.

MAURICE DAUMAS

PART ONE

BIRTH AND EARLY DEVELOPMENT OF TECHNOLOGY

INTRODUCTION

LOOKING CLEARLY at the birth and early development of technology means abandoning the conventional notions of history. In their place a system of references must be adopted from which not only the ideas of inventor and date of invention have disappeared but also in which the fact of invention itself is directly dependent on biological concepts. The reassuring perspectives of the old authors as they spoke of the first man who stole fire from a volcano or the inspired hunter who entrusted a part of his stock of seed to the earth to make it fruitful have been replaced by the impersonal development, against a background of geologic ages, of an evolution of implements that cuts across the boundaries of several forms of humanity different from ours, as if it were unconscious of them.

It is evident to the historian of technology that the inventor is inseparable from his time and his social context and that he acts as a spokesman for a civilization. In historical time the accent is placed on a name, a place, a date, even if they are apocryphal, whereas in the vastness of prehistoric times it is the makers of chipped flints, spread over three continents, who blaze the interminable paths of the Quaternary period. Owing to acquired culture, we continue to speak of a "first man," when we know only of anthropoids on the margin of the animal kingdom, and when the idea of "first" is opposed to the imperceptible linkage of forms.

Technology and the prehistory of the human race

Despite the unusual nature of its system of reference, prehistory is nevertheless primarily a history of technology. The temporal succession of increasingly sophisticated objects forms the only link between the hundreds of thousands of years of the infancy of the human race. Paleontology is not in a position to determine the human or animal character of a supposedly human skull; the presence or absence of handmade tools is the determining factor. During the century in which we have been discovering increasingly distant ancestors of contemporary man, anatomy has hesitated to see human beings in the creatures with low brain pans and heavy jaws that it has discovered, and it is only because they left flint implements that Neanderthal man, *Sinanthropus,* and more recently *Atlanthropus* have come to be included among our ancestors. Actually, the greatest problem arises with the australopithecines of South Africa. The anatomy of these creatures indicates that they stood upright, had hands like ours, a face shorter than that of the large monkeys, and a more developed brain. But we do not yet know whether they were capable of making tools. A fortunate discovery in 1960 led to the recognition of several crudely chipped stones near the bone remains of an australopithecine, adding a new chapter (and certainly one of the first) to the history of the human race.

The curve of humanization The numerous anatomical characteristics that distinguish our ancestral anthropoids from the other major zoological groups form a unified picture. The importance of the cerebral equipment, upright posture, mobility of the hands, and the reduction of the face and front teeth are closely linked in their evolution, and are characteristic, in varying degrees, of contemporary *Homo sapiens* as well as of Neanderthal man, *Pithecanthropus,* and the australopithecines. It is for this reason that we include these various representatives of humanity under the general term "anthropoid," distinguishing arcanthropines (the oldest group, which includes *Pithecanthropus, Sinanthropus,* and *Atlanthropus*), Paleolitic man (forms related to Neanderthal man), and the *Neanthropinae* (the present human race, beginning with Cro-Magnon man). We possess tools made by all these anthropoids, and are aware that the evolution of these tools forms a continuous, progressive ensemble. The birth of the tool is obviously linked with an adequate development of the brain, or at least of certain of its parts, but above all with the liberation of the hand through the acquisition of erect carriage. Thus the fundamental problem of our material evolution consists in the search for methods of freeing the hand. Still more precisely, the "curve of humanization" could be expressed by the transition from animal forms, in which organ, motion, and tool are combined, to human forms, in which the tool is separated from the organ and the movement. Then mechanical forms are evolved, in which the movement itself has become detachable.

THE HAND

The prehensile mammals The history of the human hand originates in the history of the first vertebrates. Primitive technological activity is linked with the acquisition of food. Beginning with the primitive fish of the Paleozoic era, a "technological field" is established in the forward portion of the body. The principal organs are the facial organs — the lips and the front teeth. In certain species the front fins were very soon incorporated into the technological field, and beginning in the Mesozoic era the vertebrates were divided into two functional groups. In one group the facial movement remained almost exclusive, and its organs became more and more specialized. In the other group the frontal technological field was divided between the facial center and a manual center in which the front limb was oriented toward gripping. In the Tertiary epoch, with the appearance and development of the mammals, the division became more and more clearly defined between animals like the ruminants, whose head, with its horns, prehensile lips, and cutting incisors, combined all the methods of technological action of the foreward field, and the carnivora and the rodents, in which the hand played an increasingly active and varied role in the acquisition and handling of food. Thus technical ability as it exists in man was present in rudimentary form in the series of the prehensile mammals.

The squatting position — In some of them, particularly in the rodents, manual ability was facilitated by the adoption of the squatting position, which frees the hand while the animal is in repose. Some mammals, for example beavers, achieved great precision in manual operations: the possibility of opposing one finger to the others, and of using the hands separately. The monkeys went a step further: their hand acts by means of a prolonged gripping action, when seated or as when moving in the trees. Among the most highly developed groups, particularly the large anthropoid apes, face and hand balance each other in their technological operations, the sitting position assuring a complete and prolonged straightening of the spinal column. The motor centers of the cerebral cortex have acquired an importance in proportion to the diversity of the movements of the face and hands, and the temporary use of sticks to reach distant objects completes the highest level reached by the monkeys.

Bipedal locomotion — We do not yet know exactly what were the conditions of the transition to the human level, but for some years now we have had inklings of the existence, at the end of the Tertiary epoch, of primates among whom bipedal locomotion was to play an important role. We are also aware of the existence, in southern Africa, of beings known as the australopithecines, who realized in almost ideal fashion the first stage of humanization. They walked erect like human beings, had a short face with small canine teeth, and their hands were completely mobile when they walked, all of which are specifically human characteristics; their brain, however, fell far short of what is attributed to a human brain. They nevertheless showed a remarkable development of the frontoparietal regions, in which the centers of coordination of movements, and particularly movements of the hand and the face, are located. These creatures, who lacked offensive canine teeth and had free movement of their hands when standing, certainly were not intelligent in the modern sense of the term. Undoubtedly, however, they already possessed psychomotor abilities that enabled them to complement the inadequacy of their natural equipment by means of objects held in their hands. Physically, in any case, they were indisputably human beings. Certainly one of the most unexpected aspects of recent discoveries is the revelation that man's technical organs were perfected well before the development of his brain was complete.

The tool precedes intelligence — By the following period the situation was defined in terms that make long paleontological explanations unnecessary: the existence of tools is now certain; their development illuminates the early technological history of the human race. Several factors, however, are of immediate interest for technology. It is quite paradoxical that the hand preceded the brain; it is equally paradoxical that in a certain sense the tool preceded intelligence. Related discoveries, in the Indian Archipelago, China, and North Africa, of beings belonging to the *Pithecanthropus* group, formerly considered by paleontology simply as crude intermediate forms between man and the monkey, have revealed beautifully made flint implements unquestionably superior to everything that a mere consideration of their skulls could have led us to suppose. This seeming paradox arises from two series of convergent facts. We consider our technological activity as being closely connected with the highest forms of our intelligence — something that is true only for man, and even then only partly true. Technical ability, even when rationally employed, far from exhausts the cerebral possibilities possessed by contemporary man; on the contrary, it requires the intervention of neuromuscular equipment the existence of which is possible at a very early stage of human evolution. What surprises us in the technical ability of *Pithecanthropus* is comparable to that which surprises us in the technical behavior of certain birds and rodents, and the traditional philosophical distinction between *Homo faber* and *Homo sapiens* may possibly correspond to a paleontological reality. Moreover, by the fact that they survive their maker, man's technological inventions tend to form a reality endowed with an existence of its own. The foundation of that reality is no longer individual but social. On it the human community acts independently of its successive generations. In the course of millennia the collection of tools, whose production is transmitted from one generation to the next, almost imperceptibly acquires minor improvements. The accumulation of these tools results in more complex forms that could not have been developed by any individual left to his own intellectual resources. To a certain extent we may suppose that the individual *Pithecanthropus* was already surpassed by his own technology.

The appearance of language

The progressive surpassing of one innovation by another raises one final question: that of the appearance of language, an indispensable foundation, in the anthropoids, for group existence and the handing down of complex techniques. While stone implements have been discovered, we shall probably never find direct evidence of verbal transmission, but the coordination of atomic and technological evidence makes possible a hypothetical approach to the subject. The minimal cerebral equipment of the primates ensured the coordination of the movements of the facial organs and the hand: the monkey prepares its food by means of carefully coordinated movements of the teeth, lips, tongue, and hands. In the transiton from the monkey to man, the cells of the cerebral cortex were multiplied, and their growth was as great for the facial organs as it was for the hand, although the face lessened in volume. It can be supposed that along with the modification in manual operations that led to the birth of the tool, a comparable modification in facial operations led the lips, teeth, and tongue to establish verbal tools. The early link between the two poles of technical capacity, uninterrupted throughout the animal series, retains an equally imperative value in the human race, and there exist numerous reasons for supposing that the verbal capacity of the first human beings was comparable to the tools they created. Moreover, such a supposition is necessary in order to reconstruct the fundamental conditions of the development of technology that correspond, throughout the history of the human race, to the development of social structures. From its birth, technology lived a life independent of that of its individual units, equal to the durability of social institutions.

PREHISTORIC EVOLUTION OF TECHNIQUES

The most striking phenomenon in the evolution of techniques is their accretion. Until the disappearance, hardly a generation ago, of the last gun-flint cutters, all the progressive achievements of the human race were still in existence. To consider only the field of pottery, all forms of production, beginning with the most primitive, can still be seen in the modern world. The gathering of mussels on the rocks coexists with harvesting combines, and wrestling coexists with atomic missiles. Consequently it is difficult to establish systematic divisions between the successive waves of the human race. Only one seems really decisive: that marked by the appearance of agriculture. Several hundreds of thousands of years led from the techniques of the hypothetical pebble culture of the australopithecines to the achievements of the last reindeer hunters; five thousand years mark the transition from the sowing of the first corn to the contemporary world. This fundamental division will serve as a framework for the separation of the *primitive societies* from the *protoagricultural* and *early agricultural societies* that forms the very brief preface to the time of written history.

The order of presentation of the various techniques within these fundamental divisions cannot be completely chronological; each technique has continued its evolution, from its birth down to modern times, by profiting from the general technological achievements of every age. Nor can it be completely logical, for the relationships among production, acquisition, and consumption vary greatly from primitive to agricultural societies.

Taking this dualism into account, our plan sets up the following divisions:

Primitive societies. The decisive step in humanization is not the acquisition of game or plant life, nor is it the utilization of shelter — possibilities that already exist at the animal level — but the production of artificial methods of action. Thus we shall consider in succession *fire, stone, wood, hides,* and *skins* among the techniques of manufacturing; among the techniques of acquisition, *hunting, fishing,* and *food gathering.* These techniques, which are found in primitive societies, continue to exist in agricultural societies; they are influenced by technological progress in general, but remain sufficiently constant to justify our discussing them in the section devoted to agricultural societies.

The techniques of consumption — *food, clothing,* and *lodging* — will be treated separately for primitive societies and agricultural societies, since this area presents particularly characteristic distinctions between the two types.

Protoagricultural and early agricultural societies. Agriculture on the one hand, and the metallurgy of the Mediterranean peoples at the end of the prehistoric period on the other, are treated in other chapters of this work, and in this section we shall make only the most indispensable references to these activities. However, in view of the fact that the entire evolution of contemporary societies is based on the two major factors of agriculture and metallurgy, it will be absolutely indispensable to consider them in the societies that paved the way for the historical period.

Techniques of acquisition combine the most significant factors of the evolution of agricultural societies; therefore *protoagriculture* and *agriculture, protopastoralism* and *pastoralism* will be discussed first. Permanent settlement on the land and its economic and technological consequences for early agricultural societies then require the examination of questions regarding the *dwelling place* and its *structure*. For it was the acquisition of agricultural techniques and a fixed place of abode that led to the development of new techniques of manufacturing (*basketmaking, weaving*), and of the techniques of fire (*pottery* and primitive *metallurgy*), that marked the beginnings of contemporary societies.

CHAPTER 1

PRIMITIVE SOCIETIES

FIRE

FIRE HAS ALWAYS been among the principal criteria of the existence of human beings. The myth of the conquest of fire is found in most mythologies, and as soon as the idea that man was born in a very distant past became common, hypotheses on the discovery and domestication of fire were formulated. Volcanoes, lightning, and spontaneous forest fires are usually considered as the sources where the first fire was captured, and it is believed that its preservation then became, during long millennia, the object of primitive man's jealous protection.

The origin of fire among human beings Although it is impossible to discover information likely to contradict this story, which still has a place in contemporary popular literature, there are few definite facts known about the origin of the use of fire by human beings. The truth is that, except for quite recent periods, we cannot hope to find remains of hearths outside of the caves that were inhabited by human beings; elsewhere their traces have been swept away by water and wind. Geographically speaking, however, caves used as dwelling places are very unusual, and in addition they were frequented by human beings only in regions and at periods when they offered a refuge preferable to every other. Added to these difficulties is the fact that only ashes or charcoal, or stones reddened by contact with the glowing hearth, can be found; the methods that made it possible to preserve or obtain fire have disappeared, and it is only through historical sources or the example of contemporary primitive peoples that we can imagine the various methods by which fire was obtained from the rubbing together of two pieces of wood. Artificial production of fire is certainly very ancient, but owing to lack of evidence we can consider only the presence or absence of fire in the few very early sites that were favorable to its preservation. Thanks to geological circumstances, however, we possess certain valuable evidence.

Sinanthropus and fire The australopithecines, whose remains have been discovered in caves but whose implements are still a subject for hypothesis, seem not to have known fire, for no charcoal remains have been discovered in the conglomerate that includes bone remains. Among the four known varieties of archanthropines, *Pithecanthropus* of Java, *Atlanthropus* of Algeria, and Heidelberg man in Germany have been found without any traces of dwellings, but the fourth — *Sinanthropus* of the Peking area — was discovered in a cave. The bone remnants of this very early hominid, the frontal area of whose brain was still very small, were found together with abundant flint implements and remains of a variety of game animals. A very long occupation of the Choukoutien cave by *Sinanthropus* is indicated by the presence of large hearths in a number of strata — a very important piece of evidence. Thus it is certain that at least one of the representatives of the oldest group that manifests human technological behavior possessed fire and utilized it regularly. It is not possible from available evidence to say whether he used it only to warm himself and frighten away wild animals or also to cook his food. It is still more difficult to say whether he knew how to make it; nothing indicates that there was a volcano nearby. On the other hand, *Sinanthropus* occupied the Choukoutien cave periodically for thousands of years; the successful preservation of fire, and the search for it when it was lost among neighbors who were certainly very distant and probably hostile, raises difficult problems. But its possession alone already constitutes evidence that the extraordinarily primitive anatomy of its possessor renders truly astonishing.

Neanderthal Man The interval separating the testimony of *Sinanthropus* from the later evidence includes a major part of the Quaternary era, for no evidence is found in caves until the almost recent age of Neanderthal man, between 100,000 and 40,000 years ago. Although very different from contemporary man, Neanderthal man (traces of whom are found in several hundreds of sites) led an existence comparable to that of the last Fuegians. His tools were quite varied, and it is certain that he had a religious life. Traces of his hearths can be found wherever geologic circumstances are favorable. It was recently realized that he gnawed the spongy, marrow-impregnated tissue of certain bones, an operation that is possible only if they have first been roasted; this fact proves that in this period fire was used for cooking purposes. Though we do not know whether Neanderthal man made his fire, given the advanced state of his other techniques it is probable.

The oldest hearths Toward 30,000, at least in the Occident, there appeared the first cultures with human beings similar to us, beings whom for the sake of convenience we have named after their first known representative, Cro-Magnon man.

It is arbitrary to consider as one unit the early Chatelperronians and the late Magdalenians, who were as different from each other as the Gauls were from modern Frenchmen. Nevertheless, the most primitive among them already possessed technological equipment that had nothing in common with that of Neanderthal man, and as regards fire we may lump together for the period of the Reindeer Age what very little evidence we possess. The truth is that although hundreds of hearths have been discovered in caves, the scientific inadequacy of most of the excavations has resulted in the neglect of structural details that could have furnished us with information about the manner of their preservation and use. Nothing in an examination of the tens of thousands of chipped flints that have been discovered from this period leads us to suppose that flint was used to produce fire: the flint tinderbox requires the use of a piece of metal, or at least metal pyrite, and while we sometimes find the latter in prehistoric sites it does not seem to have served this purpose.

The combustible material used was not limited to wood, for fragments of bones that had been broken to extract the marrow, and spongy bones such as vertebrae, were burned in the hearth. The construction of the latter was most often extremely simple; sometimes it was surrounded by a ring of stones. Until now no device justifying the supposition that utensils were placed on the fire has been found. Further, neither pottery nor metal existed at this period. It is possible, on the other hand, that liquids were warmed by means of stones that were first heated in the fire. This method was still being used quite recently by several American peoples, but no positive proof of it exists: such proof would require careful study of a large number of hearths. Occasionally a hearth appears to have had a damper. One such was discovered in the Ukraine, on the site of a hut constructed on the loess: two small channels dug in the loose soil carried fresh air under the hearth.

Culinary use of Reindeer Age hearths is suggested by the use of bone remains as a combustible material, but it is difficult to find direct evidence. In this regard the dwelling sites discovered in the loess of central Europe and the Ukraine furnish more details than the flooring in the caves of western Europe, in which the shifting of the gravel quickly effaces all traces of the layout of the hearth. In the Ukraine there have been several discoveries of hearths that were surrounded by holes about the size of a cooking pot; by lining the bottom with hot coals these holes could have been used as ovens for steaming.

The Magdalenians utilized a quite remarkable method of heating, which consisted of covering red-hot coals with a pile of large rocks that conveyed heat. This structure was discovered intact in the rock shelter of Mouthiers (Charente) and the grotto of Saint-Marcel (Indre). Beginning with the Chatelperronian period, men knew how to change the color of ferruginous ochers by calcination. In the Reindeer Grotto at Arcy-sur-Cure (Yonne), a series of small hearths was discovered that had been used solely for the manufacture of violet ocher by means of oxidation.

STONE

Under normal circumstances flint implements are indestructible. The extraction of a cutting object from a block of stone requires a violent, precisely aimed blow that is only on very rare occasions accidentally achieved by nature, and the examination of a sufficiently large number of objects permits us to discern deliberate human effort, even in the case of very worn objects. Thus everything constructed of flint or hard rock by human beings from the very beginning of the race is still in existence; these remains, which are to be found all over the world, would probably have to be counted in the hundreds of billions.

The historical interpretation of these objects is based on stratigraphic geology, in accordance with the very simple theoretical principle that the oldest objects are buried more deeply than later ones. For more than a century stratigraphic chronology has been resolving the difficulties that separate theory from reality, and we possess, at least for certain parts of the world, series of implements that are precisely dated. For our present purpose it is unnecessary to add that several related sciences (zoology, paleobotany or the study of human remains) have combined to supply a climatic context and a critique for this story written with stone implements. Information on these implements, which obviously constitute only a part of the technological possessions of prehistoric man, is abundant and exact, and forms the basis of prehistoric research and, as regards the history of technology, the only unbroken evidential link between *Sinanthropus* and ourselves.

Evolution of methods of production

When we examine evidence arranged chronologically from early times until the popularization of the use of metals, the most striking fact is the perfect continuity of technological traditions between the various successive waves of humanity during hundreds of millennia. This continuity makes contemporary man completely one with the predecessors of *Pithecanthropus* himself. In the contemporary world it seems normal to separate the state of technological evolution from racial or linguistic questions, but it is obviously less easy to project this attitude into the geological past and to separate the evolution of techniques from that of the human species. It is obvious that the two lines of evolution are parallel and that the techniques of the most primitive human beings are the most elementary; but these lines of evolution are actually separate; certain fundamental discoveries were made by Neanderthal man, whereas once contemporary man had appeared on the earth more than 30,000 years were required for the invention of primitive agriculture.

As regards the production of stone implements, this evolution of manufacturing techniques can be divided into five stages, leading from the earliest flaked stones to the implements that, toward the year 3000 before our era, copied the first metal tools.

First stage. The first stage corresponds to the first traces of human industry of the old continent — the Chellean (also called Abbevillian) and Clactonian "pebble-tool" culture, which covers the entire first third of the Quaternary era, and consequently lasted longer than those of the other stages.

It is characterized by a single striking operation that consists of a perpendicular blow on the edge of a pebble or slab of rock in order to create a cutting edge. This is an action similar to the one used in breaking an almond, and one uses it instinctively when idly trying to chip a pebble. The resulting products are thick and heavy, with short, irregular cutting edges (Figure 1).

Second stage. The second stage appears so far to have developed in the western portion of the old continent: Africa, Europe, and western Asia. In Europe it corresponds to the various phases of the Acheulean Age, and covers the middle portion of the Quaternary era. Its characteristic implement is shaped like a more or less flattened almond about the size of an open hand. Called a *biface* or *coup-de-poing* (though the latter term is obsolete and technically erroneous), it has a straight cutting edge, and seems to have been basically a knife.

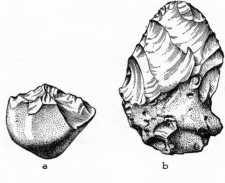

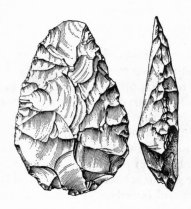

FIG. 1. (*a*) Pebble broken to form a cutting tool. This tool, known as a "chopper," is the oldest deliberately fashioned implement. It appears as a complementary implement in various periods. (*b*) The oldest bifaces, still uneven in shape, develop the characteristics of the primitive chopper, and taper toward the point.

FIG. 2. In the Acheulean biface the entire surface is shaped; the outline of the implement is even, and it has a straight cutting edge. The large concentric flake scars of its surface and the dissymmetry of its profile herald the Levalloiso-Mousterian core used specifically for the striking off of flakes of predetermined shape.

The biface was made in two linked series of operations. The first series was taken over from the first stage, and consisted of removing extraneous material from the core to make it more or less even in shape. The second series, on the contrary, consists in using the parings from the first series as striking implements to attack the stone. The blows were not perpendicular on its frontal face, but were made as obliquely as possible, producing long narrow flakes that permit the "peeling" and exact shaping of the core (Figure 2).

Third stage. The third stage represents a major step in the history of the human race, a step as important as the invention of agriculture or artificial propulsion, for it served as the foundation for the conditions of technological development until the appearance of metallurgy.

The industries corresponding to the third stage are found in the same regions as the first and second, but are more widespread in the Asian area. In the West they are designated as "Mousterian" and "Levalloisian," and taken together they constitute a homogeneous "Levalloiso-Mousterian complex." Chronologically they overlap in the second part of the Acheulean period, and reach their peak between 100,000 and 40,000. Until now the implements discovered with Neanderthal remains belong to this stage, but it is almost certain that the important step represented by the third stage was surpassed by Neanderthal man's predecessors, or by the oldest among them.

Production begins with a core, or nucleus, of raw material, as before, and its end result is a kind of biface that is dissymmetrical in thickness and has the shape of a tortoiseshell. To obtain this dissymmetry two series of actions are successively brought into play: the series of the first stage, which form the sharp slope, and the series of the second stage, which produce the profile of the flat face. After this there is a third series of actions that uses the biface not as an implement to be fashioned, but as a nucleus from which large, flat flakes that

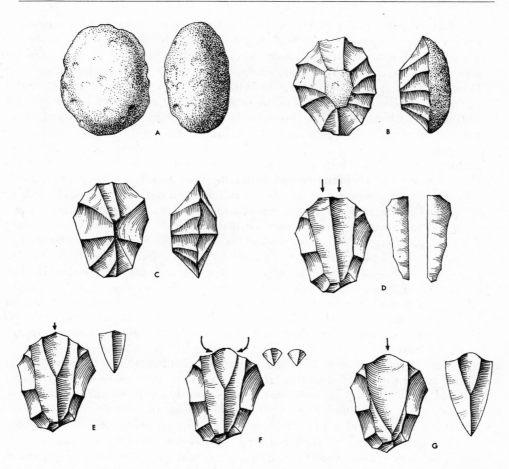

FIG. 3. Cutting of a Levalloiso-Mousterian point: (*a*) Original lump of flint. (*b*) First series of flakings (perpendicular blows). (*c*) Second series of flakings (oblique blows); the core has acquired the appearance of a very thick biface. (*d*) Striking off of two preparatory flakes. (This stage is the starting point for the cutting of blades in the following period.) (*e*) Striking off of a trimming flake from the top of the future point. (*f*) The edges of this trimming flake permit the centering of two small flakings on the striking angle; the profile of the base of the tip has been formed. (*g*) A final blow completes the flaking of the finished product.

will themselves be implements will be struck. The succeeding operations constitute preliminary paring down of the future flake, determining of the striking plane, and striking off of a flake of predetermined shape. These require a well-developed feeling for the material, some preconception of the desired result, and a manual skill that could not be deduced from an examination of the maker's cranium. Several types of cutting implements were obtained by this method, and a preconception of their shape was necessary right from the very first blows. The pebbles that served as hammer stones corresponded to very precise characteristics of density and volume for each stage of the operations (Figure 3).

Fourth stage. The duration of the fourth stage was comparatively very short, since it began between 30,000 and 40,000 and ended around 3,000 years before our era in the most advanced regions, where stone was replaced by metal. On the other hand, it corresponds to numerous cultural divisions included (for the West) in the Upper Paleolithic, the Mesolithic, and a part of the Neolithic periods. It reached its peak in Europe between 30,000 and 10,000 before our era, during the Chatelperronian, Aurignacian, Gravettian, Solutrean, and Magdalenian periods, which constitute the Reindeer Age.

From the technological point of view, the fourth stage repeats the innovations of the third. The core is treated as in the methods of the third stage by the three series of preparatory and extracting movements. The nucleus thus obtained is longer, however, than the Levalloiso-Mousterian "tortoise." Nothing is retained of the earlier methods except a kind of flake of predetermind shape — long and narrow, with parallel sides — called a "blade." A nucleus supplies many more blades than flakes. The method of obtaining implements consequently remains closely linked with earlier traditions; its originality consists in systematizing the flaking of uniformly light products. Perhaps the most important fact is that these blades are simply a basis for genuine implements. Just as the Acheulean biface became the core from which the Levalloiso-Mousterians obtained their flake tools, the flake in the form of a blade became the source of implements as such. The blade was then carefully fashioned into scrapers, gravers, drills, and thin knives.

Fifth stage. The fifth stage was short and worldwide, and its products, included under the term "Neolithic," form a narrow fringe around the early metallurgical period. Many of these implements simply continue the fourth stage, but this stage also witnesses the beginning of the use of such nonvitreous stones as jadeite, diorite, and schist, which are shaped by chipping; their cutting edge or their entire surface is then polished. These new implements were for the most part axes or adzes used for working wood, and most of them were produced in regions surrounding the metallurgical centers where we find the first copper or bronze axes. It is still difficult to determine the exact relationship between the polished stone axes and the first metallurgical products, which come soon afterward, but polished stone objects appear after Egyptian metal in Mesopotamia (except, perhaps, in the eastern part of the Mediterranean basin); they seem to have played the role of replacements during the protometallurgical phase. The same fact has been noted in the Far East, and seems possible in America. As for western Europe, where the evidence is most abundant, the protometallurgical period (3,000 to 1,500 before our era) shows, in the case of both fourth-stage products and polished stone implements, a search for dimensions larger than those of the preceding periods.

This unified evolution of flint techniques and the linkage of the various stages demonstrate one of the most persistent factors in the history of technology; namely, that innovations are born by the addition of new operations to earlier achievements, which then serve as substrata. Thus the actions of *Pithecanthropus* persist in the preparatory phase of the work (Figure 4) right down to the very last flint knapper. The significance of this evolution emerges in the relationship which is established in every period between the manufacturing process and the technical efficiency of the implement obtained.

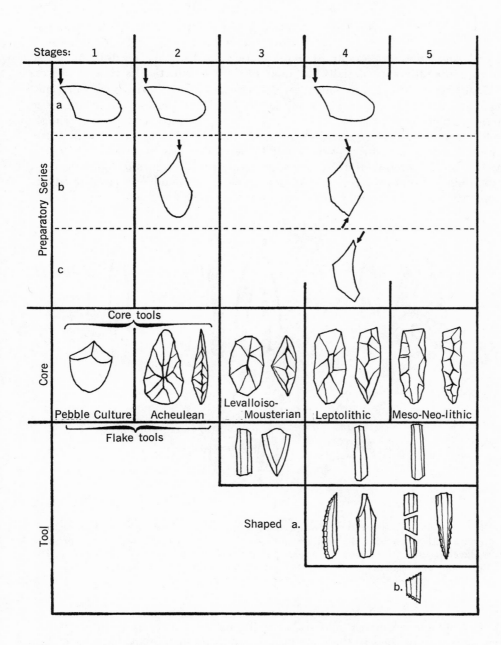

F̲ɪ̲ɢ̲. 4. Table showing the progressive improvement of stone techniques, from the "pebble culture" to the Neolithic period.

Top: Development of series of actions whose effects on the core are cumulative from stages 1 to 3. *Center*: Development of the core, at first an implement in itself, then a source for other implements. *Bottom*: The flake and the blade, struck directly from the core in stage 3, then a source for the shaped implements in stage 4(*a*); the blade may itself become the core for the striking of geometrical segments shaped into implements (*b*).

Economic evolution of implements

A series of chipped-flint implements arranged in chronological order shows that in the great majority of cases the technological effect sought was a cutting edge that was longitudinal in relation to the piece of flint. In other words, temporally and spatially the most uniform production is that of knives.

Another fact becomes evident in this chronological arrangement: descending from the beginning stages down to the Neolithic period, the average size of the implement diminishes. The heavy bifaces are succeeded by increasingly smaller and thinner ones, then by flakes, and then by blades, and we witness the proliferation of microliths — very small blades sometimes less than a half-inch thick. The fact of this decrease in weight and reduction in size has long been recognized by prehistorians; it did not, however, preclude the continued existence of larger dimensions for certain implements (Figure 5).

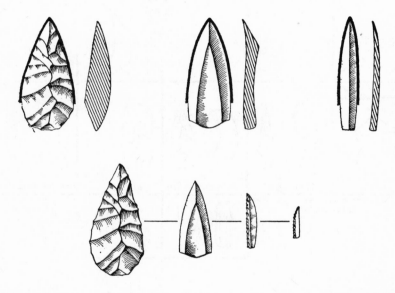

Fig. 5.   Schematic representation of the economic development of the Paleolithic period.
Top: While preserving the same amount of cutting surface, the implement progressively develops from the heavy Acheulean biface into the lighter Levalloiso-Mousterian pointed flake and the Leptolithic blade. *Bottom*: Between the Acheulean and Mousterian periods, the size of cutting implements decreases.

For obvious geometric reasons the utilizable cutting portion of the implement does not decrease in the same proportion as its surface and bulk, and thus its relative efficiency is in inverse proportion to its decrease in size.

A table showing the total length of utilizable cutting edge obtained at different stages in the technological evolution from 2.2 pounds of flint illustrates the evolution of the relationship between efficiency and the volume of material converted into implements:

FIRST STAGE:	Abbevillian (bifaces)	2 ft.
	Clactonian (flakes)	2 ft. 10 in.
SECOND STAGE:	Acheulean (bifaces)	3 ft. 11 in.
THIRD STAGE:	Mousterian (flakes)	13 ft. 1 in.
FOURTH STAGE:	Aurignacian (blades)	29 ft. 6 in.
	Solutrean	36 ft.
	Magdalenian (blades)	56 ft. 8 in.
	Magdalenian (microliths)	216 ft. 5 in.
	Mesolithic (microliths)	328 ft.
FIFTH STAGE:	Neolithic (blades)	24 ft. 6 in.
	Neolithic (paring knives)	3 ft. 3 in.
	Neolithic (axes)	1 ft.

The direction of the transition from one series of operations to another becomes apparent in this improvement of the relationship between the implement and its raw material. The search for flint, indispensable for the survival of a group of hunters, dictated the group's installation in mineralogically favorable regions, or at least its periodic return to the sources of this material. Beginning in the fourth stage there occurred a considerable liberation; a few pounds of blades represented months or years in territories in which flint was absent.

The adoption of a permanent abode by the agricultural peoples in the fifth stage and the introduction of heavy implements for woodworking reveal, in contrast, a temporary return to formulas of production that were costly in terms of raw material.

Evolution of equipment The progressive development of industries can be grasped under another aspect: that of the enrichment and diversification of implements.

The implements of the first stage are limited to *choppers* (pebble tools with a cutting edge crudely shaped by one or two flakings) and, in certain areas, to the early bifaces, which were very heavy and had an uneven cutting edge.

Second-stage implements include the *chopper* of the preceding period, lighter bifaces with straight cutting edges, and the first flake tools: the *racloir* (a knife with a concave cutting edge), *tips* (a more tapered form of knife), and *hatchets,* whose use is still unknown. Moreover, the existence of woodworking is revealed by the appearance of implements fashioned into *scrapers.*

The implements of the third stage add a profusion of flake implements. *Racloirs* and tips are complemented by *knives,* one edge of which has been flattened to form a back. The scraper was joined by the first *gravers* for working bone or deer antler, and *raclettes* (scrapers for fine planing of wood).

The implements of the fourth stage, considering only the Upper Paleolithic Age (from the Chatelperronian to the end of the Magdalenian periods), present a great number of new shapes obtained from flakes and blades. We can count some thirty major forms of *racloirs* and various knife blades, about ten forms of microliths, twenty-five types of scrapers, some twenty types of gravers. Drills were added to the *raclettes.* The first sandstone polishers for sharpening tips and needles appear, so that in all more than a hundred types of flint implements characterize the industries of this period. In addition, the bone, ivory, and horn industry led to the creation of some thirty types of implements or new weapons. From this point on, an average technological level for preagricultural groups has been established; the Eskimo equipment, for example, retains approximately the same proportions.

BONE AND ANIMAL SUBSTANCES

The use of animal bones, elephant or mammoth tusks, and deer, reindeer, or moose antler seems so natural that we could expect to find numerous evidences of it from the very beginning of human industry. But a cause that is purely physicochemical in nature immediately lessens this evidence. Bone remains have disappeared from most of the prehistoric sites, and it requires exceptional conditions of chemical equilibrium to prevent the calcium in the remains of living beings from returning to the common body of matter. These requirements are most frequently met by clay soils, particularly those in caves. Despite the massive destruction of bone remains, the chief stages of this industry are known, and we possess abundant animal remains connected with the australopithecines, *Sinanthropus,* and *Atlanthropus;* very abundant remains accompany Neanderthal man, while those discovered with the implements of the Upper Paleolithic period are still more numerous.

Erroneous interpretations The fate of bone remains in a prehistoric dwelling place varies considerably, depending on the origin of these remains. As a result of the liberation of chemical particles by the soft tissues, a cadaver abandoned to natural decomposition leaves a skeleton that is threatened with more rapid disintegration than occurs with food remains and bones from which the flesh has been removed. Implements, carefully cleaned and dried in the course of their use, easily become fossilized. Thus the best selection is found among the products of human industry. These fortunate conditions emphasize the almost total absence, during the major portion of the prehistoric period, of implements made from animal substances. There has been much speculation about the use of the jawbones of cave bears as offensive weapons, pointed bone flakes as tips for javelins or spears, and portions of skulls as cups. Most of these theories however are not substantiated by experience. Aside from the absence of decisive evidence of their use, these supposed implements would result in mechanical absurdities, that the inventors of a stone industry would not be likely to construct.

The explanations suggested for bones that have been intentionally broken and bones that show traces of polishing are not convincing. The deliberate breaking of long bones is found continually throughout the prehistoric period; all the bones likely to have contained a scrap of marrow have been crushed by a blow from a stone and carefully emptied of their contents. It has been demonstrated that a violent blow on a long bone whose cavity is filled with marrow and blood causes the formation of large splinters shaped almost invariably like triangles or butterfly wings. These fairly regular pieces have sometimes been taken for deliberately fashioned objects, although traces of any work done on them are extremely rare or nonexistent.

Broken bones with one end beveled and apparently carefully polished have also been discovered in caves. This polishing is as a matter of fact the result of an artificial activity: the constant passage of men or animals over the bones, which were embedded in the clay but partially protruded above the ground. Among numerous samples, there are very satisfactory examples of implements, but as a whole the polished areas suggest completely improbable polishing tools or implements for skinning hides. Moreover, it is easy to discover, by carefully skinning the ancient cave floors that were trampled by living beings, numerous bones that might be polished by the passage of these beings. Once these "artifacts" have been eliminated, we are left with very few really indisputable cases prior to the Upper Paleolithic period.

Earliest traces of use The South African sites of discovery of the australopithecines have raised the question that these very early anthropoids used bones. Among the animal remains are those of baboons whose skulls contain hollows that could have been made by a blow from a weapon used by the australopithecines, for whom the baboons may have been rivals or game. It has been thought that this weapon could have been a long antelope bone — the humerus or the femur — grasped in the middle. This hypothesis is not completely convincing, though it is possible that antelope horns were used as weapons.

Sinanthropus, or Peking man, a close relative of *Pithecanthropus* of Java and *Atlanthropus* of Algeria, left numerous remains of his game in the Choukoutien cave. These remains were broken in order to extract the marrow, as is the case in all the succeeding periods, and there are no fragments that show really unquestionable traces of systematic manipulation. Deer antlers, on the contrary, were definitely utilized; investigators have discovered large stag antlers stripped of their tines by flexion or by a blow, and traces of rudimentary sawing with a stone implement are sometimes visible. The very early use of deer horns is not surprising, for after very simple preparatory work (for example, cracking), they supply clubs, picks, and daggers — the perfect complement for flint knives.

In the earliest stages of the human race — those corresponding to the australopithecines and *Pithecanthropus* — minimum survival equipment must have included weapons in addition to cutting tools, which could be of no assistance for hunting or defensive purposes. Among contemporary primitives the work represented by the making of a club from a wood sufficiently dense to serve as a genuine weapon suggests that the early hunters chose the antlers of large deer, which had the desired weight and required very little preparatory work. Though in the third stage of the flint industries (the Levalloiso-Mousterian), the already considerable development of flint implements suggests that there was a corresponding development of bone industries, the archaeological facts do not confirm this view. We discover the same evidence as in the first and second stages, that is, deer antlers sometimes carefully broken or divided into sections. Antler tines have been found whose tips seem to have been polished by prolonged use, and which may have served as "polishing tools." It is very difficult to establish this with any certainty, because deer themselves have the habit of polishing their horns by rubbing them against trees.

From time to time Neanderthal man's dwelling places have revealed large bone fragments that have been deliberately carved like flint. This is an infrequent practice that cannot be regarded as a genuine technical innovation. Consequently, taken as a whole the first three stages of the evolution of human industries are very poor in evidence of the working of bone.

Only at the very end of the Mousterian period do the first genuine bone implements appear, and even then they are very rare. They include various types of awls, which were probably used in the working of hides. The deficiency of the bone industry in the third stage is understandable. The very early use of deer antlers was due to the fact that almost no preparation was needed. Making genuine implements from the same material, in contrast, requires equipment that was not available until the end of the Mousterian period, when flint gravers became common. Consequently the third stage represents a long period of stagnation in the use of bone, the utilization of antlers having reached its limits in a single step.

Developments in bone implements

The fourth stage, on the contrary, is marked by the sudden explosion of bone implements. Beginning with the Chatelperronian period, we find many pierced teeth, bone pendants, javelins made from mammoth tusks, reindeer antler, or bone, and later harpoons with fine barbs and slender needles with eyes. Sections of reindeer antler are carved into splitting wedges, and ribs from horses are used in their natural state as small picks or are cut into spatulas. Cro-Magnon man, who already resembles contemporary man, and his successors appear to have possessed a much wider variety of implements than did Neanderthal man; they are surrounded by objects that are already almost familiar to us.

The most important factor for the history of technology was to discover the implements used for the working of bone. This meant searching among the flint implements whose appearance coincided with that of bone tools for those that might fill the requirements of working with bone. It seems certain that the graver and the scraper filled these requirements. It is a fact that these two implements slowly began to proliferate in the Mousterian period when the first bone awls appear, and beginning with the Chatelperronian period they increase to the point where they sometimes form the majority of the stone implements. Probably *raclettes* and notched or denticulate blades or flakes also played a role in boneworking, but since they were already numerous in the Mousterian period, and seem to apply more particularly to woodworking, only a complementary role can be attributed to them.

The preparatory working of objects was based on the use of the graver, which acted as a very narrow plane for carving deep grooves. By means of appropriately positioned grooves, the workman detached a piece of the raw material, which was then shaped by filing with the scraper, much as we might use a knife blade held at a cross angle. Very fine gravers might be used for details or for piercing holes. Notched implements were used to finish off cylindrical objects such as spearheads. Occasionally an implement was improvised from flakes of all shapes, selected on the spot from among the waste pieces. But the basic equipment consisted of the graver and the scraper, and these were so important that we frequently encounter flint blades that have been shaped into a graver at one end and a scraper at the other. Wedges carved from reindeer antler must have been used to detach the pieces marked out by the chiseled grooves.

Peak period of boneworking

It was in the Magdalenian period, from 12,000 to 8,000 before our era, that the working of bone reached its apogee, with foot-long javelins as thick as a pencil, sewing needles, and harpoons with long barbs. The making of javelins can be reconstructed with a fair degree of certainty. The worker selected the slightly curved shaft of the antler of a male reindeer killed during the season when the horn is most compact. With the graver he outlined two parallel furrows that cut through the hard cortex of the antler on the bias to the spongy center tissue. It was then possible carefully to lift out a long, curved, more or less square wand, which was then heated and straightened while still warm with the help of a reindeer-antler implement with a hole in it (long thought to be a *"bâton de commandement"*). Probably after this operation, the wand was worked with the scraper and filed down into cylindrical shape, with a pointed tip and a base beveled on both sides. Such spears must have had a tendency gradually to return to their original curve, for they are frequently discovered in this state, but heating them was all that was required to restore their straight shape (Figure 6).

Needles were made by the same method from the tusks or bones of mammoths. Once the grooves were outlined, it seems that the eye was pierced before the fine rod was detached. This was then filed down and polished on a small sandstone polisher. Such needles could be as little as one millimeter in diameter, which suggests the precision of the sewing tasks for which it was used.

The regression

After the year 8000 in western Europe, the disappearance of the large animals, particularly the reindeer, caused a decrease in boneworking. The traditions were continued, however, for quite a long time among the deer hunters of the the Maglemosian culture from England to Scandinavia. The graver was still common, and long wands cut from deer antler with the same technique as the Magdalenian reindeer-antler wands served for the making of long barbed spears.

The appearance of agriculture in the Neolithic era brought new uses for bone. Deer antler found a use in the making of handles for hafting polished stone hatchets (Figure 7). Various portions of the antler were shaped into picks and wedges with handles; moose-antler horn supplied hoe blades, and even cups were carved out of the hollow base of large deer antlers.

Bone awls certainly played an important technical role, for they are very numerous and were the object of a very systematic method of production. The metacarpal or metatarsal bone of a stag or a roe (whose tissue is extremely compact) was chosen. The head of one of the ends was sawed off by means of a circular groove; then the body of the bone was split, using two longitudinal

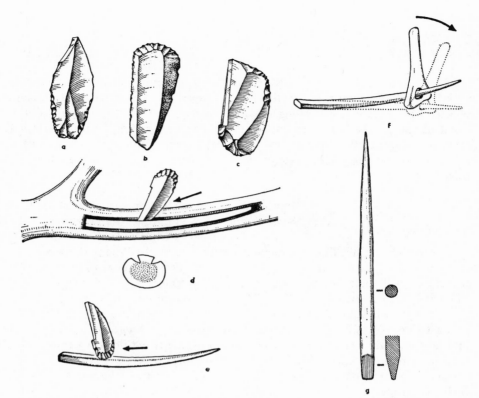

FIG. 6. Making a Magdalenian spearhead: (*a*) Graver. (*b*) Scraper. (*c*) Composite implement (graver-scraper). (*d*) The outline of the spear is detached from the reindeer antler by means of deep incisions made with the graver. (*e*) The rough piece is trued and filed down with a scraper. (*f*) Piece is straightened while warm with the help of a "shaft-straightener" — a stick with a hole in it, made of reindeer antler. (*g*) The finished object.

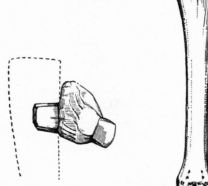

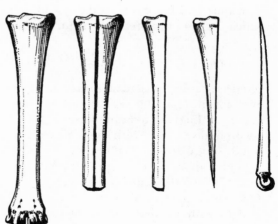

FIG. 7. Ax. The polished stone blade is mounted in a locking ring of deer antler; the tenon is fixed in a wooden handle. Neolithic, France.

FIG. 8. Production of awls carved from metacarpal bone of ruminant. *Left to right*: the original bone; a section of the bone marked with longitudinal grooves preparatory to splitting; the detached piece worked into an awl; completed polished awl; awl obtained by cutting the bone from the other end.

grooves. The resulting splinter, whose joint often formed the head, was sharpened to a point with the sandstone polisher (Figure 8). The technique of stone polishing seems generally to have influenced boneworking in the Neolithic period, for a number or implements fashioned from ribs, vertebrae, or fragments of long bones were shaped by abrasion on a rough stone.

WOOD

The origins of woodworking are particularly obscure, for until a very late period (almost until Neolithic times) direct evidence is practically nonexistent. Preservation of wood is possible only in peat bogs or accumulations of vegetable matter on lake bottoms, and with the exception of the Magdalenian period in northern Germany no site of this type is found before Maglemosian and Neolithic times. Thus we can study the problem only indirectly, that is, on the basis of the stone implements that could have been used for wood-carving or that may have had to be hafted in wood.

Doubts of technology Regarding the first and second stages (those of the pebble-tool and biface industries), no sound hypothesis can be formulated. We can imagine *Pithecanthropus* brandishing a club or a spear, but there is not the slightest supporting evidence from technology. Hafted bifaces have familiarly been depicted as axes, halberds, and pikes; but everything seems to indicate that bifaces were skinning and butchering knives, and putting handles on them produces results that defy their effective use. Cutting stones and deer-antler horn are the only objects remaining from these stages.

The first wooden weapons The third stage — the Levalloiso-Mousterian — contributes somewhat more substantial evidence. The existence of wooden weapons is definitely established by the discovery in Palestine of a Neanderthal skeleton whose pelvis had been pierced by a pointed weapon. The cast of the hole revealed that it had been made by a wooden spear — or javelin head. This proof that the Mousterians used wooden javelins is confirmed by an examination of the flint industry. Particularly during the second half of the Mousterian period there were many flakes of all sizes, the edges of which have identations made by a scraping implement. These implements have been variously termed *raclets, coches,* and notched and denticulate tools (Figure 9). Experience confirms the obvious assumption that they were used

Fig. 9. (*a*) Notched and denticulated flakes, Mousterian.
(*b*) Notched blade, Upper Paleolithic.

a   b

as scrapers for materials like bone or wood. The rarity or absence of shaped bone objects leaves woodworking as the only possibility. The size of the *raclettes* and the profile of their cutting edge corresponds empirically to the paring of thin chips several inches long. Moreover, the series of semicircular notches produced by flaking the flint, or evened off through careful retouching, indicates that the implements were usually no more than one inch in diameter. All these elements point toward the use of javelins in the Mousterian culture, or at least in its period of greatest development.

It is difficult to establish the shape of these weapons, although the Australian aborigines in particular, and numerous Oceanic and American peoples who do not have the use of metal, produce excellent wooden javelins with heads that are simply sharpened to a point. The only important question is that of the possible hafting of flintheads. Certain large triangular flintheads, thin at the base and sometimes eight inches long, could have been fastened to the end of a spear or a pike. They would have been too fragile for a throwing weapon like the javelin.

Utilization of wood for various purposes

There is no doubt whatsoever about the importance of woodworking in the fourth stage (Upper Paleolithic and Mesolithic). It is difficult, on the other hand, to guess for what objects wood was used. The considerable quantity of spearheads made of bone, ivory, or reindeer antler (flint in the Solutrean period) implies the making of thin cylindrical shafts for which *raclettes* and notched blades were used. Given the very high technical level reached in this period, the existence of bark utensils, wooden platters, palings, and frames for bedding or for the drying of garments can be conjectured. No evidence, however, even indirect, has yet been uncovered. A few wooden posts have been found in northern Germany, in the only Magdalenian site where wooden objects had been preserved. Here sites of circular tents, which must perforce have had wooden poles, have been discovered. We may also consider as very probable the existence of wooden spear-throwers — rods provided with a hook at one end that were used for hurling spears. Examples of these, made of reindeer antler, have in fact been found in a certain section of the Magdalenian area of southwestern France, Switzerland, and southern Germany. Their scarcity, and the large area represented by the regions where the same spears without such reindeer-antler throwers are found, suggest that these devices were in most cases made of wood.

One problem is common to the third and fourth stages: the question of the methods used to fell and cut up large trees. Javelin-making required tree trunks at least four inches in diameter, to provide a sufficiently thick wood and the necessary length when split. In the third stage, heavy scrapers or hatchets could have served as hand axes, but the flint implements of the Upper Paleolithic period are uniformly light in weight, and even among the waste products of the flaking operation (some of which are the size of a fist) we have not found enough sufficiently uniform types to permit the formulation of a hypothesis. Carbonization by means of a small fire made against the trunk, and the gradual scraping of the burned portion, are the only elements that can be posited with any degree of probability.

The ax   The situation changes completely in the Mesolithic and proto- and para-metallurgical periods of western Europe. A new implement appears: the *ax*. Its production involves a relatively large quantity of material, and of necessity presupposes a hafting device. The term "ax" is, however, somewhat inaccurate, for the stone blade was hafted sometimes in the axis of the handle (like the modern "ax"), sometimes perpendicularly, as in the case of the tool called an adze. The discovery of wooden handles that have been preserved confirms the existence of these two implements, but since in most cases the handle has been destroyed, it is impossible to determine the original nature of the stone blade.

Beginning in the Maglemosian period, around 7,500 years before our era, discoveries in peat bogs reveal large trees felled with the hatchet or the adze, their bases patiently whittled down into the shape of a pencil. Beginning in this period, and especially during the Neolithic stage in Switzerland and southern Germany (around the year 2000 before our era), evidences of wood sawing multiply (Figure 10). Traces of postholes have been discovered in the Azilian period (toward 7000). Outlines of huts indicated by similar holes are found almost everywhere, and in peat bogs from Denmark to Switzerland we find the posts themselves. The first evidences of canoes hollowed from tree trunks and the first wooden utensils also date from this period.

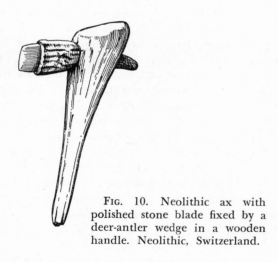

FIG. 10. Neolithic ax with polished stone blade fixed by a deer-antler wedge in a wooden handle. Neolithic, Switzerland.

FIG. 11. Beaver mandible used as wood cutter. The corners of the jawbone have been broken off and the cutting edge has been sharpened. Neolithic, Switzerland.

Small implements for woodworking dating from the end of the Upper Paleolithic to the Bronze Age have not yet been identified with precision. The presence of numerous scrapers is probably related to this problem, but the only implement whose usage can be determined with certainty is a graver made from the lower jawbone of a beaver. Its large incisor has been sharpened and it has been broken at the corners to provide a place for the hands (Figure 11). This implement has been found in a large number of Neolithic sites, down to the Iron Age.

HIDES AND SKINS

Hides and skins have left even fewer traces than wood. Since they have been destroyed even in the peat bogs and lakes, at least prior to the Bronze Age, their use can be established only very indirectly. This indirect evidence may take the form either of implements that were definitely used for the working of skins or of objects implying the use of skins or of the bone remains of animals that retain traces of the use of their fur or hide.

The first of these clues — implements used for the preparation of skins — is so far very weak. Any cutting implement could have been used to skin an animal or scrape its hide. Scrapers were so named precisely on the supposition that they were used to scrape the skins of wild animals. It appears, however, that they served as general-purpose knives, since they disappear after the Mousterian period, although the use of skins certainly continued. Among the bone remains of bears found in Swiss caves, fragments with polished broken edges have been found that may have been used in the preparation of fur skins of plantigrade animals; but, as we mentioned earlier, these are natural phenomena, not tools. While several of these prehistoric implements could have been used in the preparation of skins, in no case has this use been confirmed.

Implements for the working of skins   The testimony of objects implying the use of skins is more eloquent. None have been discovered for the period between the beginning of human industry and the end of the Mousterian period. After this, the first bone awls appear. Unless we suppose that they were used in basketmaking, a difficult hypothesis to prove, these awls were probably connected with the use of skins. The same is true in the Upper Paleolithic period, when awls increase in number and become very fine (Figure 12). The appearance of needles toward the Solutrean period confirms this point of view by indicating the existence of sewing.

From the Magdalenian period there exist pieces of reindeer antler shaped into a wedge at one end and polished by use. Some of them seem, because of their curved shape or their narrowness, to have been designed for softening skins rather than for splitting wood. Large mammoth bones, particularly femurs, have been discovered bearing numerous intersecting curvilinear incisions that can be explained if we suppose that these bones were used as blocks for cutting pieces of clothing or shoes.

The tools of the following period (Mesolithic and Neolithic) are not well known. The needle disappears, but the awl is still found in large numbers. It is difficult to distinguish among the bone or reindeer antler wedges those that were used for splitting wood from those that may have served for the scraping or softening of skins.

Traces of skinning on animal remains   The most consistent evidence is that supplied by the remains of furred animals. The dismemberment of animals with flint knives leaves traces on the bones in the form of fine nicks that can be discovered with careful examination. An inventory of these marks, which are found at the joints, permits us to recon-

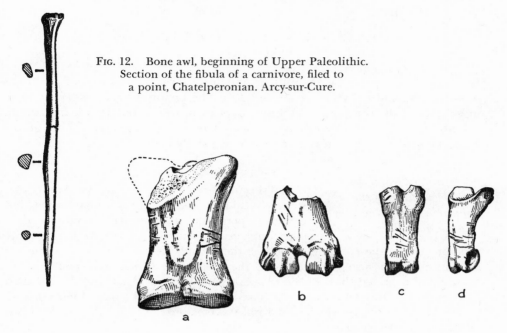

FIG. 12. Bone awl, beginning of Upper Paleolithic.
Section of the fibula of a carnivore, filed to
a point, Chatelperonian. Arcy-sur-Cure.

FIG. 13. Bones showing traces of dismembering. Mousterian.
(*a*) Phalanx of a horse. The marks from the flint knife used to cut the skin around the foot are visible around the bone. (*b*) End of a reindeer shank. The incisions resulting from the cutting away of the skin appear on the front face of the bone. (*c*) Reindeer phalanx with incisions resulting from disjointing. The marks are located on the inner face, at the location of the tendons. (*d*) Phalanx of cave bear, showing incisions resulting from cutting of the skin. Arcy-sur-Cure.

struct the manner in which the Mousterians dismembered reindeer and horses. The operations for the dividing of the limbs never varied, since the same nicks are always found at the same places in all the animals. On the reindeer in particular, in addition to the marks at the joints, marks of cuts made on the limbs preparatory to skinning (Figure 13) have also been found. Consequently we are certain that Neanderthal man already carefully separated the skin from the slain animal, an operation that is not performed by all the contemporary hunting peoples.

The same marks are found on bones of the Upper Paleolithic and, in the case of certain animals, Neolithic periods.

The skinning of carnivores, such as the lion, wolf, bear, or fox, is generally done by abandoning at the end of the paws the last phalanges and the claws. Statistical research shows that precisely these parts are encountered unusually often in the Upper Paleolithic period. Moreover, incisions on these phalanges are frequent. Thus we are certain that skins, whole and with their claws, were prepared and used in dwellings. In the Neolithic period, we also find bear phalanges, singly or in whole paws, which came from similar skins.

Hypotheses advanced about preparation and tanning, particularly the marrow and the brain, are, however, insufficiently supported by proved facts.

TECHNIQUES OF ACQUISITION:
HUNTING AND FISHING

It is perhaps in the areas of hunting and fishing that "scientific folklore" has been best developed. While we are certain of the important role played by these techniques in the economy of prehistoric societies, we are far from being able to describe them exactly. Spectacular legends have grown up, often born of the discreet suggestion of an investigator and gradually transformed into certainties.

Hunters from the beginnings of the human race to the end of the Acheulean period

The earliest accessible evidence is that of the australopithecines of South Africa. We know that strictly speaking these were not human beings but creatures possessing the human characteristic of erect carriage. Their front teeth were hardly more prominent than those of human beings; their hands were similar to ours; and, at least in their physical characteristics, they represent *Homo nudus et inermis:* Nature had deprived them of the customary methods of attack and defense used by mammals. Although no weapons have as yet been found among their remains, given the conditions in which they lived it seems probable that they had the use of some offensive weapon.

Their remains are found in several grottoes in South Africa, which is the most favorable milieu for the study of traces of intellectual or technical activity. Unfortunately, these remains lie with thousands of other bone remains in extremely hard soil, and laboratory examination of blocks of rock extracted by means of explosives does not permit the discovery of small details of the original disposition of the remains. Thus we do not know whether these caves were the dwelling places of the australopithecines or lairs of wild beasts who dragged them there or accidental ossuaries. If we adopt the first hypothesis, the australopithecines may have lived in groups and their diet may have been at least partially carnivorous, for they are surrounded by the remains of animals who do not normally live in caves: birds, antelopes, and particularly baboons. Even more odd is the presence of the panther among this presumed game. The well-preserved remains of a certain number of baboon skulls have been studied, and it is believed that they were smashed in by a blow comparable to that of a club or a fist armed with the humerus of a gazelle. The australopithecines are thus supposed to have been able to overpower the baboons, who have enormous canine teeth, travel in packs, and are now considered to be more dangerous than most wild animals. These facts have not yet been sufficiently proved, but that the australopithecines could have been partially carnivorous is not in itself very surprising, for all our information on prehistoric man, even in his most primitive state, is of the same type.

For the first and second stage, our information relates to *Atlanthropus* of Ternifine in Algeria and *Sinanthropus* of Choukoutien near Peking, both similar to *Pithecanthropus* (who has not been discovered in his dwelling place).

The remains of *Atlanthropus* of Ternifine were found not in a cave but in the deposits of a spring near which they lived. Lying together with them were their flint implements and the vestiges of a large number of animals. These vestiges must include the game animals of *Atlanthropus* and the cadavers of animals who died nearby of natural causes. *Sinanthropus*, in contrast, is found in his dwelling. We have already seen that he left hearths, an enormous quantity of flint implements, and deer antlers that were probably used as clubs and rudimentary pickaxes. The animal remains are broken; they undoubtedly come from game animals, among which are included, in addition to deer, relatives of the fallow deer who were still living in Europe after the Reindeer Age, horses, elephants, rhinoceros, beavers, and several large carnivorous animals. No information can at this point be supplied about the methods used to overpower these large animals.

The Levalloiso-Mousterian hunters   Information about the Mousterians, who peopled Europe between 60,000 and 40,000, is extremely abundant. It is known that hunting played an important role in their acquisitive activity. In addition to evidence of their methods of dismemberment, we have been able to make an exact reconstruction of their technique for cracking bones in order to extract the marrow.

The Mousterian game animals are also known. The principal game animal was the horse. In varying quantities, depending on region and variations in climate, the horse was followed by the ox, the bison, the reindeer, the stag, the fallow deer, and the boar. The fate of the mammoth has not been definitively established. From time to time we find tusks or bone fragments, but these may possibly have been removed from carcasses of mammoths who died a natural death. The size of this animal is perhaps the only reason for the absence of numerous bone fragments in caverns; mammoth hunting must have been no more difficult than hunting the lion or the aurochs. The Mousterians probably hesitated to attack the adult rhinoceros, but must have learned to separate a young rhino from his mother, and in general more fragments of young rhinos than adults are found. The hyena and the wolf were hunted and eaten. As for the bear and the lion, whose remains may possibly come from animals who died in the caves of natural causes, the situation is less clear.

Following excavation in the natural bone sanctuaries of cave bears in the Swiss and Austrian caves, it was believed that there existed a real "civilization of the cave bear." The Mousterians, profiting by the winter hibernation of the bears, are supposed to have stretched nets in the cave to capture the beasts, who desperately clawed at the cave walls through the holes in the net that imprisoned them. Once the animals were killed and eaten, their skeletons are supposed to have been used to make various tools for the working of their hides — tools of odd shapes, with some of their edges carefully polished; the jawbones, bristling with canine teeth, made formidable bludgeons. Moreover, or so the story continues, a religious cult centered upon the remains of these dangerous animals: the large bones were piled in heaps along the walls; the skulls were placed in small cases made of flat stones.

This reconstruction of the life of the bear hunters was based entirely on errors of interpretation made in the course of excavations carried out under inadequate scientific supervision. The reconstruction of the cave bear's winter customs showed how the legend of the Mousterian hunters was able to develop. The claw marks on the walls do not correspond to the traces of a desperate struggle: all animals who frequent caves claw the walls when exploring in the darkness the nooks in which they are looking for an opening or a hiding place. From the large fossil bear to the mouse, every cave contains innumerable traces of this behavior. The bears wandered around in the corridors and unconsciously pushed back toward the walls the large bones of their predecessors, which formed a seemingly deliberately arranged row of femurs or tibias. When they arrived at the point selected for hibernation, they cleared a spot on the ground by hollowing out a circular sleeping place; this caused the accumulation of a circle of large skulls and bones. Most of the skulls disintegrated and were destroyed, but some rolled into out-of-the-way corners or slid between the slabs of stone; these are the only ones we now find intact. As for the "implements," they were made, as we have already noted, solely by the rubbing action of the feet of generations of animals and men.

If we abandon the legend of the bears, direct evidence regarding the hunting methods of the Mousterians is almost nonexistent. The use of reindeer or deer-antler horn permits us to suppose that clubs existed, but, as we have already noted, the Mousterians certainly made javelins and perhaps armed spears with large flint blades. These two weapons — the pointed javelin for striking from a distance, and the spear with a wide cutting blade for close-quarter attacking and for killing wounded animals — are still the arsenal of hunters of lions, elephants, and other large game animals in Africa.

Nothing is known of the existence of traps in the Mousterian period. Beginning in the first industrial age, we discover, first in Africa and later in Europe, lumps of stone carved into spherical balls about the size of a fist. These balls are quite numerous in the Mousterian sites of Europe. They appear to be related to hunting rather than to any other technique, and it has long been thought that they must be equivalent to the *bolas* of the South American Indians. These are balls attached by twos and threes to leather thongs, and used as slings to stop the animal by impeding its movement. No precise observations have been made of the Mousterian *"bolas."* They have been found sometimes in groups of two or three, sometimes isolated, sometimes heaped up in piles.

The hunters of the Reindeer Age

In the Upper Paleolithic we find ourselves among human beings very similar to ourselves, whose culture is much richer than anything created by their predecessors. We may therefore expect to discover more precise evidence of their behavior as hunters and fishermen. Although the scientific legends about their hunting methods are concerned especially with trapping, there exists no material evidence to prove that the men of the Reindeer Age practiced this skill. Knowledge of its techniques is completely probable, but the pits, which were dispersed at a distance from the dwellings, and the traps, have disappeared without leaving traces. The frequently numerous remains of birds, marmots, and white foxes found in dwelling places almost necessarily imply capture by means of traps or snares, but again the methods themselves are still unknown to us. Certain signs painted or engraved on several animal figures (Figure 14) drawn by artists in the caves of southwestern France and northern Spain have been interpreted as representations of traps. In fact, however, they are undoubtedly symbolic representations of a feminine nature. Other feminine representations, called "claviform signs," have also passed for boomerangs. If the Magdalenians did trap animals, they have left no pictures of their tools.

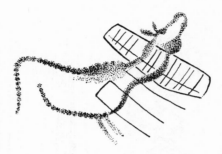

Fig. 14. Sign superimposed on the picture of a doe, often considered to be a representation of a trap. This is a chance combination of two subjects discovered singly in tens of examples. Grotto of la Pasiega, province of Santander (from H. Breuil).

A persistent explanation was born of the discovery, at the end of the nineteenth century, of the remains of several hundred horses at the foot of the escarpment of Solutré. It was believed that the Soultreans had the custom of driving herds of wild horses over the edge of the rock. The hypothesis of herds driven by beaters toward cliffs should not be rejected; we have examples of this among recent hunting peoples. The case of Solutré, however, cannot be taken as proof, for the pile of horses is located at the site of a village, around hearths, and the remains found are similar to cave finds in that they have been carefully broken to extract the marrow. There is another consideration: the horses would have had to make a gliding flight of about a hundred yards before landing in this spot.

If we reduce the Upper Paleolithic hunting techniques to what is definitely known, we are left with the fact that they hunted with javelins and that the late Magdalenians used a barbed weapon that has been given the name of "harpoon." Some of these weapons could have been used for fishing, for fish debris has been found among the food remains, and works of art frequently include pictures of salmon and pike.

The javelins were of a type usually light in weight and armed with a bone or reindeer-antler head. The oldest javelins are either cylindrical or leaf-shaped (Chatelperronian, Aurignacian, and Gravettian periods). In the Solutrean period, certain bone heads were replaced by very beautiful flint heads referred to as "laurel leaf" or "polar leaf." The series of reindeer-antler javelins reappears at the beginning of the Magdalenian period (although actually it had continued uninterrupted during the Solutrean period), and the forms become increasingly cylindrical, longer, and more narrow (Figure 15), and frequently have one or two long, deep grooves often believed to have contained a poison. Though as yet unsupported by facts, this hypothesis should not be entirely discounted. A javelin of this type was recently found in which the groove was filled with very fine flint splinters, probably held in place by a resin, that form a finely barbed cutting edge.

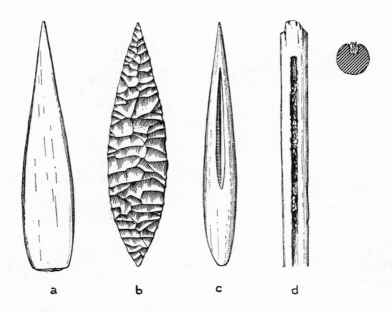

FIG. 15.   Evolution of spearheads in the Upper Paleolithic period.
(*a*) Bone spear, Upper Aurignacian. (*b*) Solutrean flint spearhead ("laurel leaf"). (*c*) Early Magdalenian reindeer-antler spearhead. (*d*) Fragment of a spearhead comparable in size to Figure 6*g*. The groove contains fine flint splinters. From the grotto of Saint-Marcel, Indre. (Excavations of Dr. Allain.)

Javelin-throwing in the Reindeer Age raises interesting questions. Given their diameter — around fifteen millimeters — we can only imagine that they compensated by their speed for what they lacked in mass on account of their lightness. In the modern world this problem of ballistics is solved by the bow or the projector. The bow may have existed in the Upper Paleolithic period, but unfortunately direct evidence is completely lacking. Though a few drawings on cave walls have been variously interpreted as feathered arrows or harpoon tips, these are probably abstract signs. Feathering would not, however, exclude their use as javelins, for the modern weapons whose ivory heads bear the strongest resemblance to those of the Magdalenian weapons are certain small feathered harpoons of the Eskimos of southern Alaska.

The second solution, common in recent centuries to a large area of America, New Guinea, and Australia, is that of the thrower, made of a small stick or rod about a foot long (or slightly longer), equipped with a cupped hook at its free

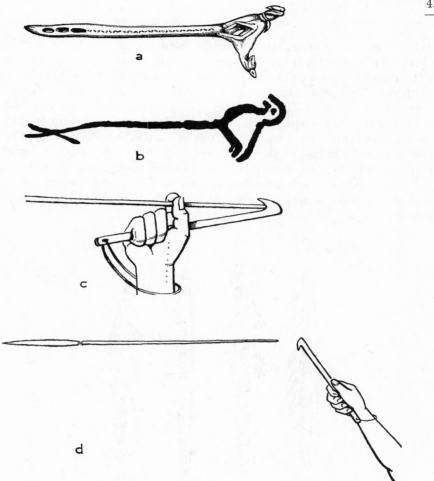

Fɪɢ. 16. Object made of reindeer antler that very probably served as a spear-thrower. The numerous prehistoric spear-throwers are decorated in various ways, but invariably have the same functional characteristics: a hook or bulge at one end and a perforation in the other. The function of the hook is illustrated in *c* and *d*. The perforation (or sometimes three perforations) could hold either a bar for the hand (a similar device is found among various contemporary peoples) or a thong that bound the thrower to the wrist and left the right hand free after throwing. Item *b* shows an object painted near the picture of a man in the grotto of Lascaux; it apparently represents a spear-thrower decorated with a bird (an actual example exists from the Magdalenian period), equipped with a device for holding it.

end and often with rings or notches so that it can be held in the hand by its other end (Figure 16). The javelin is laid on the thrower, with its base resting on the hook, and the action of throwing by raising the thrower to vertical position hurls the weapon with considerable speed and precision. Such hooked rods, carved from reindeer antler, have frequently been found in Magdalenian dwelling places, and have long been believed, with great probability, to be spear-throwers.

Harpoons carved from reindeer antler were in use in the second half of the Magdalenian period. They consisted of a cylindrical core, originally barbed on only one side, with a bulge (or sometimes even a ring) at the base for attaching a thong. They seem identical to the numerous barbed male harpoons still used in numerous regions of the world for the capture of large fish. We may thus assume that a portion of the Magdalenian "harpoons" were in fact harpoons and were probably intended for use in salmon or pike fishing. The very late Magdalenian harpoons, however, acquired enormous angular barbs not in keeping with their supposed purpose. The terminal point is often blunted or the barbs are reversed, pointing toward the tip. Such objects may have been used in groups of twos and threes to decorate the end of pronged harpoons (Figure 17).

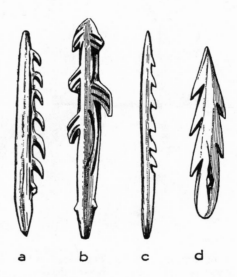

a   b   c   d

FIG. 17. (*a* and *b*) Magdalenian "harpoons" of reindeer antler, with single and double rows of barbs. (*c*) Maglemosian arrowhead, deer antler. (*d*) Azilian harpoon, deer antler.

The Mesolithic and Neolithic hunters

The Maglemosians had long deer-antler javelins, barbed on one side, and remarkably shaped. Being of a standardized type, they could have had several uses — as heads for javelins or harpoons. The Azilians possessed a flat harpoon with an eyelet for the attachment of a strap, derived from the Magdalenian harpoons with eyes of northern Spain. Hardly any other information is known about the Mesolithic hunters. The boneworking industry experienced considerable regression during this period: the flint industry consisted partly of microliths of geometric shapes that could equally well have armed arrows, javelins, or sickles for grain reaping.

The first agricultural peoples indisputably did hunt with the bow. Beginning with the Neolithic period, we discover numerous arrowheads, and even bows have been discovered at lacustrine sites, where fragments of fishing nets have been found under the same conditions. Wild animals still represent approximately 20 percent of the alimentary bone remains among the Swiss agricultural peoples of around the year 2000 before our era.

FOOD GATHERING

The fact that only the bone remains of game animals have been preserved in the prehistoric geological layers should not cause us to underestimate the importance of vegetable products in primitive man's diet. The Eskimos themselves, although ill favored from the botanical point of view, make the best possible use of berries, tender shoots, marine algae, and predigested lichens removed from the stomachs of reindeer. Among the hunting peoples of warmer regions—the Australian aborigines, the bushmen, and the pygmies — food gathering is quite important. The life economy is maintained in a constantly precarious balance between the produce of the male hunting activity and the gathering of plants or insects, which is the women's job. Unfortunately, agricultural produce is extremely perishable, and no traces of it have been found for the prehistoric period as such. Not until the Neolithic period do we find indications of such products in the lacustrine settlements, while for the Bronze Age the analysis of the contents of the stomachs of mummies found in bogs in Denmark enables us to understand the very important role played by wild grains.

Thus only very indirect evidence links food-gathering activity to the prehistoric hunters. The first evidence is paleontological: the teeth of the entire primate group from the very beginning were adapted to the consumption of a variety of pulpy foods. Depending on the species, from a functional point of view this dental equipment resembled that of the insectivores, the fruit- and grain-eating rodents, and badgers and bears with a mixed carnivorous and vegetarian diet, and is as far removed from the diet of the grass- of leaf-eaters as it is from that of the exclusive meat-or bone-eaters. The anthropoid group in general possessed these characteristics, beginning with the Australopithecines and continuing down to contemporary man. By his anatomical constitution man has a mixed diet. In the modern world this diet has been greatly modified in one way or another, but as regards the past it is easier to imagine peoples completely specialized in the gathering of plants, insects, and shellfish than in hunting alone.

The division of labor　　When did the tasks of hunting and food gathering come to be divided between men and women? It is impossible to determine this factually, but contemporary or recent examples suggest that this division corresponds to a functional arrangement directly linked to the primitive economy. For various physiological and psychological reasons, the division of activity finds its natural equilibrium in the assignment of hunting to the men and food gathering to the women. Even though the ac-

Plate 1.

a. Italy (Lombardy, Val Camonica), First Age of Metals.
 Agricultural scene. *Photo Emmanuel Anati*

b. Italy (Lombardy, Val Camonica). Rock engraving, end of
 the Bronze Age or the First Iron Age. At right is the smith,
 who seems to be beating a piece of metal on an anvil.
 Photo Emmanuel Anati

tivity of food gathering requires considerable effort ranging over long distances, its rhythm remains compatible with the mobility of the women (a group that includes preadolescent children). Thus, although no material evidence exists, it can be presumed that as soon as a community settles down into a mixed economy of the primitive type, this division tends to appear.

Fossilized plant remains There is a final method of research, still largely unexplored, which consists in establishing a list of plants whose alimentary use was possible in each climatic period. The identification of charcoal found in prehistoric hearths sometimes permits investigators to explain a small portion of the problem: the presence of charcoal from chestnut or walnut wood, for example, is an indication of the possible use of these nuts. The search for charcoal is uncertain for two reasons: first, because wood charcoal is preserved only under conditions of chemical isolation that are very seldom present, and second, because a large number of alimentary plants have no combustible value. Much more substantial information is gleaned from the study of fossilized pollen. Long preservation of pollens and spores is even more frequent than the preservation of bones, and very often in the prehistoric sites where the industry and the fauna permit the reconstruction of a portion of the behavior of the prehistoric inhabitants, the state of the contemporary flora can accurately be determined.

Plant foodstuffs Comparison of the lists thus established with the known alimentary habits of various peoples of historic times indicates that the number of plants capable of contributing to the food supply of Paleolithic man is considerable. In addition to hazelnuts, acorns, chestnuts, and berries, whose consumption may seem obvious, the inner bark of conifers and young willow and birch shoots could have been eaten. The herbaceous plants furnished numerous resources: dryas, anemones, nettles. The small chenopodium, violet, and starwort seeds, water-lily and water-caltrop seeds, the pulpy roots of Solomon's seal and goose grass, were definitely eaten in the historic or protohistoric period, as were ferns and mushrooms. The few species mentioned here relate only to the temperate regions of Europe, less favored by comparison with the warmer areas, where the cereal grasses, tubers, and fruit trees furnished a much larger contribution.

Until now it has not been possible to distinguish among the prehistoric implements those that could be connected with food gathering. The mills and grinders of the Upper Paleolithic period are connected with the preparation of colors (ocher and manganese), and not until the Mesolithic period can the existence of knives for the harvesting of cereal grasses be presumed. This question will again be considered in connection with protoagriculture.

PREPARATION AND CONSUMPTION OF FOODSTUFFS

Until upward of 4,000 years before our era, that is, until the appearance of the first pottery, evidence of the preparation and cooking of foodstuffs is practically nonexistent. The few known facts, such as the roasting by Neanderthal man of the ends of large bones in order to eat the spongy tissue impregnated with marrow, the probability of a ventilating system in several Upper Paleolithic hearths, and the probable existence of small ovens dug in the ground for the steaming of foods, have already been mentioned in connection with fire. The existence of leather bottles and utensils of bark or wood is very probable, at least in the Upper Paleolithic period. It is also possible that heated stones were used in these vessels for the cooking of foods; precise observations during excavations in certain dwelling places where the objects have not been disturbed could help solve this problem.

Comparison with contemporary primitive peoples whose way of life is comparable to that of the prehistoric hunting peoples furnishes information only on what was possible. This information shows that there does not exist a type of behavior sufficiently uniform to be projected with certainty into the past. The Australian aborigines and the Fuegians did not know the practice of cooking in utensils; animals or plant foods were roasted over a flame or in the ashes. Beginning at least in the Mousterian period, the same practices can be attributed to prehistoric man. The bushmen use the same methods, but in addition utilize imported utensils for the making of stews, and grind meat in wooden mortars.

The Eskimos have two diametrically opposed methods of consumption. A major portion of their meat and almost all of their vegetables are eaten raw, the meat being fresh or gamy. The other method consists of stewing in pottery or stone utensils, possibly utilizing the blood of the animals. They are unacquainted with roasted meat; wood is almost nonexistent in their country, and all hot food is prepared over oil lamps.

We possess numerous oil lamps from the Magdalenian period, but nothing leads us to suppose that they were used for anything but lighting. If we attribute to the Magdalenians a method of cooking other than roasting, it can hardly be conceived as anything more than the so-called "Polynesian" oven. This is a kind of pit in which coals are covered with a layer of cinders. The food is arranged in leaves of leather or bark utensils and finally covered with a layer of earth. A discovery made in the Ukrainian loess seems to confirm the existence of this method (Figure 18).

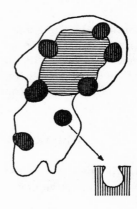

Fig. 18. Plan of the location of a hearth in an upper Paleolithic dwelling in the Ukraine. The main hearth (horizontally hatched area) is surrounded by a series of small pot-shaped pits that were probably used as ovens.

Meat pastes prepared in stone or wooden mortars are quite common, particularly in North America. This practice is necessary for the complete consumption of shelled animals; it probably was practiced in the prehistoric period, but cannot be proved, for pebbles were often used as hammers in dwelling places, in addition to serving other functions of which the grinding of meat leaves the fewest visible traces. The famous Neanderthal skull of La Chapelle-aux-Saints reveals that when its possesser died he had long been practically toothless. We may suppose that, like contemporary primitive people who find themselves in the same situation, he ground up his food by hand, with a hammer.

Minor equipment connected with food is very scarce. Awls could have been used to extract marrow from bones. From the Magdalenian period there exist flat bone spatulas, very carefully decorated and often polished by long use, but nothing proves that they were used in food consumption. In the same period, and apparently closely related to those spatulas, there exist real spoons with very slight hollows. Their rarity (only three or four have been found) suggests they were not in common domestic use. However, objects of the same type have been found in the early stage of the Mesolithic period — the Maglemosian. Beginning in the Neolithic period, genuine spoons appear, linked with the spread of pottery vessels.

CLOTHING

This is one of the subjects in which science-fiction has taken the greatest interest. Ape-men, sometimes draped in skins carelessly knotted over their shoulders, people the illustrations in popular books. Our actual knowledge is extremely limited, and it would be very difficult to prove to the artists who make reconstructions that Neanderthal man did not tie his wolfskin by its paws around his waist.

The pelt of the primates was not inferior in protective qualities to that of the other furred animals. Only the large anthropoids, especially the chimpanzee and the orangoutan, have a fur pelt that is sparse and lacking in down, but these are tropical monkeys unrelated to our ancestors. We imagine that the australopithecines, *Pithecanthropus,* and *Sinanthropus* were already much less hairy, like the orangoutan.

The only definite fact is that Neanderthal man carefully skinned wolves and reindeer, which proves that in winters comparable to ours he felt the need of additional protection. Neanderthal man is the very last arrival in the chain of ancestors of contemporary man, and it is reasonable to suppose that his hairy condition already tended to resemble our own. His garments can only vaguely be imagined, for his remains contain nothing that could have served as sewing equipment. The first bone awls appear at the end of the Mousterian period, which may justify the supposition that holes were pierced in the edges of skins to attach them together. The Mousterians, however, may have worn clothing much like that of the Fuegians, who lived practically naked all year round, protected from the wind, when necessary, only by a fur cape made by joining skins with tendons, with the assistance of a bone awl, or simply by sheltering under a seal-skin turned so as to break the wind.

This situation seems to have changed considerably in Upper Paleolithic times, especially after the Solutrean period, when needles with eyes appear. The climate appears to have been harsher than in the Mousterian period, and the survival of the hunters in winter is hardly conceivable without the protection of garments somewhat like those of contemporary Arctic peoples, from the Laplanders to the Eskimos. Several engraved slabs, and particularly a bas-relief from the beginning of the Magdalenian period depicting a male bust, show hunters wearing a garment with sleeves reminiscent of the Siberian parka or the Eskimo *anorak* (minus the hood). Information about their other garments, which may have included leggings and moccasins, is at present totally lacking.

Also lacking, unfortunately, is information about the succeeding periods; neither the Mesolithic nor the Neolithic sites have furnished the slightest evidence of clothing. The existence of weaving in the Neolithic period must have transformed clothing and caused it to resemble that of contemporary agricultural peoples. Discoveries in the peat bogs of northern Europe show the women at the beginning of the metallurgical period wearing long chemises with sleeves, and skirts caught at the waist by a belt, the men dressed in tunics, capes, moccasins with leg thongs, and leather breeches — a costume characteristic of a part of northern Europe until modern times.

THE DWELLING PLACE

The true importance of the caves
The cave is usually considered as the home of primitive man, who was constantly disputing this dark, generally damp retreat with the bear and the lion. Nothing could be less exact, for the caves were dwelling places only for a very small minority of men whom we would not in all cases consider as exceptional.

Caves are an unusual geological accident found only in certain types of terrain that cover a fairly limited surface of the globe. Their interest lies less in their habitability than in the fact that the remains of the men who lived in them were preserved as if in hermetically sealed jars. The neutral clay soil permitted the fossilization of bone remains, and objects that make it possible to reconstruct the existence of our ancestors are found, under sometimes miraculous conditions, right where they were abandoned.

Where the loose, exposed soil has made possible a rapid burying of debris, as in the loess of central Europe, we find something still more eloquent: the remains of huts or tents. Thus the cave is simply an accident in prehistoric cultures, just as the troglodytic dwellings of certain calcareous regions are a very minor aspect of contemporary French architecture. They are, however, the source of the major part of our archaeological information. As is the case with modern man, the normal living condition of prehistoric man was the construction of shelters whose development must have varied, depending on period and climate; the caves played only the role of chance finds.

For the entire early Quarternary period, we have hardly any evidence other than the cave of Choukoutien, if we omit the australopithecine grottoes, in which living conditions seem incapable of being reconstructed. *Pithecanthropus* has been found in beds altered by streams; *Atlanthropus* is found in the sandy areas around springs; and Heidelberg man was discovered in fluvial gravels. Though they could have been transported from some distince by rivers if they did live on the riverbanks, no traces of their dwelling places could have survived. The dwelling of *Sinanthropus* in the Choukoutien cave has revealed no characteristic structures, only layers of bone debris from the meals of the inhabitants (or of the animals who may at times have replaced them in the cave), and hearths. What is known of the Mousterian dwellings, which date from a much later period, does not encourage us to image a very highly developed domestic organization among the archanthropines.

Traces of open-air dwellings From the period characterized in a major portion of Eurasia and Africa by the Levalloiso-Mousterian industries, there is an enormous dispersion of open-air dwellings. The vast majority of our information is obtained, not from grottoes, but has been discovered in layers of loess, in alluvial beds, or in areas swept by rain or wind. Under these conditions it is difficult, even when the vestiges of industry have not been disturbed, to uncover anything other than small areas. Such inhabited areas are known by the hundreds in Europe, as in the Sahara and the Near East. The frequency of the finds denotes a concentrated occupancy over a more or less long period of time. The destruction of everything not made of stone, and the elimination of the traces of hearths, make every reconstruction of dwellings illusory. It is certain that in the warm regions these could be reduced to practically nothing. Neither the few strips of bark supported on sticks that sheltered the Australian aborigines, nor the circular huts with frameworks of sticks covered with leaves or grass built by the bushmen, the African pygmies, and the Fuegians, leave much trace after they are abandoned, except for the marks of a fire and some scattered debris.

Evidence from the caves The only coherent information is that supplied by the caves. Since the grottoes furnished a natural roof, superstructures for the shelter are lacking or are unusual. Thus a dwelling in a grotto usually possesses a more "primitive" character that is not necessarily an index of the technical level reached by its occupant. The very variable configuration of the caves, the presence or absence of drafts, the presence of flowing water at certain points, exposure to sun in various portions, and the possibilities offered for installing a shelter or for protection against animals make each case a special one. Lastly, the flooring of the grotto, which was sometimes clay, sand, or dust, sometimes shifting gravel, modifies the conditions of layout and preservation of the infrastructures of the dwelling. The hearth, ringed by rocks in one case, may be simply dug out in another; traces of posts will disappear in gravel, whereas in a clay soil they will be preserved.

In the caves of western Europe (the only ones for which some facts have been collected), the floor usually consists of gravel more or less supported by clay, which forms a waterproof base. This purely natural surface has in many cases been greatly worn by the passage of men and animals. The ground is littered with bone remains of all kinds; in most cases the carbonaceous remains of the hearth have disappeared, but small areas are sometimes found in which the stones have acquired a characteristic red coloring.

A single group of dwellings — that of Arcy-sur-Cure (Yonne) — has been studied in detail, and its study permits the approximate reconstruction of the dwellings of two groups of Mousterians, visibly dating from the same period. One in a fairly large hall lighted by daylight, the other, on the contrary, in a narrow gallery joined by a trench to a small cave opening onto daylight. In neither case did the ceiling height reach five feet. No deliberate construction is apparent, but the center of the floor, an area about ten to thirteen feet in diameter in the first grotto, was relatively well cleared, the cumbersome bone remains being pushed toward the walls (Figure 19). Small hearths were used, but

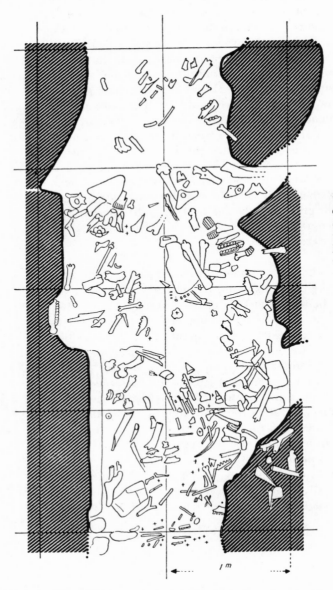

Fig. 19. Plan of a Mousterian dwelling in the Reindeer Grotto, Arcy-sur-Cure (Yonne). Broken bones litter the ground. The largest ones have been pushed toward the walls.

1 m

nowhere do they show signs of intense and prolonged use. In the gallery, where the darkness is total, reddened stones are more abundant. A small mammoth tusk placed in a corner, and a heap of calcareous stones and large fragments of sawed-up bones, are the only indications of interior arrangement. The presence of the signs mentioned above, testifying to the use of reindeer and wolf skins, permits us to suppose that such skins were used as sleeping mats. Rush pollens and fern spores, perhaps of natural origin, raise the possibility that these plants were spread on the ground to protect the fur pelts from the damp, uneven ground.

The Upper Paleolithic period As in the preceding stage, open-air dwellings were the normal form of habitation, but whereas the Mousterians have left no evidence of their buildings, the men of the Reindeer Age have on more than one occasion revealed to us the infrastructures of their huts and tents.

The most remarkable of these vestiges of dwellings have been found in central and eastern Europe, in the layers of loess. There bone remains are preserved and the smallest details of the original ground layout have remained visible. Each year the Russian excavations uncover new examples of buildings that on the whole must not have differed from those being constructed in western Europe during the same period.

An early form has a circular base, slightly more than six feet in diameter, formed of mammoth skulls or large bones intended to ensure the stability of a small dome-shaped building. Large skulls of more than six feet, embedded in the ground, formed the dome itself, which could have been covered either with skins, bark, or sods. Scapulas or pelvic bones placed on the covering seem to have assured stability against the wind (Figure 20). In certain cases the interior was belowground, hollowed out in the form of a pit, and it seems that large bones were sometimes used to support the sides of the opening. The structure must have been reminiscent of an igloo, or even more of the houses of whale ribs and jawbones constructed by the Eskimos several centuries ago. Around the dwellings, pits were dug in which the bones of game animals were piled — a practice whose utilitarian nature is not clear but which may have a religious significance, for modern examples of accumulations of this type have been found. In some cases the hearths were placed outside the dwellings, in others, inside. These hearths were often shaped like small pits, and were sometimes supplied with dampers.

Various types of dwellings existed during the Upper Paleolithic period on the loess of eastern Europe. One of the strangest is a dwelling perhaps ninety eight feet long, which was probably covered with mammoth hides. Inside, a line of about ten hearths seems to have corresponded to partitioning by family. Details of furnishing are not precisely known; however, it seems that in circular dwellings a sleeping pallet, probably made of skins placed over a mattress made of plant matter, was placed against the outer wall.

In the Magdalenian sites of northern Germany, stone circles marking the location of tents have been found comparable to those of most of the Eurasian and American peoples of the northern regions. In the other European coun-

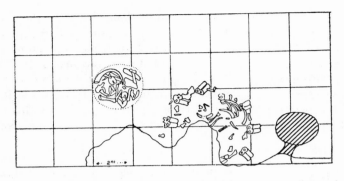

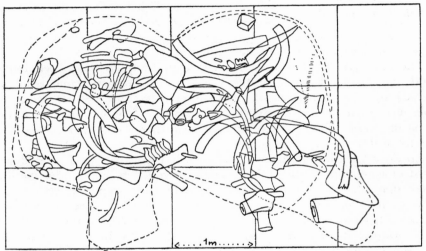

Fig. 20. Plans of Upper Paleolithic dwellings in the Ukraine.
(*Top*) The Dobranitchevski group, including a circular hut surrounded by fragments of mammoth skulls (the front portion has been destroyed), a hearth outside the hut, and, behind it, a ditch for bones. (*Below*) Plan of a pit with two sections, Kostienki. The entrance (*below, right*) seems to have been supported by two mammoth scapulas; inside, large tusks supported the covering.

tries, particularly in France, no such observation has yet been made. On the other hand we possess, despite the often very rudimentary character of the structural observations made by western-European investigators, some information on the organization of the cave dwellings during the Reindeer Age.

Improvements of the ground

The most striking fact is the treatment of the flooring. Beginning in the Chatelperronian period, the damp clay surfaces were smoothed out and covered with flat stones sometimes arranged in a fairly regular pattern (without, however, forming a true flooring). The cleanliness of the dwelling is another contrast with the Mousterian dwelling: large bone remains are eliminated, except when they are used as a source of raw material or as a worktable. Remains of food, when they did not disappear as combustible materials into the hearths, were thrown outside. Unfortunately for the archaeologist, we do not find, as is the case in the Soviet Union, pits of bone remains and waste material.

The cave floor presents another, universal, characteristic: the presence of red ocher in sometimes extraordinary quantity. Even if we accept the idea that all the objects, all the skins, and all the bodies of all the inhabitants were thickly coated with ocher, and this over a period of centuries, it is difficult to understand why certain cave floors eight inches thick are composed almost solely of ashes and ocher. It seems that ocher was spread over the ground, probably in small quantities, but repeatedly.

Interior layout   Although few infrastructures have been discovered, rocks lined up to form the base of an enclosure have been found in shelters under other rocks or at the entrance to grottoes. The interior layout must have frequently included structures that left traces in the ground, but the only observation made (at Arcy-sur-Cure) shows cylindrical holes dug in the clay soil; these served for the embedding of mammoth tusks that formed the framework of a screen or a tent constructed at the entrance to the grotto (Figure 21). Crevices or niches in the walls were used as shelves

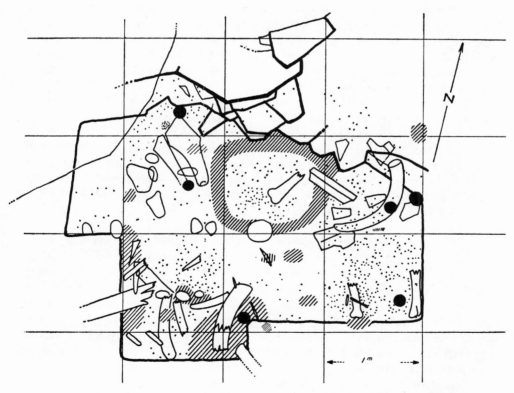

FIG. 21. Part of the floor of a dwelling, beginning of the Upper Paleolithic (Chatelperronian), in the Reindeer Grotto at Arcy-sur-Cure. The hatched portions indicate the location of hearths or cinder remains; the dots represent the location of flint implements. Note the six postholes and the mammoth tusks.

or closets, and objects have sometimes been found in them. The discoveries of caves blocked up after their abandonment, or which had not been entered for thousands of years, are relatively frequent. Their discovery generally results in a disorderly collection of the irreplaceable evidence they contain, as in the sanctuaries of Lascaux, Gabillou, and Labastide. Stone lamps, small hearths, and fragments of ocher for paintings were still in place, and in the last-named grotto engraved plaquettes of bisons or horses were still in the position in which they had been abandoned 15,000 years ago.

The best studied of the Magdalenian dwellings is that of Saint-Marcel (Indre), consisting of two grottoes that were successively occupied. One of them was discovered absolutely intact. In an alcove in the rock were found reindeer antlers ready to be made into javelins, and a reserve supply of javelins, still tied together in a bundle. The crevices in the lower grotto still contained javelins thrust in one next to the other, with flint scrapers standing on the ledges. The hearths, covered with heaps of stones that acted as heat conveyors, were found just as they had been extinguished. It seems, in addition, that the walls were decorated with massacres of wild oxen formed of the horns and portions of the skull, for several of these enormous remains were found in the position they would have taken in falling off the walls.

The lighting system of Upper Paleolithic man is known from numerous lamps made of stones hollowed out either by nature or by human hands; sometimes signs of combustion are still found on their edges, and they must have contained a wick fed by soot. These lamps, which provided a very satisfactory light, permitted the Magdalenians to examine the fastnesses of the caves and to execute their cave drawings and engravings.

A final detail of the layout is still a subject of discussion. We occasionally discover in the walls of rock shelters rings engraved in the rock itself. These rings seem to have been intended for the attaching of curtains, and it is thought that they could have been used to close the front opening of the shelter with skins. However, they have been found only in shelters in which the wall was carved with animals in bas-relief, and their position generally does not correspond with the most favorable use for a dwelling place. More probably, they were connected with religious practices.

Habitation in the Mesolithic period

Remains from the Mesolithic period are scarce. There is a Maglemosian site in Yorkshire, where the Maglemosians settled on the edge of a swamp, building a flooring of branches and pieces of birchbark to protect themselves from the dampness. Trees were felled, but no details are yet known about the dwelling itself. Almost nothing is known of the Azilians. The only detail relating to their housing comes from a cave in Cantabrian Spain, where postholes dug vertically inside the cave justify the supposition that they, like Paleolithic man, sometimes erected additional structures inside the caves.

Except for the remains in the Soviet Union, the history of primitive dwellings is still almost entirely based on suppositions. The scarcity of information results partly from the destruction of the fragile structures of the huts or tents,

Plate 2.
a. Switzerland (Lake Neuchâtel, Auvernier). Woven straw comb
found on the flooring of a lacustrine hut.
Photo Leroi-Gourhan
b. Grotto of Montespan, Ariège. Engraving on clay of a horse.
The picture has been disfigured by blows. Magdalenian
(12,000 to 10,000 B.C.). *Photo Leroi-Gourhan*

but still more from the fact that, especially in France, the goal of the excavations, even when properly done, is too often the discovery of objects characteristic of each age, and they are less directly concerned with the traces of the actual existence of human beings — traces whose discovery and recording require a great deal of time and care in the·examination of the flooring.

CHAPTER 2

THE FIRST AGRICULTURAL SOCIETIES

THE HISTORY of the revolution that occurred in human societies between 5,000 and 3,000 years before our era can be understood only if the study of the technological evolution is supported by a study of geographic and economic conditions in which this evolution occurred.

PRIMITIVE ECONOMY AND TECHNOLOGY

When we consider as a group the prehistoric societies (from the early stages to Mesolithic times) and recent or contemporary societies whose economy has continued to be based on the exploitation of natural resources, it is evident that they belong to a similar economic and technological system. While the identification of prehistoric equipment or practices based on comparison with contemporary primitive peoples has many pitfalls and frequently leads to errors, the parallels in the structure itself reveal a general similarity comparable to that which may exist (ignoring all ethnographic details) between the Sudanese peasant, the Chinese peasant, and the French peasant of the medieval period. A consideration of this technoeconomic infrastructure of the primitive world is of prime importance for an understanding of the causes of the subsequent revolution.

The technological equilibrium

Even when natural plant and animal food sources are abundant (which is far from having been generally true), the density of a primitive human group cannot surpass a certain level determined by the range of activity of the food-gatherers and the density of the supply of edible plants and animals. Under the best known conditions, a society can temporarily include several hundred individuals, but normally it achieves a balance with its surroundings when it consists of permanent groups of several families at most, the various permanent units forming a loose network over a wide territory. Each unit is characterized by diversified technological achievements and an economic autarchy that in theory is absolute. In actual fact, there is an exchange of certain products or objects that are rare or cannot be found everywhere in the territory. These exchanges play an important role in the evolution of techniques and institutions, but this role follows the general rhythm of development of the entire community, a rhythm that slows down the closer the conditions of the natural surroundings match human consumption. Thus the primitive economy is by turns almost closed, stable, or superabundant for short periods such as a successful hunt or the harvest period of a product that exists in superabundance in nature. Except among the Eskimos, who are restricted by climate, stockpiling is practically non-existent.

Since the limited family group forms the functional basis of small communities, the entire technical activity is concentrated among several individuals of both sexes, all absorbed in operations of acquisition, production, and consumption that have been rigorously tested over millennia and the use of which is balanced rather than rigidly fixed.

Causes of their evolution   These techniques evolve and are perfected over a very long period, as appears throughout the long development of the Quaternary period. But their modification is invisible to the people involved in it. It seems that the stimulus for the modification of techniques cannot come from a direct relationship between acquisition and production: the gradual perfection of the javelin leads to an admirable harmony between the use of bone and the killing of game animals, but not to metallurgy or a radical economic transformation. The intervention of an external influence is indispensable for initiating a movement in a new direction. This influence may be negative (for example, the growing scarcity of a portion of the resources; in this regard changes in climate have been able to play an important role), or positive (such as contact with societies of a different economic type, or the appearance, for climatic reasons, of new resources). In the change that occurred at the end of the prehistoric period, human factors could not play any role, since the entire world was in a primitive economic state, and climatic changes must be considered first and foremost. The situation will be different once the change of economic system has begun and is overtaking the world despite climatic conditions (excepting only regions that are isolated or subject to temperature conditions incompatible with the new techniques of production).

The importance of climatic stimulation varies according to latitude. Changes in the Torrid Zone could not have been sufficiently intense to affect a perceptible transformation in their resources. The more northerly zones were excluded by virtue of the fact that a moderating process involving the liberation of surfaces covered by glaciers could draw the primitive populations from south to north but, for lack of a population already settled there, could not create the conditions of a favorable environment. Thus the favored zone was limited to a climatic strip that included the temperate regions of the Old and New worlds, a strip that in the Old World followed the parallels, and in the New World corresponded to areas delimited by the mountain ranges of the Rockies and the Andes.

THE "MESOLITHIC" PERIOD

The term "Mesolithic" was invented when it was realized at the end of the nineteenth century, that for the western regions of Europe a considerable time lapse separated the Magdalenian (the last stage of the Paleolithic period) and the Neolithic, characterized by the appearance of polished stone implements and agriculture. The technological notion of "Mesolithic" has been greatly

weakened by subsequent research, the division between the Magdalenian and
the Maglemosian or the Azilian appearing less clear-cut than was at first thought.
If in its early stages the "Mesolithic" period has lost precision, it has become
completely blurred in its terminal stage, when such cultures as the Tardenoisian
or the Campignian now appear to have closely overlapped in the early agricultural periods. These distinctions that characterized a portion of the prehistoric
period become still more difficult to maintain when we consider the rest of
western Eurasia. Nevertheless, the terms "Mesolithic" and "Neolithic" are retained because, provided the necessary distinctions are made, they characterize
a period of transition and intense, rapid transformation that constitutes the
short preface to modern times.

The influence of climate The most important factor is climate: beginning
around 8,000 years before our era, a warming
process created botanical conditions similar to contemporary ones. The Mediterranean regions were still scarcely affected by the progressive desiccation, and
were to constitute an environment favorable to a new economic adjustment.
Such warming periods had intervened on several occasions in the past, but at
stages in the evolution of human beings and human technology that were not
conducive to a decisive development. By the time of the last warming period,
in contrast, the hunters and food-gatherers had attained a considerable degree
of maturity in technological matters, and only a few geographic coincidences
were required for the completion of the transition to agriculture and pastoralism.

The center of the evolutionary movement of the Eurasian temperate regions appears, in the light of our present knowledge, to have extended from
Egypt to the Caspian Sea. Between 7,000 and 5,000 years before our era, the
Near East was occupied by populations whose economy was still primitive but
among whom the gathering of plant foods had surpassed that of the large mammals. In these regions, where the ancestors of present-day wheat and barley grew
wild and, to the east, the wild sheep and the wild goat were the principal game
animals, the Mesolithic communities had to base their technoeconomic apparatus on the resources that were to become, several millennia later, the foundation
of the agricultural economy.

Influence of
intercultural relations The systematic exploitation of wild plant products
as a primary nutritional base spread in a relatively
short time over a major portion of Eurasia and
Africa. It was certainly favored, around the Mediterranean Sea and in Europe,
by the botanical evolution and, for France in particular, by the growing scarcity
of large game animals. Intercultural relations, however, definitely played a major
role. The evolution of tools seems to have taken place simultaneously over a
very large territory. The uncertainty of absolute dates for each cultural group
makes it still difficult to determine the true nature of this apparent synchronism.
The appearance of the ax and the adze, which imply a total revolution in woodworking, the increasingly geometric character of the microlithic flint implements,
the colonization of sandy mounds near rivers, could have occurred separately
in various regions, since the traditions of microlithic flint chipping were already

in existence and corresponded to the normal evolution of production, while the extensive forest area led to the appearance of bulky tools and colonization on its outskirts, near water (Figure 22). It is much more likely, however, that phenomena of cultural osmosis, made possible by the community of new needs, played an important role; not until the dating of the various Mesolithic cultures has been more exactly established will we be able finally to define them.

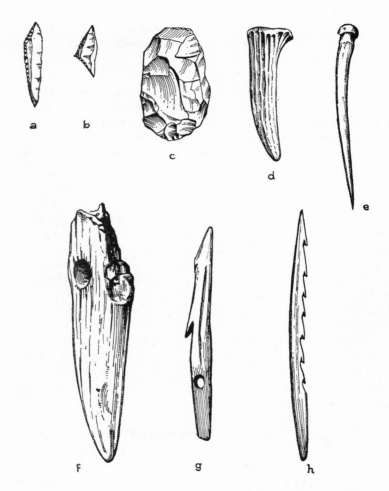

FIG. 22.   Maglemosian implements from Star Carr, Yorkshire.
(*a* and *b*) Flint implements. (*c*) Ax or adze, chipped flint. (*d*) Wedge or pickax, deer antler. (*e*) Bone awl. (*f*) Implement of moose antler, with a hole for a handle. (*g*) Harpoon, deer antler. (*h*) Arrow tip, deer antler. (From John G. D. Clark, *Prehistoric Europe*.)

The origin of the transition to agriculture begins to emerge through the study of the Mesolithic cultures. Among them it is the cultures whose natural resources correspond to the resources of the earliest farmers that deserve our greatest attention. This leads us to search for the key to the agricultural problem in the Near East of the ninth millennium.

PROTOPASTORALISM AND PROTOAGRICULTURE

The birth of agriculture and pastoralism can be conceived as sudden, separate phenomena. But the evidence supplied by archaeology and ethnology tends to make such a hypothesis improbable. Rather it indicates the gradual emergence of agriculture and pastoralism from within a cultural ensemble in which favorable conditions were present together during a sufficiently long succession of centuries. The combining of the conditions of a gradual transition from hunting to cattle raising, or from food gathering to agriculture, is a matter of chance; nevertheless, the testimony of several recent ethnic groups who were placed in conditions comparable to those of the Mesolithic peoples of the Near East is of some use to us.

The migrations of the reindeer The best example, as far as pastoralism is concerned, is that of nothern Lapland. The reindeer flocks migrate from summer to winter along the short valleys perpendicular to the sea, alternating between the mountains and the coast. The ethnic groups in Lapland occupied every one of these valleys, and consequently were included in the system of migration of a given reindeer population, whether wild or domesticated. Convincing proof exists for the two forms of exploitation of this animal stock in its alternating migratory movement within the partitioned valleys; in the course of recent centuries, depending on the region, hunting or cattle raising, or both of them on different territories, have been practiced. What is striking is the fact that whether the packsaddle or the sleigh is used, the methods involved in both cases are identical: the assistance of the dog, and familiarization with the movements of an animal population that, whether game or livestock, periodically moves about within a limited area. In only one other area of the immense arctic region do the migrations of the reindeer, whether wild or domesticated, have the character of limited movements between the sea and the mountains: in the territory of the Chukchis in eastern Siberia, where exactly the same forms of husbandry exist. These consist in enclosing and protecting, while following its seasonal movements, a herbivorous, gregarious animal that is compelled by geographical conditions to change altitude seasonally and consequently within a limited geographical area.

Proper geographical conditions makes such forms of protopastoralism in the area between the Caucasus, Persia, Turkey, Mesopotamia, and the coastline of Syria, Lebanon, and Palestine very probable. The raising of sheep and goats has continued through the centuries in the folds of the natural topography. Unfortunately, it is difficult to uncover archaeological proof, since the transitions from hunting to domestic animal breeding are imperceptible, and nothing in the skeletons of the animals discovered in the course of excavations indicates whether they were game animals or livestock. On the other hand, such an explanation of the facts does correspond to the normal characteristics of the cultural evolution.

The gathering of wild rice

The case of agriculture is not greatly different. It implies the existence of a plant crop that can be harvested over a relatively limited area and stored for a period of a year. Except for those plants that because of their very small size or their dispersion would require a disproportionately large harvesting effort, the plants whose fruit or seeds answer to these requirements are fairly numerous. In the regions under consideration, however, the wild grains have the maximum number of advantages. In the contemporary world there are numerous examples of food-gathering peoples; the best example is that of the "wild-rice Indians" who as late as the nineteenth century occupied the region to the north of Chicago. Wild rice (*Zizania aquatica*) covered vast stretches of swampland, and furnished the Indian populations with an important dietary supplement. Three groups in particular used this grain: the Dakotas, the Ojibwas, and the Menominees. What is important is the forms of exploitation of each of the three groups, which illustrate the gradations between simple food gathering and an already organized system of agriculture. The Dakota Indians belonged to the Sioux group, who were hunters of bison, and food-gatherers. Pushed back toward the wild-rice area, their exploitation was limited to food-gathering expeditions at the moment of the maturing of the grain, operations that frequently took place at the expense of other tribes, which had protected the grain from the birds. The Menominees belonged to the Algonquin group, and were hunters and food-gatherers in forests and swamps. Fishing, hunting swamp birds, and gathering of wild rice on the one hand, a few forest animals, forest berries, and maple syrup on the other, assured them an economic cycle sufficiently stable to make them practically a sedentary people. During its growing season the rice was protected by tying the heads in bundles, and the various sections of the swamp, like maple trees, were considered private property. The Ojibwas belonged to the Chippewa people, and had long cultivated Indian corn, although hunting and food gathering played a considerable role in their economy. Among them wild rice was given the same care as among the Menominees, but in addition a portion of the harvest was put back into the soil as seed.

These three recent examples of gradual adaptation of a wild grain probably do not duplicate exactly the line of development of early Mediterranean agriculture, for the Ojibwa rice sowers had already received knowledge of agriculture from other sources, and the Menominees probably were not the inventors of the protective binding of the heads. They do, however, furnish a series of images that can be transposed to the first farmers of Mediterranean wheat.

CHAPTER 3

THE TECHNOLOGICAL COMPLEX
OF THE NEOLITHIC PERIOD

FOR WESTERN ARCHAEOLOGY the Neolithic period corresponds to the remains discovered in lacustrine settlements in Switzerland, Germany, and France. Numerous other factors have been added to this traditional picture, but in essence it persists, whether on a conscious or unconscious level. Neolithic man was a farmer with oxen, sheep, goats, pigs, and a dog. Wheat and barley formed the basis of his agrarian economy; his technical equipment consisted in particular of the polished stone ax, basketmaking, weaving, and pottery — a totally new set of equipment in comparison with that of the last Magdalenians. We must realize that this picture of European Neolithic agriculture is more than 3,000 years later than that of the Near East, and 1,500 years after the appearance of metal in the Mediterranean area, and that its livestock and plant life consisted of imported species foreign to Europe.

In the Near East the contrast between Mesolithic, Neolithic, and the Age of Metals is almost imperceptible: Mesolithic man passed from food gathering to protoagriculture, settled down to being a farmer, and almost immediately became a metalworker. His equipment, unlike that of man in western Europe is not in striking contrast to that of still-primitive predecessors: element by element it improves in imperceptible transitions. The sickle for grain harvesting existed before it was possible to speak of a genuine agriculture (Figure 23), for the food-gatherers had already used it; the sheep and the goat were already in ex-

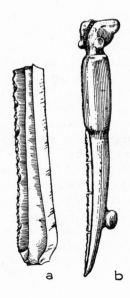

FIG. 23.   Flint sickles.
(a) Notched flint blade. Byblos (Lebanon), Neolithic.
(b) Bone sickle with short, notched flint blades. Kebbara (Israel), Mesolithic-Natufian.

a   b

istence as game animals, and the only moment that can be safely singled out is the period for which excavations reveal groups of people who are already settled on the land, organized into dense and coherent communities, in an economy that is both agricultural and pastoral. This point was reached between the years 7000 and 3000 in the area stretching from Egypt to the Caspian Sea and the Persian Gulf, in the course of an accelerated evolution that already indicates the existence of cities before 3000 B.C. It was from this large center that the agricultural evolution directly or indirectly spread over all Eurasia. The contrast between the explosive nature of the agricultural transformation and the sluggishness of the development of the Paleolithic culture over several hundreds of millennia is striking, and we must analyze the reasons that abruptly led the various populations into a series of technical innovations, the most important of which were metallurgy and pottery.

PERMANENT SETTLEMENT

It is impossible clearly to separate the nomadism of groups with a primitive economy and the immobility of the groups engaged in farming and cattle raising. Among numerous contemporary peoples, agricultural activity may include periods of the year when the community temporarily moves to a new area, or a system that details only a portion of the population to agricultural or pastoral activities, or an agricultural system that requires the relocating of the entire community after a certain number of years. All these forms, however, require that the community have a principal center of residence for a period of at least several years.

Appearance of groups of dwellings In the environment of the Torrid Zone, the nature of the products harvested and the alternation of harvest periods eliminate the need for stockpiling of food. But in the Mediterranean area, with its marked climatic contrasts, an economy that is even partly based on cereal products imposes conditions of immobility around the source of food. Toward 7000 in northern Iraq, 6000 in Mesopotamia, Syria, and Lebanon, 5000 in Egypt, and 4000 in the area from the Sudan to Baluchistan, we discover groups of dwellings, constructed of durable materials such as baked brick, sometimes with floors coated with clay or plaster, that suggest permanent settlements. The existence of the pig at Byblos suggests the same conclusion. From this point on, examples proliferate, and in the thousand years that follow, the influence of the agricultural Neolithic culture spreads like an oil stain over a portion of Africa and all of cultivable Eurasia. Long before primitive agriculture had spread to its borders, the Mediterranean heartland had already made the transition to the next stage and had begun the great metallurgical expansion, but this "border Neolithic" played a protracted role, and represents, particularly for Europe, a lengthy portion of protohistory.

The different periods of the countries involved have been variously named. In France, for example, there are "Campignian," "Cardial," "Danubian," "Chassian," and so on. The names are indicative of a general structure of Europe already comparable to its present structure, with several clearly defined areas that can be subdivided into large archaeological areas. The whole is in a state of rapid evolution.

The Neolithic villages of Europe   The structure of this protometallurgical society of the Neolithic period is indirectly apparent in the forms of its economy and its dwelling places. There exist genuine villages of huts, either rectangular or round, whose grouping, layout, and similarity of proportions suggest the juxtaposition of more or less identical family units (Figure 24). Traces of defense organization are not particularly pronounced; and the buildings are more often located near water than on hilltops, as will a little later be the case. There is nothing in the architecture to indicate a very pronounced social stratification. Living on the edges of forests, on the arable lands around valleys and lakes, raising their oxen, sheep, cows, goats, and pigs, living in rows of huts with thatched roofs and walls of wattling covered with clay, working the soil with hoes, the peasants of the various Neolithic cultures of Europe are reminiscent of the peasants of Africa, America, or

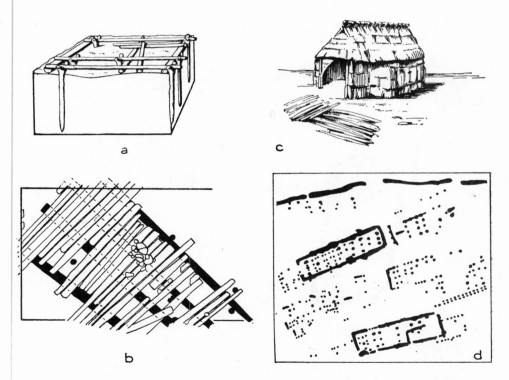

a

c

b

d

FIG. 24. Types of construction during the Neolithic age in western Europe. (*a*) Foundations of a "lacustrine" hut (from R. R. Schmidt); (*b*) Flooring and hearthstones of a hut similar to "*a*" (from D. Viollier); (*c*) Reconstruction of the general appearance of such a hut (from H. Reinerth); (*d*) Ground plan showing postholes and outlines of huts of a German Neolithic village. (From W. Buttler and W. Haberey.)

Oceania. Their social organization will probably never be known, but what remains of their material culture reveals neither palaces nor buildings recognizable as temples; their burial places contain no traces of major social distinctions between the individuals buried there. This does not preclude the existence of a hierarchy, for the necessities of collective defense or work existed; it is even possible that there were slaves and masters, if only because of the influence of the Mediterranean agrarian capitalism that was already in the process of development. The absence of the metalworking industry, however, maintained society within relatively narrow limits of production.

The lacustrine pseudocities   The European Neolithic period has long been considered the age of lake cities, and firmly rooted illusions still persist about this last remaining "scientific legend" of the prehistoric period. We now know that villages built on piles above lakes never existed. The huts, made of wooden posts embedded in the mud, were built on the riverbank, not over the water (Figure 25). The level of the lakes having changed, the rotted posts at ground level were submerged, and created the illusion of lacustrine settlements.

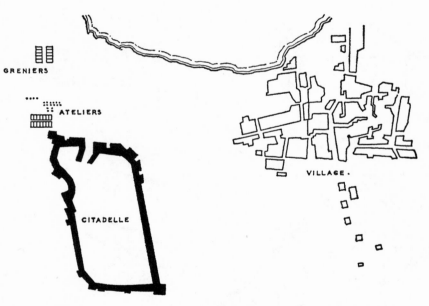

FIG. 25.   Plan of the stronghold of Harappa (Indus Valley), showing the location of the granaries and the artisans' quarter (from Wheeler).

The process of simultaneous evolution   What seems to have been for a long time the cause of a certain confusion in the archaeological synthesis is the fact that the phases of the general development from the Paleolithic period to the Bronze Age are not exactly the same in the birthplace of the agrarian structures as in its peripheral regions. Encouraged by the excavations in western Europe, the logical mind sees a progression, beginning with the Mesolithic cultures, of a primitive agriculture accompanied by a special stone-tool culture (in this case, polished stone). Then would come the appearance of metal: first copper (because it is a pure metal), then bronze (an alloy), and lastly iron. The truth is that while this development did occur, it was in the peripheral area, and it was less characteristic of the stages attained by technology at a given moment of history than of the gradual adaptation of regions distant from the center of origin. In the center itself, the various stages unfolded almost simultaneously: the Mesolithic was still in an evolutionary stage when protoagriculture and protopastoralism were already turning this evolution into a Neolithic culture; agricultural stabilization was hardly completed when the foundations of metallurgy were already in existence. The eastern Mediterranean area already possessed the beginnings of cities at a time when the western area had hardly had an opportunity to learn of the existence of agriculture. Only because of the lag caused by the expansion toward regions where the new state of affairs was not an evolution but a revolution does the logical stratification become apparent.

This phenomenon of adaptation in stages, with a polished stone phase far in advance of the popularization of metals, depends more on cybernetics than on history. It is found not only all around the Mediterranean heartland but also around the area of Chinese influence and the great American civilizations. Later, when ironworking is already an established fact over a large portion of the ancient continent, the same phenomenon will characterize part of agricultural Africa, Oceania, North and South America, forming an aureole of "Neolithic cultures" around the most highly developed cultural areas. The observation of these universal facts has led certain theoreticians to formulate the hypothesis of a "universal Neolithic civilization," a hypothesis that does not withstand an examination of the facts. Each "Neolithic culture" is formed as a direct function of the center that supplies it with new inventions. Thus we must reverse the logical development of the evolution, beginning with the metalworking cities of the Near East and ending with the villagers polishing their axes in the peripheral areas.

TOWARD THE CITY

Agricultural immobility, even when only partial, determined the appearance of a system of social and economic relationships profoundly different from those of the system of primitive economy. Food stockpiling played a decisive role in agricultural settlement. This role was expressed in the increase of the density of the group, whose functional base is no longer the limited family group of the couple and its immediate descendants or ancestors. The new base is the wider family group, the permanent aggregation of a family life. On the technoeconomic level, the increase in the members of the functional group ensures a greater latitude in the division of labor; certain members are able, on a part-time or full-time basis, to perform work not directly connected with the search for food. The appearance of techniques whose economic utility is not immediately apparent (for example, metallurgy) was possible only after the group was able to distribute among its members the burden of supporting the specialist. Here, again, this did not happen suddenly, and the artisan was not liberated overnight from food-gathering tasks, but some hours of artisanal activity became accessible to some of the members of the group (and to some more than to others).

Other factors played an important role, for example, the change in distribution of periods of food production and periods of intense mental activity. The high point of this change occurs among primitive peoples during hunting, among farmers on the margin of the stereotyped operational sequences of working the land. Equally important was the "creative urge" determined by the development of the new economic situation: invention centers upon methods of defense of the agrarian equipment (fortification and metallurgy), storage (basketweaving and storage pits), and the preparation of grain (grinding mills and pottery).

A network of needs and possibilities is consequently created in the functional group that constitutes the agricultural village. It is not limited to a single village, but includes geographical areas that increase in size over the centuries and lead to the interacting of a more or less large number of similar villages distributed over the areas of agricultural exploitation. This network of relationships, which is diametrically opposed to the primitive apparatus, is a dynamic one: the density of population is not limited by the density of its resources, but on the contrary the volume of the resources is a direct function of the increase in population. Since the number of extra-alimentary relationships increases in relation to the number of human beings and the "humanized" area, technology in particular is caught up in a rapid evolution.

Thus a direct relationship exists between the forming of a dwelling place, the increase in the density of the population, new techniques, and the social configuration, so that all these elements must be examined simultaneously in order to understand the technological evolution.

The cities   Between 4,000 and 2,000 years before our era, in the area stretching from Egypt to the Indus Valley, the agricultural system acquired the form that leads directly to the modern civilizations. The basic human groups assured the alimentary exploitation of the agricultural-pastoral areas. They became organized into more or less large territorial units, of similar ethnic characteristics, linked together by the community of the social, economic, and defensive organization. This bond was assured by a structure in which social specialization intervened to separate the defensive and religious authority from the peasant community. This separation is revealed in the appearance of areas of concentrated habitation, equipped for defense, which correspond to a new regulatory organ of community activities. It is the cities that house a major portion of the food supply, the military and religious leaders, and the most dynamic members of the artisan class (Figure 26). The functional apparatus gradually became such that a separation both social and territorial occurred between the rural majority engaged in food production and the urban minority engaged, on a level both secular and religious, in the management of the community wealth.

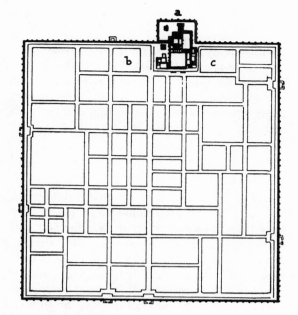

Fig. 26 Plan of the city of Khorsabad (Assyria). The city, geometrically laid out, housed government employees, artisans, and merchants. The palace buildings (*a*), major temples (*b*), and granaries (*c*) were grouped together.

This process is *civilization* in the strict sense of the word: a special form of the development of human societies that brings the city to the fore. From this point on, the units of civilization were brought together into a narrowly hierarchical system with a capital city, topographically structured within its ramparts, which housed the grain supply and the wealth, the king and the military dignitaries, the priests, craftsmen, servants, and slaves. This urban microcosm was linked with its food supply by means of a network of administrators. Below this was the peasantry, most often in a state of servitude, divided into rural units that preserved the Neolithic structure.

The countryside   At this point in the evolution of technology, the separation between the city, which is the source of innovations, and the countryside, which only later benefits from a portion of the technical progress, was about complete. The last to benefit from metal during the Bronze Age was agricultural equipment, which was still in too short supply to reach the peasant masses.

Thus, on the technoeconomic level, during the period of their early development (and in many regions right down to modern times) civilizations have a dual structure. The city possesses and promotes all the technical acquisitions. The countryside continues the basic "Neolithic" structure, that is, a communal habitat with identical family structures, agriculture and cattle raising, the techniques of basketweaving and weaving, which ensure the storage and handling of produce; clothing made from cultivated plant materials, and pottery shaped and fired by simple methods.

These two strata within the same ethnic system correspond to a technoeconomic formula of equilibrium, and they explain the existence of the Neolithic periphery that forms a growing border around the civilizing center. The first achievements often persist for centuries before reaching urban civilization. They occur in the rural stratum, accessible without the installation of the urban establishment. Assured of survival from their birth, thanks to local resources, cereals, domestic animals, basketmaking, weaving, and potmaking were established from China to Great Britain. It is not a matter of a beginning, or of primary structures, but of a partial borrowing, complemented, moreover, as regards armament and tool equipment, by polished stone objects that are substitutes for a still-inaccessible metallurgy.

POTTERY

There is as yet no evidence that would establish an exact place and time for the birth of the first molded and baked clay utensils. In light of our present knowledge, however, it seems that the first pottery cannot be older than 7000 B.C. In the old continent, it seems possible that it could have appeared in the Near-Eastern area of protoagriculture. This is quite surprising, for the raw material exists everywhere, and particularly in caves, where the Paleolithic hunters constantly came in contact with it and even molded it into figurines. Moreover, in the case of hearths located on clay, the surface beneath them is sometimes found to have become baked; many types of clay can be baked in an ordinary fire and without special preparation. There was no real technological obstacle to the birth of pottery, but its absence is no stranger than the universality of its diffusion after the beginning of early agriculture, as if there were an organic link between cereals and pottery.

The explanation most frequently advanced is that the nomadism of the hunter–food-gatherer groups was ill-suited to the long process of preparation and drying of vessels and to their fragility. Without being preponderant, this factor must have played a role. The principal reason for its development may have been the need for various types of utensils created by the development of harvesting, handling, storage, and cooking techniques for cereal grasses. A portion of these needs was satisfied by basket or bark vessels, the others by utensils of dried or baked clay. In addition to the growing need for containers as a result of the use of grain, there was the making of houses from hardened clay — an indirect result of permanent (or even semipermanent) settlements on the land. It is possible, and even probable, that the use of dried clay in the construction of walls, storage pits, hearthstones, and floors preceded pottery making as such, and that familiarization with sculptural materials made possible the associations of ideas conducive to the production of baked clay vessels. In contrast, "sun-baked" pottery exists, needless to say, only in the imaginations of a few theoreticians who are ill-informed about the temperatures of several hundred degrees needed to alter the composition of clay and permanently harden it.

Inventions are born through a series of coincidences. Their exact date of birth can never be pinpointed, for they always emerge in a technological environment that is prepared for them. Pottery was born in a more or less stable environment of proto or primitive farmers, who shaped clay and stored grain. In the ancient world this environment corresponds to the Near-Eastern area from the seventh to the fifth millennia. In the New World, very similar conditions seem to have existed from California to Peru at a perhaps slightly later period.

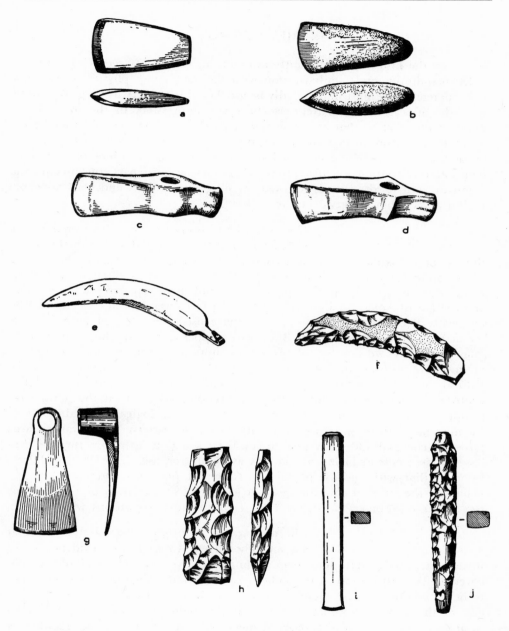

FIG. 27. Mediterranean and Neolithic metallurgy in western Europe. (Copper or bronze objects on the left are *anterior* to the carved or polished stone objects in the right-hand column.)

(*a*) Metal ax or adze, Cilicia, around 3500 B.C.   (*b*) Polished stone ax or adze, Switzerland, around 2000 B.C.   (*c*) Metal ax head, Caucasus, toward 2000 B.C.   (*d*) Stone ax head, Silesia, toward 1500 B.C. (*e*) Metal sickle, Syria, around 1500 B.C. (*f*) Flint sickle, Netherlands, Bronze Age (after 1500 B.C.). (*g*) Copper hoe, Susa, Iran, around, 2500 B.C. (*h*) Flint hoe, Neolithic, France. (*i*) Copper chisel, Sialk, Iran, before 3000 B.C. (*j*) Flint chisel, Denmark, after 1500 B.C.

METALLURGY

Based on the information presently available to us, pottery seems to have preceded metallurgy, and familiarization with the methods of physical transformation offered by fire must undoubtedly be attributed to pottery making. Beginning with the appearance of pottery, we have the impression, in the southeastern Mediterranean area, that experimental research into the possible uses of fire as an agent of transformation had begun. This impression is correct only insofar as the behavior of the pottery kiln offers numerous occasions of correspondence and associations of ideas. Research is for the most part unconscious and empirical, but it corresponds to a reality, since metal, lime, and glass appeared within a very short interval of time.

Copper   The Near East seems to have discovered metalworking around the year 3000 b.c., with the reduction of copper oxides into metallic copper. The idea of reconstructing the process of invention has tempted numerous investigators, who have often suggested the gathering of nuggets of native copper as a possible first step. Among the Eskimos and the North American Indians, there are in fact several primitive groups who by means of cold-working and polishing have used fragments of native copper for making ornaments or tools. But native copper is rare in the Near East region, and there are few logical reasons to support its use. Actually, the chain of inventive associations that may have made it possible to proceed from the natural metal to the reduction by fire of oxides apparently unrelated to it is contrary to everything taught us by historical technology about the normal development of new methods. In contrast, it is archaeologically certain that at the moment of the appearance of the first metal the eastern Mediterranean area possessed charcoal and knew how to construct a high-temperature furnace, possessed lime whose reducing action was available for use, and, lastly had long been familiar with the use of copper oxides for pigments and eye cosmetics. Whatever the series of coincidences that led to the combination of all these factors in the creation of metal, favorable environmental conditions were present.

Bronze   The appearance of bronze poses a problem that has yet to be solved. Archaeological evidence is still inadequate, and to date few chemical analyses have been made of objects discovered. Recently, however, it has been realized that numerous objects thought to be bronze are actually copper and that, on the other hand, in the area to the east of the Syriac-Palestinian regions, bronze could have been the first metal to appear. In the primitive stages of metallurgy, it seems that the question of deliberate alloys does not arise and that the presence of arsenic or tin in the metal obtained is determined by local mineral conditions. Alloying operations were only gradually systematized, after metalworking had definitely become an indispensable auxiliary to the central government. Metallurgical problems were from then on closely linked to the development of the urban system, the formation of empires, and economic expansion in the direction of sources of minerals. It was after the appearance of metal that Egypt, the Levant, Mesopotamia, and the Indus Valley experienced the establishment of the political and economic systems mentioned earlier. In this system the city, bronze, and the use of writing appear as the instruments of a new social order.

The influence of the centers of civilization We must once again go back in time in order to determine the fate, between 5000 and 1500 B.C., of the European and Asian areas peripheral to the Mediterranean centers of civilization. Civilization (that is, the first cities, bronze metalworking, and possibly writing) reached the outskirts of Europe and the Far East within two or three centuries of each other), assuring the gradual formation of secondary centers of civilization that would later replace the early Mediterranean civilizations. By 1500 B.C. the structure was complete. During the preceding centuries the peripheral agricultural world had experienced a period of adaptation by way of the various phases of the Neolithic, and on the whole the situation in China did not differ from that of western Europe. Contacts with the Mediterranean and Near-Eastern world were sufficiently close to ensure the appearance of copper, toward 2000 B.C., in very small quantities in a number of areas, but, as we have already noted, this peripheral world was organized along the lines of the primitive agricultural society, and cutting objects were derived from stone, not metal.

Mesolithic traditions continued to govern all the objects connected with the traditional techniques, but the new tools born of agricultural needs — the ax, the adze, the hoe — were made, as in the poor Mediterranean communities, of hard stone that could be polished to a cutting edge. It was these "Neolithic" (actually parametallurgical) implements that made possible the agricultural colonization of most of the ancient world, including its most remote areas, where they have sometimes continued to be made right down to our own age.

The metallurgical influence of the Mediterranean area made itself felt in Europe in another form. While the exportation of large metal objects, particularly axes and daggers, was still impossible, a knowledge of these objects is apparent from the efforts made in less favored areas to imitate with poorer materials the objects their economic situation did not yet permit them to acquire. As soon as the first traces of copper appeared in the West, axes of greenstone, comparable to oxidized copper axes, or yellow flint like the new copper and bronze materials, enjoyed great favor. The same is true of the large daggers, carefully fashioned from flint and polished to give them the appearance of the Mediterranean weapons. There were numerous sources of this flint, but the largest center so far discovered was that of Grand-Pressigny (Indre), where the production of large daggers that were imitations of metal versions has left evidence in the form of thousands of lumps of flint, very skillfully prepared for the extracting of blades ranging in length from seven to eleven inches, and then abandoned. The same phenomenon of imitation is discovered around the first centers of metallurgy in China, and thereafter, down to modern times, on the oceanic or arctic fringes of the Eurasian world.

Beginning in 1500 B.C., the human race entered another phase, and the prehistoric modes of economy and technology persisted only in the most remote regions.

BIBLIOGRAPHY

There is a growing literature on prehistoric technology, dealing with materials from the origin of the human species (which many anthropologists associate with tool making and tool using) to the development of writing, which marks the transition to historical times. In addition to the specific items mentioned below, the reader is referred to anthropological journals that publish research on prehistoric materials and on primitive peoples still living in a Stone Age culture today.

AITCHISON, L., *A History of Metals* (New York, 1960).

ATKINSON, R. J. C., "Neolithic Engineering," in *Antiquity* (December 1961).

BRAIDWOOD, ROBERT J., *Prehistoric Men* (5th ed., Chicago, 1961).

––– and WILLEY, G. R. (eds.), *Courses Toward Urban Life* (Chicago, 1962).

BUTZER, K. W., *Environment and Archeology* (Chicago, 1964).

CHILDE, V. GORDON, *What Happened in History* (New York, 1946).

–––, *Man Makes Himself* (New York, 1951).

CLARK, GRAHAME, *The Stone Age Hunters* (New York, 1967).

––– and PIGGOTT, STUART, *Prehistoric Societies* (New York, 1965).

CLARK, J. G. D., *Prehistoric Europe: The Economic Basis* (London, 1952).

COGHLAN, H. H., *Notes on the Prehistoric Metallurgy of Copper and Bronze in the Old World* (Oxford, 1951).

CORNWALL, I. W., *The World of Ancient Man* (London, 1964).

DART, RAYMOND A., with CRAIG, DENIS, *Adventures with the Missing Link* (New York, 1959).

FORBES, R. J., *Metallurgy in Antiquity* (Leiden, 1950).

FORDE, C. DARYLL, *Habitat, Economy, and Society: A Geographical Introduction to Etanology* (London, 1945).

HAWKES, J., and WOOLLEY, SIR LEONARD, *Prehistory and the Beginnings of Civilization* (Vol. I in the UNESCO History of Mankind, New York, 1963).

LEAKEY, L. S. B., *Adam's Ancestors: The Evolution of Man and His Culture* (4th [1953] ed., New York, 1960).

–––, *Olduvai Gorge*, 2 vols. New York, 1965–67).

LEE, RICHARD B., and DEVORE, IRVEN (eds.), *Man the Hunter* (Chicago, 1968).

LEROI-GOURHAN, ANDRÉ, *Evolution and techniques*, 2 vols. (Paris, 1948–50); 2 vols. (Paris, 1949–50).

–––, *Prehistoric Man* (New York, 1957).

–––, *Treasures of Prehistoric Art* (New York, 1967).

LIPS, JULIUS, *The Origin of Things* (London, 1949).

MELLAART, JAMES, *Catal Hüyük: A Neolithic Town in Anatolia* (New York, 1967).

MUMFORD, LEWIS, *The Myth of the Machine: Technics and Human Development* (New York, 1967).

OAKLEY, KENNETH, *Man the Tool-Maker* (London, 1952).

PIGGOTT, STUART (ed.), *The Dawn of Civilization* (New York, 1961).

–––, *Ancient Europe: From the Dawn of Civilization to the Classical Age* (Chicago, 1965).

–––, "The Beginnings of Wheeled Transport," *Scientific American*, July 1968, 82–90.

QUENNEL, C. H. B. and M., *Everyday Life in Prehistoric Times* (New York, 1959).

ROWLETT, RALPH M., "The Iron Age North of the Alps," *Science*, Vol. 161 (July 12, 1968), 123–134.

SEMONOV, S. A., *Prehistoric Technology* (London, 1964).

SHEPARD, A. O., *Ceramics for the Archaeologist* (Washington, D.C., 1956).

TYLECOTE, R. F., *Metallurgy in Archaeology* (London, 1962).

UCKO, PETER J., and ROSENFELD, ANDREE, *Paleolithic Cave Art* (New York, 1967).

WHITE, LESLIE A., *The Evolution of Culture: The Development of Civilization to the Fall of Rome* (New York, 1959).

CHAPTER 4

BIRTH AND EARLY DEVELOPMENT OF AGRICULTURE

For long periods of prehistory human beings were able to provide for their subsistence only through hunting, fishing, and food gathering. The belief that the hunting peoples, the gatherers of wild plants and fruit, and lastly the farmers followed each other in succession is inaccurate. Food gathering certainly did not disappear with the first attempts at cultivation, and hunting and fishing continued to be practiced even when cattle raising made it possible to obtain a more steady and abundant supply of meat. It is true, however, that the introduction of agriculture into human activities revolutionized the civilizations of prehistory, and it must be considered as one of the greatest events in the development of the human race.

The hypothetical birth of cultivation
The origins of cultivation remain obscure. Prehistorians have discovered only that the moment of formation of the first centers of a genuine agriculture must be pushed back to the Neolithic or Mesolithic age. Deliberate planting or sowing of seed with a view to a preconceived harvest undoubtedly came about only through a combination of auspicious accidents, awkward trials, and a succession of unexpected failures and fortuitous successes of which there is no record.

The often suggested hypothesis that plant cultivation was born of the observation of the plants growing in the compost piles near the primitive cabins, or accumulating around the entrance to the caves where the primitive families found their precarious shelter, remains unproved. Evidence supplied by the ethnography of the most backward peoples can furnish only analogies that may be deceptive. It merely permits us to speculate about the primitive techniques, and this almost always in biological environments very different from the ones in which we believe agriculture could have been born. The first attempts at cultivation very probably occurred in the homelands of the wild plants considered to be the ancestors of our cultivated plants. Thus wheat and barley were first cultivated in the Near East, while rice was born somewhere to the northwest of Indochina. As for corn, almost all botanists recognize it as a plant native to Central America and the Andean area.

Wild species or varieties did not spread without being profoundly transformed as a result of the ecological conditions they encountered. Millet and sorghum have an important place in the nutrition of the Old World. They were widely cultivated in Europe beginning in the Neolithic period, but lost ground before the bread grains (the former were consumed chiefly in the form of gruel).

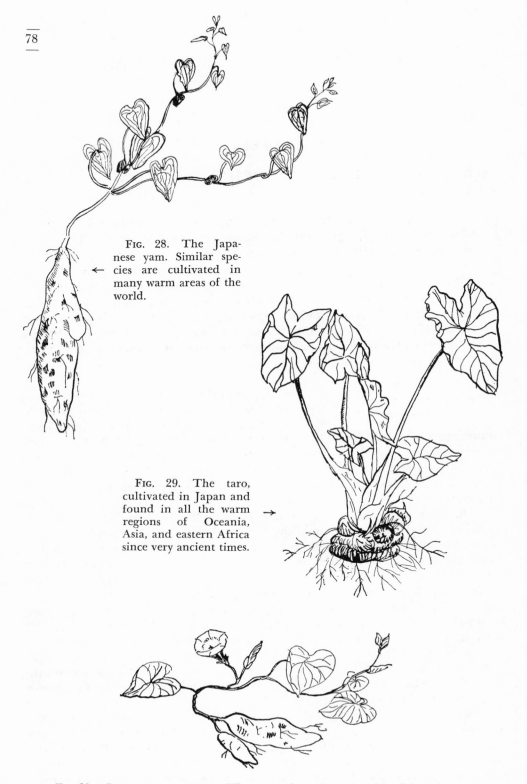

FIG. 28. The Japanese yam. Similar species are cultivated in many warm areas of the world.

FIG. 29. The taro, cultivated in Japan and found in all the warm regions of Oceania, Asia, and eastern Africa since very ancient times.

FIG. 30. Japanese sweet potato. The potato is a tuber vegetable of American origin, now cultivated in all tropical areas.

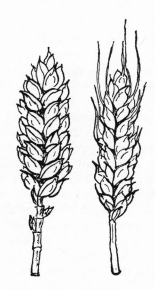

FIG. 31. Millet, cultivated in Europe beginning in the Neolithic period, and still used in Russia and China.

FIG. 32. Cultivated Chinese barley. The wild species most closely related to it grows in the eastern valleys of Tibet.

FIG. 33. Forms of wheat found in the Swiss lake settlements. They are still being cultivated, one in China and the other in Afghanistan.

FIG. 35. Improvement by cultivation. At right, early cultivated corn found in excavations in New Mexico. At left, an ear of modern hybrid corn.

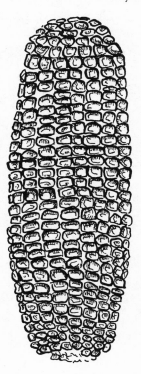

FIG. 34. The oldest species of wheat cultivated in Egypt and the Near East, Europe (during the period of antiquity), and the Middle Volga (in modern times). Sometimes called by its German name, "emmer."

The large-grained wheat variety (emmer, or *Triticum dicoccum*) and the cultivated einkorn (*Triticum monococcum*) seem to have been more widespread in prehistoric Europe than the soft grains that were later to replace them; the dense-eared variety of barley with six rows (*Hordeum hexastichum*) was gradually supplanted by the four-rowed variety (*Hordeum vulgare*), and so on, for numeorus examples. Through natural adaptations and hybridizations a plant base was gradually formed on which agricultural activity could be exercised.

Each climatic zone thus acquired its productive potential and its own techniques. Throughout the history of our cultivated plants we find this play of biological factors combined with the collaboration of actions dependent on man himself. We note, for example, that secondary cultivated plants such as oats (*Avena sativa L.*) and rye (*Secale cereale L.*) were originally mixed with wheat and barley. Favored by the harshness of the climate, which they withstood better than the other cereals, they ultimately became established as separate cultures. We first discover the presence, in Switzerland, of a separate oat culture toward the end of the Bronze Age, while the rye culture does not appear in Germany until the Hallstattan period.

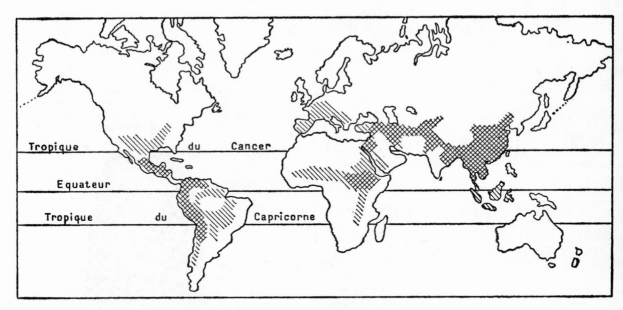

FIG. 36. Areas of origin of cultivated plants. The density of the hatchings is in proportion to the number of species under cultivation.

*Techniques of sowing:
the hoe*

In any event, each group of cultivated plants has its own requirements which lead to special techniques discovered only through long, groping research. As a group they constituted a biotechnical system that imposed its needs on every farmer, and determined the rhythm of life among each group of farmers. The practices of the cereal farmers, for example, resemble each other very closely, no matter where they are observed.

Grain cultivation, in however elementary a form, can be practiced only in properly prepared ground (the seedbed). The adaptation of stone implements to these operations, as simple as they may have been, must have created difficulties so overwhelming that it is difficult to understand how they could have been used, except perhaps for the preliminary work of ground clearing. Experience shows that the ax and the adze of the Neolithic peoples were at most suited to the felling of trees or the cleaning out of underbrush. Fire came to their assistance in clearing land in the temperate regions, where the forest of foliage did not have the overpowering vigor of the equatorial forest, and the winter hiatus of plant growth favored their attack. The Germans call this practice *Brandwirtschaft;* it was practiced by land clearers until modern times, and still exists in its attentuated form of grubbing and burning of weeds. Almost everywhere in the tropics forest and brushfires are still used.

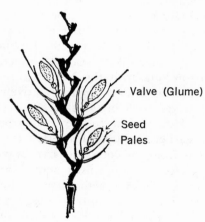

FIG. 37. Diagram of an ear of grain. The black portion is the axis of the ear (the rachis). On each side are the spikelets, enclosed within two valves (the glumes) that contain one or several seeds, each enveloped in two pales.

← Valve (Glume)

Seed
← Pales

FIG. 38. Threshing of the harvest before the development of agriculture. Two Ojibwa Indian women (Canada) in a canoe are threshing the wild aquatic grain by striking (but not cutting off) the heads.

However, until the discovery of metal made possible the utilization of solid cutting tools, the primitive farmers were most probably forced to seek out preferably unwooded areas for the establishment of their fields. Although all of Europe had been gradually covered over with forests after the Glacier Age, some natural clearings certainly existed. In any case, there were certain types of soil in which the forest subsisted with difficulty. Perhaps the calcareous regions of the Mediterranean Midi never supported anything more than a precarious forest growth that was slow to develop and even slower to grow back once it had been destroyed; the climate contributed to this condition. The alluvial soils that spread over the low plateaus and plains from the Paris basin to the Russian or Siberian steppe were covered with a relatively sparse vegetation, while those of Mesopotamia and Egypt, deposited by the rivers that periodically flooded — the Tigris, the Euphrates, the Nile, and certain Mediterranean streams — offered cleared areas along the riverbanks whose productive capacity was renewed annually. It is not surprising that some of the oldest centers of diffusion of agricultural practices sprang up in these areas. Similarly, the boggy areas of continental Europe have yielded evidence of groups of dwellings that can hardly be explained without the existence of agricultural practices whose effects were already a matter of common knowledge.

Once the ground had been cleared, it had to be broken and loosened so that it could receive the seed intended for it. It is possible that the oldest attempts at sowing were made in simple holes dug with a digging stick. Its use can be seen in the planting of taros, yams, sweet potatoes, and other root and tuber plants in Africa and in the Pacific islands. The F. Girard Mission exhibited several striking examples of these, discovered in Borneo, in 1956 at the Musée de l'Homme in Paris. There is nothing to prevent us from thinking that the grain cultivators utilized a similar technique. It would be equivalent to the sowing in seed holes practiced by our gardeners with the help of the dibble.

It seems likely that this practice was very soon replaced by the use of a furrow dug with the hoe. Pictures of this technique from Mesopotamia and Upper Egypt date from a relatively late period, but this work, which was done by slaves, probably goes back to a time very close to the birth of agriculture,

Hoes were sometimes made from the horn of a large deer, as we have seen in an earlier chapter. Most often they were made of wood, and thus were taken right from the forest; a large branch with a smaller, protruding branch made a perfectly usable instrument. The tip could be hardened in the fire, as was done in the case of the digging stick. The work the hoe made it possible to accomplish was easy only in loose soil; even then it was still crude, and exhausting as well: the fellahin depicted in reliefs and painted pictures from the old Kingdom in Egypt work in groups, under stern overseers. They seem to have worked to rhythm, as is often the case in tropical Africa for the land clearers and harvesters.

The plow   Under these conditions it was quite natural to think of using traction — whether human or animal — to make this preparation of the soil more efficient and less difficult. Not until long after the domestication of animals was the draft animal used for traction; the first domesticated animals were used only as food for the hunters of the earliest prehistoric periods. Much patience and intelligent observation, combined with primitive man's natural familiarity with animals, was undoubtedly necessary before he was able to enclose certain animals within shafts, yoke them to a beam, and fix a stable load to their backs. Nothing is known about this search for animal assistance. Contemporary breaking-in of animals captured in a wild state, like that of the elephant, undoubtedly presents only analogies with what domestication of the horse or the ox for draft purposes could have been like.

FIG. 39. Plowing of fallow land, Sunda Isles. The farmers turn the clods with digging sticks, without using their feet.

FIG. 40. Land cultivation among the Maoris of New Zealand. The large wooden spades are pushed into the ground with the foot.

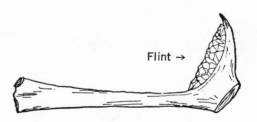

Flint →

FIG. 41a. Peasant hoe from ancient Egypt (from a fresco in the Louvre Museum).

FIG. 41b. Sickle with flint blade, in a deer-antler frame. From Denmark, around 2,000 years before our era.

A result of incorporating the animal into man's agricultural activity was the utilization of an instrument known to the Romans as the *aratrum* — the plow in its oldest known form, without wheels or frame. Whether or not it was derived from the hoe, it was distinguished from the latter chiefly by the fact that instead of plowing the soil with a striking movement, it opened it by acting on a horizontal plane. Its principal part, which was made of wood like the rest of the instrument, was a pointed plowshare probably hardened in the fire, sometimes made more efficient and durable by a piece of flint before it was provided with an iron shoe.

Pictures of the earliest plows, for example those in the rock engravings in the Maritime Alps or southern Sweden, date at the earliest from the beginning of the Bronze Age. They lead us to believe in the utilization of an *aratrum*-spade and an *aratrum*-hoe to which two oxen were harnessed. The instrument seems to have been all in one piece, and perhaps all that was needed in order simultaneously to obtain the beam, handle, and plowshare was a search in the forest for a suitable branch. This is the *aratrum* described by Hesiod together with another, heavier plow made of several pieces assembled together, to which Homer had already alluded.

The plowshare, which was always symmetrical, opened a shallow furrow with two edges like the lips of a wound. By marking out a line and economizing on the farmer's work, it made possible the sowing of large areas which could have been done with the hoe only with a tremendous expenditure of labor. In this sense, at least, it is not an exaggeration to say that the *aratrum* created the field, with all its consequences, consequences such as relatively permanent settlements, the construction of more numerous and more thickly populated villages of farmers, and, to a certain extent, the development of cities.

The *aratrum* was utilized over an immense area extending from the Far East to the seas and oceans of Europe, including those parts of Scandinavia where agriculture had been able to gain a foothold, that is, at the edges of the conifer forest. Our present knowledge, however, leads us to locate its earliest use in ancient Mesopotamia and Egypt. From here it succeeded in spreading throughout the Mediterranean world, then extended toward the north and the

FIG. 42*a*.   Prehistoric *aratrum* from Walle, Germany. From a photographic reproduction of a rock drawing.

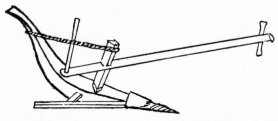

FIG. 42*b*.   Mediterranean *aratrum* used by the Benis Ouriagel, northern Morocco.

FIG. 43.   Chinese plow with plowshare and small cast-iron moldboard.

east. However, it could have been invented in several areas simultaneously, although "intercultural relations" certainly played a major role in the dissemination of techniques, as has been shown by Mr. Leroi-Gourhan.

The *aratrum* was later replaced by the plow (French *charrue,* believed to be from the Celtic *caruca*), which had a set of wheels and moldboard. With this development we are already in the historical period. The *aratrum* continued to be the preferred instrument in southern Europe and in the Near and Far East for light work on dry land, where the repetition of furrows on the same field ultimately became a method of retaining the moisture in the soil.

Techniques of cultivation   Needless to say, the farmers who used the *aratrum* did not invent this technique, any more than the techniques of broadcast sowing or harrowing, all at once. Hesiod still talks of planting seeds by plowing; a young worker had to break up the sods that had not been broken up by the plow, so that all the seeds would be covered over. This was the method formerly used by the peasants for heavy soil, where they employed a wooden sledgehammer for this purpose. The first settlers could be satisfied with a very superficial harrowing done by passing a simple thorny branch over the calcinated, dusty, ash-covered ground. Medieval iconography reveals that this method was still being practiced in the Middle Ages; the triangular harrow and the rolling harrow were not invented until later.

Such rudimentary methods condemned production to an ineffectiveness characteristic of all primitive agricultural societies. However, except in areas whose fertility was periodically renewed by river floods, the land undoubtedly was very quickly exhausted. The inevitable result was an agricultural nomadism, which must not be confused with pastoral nomadism, and which decreased or disappeared only when cattle raising made it possible to revitalize the productive capacity of the fields with manure. This was undoubtedly the starting point of the alternating periods of cultivation and fallowness that characterized early crop rotation and fallowing. Many centuries must have passed before they were systematized. All that can definitely be said is that the climate — and sometimes the fertility of the soil — made it possible in some areas to adopt a more varied system of fallowing than the simple succession in "two-quarter rhythm" of cultivation and fallowness. It seems likely that certain peoples of protohistorical northern and central Europe were already familiar with three-course rotation, which was based on the succession of a winter grain, a spring grain, and a fallow period.

Groups whose agriculture was based on root or tuber plants apparently had fewer problems to solve. These were also the most retarded of the agricultural peoples, and our knowledge of the origins of their farming activity is no greater than our knowledge of the grain growers. All that can be said is that their agriculture was not very far removed from food gathering: a hole or a furrow for the rhizome or tuber was dug, and this practically completed the entire agricultural operation. Only over a long period of time did certain of these primitive peoples learn, at least vaguely, the rules for the mixing of cultures, as in the *lougans* and gardens of tropical Africa.

Harvesting techniques Among these same peoples, harvesting was often a simple food-gathering operation. The natives of Oceania dig up edible roots, both the wild and cultivated varieties, with a digging stick. The harvesting of grain requires more attention. When mature, the grains break away naturally from the ear. Experience acquired on previous food-gathering expeditions told the early farmers when to gather them.

They may have practiced hand-harvesting, as is still done among the Ojibwa Indians (see M. Leroi-Gourhan's earlier discussion of this practice with regard to wild rice). The peoples of Indochina harvest various types of echinochloa in a similar manner, and people may in the same way have become familiar in the Far East with the rice that grew wild in the streams of Upper Burma and the rivers of Indochina. Perhaps, too, the gathering of aquatic plants — the common reed grass (*Phragmites communis*), the buckbean or bogbean (*Menyanthes trifoliata*), the seeds of the water lily (*Nuphar luteum*), the water chestnut (*Trapa natans*), and manna grass (*Glyceria fluitans*) — which complemented the diet of the fishermen of central, eastern, and nothern Europe and, beginning in the Neolithic period (toward the fourth millennium before our era), oriented them toward farming.

Beginning in the Neolithic period the grain farmers of the ancient world seem to have used a chipped flint implement with which they could detach the ear of grain from the stalk. Prehistorians have named this primitive tool the "tranchet-sickle," and claim that it is polished in a special manner. In this case it may be the ancestor of the sickle, made of terra-cotta and armed with flint teeth, like those that have been discovered in excavations in Mesopotamia on the site of Obeid (beginning of the third millennium), or from a curved piece of wood, as can be seen in engravings or pictures from Upper Egypt. Harvesting knives discovered at the site of Hitzkirch (near Lucerne) and Polada (in Italy) are already more complex in structure, and help us to imagine the action used in harvesting: the ears were grasped in the left hand and "sawed off" with the tool.

This is approximately what was done with the sickle. Metal (copper and bronze to begin with; the iron sickle appeared in the La Tène I period) made it possible to give it the most suitable curve and, by adding a projecting point to one end of the blade, a handle; we can follow the progress of the large sickle, made possible by the discovery of metal, from Styria to the plains of the Saône. The early sickles were often toothed, perhaps because there was no metal sufficiently durable or capable of taking a sharp cutting edge. Toothed or straight, the sickle sawed rather than cut (in certain regions of France the word "sawing" is still employed for "harvesting").

No serious efforts seem to have been made to enlist the help of animals in this difficult task. Pliny the Elder mentions an extraordinary Celtic cart, armed with a toothed blade and pushed by an animal, which harvested the ears of grain along with their stalks. A fourth-century agronomist describes it again, and declares that its use has become more common in northern Gaul. Archaeology has uncovered only one example, and it must be recognized that this "harvester" did not satisfy the wheat farmers, since it was not used outside its birthplace (Celtic Belgium).

*Storing of harvests;
cooking*

Once food has been harvested, it has to be preserved. Neither the hungry prehistoric food-gatherers, nor even the early farmers with their clubs and hoes, had to devote much attention to this matter. The problem arose, in contrast, when cultivation on a larger scale and more abundant production made it possible to depend almost entirely on crops to supply nutritional needs. The long cold winters and the warm, damp spring periods required precautionary measures for the preservation of the harvests. While fruit, obtained exclusively by random picking (no orchards are known prior to historical times), was undoubtedly consumed immediately, certain pulses — beans, lentils, and peas — produced seeds that could easily be dried. Beginning in the Neolithic period, beans became an element of food supply in the Near East, the Mediterranean countries, and even central Europe; lentils and peas did not appear until the Iron Age.

Cereal grains could not be stored as easily. They had to be harvested at the exact moment of ripening, and their excess water removed. In most cases they undoubtedly had to be subjected to additional drying after threshing, which was done with the feet or by striking the ears against a stone or a piece of wood, or perhaps with rods, as was later done with the flail. Often the storage place was simply the floor of a hut improved with a crude flooring; this was a type of silo similar to the earth-covered caches that can still be seen in the steppes of North Africa. Silos became numerous beginning in La Tène I. The earthenware vessels — large jars and dolia — already familiar to Bronze Age man were also frequently used for this purpose. Thanks to the impressions made by the grains in the walls of these jars, and thanks also to the grains themselves that have thus been preserved, the excavations in the foundations of the lake dwellings have been our best source of knowledge about the cultures of the Bronze and Iron ages. Later came wooden bins, then granaries that, given their ingenuity and diversity, must be posterior to prehistoric times.

The grains found in this way in Switzerland, Scotland, and elsewhere are often roasted; they include wheat and barley from the La Tène period (the varieties on display in the Musée d'Art et d'Histoire in Geneva, for example). This practice of roasting guaranteed their conservation over a long period, and at the same time made digestible the starchy matter thus transformed into dextrin. Water could be used in food preparation; it made the toxic roots and tubers, manioc, taros, and so on, safe for consumption, and in particular made possible the preparation of gruels and flat cakes before the art of genuine bread-making was known.

The grain was probably first crushed with a pestle, then with a quern made from hard stone — a simple stone ball rubbed by hand over a solid base that was either flat or slightly concave. A great number of these querns have been found, perhaps invented "somewhere in the Near East." They were utilized in all the grain-producing countries until the appearance of the millstone, which came into use, according to John G. D. Clark, "in the area of the La Tène civilization, during the second century B.C.," and which "crossed the Channel towards 100 B.C." The practice of pounding the grain, he maintains, did not completely disappear: the sight of women pounding rice in monsoon Asia or millet in tropical Africa reminds us of the slaves of antiquity crushing wheat, and is certainly a reproduction of the actions of the pre- or proto-historical cereal farmers.

Cattle raising for food purposes
"The transitional stages from hunting to domestic cattle breeding are imperceptible," declares M. Leroi-Gourhan. This is the same as saying that the introduction of domesticated animals into the life of man, and particularly into his agricultural activity, is almost impossible to determine. Here we are dealing with a symbiosis resulting from the proximity of man and the animal, still very close to each other by virtue of the conditions of their lives and undoubtedly because of a certain sociability. This relationship later disappeared because man used force and thereby aroused fear in the animal.

Beginning in the Neolithic period, the dog is found everywhere as a companion. The ox and the horse, the sheep and the goat, the pig and the fowl were sought and domesticated for their meat, milk, hide, fur, wool, horns, and bones. The use of some of these animals for carrying loads, pulling, and transporting human beings is dependent on other techniques of cattle breeding, as we have already seen, and occurs in an advanced stage of domestication.

Naturally, the importance of each of the domesticated species varied from place to place, according to the type of vegetation, the nature of the seasonal changes, the topography, hydrology, and so on. It has been noticed, for example, that the pig, which was particularly abundant at the beginning of the Neolithic period, seems to have declined as cultivation became more extensive and the forests were more energetically cleared. Conversely, in the same period an increase in sheep and goats can be seen. It is also known that the climate did not remain unchanged between the Neolithic period and the Iron Age. The damp climate that followed the Glacier Age gradually gave way to a drier climate, and perhaps the sheep and the horse adapted to it better than the bovine animals.

Lastly, aside from local ecological conditions, the appearance of animals in the various parts of the area settled by the human race was accomplished in stages that are not necessarily the same for all animals. Thus the domestic horse is not found in the Swiss sites until the end of the Bronze Age; in contrast, it was raised in Great Britain beginning at the end of the Neolithic period. Conversely, the small short-horned ox (*Bos taurus longifrons*), which appeared in Switzerland and Denmark in Neolithic times, seems not to have been introduced into Great Britain before the end of the Bronze Age.

The methods of cattle raising, on the other hand, reveal less diversity; they were dependent on the conditions of the animals' lives. We have seen how the reindeer migrations required seasonal moves on the part of the peoples who guarded these animals for purposes of their own. The transhumance of sheep in the plains areas near mountains (for example, in a large part of the Mediterranean area) had to a certain extent the same character of natural spontaneity. The same is true of the pasturing of pigs in the forest, sheep in open areas and fallow lands, and oxen or horses in the more humid meadowlands. The search for food for the animals was the foundation of a cattle-breeding industry that did not go very far beyond a simple guarding of the animals. Not until very much later did the farmer and cattle raiser have the desire or the knowledge to open up grazing lands for his livestock; at best he was satisfied to supply its food simply by random food gathering, when the season did not permit the animals to find their own food. Under these conditions we are not surprised to note that the slaughtering of animals took place preferably at the beginning of winter, at a time when provisions became less abundant.

Domestication was nevertheless a major element in the development of agriculture. It led to the achievement of the complex that associated man, plant, and animal; it can be regarded as the starting point of that mastery of nature by man in which his instinct, intelligence, and imagination worked together. This is the moment at which *Homo faber* becomes *Homo sapiens*. For a long time to come, every achievement was to be the result of simple experience and a kind of active submission to natural laws. For long centuries and in all climates the man of the fields was satisfied with practical knowledge, and this was the treasure he handed down to succeeding generations. What he did not understand he explained by myths, and his daily activity, even the most humble, was surrounded by rites that became traditional. Rural civilization thus remained immobilized in customs, many of which date, unknown to their practitioners, from the dark ages of prehistory.

BIBLIOGRAPHY

In addition to the appropriate references for Chapter 1 — especially the works of CLARK, FORDE, and LEROI-GOURHAN — the reader is referred to the following items that reflect the growing amount of research during the past two decades in the origins and development of agriculture.

AMES, OAKES, *Economic Annuals and Human Culture* (Cambridge, 1939).

ANDERSON, EDGAR, *Plants, Man and Life* (Boston, 1952).

CURWEN, E. CECIL, and HATT, GUNDMUND, *Plough and Pasture: The Early History of Farming* (New York, 1953).

FORBES, R. J., *Studies in Ancient Technology*, Vol. II (Leiden, 1955).

FUSSELL, G. E., *Origin and Development of Agricultural Farming Techniques from Prehistoric to Modern Times* (Oxford, 1967).

HAUDRICOURT, ANDRÉ, and JEAN-BRUNHES-DELAMARRE, MARIEL, *L'homme et la charrue à travers le monde* (Paris, 1955).

LEROI-GOURHAN, ANDRÉ, *L'homme et la matière, Milieu et techniques*, 2 vols. (Paris, 1949, 1945).

MOBERG, CARL-AXEL, "Spread of Agriculture in the North European Periphery," *Science*, Vol. 152 (April 15, 1966), 315–319.

SAUER, CARL O., *Agricultural Origins and Dispersals: The Domestication of Animals and Foodstuffs* (Cambridge, Mass., 1969).

WATERBOLK, H. T., "Food Production in Prehistoric Europe," *Science*, Vol. 162 (Dec. 6, 1968), 1093–1102.

CHAPTER 5

THE FIRST STAGES IN THE
UTILIZATION OF NATURAL POWER

THE DEVELOPMENT of methods of utilizing the various forms of natural power — animal, wind, and hydraulic — is so important, first for the appearance and then for the development of a technological civilization, that the principal facts will be summarized in a general chapter, and then considered separately in accordance with the subjects of later chapters.

In the earliest periods of human existence, methods of utilizing power resembled, or were even identical to, each other from one period or region to another. By the time history began, man had learned how to use his own muscle power and that of animals, and for long millennia this was the only power available to him. In addition, during this period he had learned how to use wind power, in a crude manner and only for propelling boats.

It was not until several centuries before our era that he began to learn how to harness waterpower. Still more centuries had to pass before he learned how to power machines by wind. Owing to the lack of exactly dated evidence, the origin of most of the simplest methods, particularly that of the types of harness known in antiquity, is highly uncertain. In some cases ancient types of harness have continued to exist down to modern times, and the observation of devices still in use often makes it possible to compensate for the lack of evidence from early periods.

ANIMAL MOTIVE POWER

Domestication The domestication of animals was not undertaken with the intention of utilizing them as sources of motive power. Domestication of herbivores did not occur until after the birth of agriculture; it protected the fields from the animals' depredations, and the grain surplus supplied their nourishment. The riverbank regions of the Near East, from the Himalayas to the Mediterranean, seem to have been the first areas to domesticate bovines, using their milk and meat for food as the wild animals became more scarce. These herbivores were ruminants: the humpless bovines of the West, the humped variety from India, the buffalo from the western regions of tropical Asia, goats, sheep, camels, and dromedaries, equine animals, donkeys from Africa, wild asses and horses from Asia. Another omnivorous animal played an important role: the pig, which was domesticated in the same areas.

The camel   Among the domesticated animals a special place must be accorded in this chapter to the camel and the dromedary. They were raised for their milk, their hide, and certain other domestic qualities, but in particular they very soon became, in certain clearly defined areas, the beast of burden best adapted to the climatic characteristics of those regions. During several centuries, and perhaps for almost a millennium, the camel, by then irreplaceable in these areas, played a major role in the development of the human race.

The principal discussions about its origin and utilization have revolved around linguistic questions, since in many of the ancient languages the distinction between the dromedary and the camel was not always clearly defined. However, it seems possible to accept the idea that the animal properly called a camel — the two-humped Bactrian camel — is a native of central Asia; its fossil remains have been found as far as Siberia and the banks of the Volga. At the beginning of the third millennium it was already domesticated in the plains of this geographical area. From there its use spread to the Far East on the one hand, to the Near East on the other.

The dromedary, which has only one hump, may have been domesticated in the Mediterranean regions of the Near East, and in Arabia beginning in the early pastoral period. Its area of expansion probably extended to the southern part of Paleolithic Europe; changes in climate during the Neolithic period drove it back to Mediterranean Asia and North Africa.

The boundaries limiting the respective areas of use of the two animals met in Persia and Mesopotamia in Assyrian times. The dromedary may have been the first to be introduced into Persia, but it does not seem to have penetrated central Asia, while the camel spread through the regions where the dromedary was already known. These countries were located on the outskirts of the areas of expansion of each species, and beginning in the middle of the second millennium they simultaneously adopted the camel and the dromedary as a result of military use of the two animals.

Domestication of the camel and the dromedary in this period was practiced only by the nomadic peoples of the desert regions, for whom they had become the best method of transportation probably as early as the third millennium. However, their breeding does not seem to have been very intensive in antiquity. In Palestine, as in Egypt and Nubia, the sedentary populations of the countryside and the cities did not utilize the camel, which they had known since a very early period only through undesired contacts with the desert nomads.

Utilization of the dromedary and the camel in large numbers began only in the Assyrian armies, toward the end of the first millennium, and then in the Persian armies. The camel seems to have been used particularly as a beast of burden, the dromedary for riding. Their use increased during the Greek conquest, and especially in the period of Roman domination, and conquered the Berber countries during the first two centuries of our era, the entire western portion of North Africa during the two succeeding centuries.

Prior to the beginning of the Islamic conquests, therefore, their use was general but not intensive. Their breeding remained limited, with periods of more intensive development corresponding to the temporary development of commercial exchanges across the desert. Only in Arabia did their use answer to a special and continuing need; trade between Yemen and northern Arabia was the principal motive. The breeding of dromedaries was considered by the Bedouins as a noble occupation. The animal was used in large numbers for the commercial caravans from Mecca to Petra and Palmyria, as well as for military needs. The camel thus became used more widely beginning in the seventh century, simultaneous with the expansion of Islam. More exact information is given on this subject in the chapter on the Moslem world.

HARNESSES

The most basic work required of livestock is treading, which is still used in irrigable fields as a form of plowing, and which can be performed by buffalos or pigs. In the Mediterranean areas treading is used only for winnowing, that is, for separating the grain from the chaff after the grain has been harvested. For this purpose horses and asses in particular are used, harnessed together and turning in a circular area.

The horn yoke   The oldest type of harness seems to owe its origin precisely to the habit of attaching two animals together so that they will walk side by side. Thanks to their horns the bovine animals are the easiest to attach in this way. The yoke, in the form of a narrow wooden bar attached to the horns of the two animals (Fig. 44), seems to have appeared in Egypt, and slowly evolved from this rudimentary form; the piece of wood was carved to fit the animals' foreheads or necks, and notches and holes were cut in it

Fig. 44. Horn yoke used in ancient Egypt, as depicted in bas-reliefs. The *aratrum* must be held in the hand.

to accommodate the leather thongs used to hold the harness on one hand and the shaft of the vehicle being pulled on the other. This yoke spread through the West, in the direction of Europe, where it is found in peat bogs dating from various stages of antiquity; as late as the nineteenth century it was still being used on the shores of the Baltic Sea, in the eastern Alps, central France, and on the Iberian Peninsula.

The neck yoke   Another form of development of the yoke was the use of two wooden bars, so that the necks of the two animals were locked, as it were, in a kind of choker. Thus this type of harness could be used for animals without horns, such as the donkey, or for animals with oblique, almost horizontal horns, like the buffalo. This was the "neck yoke," so named because it rests on the base of the neck, at the bulge of the spinal column. A neck yoke with two bars in the form of a square still exists in the Balkans for bovine animals, and was once used in the Landes in southwestern France, but in most cases it has been simplified: the lower bar has disappeared and has been replaced by straps that join the sidebars together (Fig. 45).

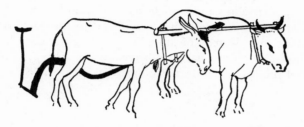

FIG. 45.   Neck yoke harness, used for both donkeys and bovine animals. From a plowing scene in North Africa.

The first horse harness   Certain animals that are saddle-backed or that move slowly could use this type of harness, but the horse, which carries its head erect and is skittish, would have been injured by the sidebars. The latter were therefore replaced by a leather band sufficiently wide to avoid cutting the animal; however, the yoke in this case rests on the windpipe, and the animal pulls by pressing the base of its neck against the collar, which impedes its breathing (Figure 46). (The ox and the ass, in contrast, rest their neck and shoulder in the angle formed by the yoke and the sidebar.)

FIG. 46.   Horse-traction harness, using a flexible collar under a neck yoke (from Assyrian bas-reliefs).

Shaft vehicles   These harnesses belong particularly to a type of vehicle that invariably ends in a shaft. Examples of such vehicles are, in Egypt, the *aratrum,* used for loosening the soil and planting seeds, and also the rectangular sledge, often pulled by a rope rather than a shaft. In the Near East, Mesopotamia, and India, the wheel was coming into use to facilitate its movement. The sledge on wheels does not appear to have been the model for our carriage; the carriage seems rather to be an outgrowth of the travois.

The travois and its evolution   The travois is a primitive vehicle composed of one or two wooden bars, one end of which drags along the ground while the other is supported by the man or animal pulling it (Figure 47). In the case that interests us here, it is the yoke that acts as the support. Thus the vehicle has a triangular shape, the two poles coming together at the place where they are joined to the yoke. All that is then necessary is for their ends, instead of dragging on the ground, to be placed on an axis with two wheels: a carriage has now been formed. This type of carriage still exists in India and the Mediterranean regions; it is often called a "carriage with a forked shaft."

FIG. 47.   The travois, first used with dogs in North America, and later with horses after they had been introduced into the region. The two poles are later taken apart and used as poles for tents.

The carriage is built in a semicircular or rectangular shape, and the shaft becomes a single piece of wood located in its longitudinal axis. The next step is the two-wheeled, shafted cart of classical antiquity. The famous chariot of the Indo-European peoples, the Hittites, and the Assyrians (Figure 46) was a carriage of this type, slung low on wheels. Despite the flaws in the horse-traction harness, because of its rapidity this vehicle, which was light enough to be drawn by two horses, gave a temporary military advantage to the peoples who used it. This chariot and the horse must have arrived in Egypt, together with the foreign peoples, at the end of the Middle Kingdom; they are seen in pictures dating from the New Kingdom. In this way the wheel was introduced into Africa.

In China, in this same period, excavations have uncovered similar carts, but the yokes utilized for the horses seem to have retained the side pieces that were abandoned in the Occident. Since only the metal casings for the vanished wooden parts remain, however, it is difficult to determine their actual use. Also in China, wheels were adapted to another type of shafted vehicle that give birth to the modern horse harness.

In regions of Eurasia where the climate, until more recent times, did not permit the introduction of agriculture with its concomitant use of bovines, the reindeer was domesticated and utilized as a draft animal as far south as the southern portion of Siberia. The study of methods of locomotion and types of harness in Siberia indicates that the horse-traction harness had no direct influence on that of the reindeer.

Dogs and reindeer   In northern Asia and North America, vast stretches of land covered with snow for a great part of the year were the setting for perfecting the use of the dog and the reindeer. The only vehicles used were light sledges that could be pulled, even by human beings, by means of straps.

In the sixteenth, seventeenth, and eighteenth centuries, when the Europeans came into contact with the arctic peoples, certain dog and reindeer harnesses were discovered that seemed to have derived directly from the use of human traction: straps attached to a belt (Figure 48) or a shoulder strap (Figure 49), and harnesses in the form of tunics (Figure 50) or simple collars (Figure 51). These harnesses have now been replaced by harnesses of another type reminiscent of those of the domestic animals of the Temperate zones (Figures 52 and 53).

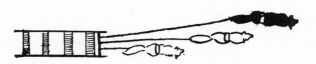

FIG. 48.   Belt harness of dogs pulling a sleigh. Found among the Ostiak people of western Siberia (from Finsch, *Reise nach West-Siberien*, cited by Montandon, pl. 1, p. 136).

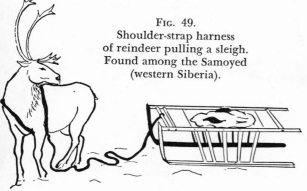

FIG. 49.
Shoulder-strap harness
of reindeer pulling a sleigh.
Found among the Samoyed
(western Siberia).

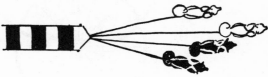

FIG. 50.
Brace harness of dogs pulling a sleigh;
used by the Eskimos of the Bering Strait
(cited by Montandon,
from Bogoras and Jochelson).

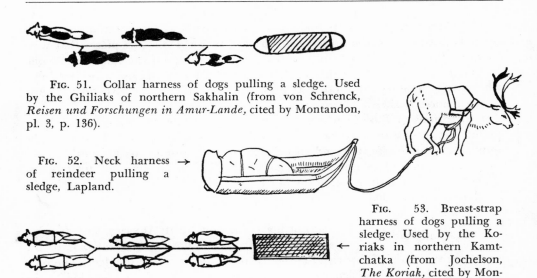

FIG. 51. Collar harness of dogs pulling a sledge. Used by the Ghiliaks of northern Sakhalin (from von Schrenck, *Reisen und Forschungen in Amur-Lande,* cited by Montandon, pl. 3, p. 136).

FIG. 52. Neck harness → of reindeer pulling a sledge, Lapland.

FIG. 53. Breast-strap harness of dogs pulling a sledge. Used by the Koriaks in northern Kamtchatka (from Jochelson, *The Koriak,* cited by Montandon, pl. 2, p. 136).

The evolution of harness in these regions was accompanied by a differentiation in the draft devices based on their functions of pulling, reversing, and supporting. In the ancient harness, these three functions were accomplished simultaneously by the yoke and the shatf; in the sledges, as we have seen, the flexible trace serves only for pulling, no support is required, and reversing is performed manually by the driver.

The shafts   It was in the Far East that the travois evolved into the shafted sledge, which in turn developed into a two-wheeled carriage with a shaft. This vehicle appeared several centuries before our era in the period (Han) of the Warring Kingdoms. The Chinese chariot of

FIG. 54.   Chinese harness (from Han-period bas-reliefs).
The two shafts are joined by a piece attached to the neckpiece.

this period has two curved shafts joined at one end, to which is attached a small yoke (Fig. 54). Draft was ensured by a breast strap similar to our breast harness, and support by the small yoke and the shafts.

The packsaddle The operation of support benefited from progress made in the field of transportation on the backs of animals.

In ancient Egypt the animal (usually the donkey) was used directly for transporting a burden placed on its back. (The only domestic animal of the Andean region, the lama, is used in the same way.) But to balance a load is only the first step; stabilizing and fastening it to the animal posed a problem as difficult as harnessing, and injury to the animal had to be avoided. The solution finally adopted, that of joining padding to a wooden framework and fixing the apparatus in place by means of straps, seems not to have been discovered until our era. Probably only at the time of the Roman Empire was the packsaddle utilized for donkeys and horses, and thereafter (in the second century of our era) the camel began to be used for carrying burdens; before then it had been used neither in the Sahara nor in Egypt.

The invention and spread of the packsaddle also made it possible to utilize two animals for the transportation of a vehicle without wheels. Sedan chairs, palanquins, and so on, carried by men, could now be attached to the packsaddles of animals (Figure 55). This method of transportation was introduced into Europe before the end of the Roman Empire.

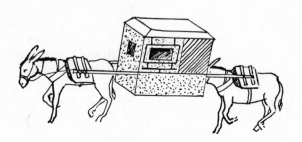

FIG. 55.   Sedan chair, China. The shafts
rest directly on the packsaddles.

Development of the packsaddle Once the packsaddle had given birth to the ridge strap, which supported the shafts, it became possible to improve the horse-traction harness. The draft point of the flexible collar was moved to the middle of the animal's back by attaching the collar to the ridge strap and the yoke to the ridge strap. This method of draft by using the ridge strap to support a yoke is still used for pulling carts in certain parts of Sicily and with mail carriages in India (Figure 56). Lefebvre des Nouettes called this "the Byzantine harness" because it appears in Byzantine illustrations.

Fig. 56.   Hindu harness (drawing made from a fifteenth-century statuette).

The saddle
Meanwhile, in central Asia the packsaddle was being transformed into a riding saddle. In European antiquity the horse was ridden bareback or simply with a soft rug. Because of its wooden framework the saddle, modified into a seat to ensure the rider's stability, prevented him from guiding his mount, and so stirrups and spurs were created. This manner of riding seems to have arrived in Europe with the Huns and particularly the Avars. It is important to note that the Arabs adopted it just prior to their Islamic expansion, whereas in Europe the saddle and its stirrups did not come into general use until the Merovingian period.

From the packsaddle to the collar
The conformation of the camel gave rise, in central Asia, to a form of packsaddle that had an empty space in the center so as to avoid injuring the animal's humps, the size of which varied with the seasons and the abundance of food. Travois or carriage shafts could be attached on either side of this circular saddle. It was apparently among the Turkish peoples then inhabiting Mongolia that the transition of the use of this harness from the camel to the horse, and the transformation of the saddle into a rigid collar around the horse's neck (Figure 57), seem to have taken place.

History and expansion
It is difficult to give an exact chronology of these events. Shafted cars and a form of breast-harness draft seem to have been known in Europe by the end of the great invasions.

Fig. 57.   Packsaddle for two-humped camel, and horse collar used by the Turkish Mongols; both objects have the same shape and are called by the same name (*qom* in Mongolian, *qam* or *qamyt* in Turkish).

This method of harnessing is still used in southern Italy in the form of the "Neapolitan harness" (the term used by the ethnographers), in which the shafts, which are quite close together, are passed over the animal and are supported directly on the saddle; the horse pulls by means of a breast strap attached directly to the cart (Figure 58).

The Italian word for shaft is *stanga,* which is Germanic in origin. Did it come to Italy with the Lombards or the Ostrogoths? In the Rhineland, Lorraine, and Champagne the breast strap was utilized in agricultural work; in French dialect it is called *warcolle,* which is partly Germanic, while the Germanic names *Ham* and *Siele* occur also in Slavic, and in the Turkish languages of central Asia the word *qaam* stands for both the modern horse harness as well as the packsaddle of the camel.

FIG. 58. Neapolitan harness (drawing made from a photograph).
The shafts, placed very high over the ridge strap, are reminiscent of the "neck-type" draft of the ancient harnesses, but here the shafts serve only for balance and for backing up; the pulling is done by the breast strap.

Adaptations of shaft vehicles

In North Africa mules and horses are harnessed by breast straps, but for pulling the *aratrum,* which has no shaft, the animals are used in teams, their breast straps pulling an underbelly rod that acts as a yoke and is passed between the legs and under the stomachs of the animals (Figure 59).

In eastern Europe, in the harness of the sleigh called the troika, the side animals are harnessed with breast straps, and the center horse has a collar and a *douga.* The camel's packsaddle and the horse's collar are connected directly to the shafts of the travois or the two-wheeled carriage, in both of which vehicles the shafts are joined together in rigid fashion. In the sleigh, on the contrary, the shafts are mobile; if they are attached to the collar there is a danger that they may come together and irritate and injure the animal. To offset this disadvantage, a wooden arch, the *douga,* is used; it keeps the two shafts apart and stretches the traces, attaching them to the collar. This results in greater elas-

FIG. 59. Donkey and camel harnessed by means of an underbelly bar. North Africa.

ticity or draft. This *douga* harness was later utilized for four-wheeled carriages to which only one horse was attached (Figure 60).

Backing up; draft by collar   The vehicle that undoubtedly penetrated into Europe together with the collar was the shafted, two-wheeled carriage or cart, pulled directly by the shafts fixed to the collar. One of the proofs that the collar harness is later than the breast strap is found in the use of the latter for backing up. The breeching, or breechband, is a strap passed behind the horse's croup and attached to the shafts; it permits the animal to stop or back up the vehicle. A study of the terms indicates that the name of the breast strap and breeching is found in Old Germanic (Silo = *Siele*), whence it passed into the Slavic (Polish *szla*, Russian *šleia*), while the German name of the rigid collar (*Kummet*) is of Slavic origin, and the Slavic word itself (*chomat*) seems likely to have come from the Turkish (*qamyt*). We are tempted to date the introduction of the collar toward the seventh or eighth century.

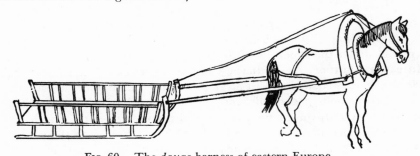

FIG. 60. The *douga* harness of eastern Europe.
The *douga* is a wooden arch that keeps the shafts apart. (They have a tendency to come together because of the tension of the traces that join the collar and the ends of the shafts.)

The whiffletree   For pulling a vehicle with a single animal, the pole was replaced by two shafts, in two- and four-wheeled carriages as well as in certain types of the Mediterranean *aratrum*. When the use of several animals together was desired, the shafts had to be replaced by flexible traces, and these were attached to an underbelly rod. Several examples have been found in central Europe and Scandinavia. Most often, however, the whiffletree was used. This was a piece of wood placed behind the animal, attached to the traces at each end, and by its middle to the vehicle or to a swingletree (Figure 61).

For pulling plows and other farm implements the whiffletree became as characteristic of the horse harness as the yoke is for the ox harness.

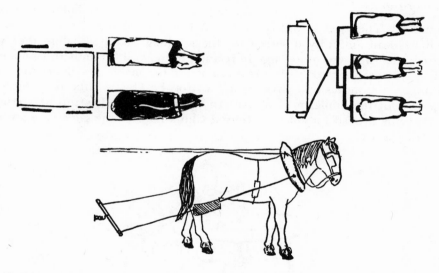

FIG. 61.   Modern harness for horses; pulling a whiffletree by flexible traces.

Ascendance of the horse   The distribution of the various types of harness in Eurasia seems to have become stabilized beginning in the twelfth century; the only change — before the steam and internal-combustion engines replaced animal power — was the ascendance first of the horse, which supplanted the ox throughout northern Europe, then of the donkey and the mule, in the south. There were a few exceptions to and regressions from this development, which occurred more slowly in rural than in urban and commercial transportation.

MILLS

Another aspect of the history of animal motive power is that of the horse-drawn mill, which is the basis for mills and a great number of machines.

Preparation of food   We have mentioned the use of animals for threshing. They were harnessed to various instruments — the *tribulum* (a board equipped with pieces of flint) or the *plaustrum* (a set of rollers) — in order to facilitate the work. Once the grains were freed from their husks, they required further preparation before being cooked and consumed. As in the case of certain tubers, grinding was necessary. Tubers were ground in graters, presses, or hand mortars. Harder grains required greater effort: the hand mortar could be used only to husk or polish them, and therefore millstones were used for the grinding. Some of the latter consisted of two stones that were rubbed together or of a stone roller moved with the hands (as can be seen in ancient Egypt or eastern Africa).

Horse-drawn mills   The stone roller attached at one end to a center post can be pulled by an animal walking in a circle (Figure 62). This type of mill is quite varied both in its form and in its use. In Mediterranean antiquity it served for the crushing of olives before they were pressed; the wheel was quite large in relation to the mill. In nothern China mills of this type are used to crush cereal grains for the making of flour; the wheel is thinner, and moves in a circular hollow trough. This was the same principle employed in the apple mills of western France, where the fruit was crushed before being pressed for cider. In central China mills of the same type are used for grinding rice.

FIG. 62.   A simple type of grinding mill:
a stone roller, one end of which is pulled by an animal.

The origin of this type of mill can be seen in certain Chinese husking practices, in which a roller is pulled over a circular area. This method may have been introduced into western Europe only in the seventeenth century.

Millstone harnesses   In every case the harnessed animal activates on a horizontal plane a pole that turns around the vertical axle of the millstone. Nowadays the animals pull this pole by means of traces, that is, by a modern harness; it is difficult to imagine what was done in ancient times, but it seems impossible to accept the idea that the modern harness was used for horse-drawn mills long before it was adapted to the pulling of vehicles. Certain Roman bas-reliefs picture what is believed to be a horse or a mule pushing the rod with its breastpiece.

Pressing mills Two types of horse-drawn mills are used to press
out liquids; they may be regarded as presses.

The first type seems very archaic in conception; it has remained confined
to India and South Arabia. It consists of a large central mortar with a round-
headed pestle the end of which is supported by a triangular packsaddle that
causes the animal to pivot. This mill is used either for extracting oil or for
pressing sections of sugarcane to extract the juice (Figure 63).

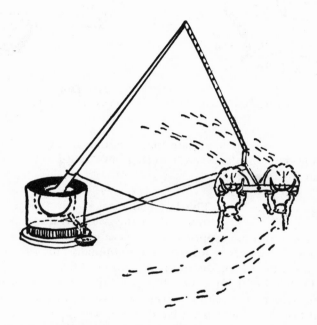

Fɪɢ. 63. Sugarcane mill, India (from *The Sugar Industry of the United
Provinces of Agra and Oudh,* by Saiyid Muhammad Hadi). The same type of mill
is used in this region for extracting oil from olives.

The second type is more recent; it was invented in the Middle Ages, somewhere in India or the Near East. Three large vertical cylinders, equipped with
teeth or projecting pitches, mesh with each other; when the center cylinder
turns, the others are pulled in the opposite direction. A stalk of sugarcane held
up to this apparatus is pulled in, pressed, and crushed by the movement of the
cylinders. This specialized device for sugarcane spread on the one hand through
southeast Asia, on the other in the Mediterranean area and as far as the West
Indies. This is the oldest prototype and ancestor of the industrial rolling mills.
It probably gave birth to a small apparatus composed of two horizontal cylinders
moved by hand, utilized for the ginning of cotton; the cylinders caught the
fibers and detached them from the seeds. This device can be found in India
and southeast Asia.

The existence of the screw (Figures 64 and 65), which the Greeks were ac-

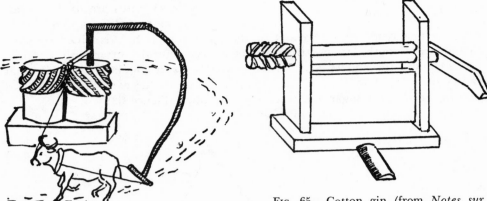

FIG. 65. Cotton gin (from *Notes sur les Tay Deng de Lang Chanh,* by R. Robert, pl. 22).

FIG. 64. Sugarcane press, southern China.

quainted with, permits us to situate the invention of these devices in the Near East in the period from the seventh to the tenth centuries, when sugarcane cultivation began to spread.

The classic mill   The last and most classic type of mill is the one invented in the Mediterranean Near East. It consists of a conical millstone capped by a concave cylindrical stone, through which the axle passes and which is turned by the animal. This cap has a cuplike depression on top into which the grain is poured. This mill may have been the model for the hand-mill, in which the grinding surface, instead of forming a pointed cone, is very flat, almost level, because it can be turned much faster by hand and because weight is replaced by centrifugal force to make the grains and flour circulate from the center out to the edges (Figures 66 and 67).

FIG. 67. The classic mill, pulled by an animal → or pushed by slaves.

FIG. 66. Crank-turned hand mill.

Human muscle power   At the beginning of human activity man himself probably performed most of the work, if only the simplest chores of pushing, pulling, and carrying, which he later imposed on the domestic animal.

The generalization of the use of animals, and then of the most diverse sources of power, did not eliminate the use of human muscle power. By way of example we shall mention several very ancient methods invented by man to increase the effects of his own muscle power. All of them are based on the principle of the lever.

The squirrel cage The method of turning a wheel by placing a man inside a circular cage, which was supported by a horizontal axle of rotation and in which he advanced, causing it to turn by its own weight, became known in Roman times. A similar mechanism can be moved from outside; the Archimedean screws for raising water were activated in this manner.

Around 1864, in San Domingo, Portugal, the discovery was made of eight overshot waterwheels buried in ancient copper mines that had been exploited from Phoenician times. These wheels must have been installed by the Romans toward the end of the third century of our era. They measured 21 feet 9 inches in diameter; the rim and the spokes were made of pine, the axle and the supports of holm oak. Around the rim were 25 wooden buckets, each 6 inches wide, 19½ inches long, and 5 inches high. The wheels were turned by men pressing with their feet on pegs inserted in the rim; the water was raised to a height of 12 feet. It has been calculated that in order to pump out the water in the mine steadily, the linear speed had to be one foot per second.

Treadles The use of treadles, that is, levers activated by foot, seems to have come from the Far East. The treadle may appear in various forms (Figure 68). One type is found in the foot hammer, a second type in the paddle noria, while the treadle of the weaving loom is a lever of the third type. The latter found many uses once the crank was combined with an arm, which seems not to have been done until around the end of the fourteenth century A.D.

There were also simple machines whose slow development during the archaic period of antiquity bore fruit during the Golden Age of Greece, and were popularized by the Romans.

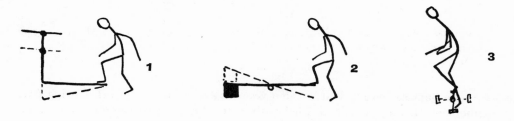

FIG. 68. Treadles: (1) Of weaving loom. (2) Of rice pounder. (3) Of paddle noria.

WATER AND WIND POWER

For historical reasons we may deal more briefly with these subjects. Although the origins of water mills and windmills are still quite obscure, we can designate relatively limited periods — quite late in time — for their respective appearances. Moreover, their development was extremely slow, the perfecting of the techniques being determined more by the quality of the materials employed in their construction than by the ingenuity of their constructors.

Wheels for lifting   It seems probable, although there is no supporting proof, that hydraulic power was first utilized to move paddle wheels designed to raise water from the stream into which they descended. This is the simplest way of using the paddle wheel, since it requires no mechanical transmission device.

In its most elementary form the ancient noria consisted of a large wheel equipped with radial paddles around the rim; it was installed on the bank of the stream in such a way that its lower portion plunged into the water, the current of which moved it. On its rim were attached buckets that dipped up the water at the lowest point of their cycle (Figure 69). It undoubtedly appeared in this form in Egypt before the period of Roman domination, in the Near East at least in the Hellenic period, between the fifth and third centuries B.C.

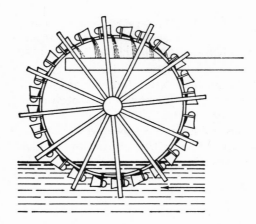

Fɪɢ. 69. Paddle wheel for raising water. It was not always a spoked wheel. The buckets must have originally been made of wood, and later of pottery.

The buckets were at first wooden vessels attached to the wheel. During part of their ascent they remained filled with water, then began to empty out when they were tilted as they approached the top. The water was collected in a wooden or ceramic trough, but a certain portion of it was lost. The only bad feature of this system was the sluggish flow into irrigation or supply conduits. Several centuries later, around the beginning of our era, pottery vessels suspended by their handles were already being used as buckets. At the top of their cycle a trip tipped them over to empty them.

Various types of norias
The distinctions between muscle power and hydraulic power that are useful in writing a history book do not, however, correspond to a historical succession of facts. Actually, the noria was very soon being turned either by a squirrel cage pulled by a man or by a crank or by a set of wheels operated by an animal moving in a circle. This latter method, which implies mechanical transmission, is perhaps later than the others. But it is not possible to say what exactly was the sequence of the inventions; undoubtedly they were all in existence by the last centuries before our era, not only in the Mediterranean Near East but also in all areas of Asia and Africa where hydrological and climatic conditions did not permit the harnessing of springs and streams. The types of norias utilized depended on the circumstances.

When it was a question of scooping water from a well or a stagnant pool, the noria must have been turned by an animal or a man. In this case the wheel was placed, not above the water to be raised, but on the ground next to it. Its axle was prolonged by a horizontal shaft that supported either the lifting wheel or an endless rope on which was hung a series of pots.

The horizontal driving wheel
It would seem logical to think that the noria led to the invention of the hydraulic mill. In truth, all this requires is the placing of a device on the end of the horizontal shaft to make it possible to move a machine (a mill in this case). However, accomplishing this seems to have been less simple.

We know that the first hydraulic mills, which were constructed by the Greeks, were conceived differently. The driving wheel was not vertical but horizontal (see Part Two, Chapter 8, Figure 54), and was placed in a millrace on which the grain mill was built. The shaft of the driving wheel extended vertically toward the top, pierced the nether millstone, and acted as a pulling shaft for the runner above it.

The first examples of these types of mills were probably constructed in the second or first century B.C. in the mountainous regions of the Near East, followed by Greece, and then Italy. But they may have been invented simultaneously and independently in various regions of the world. They were employed in Denmark at the beginning of our era, and from here, or from the south of Europe, they gradually spread as far as Ireland, where they appeared toward the third or the sixth century. Proof exists for their use beginning in the third century A.D., but there is nothing to prove that their invention spread from there throughout Asia.

These mills were not large; owing to their mode of construction the horizontal wheels could drive only millstones sufficiently large for family use. It seems probable that from the very beginning their builders thought of slanting the blades of the driving wheel more effectively on the axle. Perhaps it was even thought to direct the flow of a watercourse or waterfall against these slanted blades, and perhaps an attempt was already made to construct a distant ancestor of the modern Pelton impulse water turbine, numerous early versions of which are found at almost every stage in the history of technology (Figure 70).

The vertical driving wheel In any event, the "water mill" with vertical wheel came into existence very soon after the horizontal wheel, if not at the same time. But it seems rather to be Italian, or at least Hellenistic, in origin. Several drawings are found in the writings of Philo that suggest the use of a vertical wheel provided not only with radial paddles but even with genuine buckets. Perhaps these devices were then being made on a small scale for lesser purposes.

The only type of mill with vertical wheel known to antiquity was undoubtedly that described by Vitruvius. During the three centuries (second to first) that undoubtedly separate the two authors, the device came into use for genuine industrial purposes, together with another innovation. The pulling of a horizontal millstone by a vertical driving wheel requires the use of a device that makes it possible to reverse the right-angle movements, the driving shaft being horizontal and the receiving shaft vertical. What the constructors of the second century B.C. did not know how to do, the mechanicians of the first century seem to have learned. They invented the first model of the gear train (Figure 71).

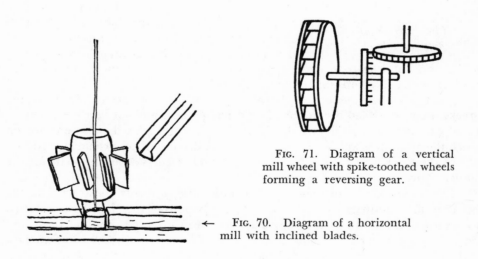

FIG. 71. Diagram of a vertical mill wheel with spike-toothed wheels forming a reversing gear.

← FIG. 70. Diagram of a horizontal mill with inclined blades.

The toothed gear This consists of a wooden disk to which wooden pegs perpendicular to the disk and arranged in a crown are attached. The combination of two toothed wheels of this type, meshing together, made possible the first right-angle gear train. To increase the strength of the device, a second wooden disk could be added to one of the wheels, in which the free ends of the pegs were embedded. This type of gear train has been known since Roman times by the name "lantern gear," and for centuries it was the only type used; it was also constructed of iron. Genuine toothed wheels made of metal did not appear before the eleventh or twelfth century; archaeological excavations have uncovered older examples, but it seems that they were used only for making machinery that later led to the weight-driven clock. Metal gear trains do not seem to have been used for industrial purposes before the sixteenth century. Even then, and until around the end of the eighteenth century, the wooden lantern gear was in common use. From this we can realize the importance of this invention, born at the beginning of our era, whose exact origin is unknown to us, and which permitted not only a reversing movement but also a reduction of effects. The entire problem of the industrial use of natural sources of energy rests on this invention.

Undershot and overshot wheels   The turning of the paddle wheel by the current of a stream flowing under it was the only method known to classical antiquity, despite several isolated ideas for other types sketched by Philo. Only at the end of the fifth century did the Roman engineers think of utilizing, in addition to the force of the current striking the paddles, the weight of the water pressing on them during their downward course. The water needed only to be brought by conduit or aqueduct over the wheel (Figure 72). This was the birth of the large installation that will be discussed in subsequent chapters.

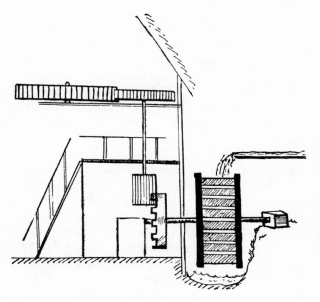

FIG. 72.   Diagram of a water mill with overshot wheel. This model can function either by the current of a river passing under the wheel or by a channel that carries the water over the wheel. The latter system is exactly the reverse of the water-raising devices. This is the first example of a reversible motor.

The use of the term "mill" for this method of harnessing hydraulic energy is due to the fact that during the early centuries wheels were often used only for turning the millstones that ground grain or olives. Only in the fourth century A.D., in fact, is mention made of the use of hydraulic wheels for turning

other machines, beginning with saws for cutting marble and millstones for polishing it. Pounders and hammers raised by camshafts and used to forge iron obtained by fusion (Figure 73) did not appear until later.

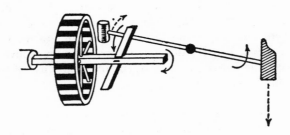

Fig. 73.   Paddle wheel, the camshaft of which activates a pounder.

Windmills   Here we shall discuss only very briefly the appearance and use of windmills, which occurred at a
relatively late period. The first windmills are mentioned at the beginning of the
Moslem era, in the second half of the seventh century, but no genuine proof
of their existence is found before the tenth century.

As in the case of hydraulic devices, and for the same reasons, the turning
mechanism was at first horizontal (Figure 74). Thus the vertical shaft turned
the millstone directly, without any mechanical transmission. Mills of this type
were certainly constructed on the Iranian plains, which are swept by steady
winds. They consisted of a circular wall with a large opening facing the prevailing wind; the turning mechanism, with its vertical axle, was equipped with
"wings," made of various materials, that must have occupied almost the entire
open space within the enclosure. But until the fourteenth century our information about their shape and construction is very fragmentary, and their distribution is no better known.

In Asia and China, toward the tenth century, windmills came into use for
irrigation or drainage. In the Mediterranean regions they are found as far as
Portugal. These are mills in which the axle of the sails is horizontal or slanted.
A variety of mill with triangular arms that could be adjusted on a stationary
axle, like ship's sails, is found in the eastern Mediterranean area (Figure 75).
The windmill with fixed arms first appeared in northwestern Europe in the
eleventh century. The body was mounted on a pivot, and could be turned according to the direction of the wind (Figure 76). Another type, which probably
appeared much later, consisted of a tower made of wood or masonry, with a
revolving cap that supported the shaft of the arms and the transmission mechanism (Figure 77). No picture earlier than the fourteenth century is known.

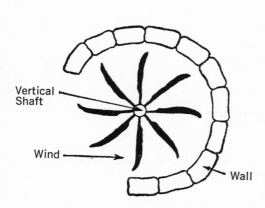

Vertical Shaft

Wind

Wall

FIG. 74. Diagram of a windmill with horizontal turning blades.

FIG. 75. Mediterranean windmill with adjustable triangular sails.

FIG. 76. Windmill with pivoting tower.

FIG. 77. Windmill with pivoting cap.

BIBLIOGRAPHY

In addition to the appropriate chapters in CHARLES SINGER *et al.*, *A History of Technology*, Vol. I (London, 1954), the reader is referred to the following:

BENNETT, R., and ELTON, J., *History of Corn Milling*, 4 vols. (London, 1898–1904) .

FORBES, R. J., *Studies in Ancient Technology*, Vol. II (Leiden, 1955).

MORITZ, L. A., *Grain-Mills and Flour in Classical Antiquity* (Oxford, 1958).

NEEDHAM, JOSEPH, and LU, GWEI-DJEN, "Efficient Equine Harness, the Chinese Inventions," *Physics,* Vol. II, No. 2, 122–162.

PUHVEL, JAAN, "The Indo-European and Indo-Aryan Plough: A Linguistic Study of Technological Diffusion," *Technology and Culture,* 5 (1964), 176–190.

STEENSBERG, A., "North West European Plough-types of Prehistoric Times and the Middle Ages," *Acta Archaeologica,* VII (1936), 244–280.

STORCK, JEAN, and TEAGUE, W. D., *Flour for Man's Bread: A History of Milling* (Minneapolis, 1952).

USHER, A. P., *A History of Mechanical Inventions* (2nd ed., Cambridge, Mass., 1954) , Chaps. 6–7.

ZEUNER, FREDERICK E., *History of Domesticated Animals* (New York, 1963).

PART TWO

MEDITERRANEAN ANTIQUITY

CHAPTER 6

MESOPOTAMIA AND THE NEIGHBORING COUNTRIES

INTRODUCTION

L ONG BEFORE Greece and Rome, and at a very early stage of her history, the ancient Mesopotamian civilization (which included the surrounding peoples in its culture) had reached an enviable level of civilization.

Geography   Mesopotamia, the "country between the rivers," watered by the Tigris and the Euphrates rivers, touches the Persian Gulf on the south, and is bordered on the east by the Elamite country and the Persian plateau, on the northwest by Asia Minor, on the west by Phoenicia. Of these regions Elam and Asia Minor, having no rivers, are the least favored; Phoenicia compensates for this lack by its location on the seacoast, which is favorable to commerce.

FIG. 1. General map of the ancient Near East.

114

Population

At the beginning of history, Mesopotamia was peopled in the south by the Sumerians, and in the northwest by an agglomeration of Semitic peoples. Surrounding her, in Elam, Asia Minor, and early Phoenicia, were peoples related by religion and language to the Sumerians, and generally known as Asianics, a term that indicates that they were neither Semitic nor Indo-European.

The protohistorical period

In Mesopotamia the historical period was preceded by a protohistorical age (around the fourth millennium), which has been divided into periods according to the names of the excavated sites that first brought to light the civilization of each period. The civilization of Mesopotamia was founded during this millennium. Architecture, metallurgy, and writing, attempts at which had been made during the preceding Hassuna and Ubaid periods, actually began (that is, definite evidence of their existence first appears) in the Uruk period.

The historical period— third millennium

The Sumerians came to power at the beginning of history, that is, in the third millennium; they temporarily lost their place, in the middle of the millennium, to the Semites from the northwest, the Agadeans (twenty-fifth to twenty-third centuries), but regained supremacy in the last centuries of the millennium with the third dynasty of Ur (twenty-second to twenty-first centuries).

Second millennium

In the second millennium the Semites again became the rulers of the country (first Babylon dynasty, nineteenth to sixteenth centuries), and the Sumerians disappeared from history. This dynasty was overthrown by the Kassites from the Zagros Mountains. They were soon driven away, and the Assyrians and the Babylonians, who had increased during this period of difficulties, struggled for hegemony. Toward the end of the millennium, Mesopotamia was raided from Elam. In the north she had to combat invaders settled there since the beginning of the millennium: the Mitanni, who soon disappeared, and the Hittites, an Indo-European stock whose power was destroyed in the thirteenth century B.C. by an invasion known as the "invasion of the Sea Peoples."

First millennium

In the first millennium Assyria conquered Babylon. This was the golden age of the empire, which ruled the major portion of the countries mentioned above; Phoenicia was frequently found among its adversaries, who were allied with Egypt. Halfway through the millennium Babylon enjoyed a return to power, with Assyria now merely a Phoenician province (the Neo-Babylonian Empire, 605–539); shortly thereafter, the Persian conquest made Mesopotamia a satellite of the Achaemenid Empire, which in turn fell after Alexander's invasion (331 B.C.).

The Mesopotamian civilization

In certain ways the Mesopotamian civilization was unique — a quality that it owes partly to the nature of the land and partly to the character of its inhabitants. The Sumerians founded the civilization; the Semites adopted it; and thereafter the Elamites and the Hittites, like Phoenicia (although to a lesser degree), were profoundly influenced by them. The Sumerians claimed that all its elements had been revealed by the gods and that nothing better had been created since then. Such a conception left room for progress only in details. Knowledge, being the object of a completed revelation and therefore sacred, could not be communicated; it was the privilege of the initiated, of priests who transmitted it, not in their writings (these contain only a series of unexplained instructions in view of the result to be obtained), but orally. These techniques can be divided into two principal groups: essential techniques of primary utility, and secondary techniques that in reality are often indispensable for complete effectiveness of the first group.

BASIC TECHNIQUES

Building Components

The fact that the alluvial subsoil of Lower Mesopotamia contained no rock, and that wood (at least hard wood) was also nonexistent led the Mesopotamians to devise ingenious substitutes.

The reed There was the reed, which abounded in the swamp-lands of the deltas of the two rivers (the mouths of which were at that time separate), and so the primitive dwelling place was the reed hut. The ditch reed (*Phragmites communis*), which can grow to a height of more than twenty feet, was utilized just as in modern times. On the site chosen, bundles of reeds were driven into the ground in two parallel rows, and were bent in such a way that their ends met and were bound together to form a row of arches. This rudimentary tunnel was reinforced by foliage and by reeds placed lengthwise. The structure could be strengthened with a clay mortar that hardened in drying.

Clay The Sumerians later discovered in the clay soil the material they needed, in the absence of stone and as a replacement for the reed, which flourishes only in swamplands. They used dried bricks: at first irregularly shaped clumps of earth, like small stones, then (after removal of foreign bodies and kneading with chopped straw to give the mud flexibility) in molded shapes that were exposed to the open air for a time. Clay was also used in a moist state with either a clay slip, moistened with water, or a very liquid bitumen, as mortar. Every four or five courses a light binding of straw or young reeds moistened with the mortar, or clay or pitch, was added; the bricks were knit together into a dense mass. Such walls withstood rain quite successfully, the burning sun of Mesopotamia somewhat less well. The bricks were of various sizes, depending on the period (approximately 12 inches by 18 inches in the early stages, 19 inches by 20 inches in the Babylonian period, and $22\frac{1}{5}$ inches by $22\frac{4}{5}$ inches in Assyrian times), and their thickness varied from 6 to $6\frac{1}{2}$ inches. Larger baked bricks were used for paving courtyards. For a brief period in the beginning the bricks were plano-convex — a form that was soon abandoned.

Bitumen Bitumen, either mixed with clay as a mortar, or as a mastic and watertight facing, came into use in the protohistorical period. Certain tablets mention the exact proportions in which it should be mixed with the other ingredients, depending on the use to which it is to be put. The surface deposits of petroleum products, naphthas, and bitumens were very soon exploited by the Mesopotamians, who were undoubtedly the first to discover their possibilities. Large quantities of bitumen were used in the construction and finishing of palaces, as well as in the construction of dikes and piers, the surfacing of bridge roadways, and the calking of ships. During early antiquity its use continued to spread throughout the Middle East, until the appearance of Roman construction techniques; its popularity then rapidly decreased, and it disappeared except in the regions that possessed deposits of bitumen. Bitumen was replaced as a mastic by pine tar, wood tar, and vegetable gums.

The roof Since the roof was flat, thought had to be given to the weakness of the palm trunks that spanned the area from wall to wall, their connecting wooden panels, and the weight of the tarred mud used for waterproofing. The result was rooms that could be as long as desired, but narrow. Moreover, the necessity for very thick walls in order to compensate for the friableness of the material precluded windows; daylight and air came in through the few doors and through a hole in the wall (or ceiling) covered with a terra-cotta grating.

Relative comfort of As for comfort, investigations have indicated the
the dwelling presence in the palaces of terra-cotta bathtubs and stoves for heating water, and so-called "Turkish" water closets — simple paving stones with holes in them, connected to a ditch. Many dwellings consisted of only one floor; when there was an upper story, it repeated the arrangement of the rooms below, opening onto a courtyard, and having a circular wooden balcony reached by a staircase of baked bricks in a corner of the house or by a wooden staircase outside. This was the typical average house. Small houses might be covered with a domed or beehive-shaped roof, built up of bricks by the corbeling method. The house was heated by a brazier, and lighted either by torches or by lamps in the shape of a cupel with a groove on one edge for the wick, which was often fed with vegetable or animal fat but also with crude petroleum, the properties of which were known to the Mesopotamians.

Palaces and temples For the construction of palaces the Mesopotamians took the individual residence as their prototype and simply multiplied it, arranging more rooms around a larger courtyard, and repeating this model as often as necessary, with communicating corridors and doors placed at irregular intervals to break the view. This arrangement is found at Mari (modern Tell Hariri) on the Euphrates (first Babylonian dynasty), and at Khorsabad near Nineveh, a palace built in the eighth century B.C. by Sargon II. The paved courtyards are sloped, and have ceramic pipes to collect water and dispose of it through baked brick drainpipes, shaped like ogive arches and laid against each other in an inclined position. The Mesopotamian land is exposed to floods because the rivers rise; palaces and temples, which were the "dwellings of God," were constructed on top of terraces built up with earth and then tamped down.

FIG. 2.   Ground plan of the palace of Khorsabad
(Botta, *Le monument de Ninive,* pp. 1849–1850, Fig. 18).

The vault   Were there no vaults other than corbeled domes? In the royal tombs of Ur, which date from shortly before the midpoint of the third millennium, the vault of the king's tomb has an arched door, which must have been very fragile but strong enough to last for the period of the funeral ceremonies. It is not a genuine arch, but consists rather of a few stones unevenly arranged in a semicircle and embedded in the mortar that binds them together. The Mesopotamians, therefore, were not acquainted with the keystone.

Brick columns The column was very rarely used. Because of the friability of clay, it could be used only ornamentally, in engaged half columns, for example. We know of such columns from the Uruk period, decorated with baked clay nails, with colored heads, sunk into the column. In the buildings of Gudea, at Tello, four assembled lower portions of columns were found on a pedestal; they were made of bricks of various shapes, alternated so as to distribute the pressure evenly. The same arrangement is found at Susa, in Elam. In both cases the columns end at this point, and we have no idea what they may have supported — perhaps a simple offering table for sacrifices.

Façades The façades of palaces and large buildings were decorated with superimposed rectangles, undoubtedly for the purpose of encouraging the plays of light and shadow that broke their monotony. Near the doors were high-relief sculptures in the form of winged bulls. Inside the rooms, areas painted with motifs that were both decorative and representative replaced the too costly bas-relief in, for example, the less important palaces at Til-Barsib, on the Euphrates. In the Neo-Babylonian and Persian empires the façades and interiors were decorated with baked bricks with glazed reliefs. An Assyrian bas-relief in the British Museum shows a winged bull being transported on rollers; men with crowbars force the bull to move, while the superintendent perched on top of the animal supervises the proceedings by means of a trumpet.

Fortifications The same principles governed the building of defensive structures — walls that were sometimes double (notably in Neo-Babylonian Babylon) and had projecting towers for taking besiegers on the flank, fortresses whose towers overlooked the other buildings (to prolong resistance), earthen embankments, or embankments with rocks to cause missiles thrown from the tops of the walls to rebound against the besiegers.

Fig. 3. Representation of a fortress (G. Contenau, *Les civilisations d'Assur et de Babylone* [Payot, 1951], p. 280).

Plate 3.
Transporting a bull of Sennacherib. Assyrian bas-relief in the
British Museum. *Photo British Museum, London*

City gates   The arrangement of the city gates is a very interesting one. The walls, reinforced at the level of the gate, were pierced with a tunnel that ran from outside to inside and opened onto one or two courtyards along the way — an arrangement that permitted the archers to shoot from the ramparts above, onto the besiegers attacking at that point.

Town planning   These buildings were not arranged haphazardly. The plan of the cities, at first circular but later quadrangular (a less economical solution), included straight, intersecting avenues that nevertheless respected the tangle of houses and tiny streets, still characteristic of the East, within the quarters thus delimited.

Multistoried towers   Excavations have revealed this layout in the principal cities excavated, especially at Babylon. There the Ishtar Gate opened onto an avenue, decorated with glazed bricks, that led to the main temple, flanked by its storied tower, or ziggurat. This construction, a prototype of the "Tower of Babel," is a stack of from five to seven stepped terraces topped by a chapel for the god; stairs or ramps led to the top. Every city had one or several ziggurats connected with the main temples; in most cases each level was painted a different color, and some were even planted with small trees.

FIG. 4.   Multistoried tower shown on a bas-relief from Nineveh
(G. Contenau, *Manuel d'archéologie orientale*, Vol. IV, Fig. 1161b).

Corrections of level   The discovery, during the excavation of palaces, of slightly convex ridgepoles has given rise to the belief that, as in Greek temples, this was a precaution taken so that from a distance this line would not appear concave. More probably it was simply a matter of preventing settling, which was quite likely to occur, given the plasticity of the clay.

Plate 4.
Ziggurat of Choga-Zambil (Elam).
Photo Roger-Viollet

Stoneworking   The working of stone, although it had no construction problems to solve, was perfected through the making of statues, the engraving of gems, and the decorating of palaces. In Assyria, where rooms were decorated with slabs of gypsum plaster worked into bas-reliefs, the stone is easy to work when it first comes from the quarry, and hardens only in drying. It lent itself, therefore, to assembly-line methods; the supervisors laid out the guidelines to be followed by the less experienced workers. In Dynasty III of Ur, a provincial governor named Gudea had a number of statues of himself executed in diorite, an extremely hard variety of granite that undoubtedly had to be imported, for no local supplies are known. It can be assumed that blocks in the desired dimensions were obtained by drilling holes in the rock, into which wooden wedges were driven; dampening the wedges caused the rock to split. The finished statue was polished, undoubtedly with sand. Although bronze was the principal metal of this period, the Mesopotamians were acquainted with various methods of tempering, examples of which still exist (for example, the golden dagger from the royal tombs of Ur). The Gudean "bricks" were artificial, unbaked stone.

Architectural plans and cartography   Similarly, architects' scale drawings made when Khorsabad was discovered show that the monuments are all noticeably irregular, as was recognized in later American excavations. An architectural plan can be seen on the knees of the statue of Gudea known as "The Architect with the Plan" (Louvre Museum).

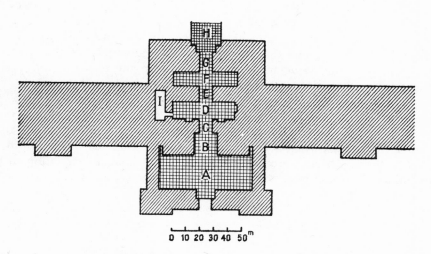

FIG. 5.   Ground plan of a gate city of Khorsabad
(Place, Vol. III, pl. 18, p. 1867).

Glyptics   The art of gem carving, whether by chasing or engraving, produced cylinder seals that are masterpieces far superior to large sculptures. Later, we shall discuss how it was used by the Mesopotamians.

The neighboring countries Early Asia's respect for tradition with regard to the Sumerian civilization was such that countries better endowed with ground resources made only limited use of them. The northern part of mountainous Assyria possessed stone and wood, but utilized the former only in bas-reliefs used to decorate palaces; they were arranged as plinths, since they could not be used as friezes (this would have torn out the portion of the wall to which they were attached).

The Hittites, who were still better endowed, used stone particularly for their fortifications, the foundations of their buildings, facings, bas-reliefs, and wall sculpture. In their buildings they used pieces of wood as a framework which they then filled with dry, irregularly shaped blocks.

Phoenicia, which was influenced by Mesopotamia as well as by the Aegean Islands and Egypt, used more stone in its construction work, without, however, attempting anything more than ordinary buildings. On the whole, construction in Phoenicia remained faithful to the Aegean and Egyptian techniques.

Achaemenid Persia Beginning in the middle of the first millennium, Persia controlled the destiny of Asia. Although Elam, despite the relative proximity of the mountains, had always preferred earthen construction, with molded bricks for ornamentation (Susa is an example of this), the only new additions to this technique under the rule of the Achaemenides were a few columns and panels of glazed bricks, done in the Babylonian technique (see p. 186). The art of construction was in part renewed at Persepolis, which is located in the mountains. A terrace was cut into the rock, and small rock terraces were built on it to support the palaces. The walls of the latter, in a last concession to tradition, were built of perishable materials that have now disappeared, but the frames of the windows and doors were cut in blocks of stone that are still in place and indicate the framework of the palace. A large stone staircase gave access to the terrace; other, smaller staircases led to the palace, which was built above the lesser terraces.

The stone columns The new element was the stone column with a fluted shaft, terminating in most cases in animal busts or in a capital corresponding to a bell-shaped base. All the technical difficulties had by now been overcome; the columns are slender in relation to their height (about fifty-nine feet), and sufficiently close together to support the roof

FIG. 6. Shaft and base of a column from Sidon (Contenau, *La civilisation phénicienne* [Payot, 1949], p. 75).

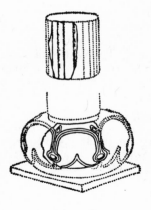

without darkening the interior, as was the case with the large Egyptian columns. The placement of these columns posed a problem. A lifting device is out of the question; probably the access-ramp method, which had long been in existence, was used. Here, again, we can only be astonished at the technical maturity that had been acquired in such a short time.

AGRICULTURE

Irrigation: canals The Mesopotamian land is fertile only when irrigated. The possibility of irrigation is offered for the most part by the Tigris and Euphrates rivers, whose annual floods would be devastating if left uncontrolled. This was the purpose of the canals, the network of which was expanded and improved from one dynasty to the next. By means of regulator sluices they supplied the land with the required quantity of water. They required maintenance, however, and dredging to prevent them from being silted up. Water was supplied to the cities by means of aqueducts, one of which brought water from a source thirty miles from Nineveh; it crossed a ravine by means of a bridge with five supporting arches and large paving stones laid on a bed of bitumen. (Remains of this bridge are still in existence.)

Bridges Pontoon bridges were usually used for crossing rivers, but the bridge, with double-arched stone piers and a wooden and stone roadbed, was also known. At Babylon, in the time of Nebuchadnezzar, such a bridge linked the city and the countryside; a few of its piers still remain.

Gardens The men of antiquity regarded the famous hanging gardens of Babylon as one of the wonders of the world. Their site is believed to have been found in a portion of Nebuchadnezzar's palace contiguous to the great avenue of processions. The garden was laid out in tiers over vaults supported by pillars; water was supplied by wells. This elevated position meant that the gardens could be seen from outside the palace, a factor that helped to ensure their fame.

Vestiges of gardens, which were sought after for their shade and freshness, can be seen in a festive hall at Assur, and at Persepolis, in the form of hollows in terraces, in which the trees which it was desired to acclimate were planted in soil imported for that purpose. A list of the various types of plants in the garden of King Merodach-baladan (eighth century) has been preserved; they seem (insofar as the translation of their names is definitely established) to have been listed according to their common properties: vegetables (garlic, onion, leek, cabbage[?], lettuce, fennel, beets, rape, radish), condiments (three sorts of mint, basil, saffron, coriander, rue, thyme, pistachio), plants of the salt-marsh variety, resinous plants (Persian ferula, balsam of Mecca), and several varieties of poppies. This was a typical ancient Oriental garden, with flower beds surrounded by drains and shaded by fruit trees, which in turn were protected from the sun by tall palm trees. Pictures on monuments depict the irrigation device known as

FIG. 7.   Series of shadoofs, Nineveh
(from Layard, *Monuments of Nineveh* [London, 1849, 1853], Vol. II, 15).

the shadoof, by which water is easily scooped up from the canal in a pot attached to a movable bar that has a counterweight, and is then poured into a basin or the irrigation runnels. Northern Mesopotamia was covered with trees typical of the temperate regions; in the central portion were fruit trees and palm trees; descending toward the south, only the palm trees persisted.

Cereal grains   Thanks to its system of canals, which brought vegetation to all its inhabited regions, Mesopotamia was the breadbasket of the ancient world. It produced barley (which was then the principal grain) and, to a lesser extent, emmer and millet. Mesopotamia is believed to have been the original home of the cereal grains. As regards the fertility of the soil, we have the testimony of Herodotus, who refused to give the production figures for fear of not being believed; the average estimate is forty to one. The agricultural tools used were the small mattock, the shovel, and the simple plow, or most often the plow with a seed drill. The wheat was stored in silos.

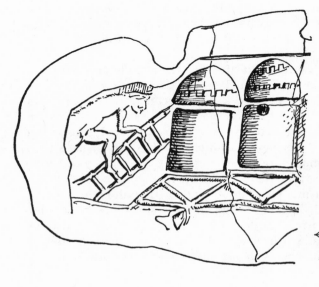

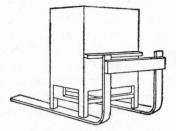

FIG. 8.   Impression made by a cylinder seal, showing granaries (from L. Legrain, *Mémoires de la Délégation en Perse*, Vol. XVI, pl. XIV, Fig. 222).

← FIG. 9.   Sledge, or *tribulum*, from the tomb of Queen Shubad at Ur. (G. Contenau, *Manuel d'archéologie orientale* [A. Picard, 1947], Vol. IV, p. 1851).

The tribulum   For threshing grain the Mesopotamians probably learned at a very early period to use the instrument the Romans called the *tribulum*: a plank with bits of flint along one edge, which was pulled slowly over the heads of grain. In the Orient the plank was surmounted by a box to seat the driver of the oxen or donkeys pulling it. The so-called "sledge" in the queen's tomb at Ur was of this shape (which, moreover, is that of a written symbol of very early date). It is still in use in the East.

THE FOOD SUPPLY

Dairy products,   The nutritional base then (as now) consisted of
vegetables, and bread   dairy products (at Mari, Mr. Parrot discovered a series of decorated molds for curdling milk), various species of cucumbers and watermelons, cereal gruels, and bread, which, judging by the pictures on the monuments, resembled the flat cakes that in modern times are cooked by being stuck on the walls of ovens for a few moments.

The Mesopotamians also used butter; a bas-relief from the protohistoric age of Ubaid depicts its production. On one side of a stable built of reeds, the farmer is milking an animal from behind. On the other side, the milk has been poured into a large jar that a superintendent is shaking constantly in order to churn the milk. The liquid is then filtered into a funnel that removes the lumps of butter; these are then packed into a jar.

FIG. 10.   Bas-relief showing making of butter (from G. Contenau, *Civilisation d'Assur et de Babylone* [Payot, 1951], Fig. 39, p. 259).

FIG. 11.   Animals in enclosures (from G. Contenau, *Manuel d'archéologie orientale* [Picard, 1927], Vol. I, Fig. 302).

*Poultry and
slaughter animals*

The most important food supply was represented by the barnyard animals: geese, ducks, a little later the hen, sheep, goats, and oxen of the type *Bos primigenius*, which seem to have been domesticated at the dawn of history, but were used especially for banquet and festival dinners.

Fish

Thanks to the lagoons and canals, fishing became a source of food supply. Excavations at Tello have yielded up the remains of a pile of fish that had been dried and disintegrated to form a kind of flour. A clay tablet lists the varieties of fish (at least eighteen) found in the market at Larsa toward the end of the second millennium. The fish were taken from the lagoons or the sea either by net or pole (or the harpoon, in the case of the large varieties), and were stored in fishponds while awaiting shipment to the market.

Fruit

Fruit — pomegranates, medlars, apricots, the peach (the "Persian" apple), figs, and especially dates — also contributed to the food supply. Almost every part of the date palm could be utilized: the trunk for its wood, the fibers for making rope, the pits as combustible material or, when ground into a paste, as food for animals, the date itself either plain or preserved in oil (at that period sesame oil was used). The olive tree did not become acclimated until around the first millennium.

Locusts; hunting

Invasions of locusts were frequent, and the Mesopotamians used them as a dietary supplement; Assyrian monuments depict servants carrying brochettes of this insect to the royal dinner table. Another, more important, supplement was supplied by the fruits of the hunt, done either for food purposes (the boar, the partridge) or for defense or sport (lion, buffalo, onager); the latter was the prerogative of the king.

Pictures show birds being hunted with bow and arrow; boar or lion hunts in which the hunter, aided by mastiffs, is on foot; lion, deer, onager, or wild-ox hunts in which the hunter, armed with the bow and lance, is riding in a chariot or on horseback, urging on the frightened flock. Primitive arrowheads were often triangular, with cutting edges; they struck the object with their long edge in order to weaken the animal through loss of blood. Falcon hunting was common in the first millennium B.C.

Fig. 12.   Boar and lion hunting. Scene on a cylinder seal from Susa (Contenau, *Manuel d'archéologie orientale* [Picard, 1927], Vol. I, p. 50; from Legrain, *Mémoires de la Délégation en Perse*, Vol. XVI, Fig. 243).

Drink   In addition to beer made from barley, the juice of the palm tree, a kind of kvass made from fruit juice, and fermented bread, the Mesopotamians used wines that, until the grape became acclimated in Assyria, were imported from Lebanon. Knowledge of alcohol gained through the fermentation of ordinary drinks was widespread. Some drinks (we are not certain what they were) left a sediment, and we possess a number of metal strainers and straws that were used to remove this sediment.

The kitchen   Kitchens were primitive; they consisted of a simple fire under a cooking pot, installed sometimes in the courtyard of the house or sometimes in one of the rooms. There was no chimney; the smoke escaped through an opening in the ceiling or the door.

METHODS OF TRANSPORTATION

The caravan   Commercial exchanges were carried on by land travel, thanks to donkey caravans (the camel did not become acclimated in Mesopotamia until around the beginning of the first millennium). Cappadocian texts dating from the end of the third millennium have been discovered to be contracts between large firms and caravan drivers. Caravan transportation was a kind of peddling. As the merchandise was sold, it was replaced by local products, which were then resold on the return trip; the profits were split.

Roads   Judging by the vestiges of the road from Khorsabad to the Tigris, the typical roadbed was slightly convex, and was filled with very ordinary, irregularly shaped limestone rocks of medium size, with somewhat larger ones along the edges of the road. The royal Achaemenid road, remains of which have been found in Asia Minor, was more carefully constructed. Main highways, which had to carry armies and war machines, were approximately forty feet wide.

Horses, chariots, and wagons   The horse did not really come into use in Mesopotamia until around the midpoint of the second millennium; it was imported from north of the Black Sea, particularly from Cappadocia. Prior to its appearance, its place was filled by a species of wild donkey, the onager, skeletons of which have been found in the royal tombs at Ur. The Kingdom of Judea exported copper and horses; at the beginning of the first millennium B.C., King Solomon appears as a great merchant doing business through his port of Ezion-Geber, north of the Red Sea. The American excavations at Megiddo have uncovered the remains of large stables for the temporary housing of the horses.

The horse was ridden with a simple saddle blanket, without stirrups; it was used almost exclusively for the war chariot, which was originally a seat on large wheels, and later a box that held from two to four occupants. The wagon, which had either two or four wheels, was sometimes pulled by oxen. The investigations

FIG. 13. Wild horse with short mane, Susa (from Jéquier "Fouilles de Suse,"
FIG. 15, *Mémoires de la Délégation en Perse,* Vol. VII).

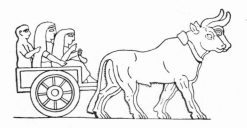

FIG. 14. Two yoked animals pulling a chariot, Assyria (G. Contenau,
Manuel d'archéologie orientale [A. Picard, 1931], Vol. III, p. 1215).

Plate 5.
Primitive chariot, Ur. *Photo G. Contenau*

of Lefebvre des Nouettes into harnessing practices in antiquity indicate that for the horse there existed only the neck collar, which strangled the animal when it pulled hard; the absence of the breech band still further diminished the usefulness of the horse as a draft animal. It is possible that the Asian (or so-called Przhevalski's) horse is one of the ancestors of the Assyrian variety.

Ships   The canals were constantly used as highways. Beginning in the historical period, the sail seems to have existed as an auxiliary; in most cases boats were driven by oars or were towed. They were constructed in peculiar shapes; there were the *kouffa,* a round basket waterproofed with a paste of clay and bitumen (it can still be seen on the Tigris); the *kelek,* a raft supported by skins to increase its ability to float; a large, sharply pointed canoe (the modern *belem*). In the swamplands temporary boats made of bunches of reeds tied together at the ends were used. All these craft are very frequently on monuments, as are inflated goatskins used for supporting swimmers.

Every large city had a regulatory agency (called the "Wharf," after the name of the place where it conducted its operations) for maritime transactions; from this we can deduce the degree of organization of river commerce. The commercial ships, often rented, that traveled up and down the canals did not have a large

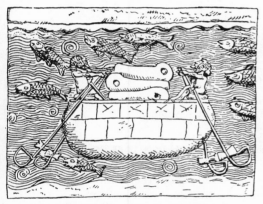

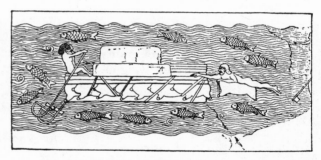

Fig. 16. The *kelek,* an Assyrian raft
(from Place, *Ninive et l'Assyrie,* pl. 43, p. 1867).

Fig. 15. The *kouffa,* an Assyrian ship
(from Place, *Ninive et l'Assyrie,* pl. 44a, p.
1867).

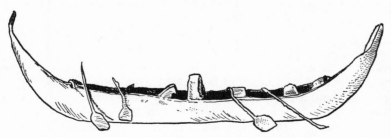

Fig. 17. Silver boat from the royal tombs at Ur
(from *Antiquaries Journal,* Oct., 1928, pl. LXII).

FIG. 18. Warship, Mesopotamia
(G. Contenau, *La civilisation d'Assur et de Babylone* [Payot, 1951], p. 298).

FIG. 19. Impression made by dried
brick seal from Susa, showing weaving
(G. Contenau, *La civilisation d'Assur et
de Babylone* [Payot, 1951], p. 208).

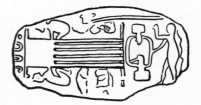

capacity; in the time of Dynasty III of Ur, mention is made of loads of from 900
to 2,500 liters of grain. A bas-relief in the Louvre depicts a convoy of cedar logs
from Lebanon lying off the coast of Phoenicia.

CLOTHING, FURNITURE, AND MUSIC

Fabrics

Our knowledge of textiles goes back to the early
days of history. In the necropolis at Susa, axes
wrapped in cloths were placed in tombs of this period as funerary offerings; oxidation has metalized these fabrics. Analysis showed that they ranged from material
similar to our canvas fabrics for wrapping and dishcloths, to the finest batistes.
They were handwoven from flax S-spun in warp and woof. We with our modern
machines have hardly surpassed these five-thousand-year-old fabrics, which were
perhaps woven on looms similar to our warp-weighted loom but lacked the comb.

Dress

The costume of the primitive period was the sheepskin, later replaced by a woolen cloth imitation (the
kaunakes), then by a draped garment with an embroidered edge. Toward the first
millennium, under the influence of the neighboring countries, the latter garment
was abandoned, and women as well as men wore long tunics (replaced, in the case
of workingmen, by short belted tunics). Embroideries were common among the
upper classes; the *opus babylonicum* was famous in antiquity. Both cotton ("the
wool tree") and flax were known.

Coiffure and shoes Among the Sumerians shaven heads and smooth chins were the custom for men. For ceremonial purposes they wore wigs and false beards (natural beards in the Assyrian period). Women at first wore their hair hanging down their backs, later in elaborate coiffures held with pins. The customary headgear in the time of Dynasty III of Ur consisted of the turban, larger for the women than for men, while the standard shoe was the sandal; the Hittite sandals had upturned tips, particularly suitable for a land of rocks and snow.

Carpets and furnishings The use of carpets in the Assyrian period is proved by a doorsill in the palace of Khorsabad: it is an imitation in sculpture of a fringed carpet.

Furniture was scanty, as in the modern Orient; it consisted of low tables, chests made for the most part of wickerwork and bamboo shoots, and low bedsteads covered with wickerwork as a substitute for the mattress.

Music Musical instruments included tambourines, the large dulcimer, hide-covered kettledrums, cymbals, reed flutes, citharas, and harps with sounding boxes. Examples of these instruments have been found in the tombs at Ur.

METAL

Ancient Assyria possessed surface deposits of silver, copper, lead, and iron, which were sufficient for primitive needs. These metals are also found in Iran, including antimony in the Anarek district to the northeast of Ispahan and, in the same region, tin. Asia Minor had gold and manganese in addition to silver, iron, copper, and lead. Lebanon in the time of Nabonidus was a producer of iron. The existence of tin in the Caucasus at this period is probable but unproved.

Copper The metal found at the necropolis of Susa, with or without a cloth covering, proved to contain 92.12 percent pure copper, without iron, sulfur, lead, zinc, or manganese, but with traces of nickel. Objects from Tello have yielded similar amounts; at Tell-el-Obeid, copper in most cases exceeds 95 percent of the total. The same results were obtained by R. C. Thompson in the cases of objects from Eridvu. Thus pure copper, and not merely an antimonial bronze, as has been claimed, was in fact in existence in the early centuries of the third millennium. Since copper did not exist in Sumer, its presence there indicates trade relations with the Caucasus and Anatolia.

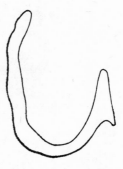

Fig. 20. Copper fishhook, Mesopotamia
(Gordon Childe, *L'Orient préhistorique* [Payot, 1953], p. 181).

Bronze

Bronze is constantly in evidence after the time of Gudea; in most cases it is tin-bronze or antimonial bronze. Elam, which was particularly noted for metalworking, has left remarkable bronze monuments dating from the second half of the second millennium B.C., the most striking being the statue of Queen Napir-Asu. This statue was cast in two halves, and was then filled with molten bronze forced in with a metal block to cause it to fill the interior completely (this part of the operation failed). The destroyers of the monument were able to remove the head and one arm; the statue, even with the head and one arm missing, still weighs more than 3,960 pounds, and is five feet tall. It must have been extremely difficult to maintain successfully the metal in each crucible at a constant temperature (they were poured out one after the other).

Gold and silver

Gold from the north and from Egypt, as well as silver and the other metals, has been discovered in large quantities in the royal tombs of Ur. The Sumerians already knew how to refine it and remove its impurities. Certain objects from the royal tombs at Ur — the headdress of Meskalamdug, for example — are very solid, but very often gold was also used simply for "gilding." At Khorsabad, for example, the bitumen-coated trunks of palm trees were covered with a bronze facing that simulated the scales of the palm tree; then gold leaf, annealed to make it flexible, was burnished and nailed to this bronze facing with small nails. Silver, more common than

Plate 6.
Meskalamdug's helmet, Ur. Baghdad Museum. *Photo X*

gold, was used in the same manner; before coins came into existence it was a unit of exchange used for the payment of purchases, as by barter. Refining of gold was necessary in view of the impurities this metal contained; King Burraburias of Babylon complained to Amenhotep IV (circa 1370–1352 B.C.) that the gold received from him contained three-quarters of its weight in impurities. The fine gold used in the objects found in the tombs at Ur was approximately 75 percent pure gold. The refining method used was a variant of our cupellation, as is indicated by the purification of five minas of gold by means of successive stages in the furnace: on the first round the gold was reduced to four minas, five

Plate 7.
Gold paten, Ras-Shamrah. Louvre Museum, Paris. *Photo Louvre Museum.*

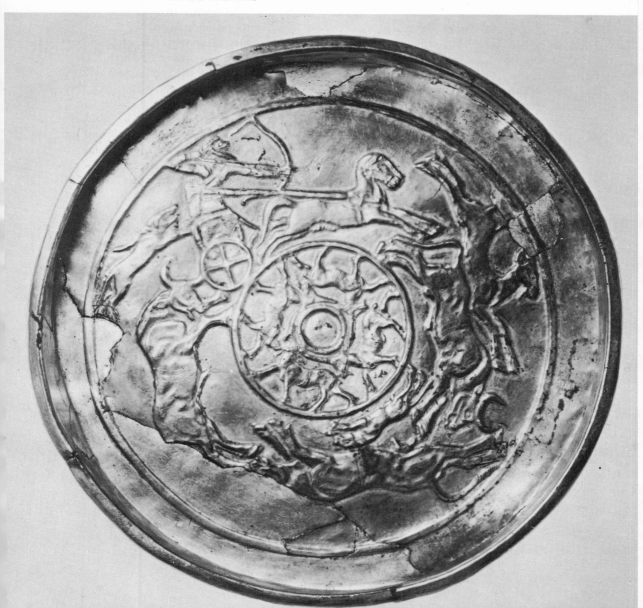

shekels, on the second to three and two-thirds minas. The Mesopotamians knew how to alloy and harden gold, which permitted it to be used for weapons.

Jewelry

We need hardly insist on the importance of gold in personal ornamentation; there are the Hittite containers of royal treasures, the royal tombs at Ur (even a modern goldsmith would be proud of the gold headdress of Meskalamdug), combs for fastening coiffures, bracelets, earrings, parts of necklaces and pendants. The Mesopotamians learned filigree work, milling, encrustation, repoussé, cloisonné, and soldering at a very early period.

Iron, tools, and weapons

Iron is found in deposits, but it can also be meteoric in origin. The two are distinguishable by the fact that the latter always contains from 5 percent to 20 percent nickel. Meteoric iron sufficed for all requirements during the early period of its use; iron deposits replaced copper only after methods of working them became known, and when abundant supplies were discovered. In this area of the world ironworking was born among the Hittites toward the middle of the second millennium. The invasion of the "Sea Peoples" in the second half of the second millennium brought it to western Asia. Pictures on monuments show Assyrian workers carrying iron (copper in the Agadean period) tools, shovels, adzes, and saws. The warehouses of the Khorsabad palace contained stockpiles of enormous iron ingots made into the customary tools — pickaxes, bushhammers, and plowshares. Iron was used (concurrently with stone) particularly for weapons — swords, lance tips, arrowheads, and maces. Moreover, the invention of new

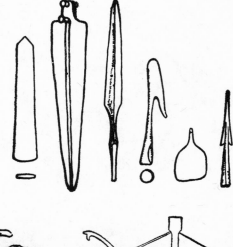

FIG. 21. From left to right: flat chisel, dagger blade, blade ending in a poker, harpoon with a cupped point, razor blade, javelin tip. (See Gordon Childe, *L'Orient préhistorique* [Payot, 1953], p. 209).

FIG. 22. Pickax, bushhammer, iron plow-share, Khorsabad, eighth century B.C. (Contenau, *Manuel d'archéologie orientale* [A. Picard, 1927], Vol. I, p. 60).

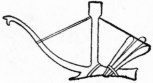

FIG. 23. Plow (from Place, *Ninive et l'Assyrie*, pl. 31).

weapons did not do away with older models: the Persians who opposed Alexander used flint, copper, and iron arrowheads. Iron was used for defensive purposes: in helmets, for the reinforcement of shields and tunics, coats of mail and "brigandines," and the poles of war chariots, which were simple boxes supported at the back by two wheels.

Siege machines; mines   Siege machines, which were moved by internal machinery, struck with iron-tipped battering rams. The Assyrians perfected a type of weapon that persisted almost unchanged in the West until the invention of firearms.

The Assyrian monuments depict sappers attempting to mine besieged fortresses. Excavations at Dura-Europos on the Euphrates, which was occupied by the Romans at the time of the Persian attack, have uncovered the mines and countermines built in this period; they are of a type that continued to be used until the invention of gunpowder. They consisted of galleries dug under the defensive works; these galleries were shored up with wood as the digging proceeded, and when the work was finished the wood was set on fire. The superstructure, deprived of support, collapsed.

IVORY AND POTTERY

Ivory   The early Sumerians used fossilized, ivory-colored marine shells from the lagoons for carving and encrustation work. Later, ivory was used — the elephant still lived in Upper Syria in the latter half of the second millennium. Remarkable specimens of this work have been discovered in recent years in Assyria, Syria, and Phoenicia-Palestine.

Pottery   Pottery existed in profusion long before the beginning of history. The decoration of very early pottery vessels is sometimes an imitation of wickerwork. The necropolis at Susa, which dates at least from 3,000 B.C., contained large goblets made on the turn-

Fig. 24. Potters. Impression made by a cylinder seal from Susa (from Legrain, *Mémoires de la Délégation en Perse,* Vol. XVI, pl. XIV, Fig. 216).

Fig. 25. Painted vase from Tepe-Sialk, Iranian plateau. (Contenau, *Manuel d'archéologie orientale,* A. Picard, 1947, Vol. IV, p. 2152).

table from a very fine, pure material; they are thin, well-fired vessels that ring when struck, and are decorated with sanguine (an iron oxide), which produces black tones of sometimes purplish-blue and even reddish hues. There was an immense production of ordinary objects; in addition, there were vases with blue-green glazes and (around the first millennium), at Tepe-Sialk near Kachan in Iran, teapot-shaped vases with very long spouts, which were imitations of metal prototypes. With these early and protohistoric vases from Susa, we are looking at objects that are already highly perfected. Numerous plaques and modeled statuettes have also been discovered. An important product of the Mesopotamian civilization, the clay tablet used for writing, is related to pottery production by virtue of the similarity of their materials.

THE CUNEIFORM TABLET AND THE CYLINDER SEAL

Mesopotamian achievements in the sciences, as well as in social affairs (historical annals, accounting and business transactions, law, and religion, among other matters), are known to us through extant Mesopotamian documents, which represent a major success in the utilization of substitute materials.

The relative scarcity of material for papermaking led the Mesopotamians to turn to clay. The primitive writing system of the Sumerians consisted of drawing the objects they wished to remember, and it is possible that the Sumerians had already invented it before their arrival in Sumer. When they decided to use small, carefully cleaned pieces of fresh clay of various sizes as "writing tablets," this system had to be modified, for circles and curved lines produced smudges. The signs were transformed into broken lines, and were no longer drawn but rather imprinted by pressing on the clay. The result was the system of little holes and hooks that is known as "cuneiform."

The Sumerians used a beveled reed as a "pen"; the impression of its fibers can often be seen on the clay. The tablet was held in the palm of the left hand; when completed, it was dried in the open air or, for permanent records, baked to a brick. In the case of a legal document or a letter, the tablet was placed in an "envelope" of wet clay marked with identifying instructions.

Major literary or religious works often consisted of several tablets; at the beginning of each tablet the last few words of its predecessor were repeated in order to facilitate their being read in the desired order. They were kept piled on shelves in the royal archives, in the temples that housed the schools of the scribes.

FIG. 26. Assyrian scribes at work. The one at left is writing on a tablet, the one at right on parchment (from Layard, *Monuments of Nineveh,* Vol. I, pl. 58).

Such documents were by definition impersonal. The scribe and the parties to a contract added to it the equivalent of their signatures, whether a stamp or (among the Sumerians and their successors) the cylinder seal, a small stone cylinder engraved with the owner's name and decorated with a picture. This cylinder was rolled over the damp clay, so that its characters and picture were reproduced in a continuous frieze.

The Mesopotamians, who traded with Egypt, were able to use papyrus, which they did not produce, but neither papyrus nor parchment (which the Mesopotamians knew long before Pergamum is alleged to have invented it) has been preserved. The Sumero-Akkadian language had two terms, one for the "writer on tablets," the other for the "writer on skins," and both types of writers at work are shown on Mesopotamian monuments.

CHEMISTRY

The techniques discussed above were of primary utility, the fruit of the daily experience of an entire people. The techniques of chemistry were developed only by a small number of people — an élite — and yet their influence on the others was often felt.

Mesopotamia's early achievements in chemistry are still unknown to us; undoubtedly many bodies whose properties were later utilized were discovered by chance, but the results indicate a long background of experiments.

Glass   Written tablets separated by long intervals of time (the first around 2,000 B.C., then a hiatus until the seventh century B.C.) once again indicate that achievements considered as definitive were accomplished at a very early stage. They concern glassmaking, and the variety of results is astonishing; in addition to ordinary glass, the tablets enumerate glass of all colors and imitations of precious stones. Several discoveries indicate a knowledge of domed kilns capable of attaining a temperature of from 1,100 to 1,200°C, as well as evidence for the use of simple tools. The knowledge required for glassmaking, the terra-cotta strainers, mortars, conical vessels with rounded bottoms and large spouts at their openings, and vessels with an opening in their lower portion (for emptying liquids) are all evidence of chemical operations. The patient investigations of R. Campbell Thompson and, more recently, Martin Levey, have reopened this area for investigation.

Paint   Frescoes, as well as the glazes of enameled bricks, are indicative of a knowledge of chemistry. Colors were obtained from plants (saffron for yellow, and kermes, red), from the shellfish *Murex brandaris* of Phoenicia (all shades of purple ranging from dark to light), and from stocks of metal ingots in the form of tools and paints intended for the upkeep of the palace of Khorsabad: sheets of lapis lazuli (blue), iron oxide or sanguine (red), copper sulfate (blue-green), tin oxide (white), lead antimoniate (yellow). Colors for enamelwork included the white and yellow mentioned above, copper without cobalt but with a little lead (blue), and copper suboxide (red).

Dyeing   Dyeing, which required a mordant, utilized tannins (nutgall, pomegranate, sumac). A hair dye (light blond?) was made from oil of cedar, alum, and the sap of the *Anthemis tinctoria*.

Perfume   Despite its very simple equipment, the perfume industry made great progress because of the demand for perfumes for religious purposes; it produced perfumed water, aromatic pomade, and incense. Extraction of perfumes by water and oil was practiced, and undoubtedly sublimation was as well known to the perfume manufacturers as it was to the metallurgists.

Soap   The elements of soap were used from the very beginning of history: oil, purified clay, ashes of plants containing soda and potassium (saltwort). Ancient tablets mention taxes on these combined substances, indicating a general use of this substitute.

CONCLUSION

As regards the basic techniques, Mesopotamia's achievement consisted, not of one or two chance achievements, but rather of improvements in all areas of activity. These improvements attained a level of perfection that remained almost unsurpassed for centuries.

The Mesopotamian civilization made particular progress in chemistry and mathematics, which it recorded for posterity in its clay tablet "books." These constitute a lasting testimony both to Babylonian thought and to the general level of the Mesopotamians' civilization.

BIBLIOGRAPHY

I. GENERAL WORKS

ADAMS, ROBERT McC., *Land Behind Baghdad: A History of Settlement on the Diyala Plains* (Chicago, 1965).
———, *The Evolution of Urban Societies: Early Mesopotamia and Prehispanic Mexico* (Chicago, 1966).
ANATI, EMMANUEL, *Palestine Before the Hebrews* (New York, 1963).
CHAMPDOR, ALBERT, *Babylon* (London, 1958).
CHIERA, EDWARD, *They Wrote on Clay* (Chicago, 1938).
CHILDE, V. GORDON, *New Light on the Most Ancient East* (rev. ed., New York, 1969).
CONTENAU, GEORGES, *Manuel d'archéologie orientale* (Paris: Vol. 1, 1927; Vols. 2, 3, 1931; Vol. 4, 1947).
———, *Everyday Life in Babylon and Assyria* (London, 1954).
FORBES, R. J., *Studies in Ancient Technology* (11 vols., Leiden, 1955).
FRANKFORT, HENRI, *The Birth of Civilization in the Near East* (Bloomington, Ind., 1950; paperbound, Garden City, N.Y., 1956).
GURNEY, O. R., *The Hittites* (Harmondsworth, England, 1952).
KOLDEWEY, ROBERT, *The Excavations at Babylon* (London, 1914).
KRAMER, SAMUEL NOAH, *From the Tablets at Sumer* (1956); paperbound ed., *History Begins at Sumer* (Garden City, N.Y., 1959).
———, *The Sumerians: Their History, Culture, and Character* (Chicago, 1963).
MELLAART, JAMES, *Earliest Excavations in the Near East* (New York, 1965).
NEUBERGER, ALBERT, *The Technical Arts and Sciences of the Ancients* (New York, 1930).
OPPENHEIM, A. LEO, *Ancient Mesopotamia: Portrait of a Dead Civilization* (Chicago, 1964).
PALLIS, S. A., *The Antiquity of Iraq: A Handbook of Assyriology* (Copenhagen, 1956).
PRITCHARD, JAMES B., *Ancient Near Eastern Texts* (Princeton, 1950–55).
WITTFOGEL, KARL A., *Oriental Despotism* (New Haven, 1957).
WOOLLEY, C. LEONARD, *Ur of the Chaldees* (New York, 1930).
———, *The Sumerians* (paperbound ed., New York, 1965).
WULFF, HANS E., *The Traditional Crafts of Persia* (Cambridge, Mass., 1967).

II. SPECIAL TOPICS

BADAWY, ALEXANDER, *Architecture in Ancient Egypt and the Near East* (Cambridge, Mass., 1966).
DAVISON, C. ST. C., "Transporting Sixty-Ton Statues in Early Assyria and Egypt," *Technology and Culture,* 2 (1961), 11–16.
DE CAMP, L. SPRAGUE, *The Ancient Engineers* (New York, 1963), Chap. 3.
FRANKFORT, H., *Cylinder Seals* (London, 1939).
GALPIN, F. W., *The Music of the Sumerians and Their Immediate Successors, the Babylonians and the Assyrians* (Cambridge, 1937).
JACOBSEN, THORKILD, and LLOYD, SETON, *Sennacherib's Aqueduct at Jerwan* (Chicago, 1935).
LEVEY, MARTIN, *Chemistry and Chemical Technology in Ancient Mesopotamia* (Amsterdam, 1958).
NEUGEBAUER, O., *The Exact Sciences in Antiquity* (2nd ed., paperbound, New York, 1962).
WILLCOCKS, WILLIAM, *The Irrigation of Mesopotamia* (London, 1917).

CHAPTER 7

EGYPTIAN ANTIQUITY

EGYPTIAN CHRONOLOGY

List of Royal Dynasties

A. Pre-Thinite and Thinite Ages, 3300–2778 B.C.
    1st Dynasty, 3300–3000
    2nd Dynasty, 3000–2778

B. Old Kingdom, 2778–2423 B.C. (the Memphite Kings)
    3rd Dynasty, 2778–2773
    4th Dynasty, 2723–2563
    5th Dynasty, 2563–2423

    End of the Old Kingdom, 2423–2220 B.C.
        6th Dynasty, 2423–2263
        7th Dynasty (*pro forma*)
        8th Dynasty, 2263–2220

    The Herakleopolitan Dynasties — First Intermediate Period
        9th Dynasty, 2222–2130
        10th Dynasty, 2130–2070

C. Middle Kingdom, approx. 2160–approx. 1580 B.C.
    1. 11th Dynasty, approx. 2160[?]–2000
       12th Dynasty, 2000–1785

    2. Second Intermediate Period, approx. 1785–1580
       13th and 14th dynasties, approx. 1785–1680. Numerous little-known kings
       15th and 16th dynasties, approx. 1730–1580. The Hyksos kings at Tanis
       17th Dynasty, Thebes, approx. 1680[?]1580

D. New Kingdom, approx. 1580–approx. 1090 B.C.
    The Theban Period, approx. 1580–1090
       18th Dynasty, approx. 1580–1314
       19th Dynasty, approx. 1314–1200
       20th Dynasty, approx. 1200–1090

E. Late Period, approx. 1090–332 B.C.
  1. The Tanite-Bubastite Period, approx. 1090–730
     21st Dynasty (Tanite), approx. 1090–950
     22nd Dynasty (Tanite-Bubastite), approx. 950–730
     23rd Dynasty (Tanite-Theban), approx. 817–730[?]

  2. The Ethiopian-Saite Period, approx. 730–525
     24th Dynasty (Saite), approx. 730–715
     25th Dynasty (Ethiopian), approx. 751–656
     26th Dynasty (Saite), 663–525

  3. The Persian-Mendesian Period, 525–341
     27th Dynasty (Persian), 525–404
     28th Dynasty (Saite), 404–398
     29th Dynasty (Mendesian), 398–378
     30th Dynasty (Sebennytic), 378–341

  4. Second Persian Domination, 341–332
     Arrival of Alexander, founder of the Macedonian Dynasty

NOTE: The 11th, 12th, 13th, 14th, and 15th dynasties are in part concurrent.

These dates are taken from the chronological list given by Drioton and Vandier in *Les peuples de l'Orient méditerranéen*, Vol. II, *L'Egypte* (Presses Universitaires de France, Paris, 1938 and 1952 [Third Ed.]).

In the third edition (1952) these authors give a double chronology, one short and one long. The latter is based on the recent work by Sharf and Moortgat, *Aegypten und Vorderasien im Altertum*, established by comparing the Egyptian with the Mesopotamian archaeology. This chronology is shorter by an average of about a hundred years, as regards the period between the Old and Middle kingdoms.

THE GEOGRAPHICAL AND HISTORICAL SETTING

In the course of a period that lasted for several millenniums, Egypt, an "island" squeezed into the hollow of the Nile Valley, and bordered on the north by the Mediterranean Sea, by deserts on the east and west, and by the infinity of the black world on the south, created a civilization out of nothing. Behind her lay the immense empty stretch of prehistory, in which the slightest technological achievement had required hundreds and perhaps thousand of years of development.

And then, relatively suddenly it seems, the invention of a new tool opened new possibilities. This tool was probably the hafted hammer or miner's pick which, by acting as an extension of the power of the hand, undoubtedly inaugurated the era of the use of building stone and the working of quarries and mines. From then on, inventions followed in quick succession.

This period of gestation seems to have begun in the predynastic age (fifth millennium B.C.), and to have ended with the Thinite kings (toward 3,300 B.C.). By the time of the first Memphite dynasties (Old Kingdom, 2778 to 2420 B.C.), when the pyramids were built, Egyptian technology had reached complete development. Then, as if exhausted by this great effort, which had spanned three millenniums, it made little further progress. From then on, the same mallets, the same copper or bronze gravers, the same methods of stonecutting and woodcutting, continued to be used.

This end of progress must undoubtedly be attributed to social life. Starting as a force of innovation by virtue of the needs it creates, society very often becomes a hindrance to progress through tradition, routine, and the development of misleading customs that lead to dead ends.

In contrast to the Greek and Roman civilizations, which were par excellence urban, Egypt possessed a purely rural civilization. Just as in modern times, moreover, her fate was in the hands of the Nile River. Egypt owes the elements of her comfort to the oily land of the Nile, with which the Egyptians molded bricks to be used in the construction of buildings. This type of construction could have continued indefinitely but for the chance discovery of a new need: the desire for eternal life. Then Egypt energetically set to work to create the components of eternity. The perishable architecture of mud, reeds, and wood was monumentalized by transposing the same elements into stone. Pharaonic Egypt remained faithful, however, to alluvium for her private dwellings and even for the palaces of her kings, which continued to be built of dried brick. But her gods and her houses of eternity had to be constructed of permanent materials.

No material was sufficiently permanent — neither sandstone nor the granite of Aswan nor the diorite from the desert. Egypt invented techniques that still amaze us: The people of the Old Kingdom succeeded in sculpturing with great suppleness of modeling the famous diorite statues of Chefren that are the pride of the Cairo Museum; they raised the stones of the Great Pyramids; they lighted the depths of the mines they exploited without being smothered by the fumes of combustion gas from the torches. With their scanty stone or copper tools, they succeeded in piercing cornelian beads, in carving statues of gigantic proportions from granite.

How these achievements were accomplished has not yet been solved. They can be explained to a certain extent by the tireless patience of the Egyptian people, who conquered the hardest materials with such unlikely methods as wearing them away. But many of their methods are undoubtedly destined to remain a mystery to us.

AGRICULTURE AND FOOD

Agricultural technology When we visit the Egyptian tombs, especially those dating from the Old Kingdom, we are struck by the great number of agricultural scenes that decorate their walls. The cultivation of the earth was — and is — the principal business of Egypt, and the Egyptian "big businessman" has always been an agricultural entrepreneur.

Methods of cultivation In Egypt, where the average annual rainfall is only 1.3 inches (Cairo has 1.2 inches a year; Alexandria, 8 inches), irrigation of the plants is dependent on the Nile. During its annual flood, which inundates almost the entire country, this river deposits an alluvium that acts as a rich fertilizer of the soil and eliminates the need for manure and even fallowing. To the great astonishment of the ancient Greek travelers who visited the Nile Valley, the Egyptians, contrary to other peóples, sowed first and plowed later. In most cases they were satisfied to send a flock of sheep over the still damp earth they had just seeded; in running about, the sheep buried the seed with their hooves — a substitute for plowing. In areas untouched by the Nile floodwaters, the Egyptians resorted to the plow and the hoe. The latter consisted of a wooden blade wedged into a handle by means of a twisted rope. In the working of the land the hoe played the same important role that the *fass*, its direct descendant, does for the modern *fellah*.

The plow was undoubtedly humanity's first step toward the use of the machine. Its appearance in Egypt dates from a very early period, since it appears in completely developed form in pictures dating from the Old Kingdom. It was always pulled by two cows yoked together with a fairly light yoke placed in front of the horns. The plow itself consisted of a wooden plowshare with a hard tip and two handles separated by a crosspiece; above the plowshare was a vertical piece of wood to which the pole was attached. This slender device had merely to scratch the loose, light soil. For harvesting, the peasants used a sickle with small flint blades inserted in a groove cut for that purpose, and held in place by a bituminous mastic. Once the harvest had been gathered, threshing was done with the help of animals. The grain was spread out in a circular area and trampled by herds of oxen or donkeys, whose hard hooves forced the grains out of the ear.

Cereal production formed the basis of Egyptian agriculture. Barley and emmer were cultivated, and, beginning in the New Kingdom, a cereal that was perhaps durra (*Holcus sorghum durra*). Food plants included the lentil, the bean, the onion, the chick-pea, the cucumber, and the watermelon. Among the fruit trees were the pomegranate, grape, fig, jujube (Chinese date), olive, the carob tree, and naturally the date palm.

Cattle raising Scenes painted on the walls of tombs testify to the attempted domestication of several species of animals. In addition to the pigs they captured by hunting, the Egyptians raised in their parks the gazelle, the deer, the oryx, the bubalus, the addax, the ibex, and even the hyena. Certain scenes show the force-feeding of an animal destined for the slaughterhouse. All these animals were penned up at night in stables and were cared for like domestic animals. But the Egyptians quickly wearied of these attempts at breaking in; they gradually abandoned this livestock and returned to the modern domestic animals. The horse appeared only in the New Kingdom, and even then was used for little more than the pulling of light war or leisure chariots.

As in the case of animals, the Egyptians attempted to raise various types of birds, such as the Numidian crane, the swan, the pigeon, and various species of

Plate 8.
Harvesting activities. Tomb of Nakt, Thebes.
Photo Roger-Viollet

ducks and geese for which large farms existed. The chicken appeared only at a very late period.

Hunting and fishing   Another type of activity filled the lives of the inhabitants of the Nile Valley, and made an important contribution to their diet: hunting and fishing. Men fished (no better method has yet been invented!) by means of lines with one or several fishhooks, pots, nets with weights and floats exactly like the models in use today in Egypt, and, lastly, by means of a harpoon with one or several barbed tips. The harpoon was also used for hunting the hippopotamus, which at that time abounded in the papyrus-covered swamps of the delta. This harpoon was naturally stronger than that used by the fisherman, and had ropes with floats, which were nothing more than bundles of papyrus. Penetrating deeply into the hide of the pachyderm, the single blade of the harpoon pulled the rope and the float with it.

Hunting in the desert was done with the help of the bow and arrow, the lasso, an instrument quite similar to the South American bola; trapping was done with a series of nets. The latter method was used for beating and capturing the wild ox, the ostrich, the oryx, the large-horned ibex, the wild sheep, and the gazelle. Very often the hunters were assisted by packs of greyhounds, magnificent half-savage animals, and sometimes (in the Old Kingdom period) by trained hyenas. The cat, which appeared in Egypt in the Twelfth Dynasty, and even the lion, also appear at hunts as assistants.

Birds were captured by means of the snare, a kind of double-net trap that was sometimes very large, and was lowered by pulling on a cable. The most characteristic hunting weapon, however, was undoubtedly the wooden thrower or boomerang, which still survives among the Australian aborigines.

PUBLIC WORKS

Hydraulics   In Egypt, where water is scarce, managing the Nile waters has always been the primordial problem and concern of its inhabitants. Without the Nile, Egypt would be only an extension of the Sahara Desert.

Practically nothing remains of the irrigation system that existed in pharaonic times. However, given the flatness of the land, major works such as bridges, dams, and stone pitchings could not have been very large, and these roles were probably filled by earthen (or similar perishable material) constructions. Only one picture of a bridge is known to us — that of the bridge which spanned the Pelusiac branch of the Nile at the site of the present El Qantarah — and even here its interpretation is not certain. Crossing from one bank of the Nile to the other must have been done by boat.

On the other hand, the network of canals dug by the early Egyptians to irrigate their lands must have been very important, since large desert borderline areas in the delta, which today are covered with sand, contain numerous vestiges of agglomerations, proof of the existence of a major system of canals.

As regards irrigation, the tradition transmitted by Herodotus, Diodorus, and

Plate 9.
a and *b* Fishing with nets. Bas-relief from the mastaba of
Akhuthotep, Fifth Dynasty. Louvre Museum, Paris. *Photo
Archives photographiques d'Art et d'Histoire*

various other authors mentions the existence of Lake Moeris, a genuine inner sea of which little now remains except Birket Karun, a mere fragment of the splendid achievement of the ancient pharaohs. Lake Moeris was undoubtedly the product of a deliberate utilization of a depression in the earth similar to that of Qattara farther north.

Lake Moeris acted as a reservoir to correct the irregularities of the flood. It was fed by a canal parallel to the Nile that rose in the vicinity of Gebel Silsileh, and, paralleling the river as far as Beni Suef, flowed through a narrow gorge into the Lake of the Faiyum. Here two parallel barrages, acting as regulating sluices, linked the Nile and Lake Moeris; a massive earthen dike was built in the region of El Bats and was linked up with them. In case of dangerous flooding, this dike was broken.

The canal of the two seas

Communication between the Red and the Mediterranean seas undoubtedly existed throughout antiquity, of course with varying success. It was accomplished through a deviation above Bubastus (now Zagazig) of the Pelusiac branch of the Nile, wide enough, according to Herodotus, to permit two triremes to pass side by side. Its path followed the valley of the Tumilat (the biblical Goshen), passed through the ancient city of Pitom, and reached Lake Timsah, near Ismailia. Then, turning south, it flowed into the Red Sea near the present city of Suez. This undertaking has been attributed to various rulers, including Sesostris, Necho, the son of Psamtik, Darius, and Ptolemy Philadelelphus. In one of the oldest written documents, the stele of Darius, who occupied Egypt toward 520 B.C., the King of Kings boasts of having "dug the canal which leads from the Nile to the sea which bathes Persia." It is more likely that he dredged an old canal that had become silted up, since steles and small forts dating from the reign of Ramses II (and therefore going back to an earlier age) were discovered near the site of the old canal. Although the physical remains of older periods have suffered from the ravages of time, it is nevertheless true, according to the Egyptian texts, that communication between the two seas was common, at least beginning in the Sixth Dynasty. A certain Khnum-Hotep, whose tomb was found near Aswan, tells us: "The Steward Khnum-Hotep says: 'Having departed with my masters, the princes and seal-bearers of God, Teti and Khui at Byblos and Punt, eleven times I "did" this country.'" The word "did" is evidently used here in the sense of "travel through."

To indicate the importance of the network of canals that existed in ancient Egypt, let us recall the statement of Herodotus (II, 17) that the two branches of the Nile, the Bolbitinus and the Bucolic, were not the work of nature, but, rather, were man-made canals.

The Nilometers

It would not be possible to end the story of irrigation without speaking about the "nilometers," the observatories for noting the height of the Nile floods. They consisted of a kind of staircase that descended into the ground to below the lowest water level of the river. On the risers of the stairs were marked graduations that served to measure the periodic overflows of the river. The nilometers communicated either directly

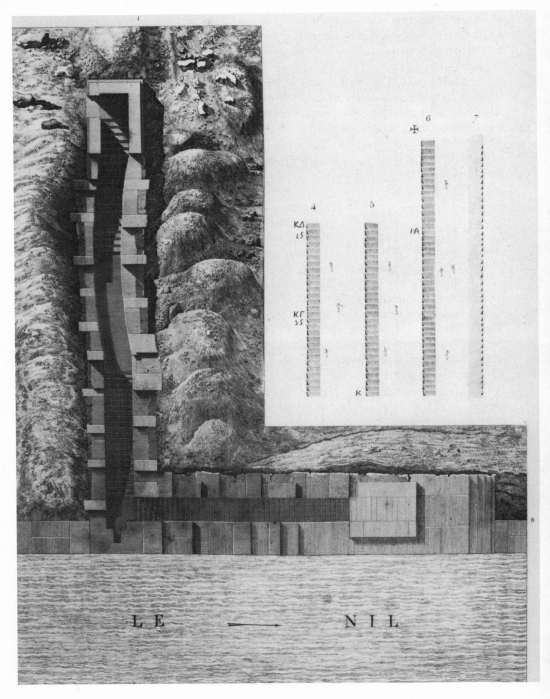

Plate 10.
Plan, elevation, section and details of a staircase nilometer,
Elephantine.

with the Nile waters (as at Elephantine), or indirectly with the seepage waters (Medinet-Habu, Dendera, Idfu). After the Late Empire, graduated columns planted in the center of a shaft (as at Gizeh) replaced the staircase type; they were similar to the column on the island of Roda, which is still used today to give the official level.

Piping The scarcity of springs in Egypt, and the location of cities and villages on the banks of the Nile or its canals, made large installations for piping water unnecessary. Even large present-day villages do not have a single pipe.

In the delta there was a system of piping inside the temple of Neuserre at Abu Sir (Fifth Dynasty), where copper pipes laid end to end have been discovered. They were made of large copper sheets rolled up like sheets of paper and held in place with an envelope of plaster.

Water-supply pipes existed elsewhere, notably in the temples at Tanis, where the sacred lakes inside the walls were fed by a piping system that consisted of pottery cones with handles, approximately $2\frac{1}{2}$ feet long and with an average diameter of $2\frac{1}{4}$ feet, sealed only by clay-filled joints. Also at Tanis (as well as elsewhere), the water needed for the life of the temple was drawn from several circular or square wells built of beautifully dressed calcareous stone; these wells descended to the underground water level (now reached at twenty-three feet), and were reached by corridors with stairs, which descended in right-angle flights to the water level. These corridors were high enough to permit the passage of a person carrying a jar on his head, while the stairs, which could be navigated by beasts of burden, had low risers and wide treads. Inside the wells, and projecting from the wall, were stone corbels that supported a platform for the use of the operators of the shadoof.

Roads In the area of communication, the construction of interurban roads (if such existed) could not have posed a problem. In this flat, dry country, nature itself favors easy solutions. Given a population that traveled on foot or donkey, without any harness, the roads must have been much as they are today: dirt paths with a simple roadbed consisting of that gray clay soil that comes from the Nile deposits and is found everywhere in the country. A simple leveling and daily watering is all the maintenance required for such roads.

The pharaonic cities, however, were linked with their necropolises by monumental routes. These causeways, which led from the plain toward the desert-like rock cliffs in which the tombs were cut, sometimes covered considerable distances.

Ancient Egyptian texts mention the existence of strategic roads — for example, the Road of Horus, the Wall of the Prince. These were probably caravan routes with small forts and customs houses rather than deliberately constructed routes similar to the Roman limes.

BUILDINGS AND DWELLINGS

Brick   Sun-dried brick is the predominant building material in the Nile Valley, and it is still employed today just as it was more than five thousand years ago. The oldest known bricks date from the predynastic (Naqada) era. They were made with Nile alluvium mixed with varying amounts of sand (depending on the locality) to eliminate shrinkage; in areas that had no sand, the alluvium was mixed with the straw, or rather the straw debris, that remained on the ground after the threshing of the wheat. This straw, however, had to be soaked so that the cellulose matter would be distributed uniformly through the mud; it was therefore left to "steep" for several days. The bricks were then made in wooden molds — simple frames open on top and bottom. The mold was placed on the ground, on a thin layer of straw debris; it was filled, tamped down crudely by hand; then shaken slightly to free the brick; the process was then repeated.

Individual homes, rich as well as poor, and even royal palaces, were all built of this material. Only temples and tombs, beginning in the Old Kingdom, were in theory built of more durable materials. The sizes of the bricks varied considerably, from $10\frac{1}{4}$ inches x $4\frac{3}{4}$ inches x $2\frac{3}{4}$ inches (Thinite period) to $17\frac{3}{4}$ inches x $9\frac{1}{2}$ inches x 6 inches (Tanis, Twenty-first Dynasty). The bricks gradually increased in size, beginning in prehistoric times, reaching maximum size between the Twenty-fifth and Twenty-sixth dynasties, then gradually decreased until the Arabic period. Sometimes they were stamped with the king's name, which makes it possible to identify them. For example, the bricks in the fortification walls of Tanis reveal that it was constructed by the Pharaoh Psusennes.

Baked bricks, although known to the ancient Egyptians, were very rarely (almost never) used; not until the Roman era did they come into common use.

Brick construction   The edifices built of dried bricks were of necessity massive. The foundations of official buildings constructed of brick were carefully laid; this was not always the case for stone buildings. The builders endeavored to reach the underlying layer of sand frequently found in Egypt. A preliminary caisson wall was built to act as a piling for preventing sandslides. The temple fortification walls were sometimes very thick (those at Tanis, constructed by Psusennes, were $55\frac{3}{4}$ feet thick). They were formed of enormous masses of rock that were contiguous and jointed but independent, with alternating projections and recesses. Both longitudinally and laterally the beds of the bricks were sharply concave, being laid in the shape of a hull — a construction method that is unavoidable when buildings are constructed in the manner of a prism whose bases are parallelograms. The probable reason for this type of construction must be attributed to the fact that the alternation of concave and convex rows neutralizes the shrinkage and expansion of the dried brick.

A layer of reeds was sometimes laid on top of the courses at regular intervals. In the walls of the mastabas and fortresses (Meydum, No. 17, Rahotep, Semneh, Tanis), branches and tree trunks were placed in the masonry, both to prevent the walls from falling apart and to facilitate their drying.

One of the curiosities of Egyptian construction is the existence, in the Middle Kingdom, of sinusoidal city walls. When viewed at an angle, these dried brick walls 3¼ feet thick are seen to consist of a combination of alternating curves. This type of construction was used because of its resistance to overturning, and its presence is warranted in sandy regions.

Cutting of hard stones   When we study the technique of stonecutting in antiquity, we are astonished at the ease with which the Egyptians succeeded in cutting and even carrying the hardest stone; the more difficult the cutting process, the more perfect the workmanship.

One of the most frequently used materials for Egyptian architecture and statuary was granite, extracted from quarries in the Aswan region. There were three varieties: rose (the most common), black, and gray.

The largest granite obelisk still standing in Egypt is that of Queen Hatshepsut; it stands 96¾ feet tall and weighs approximately 350 tons. The obelisk at Rome (Saint John Lateran) is even taller (105 feet). Remains in the temple of Karnak permit us to estimate the height of some of these "needles" as ranging between 124 feet and 157 feet; they too were carved from granite.

Near Aswan there is an unfinished obelisk still lying in its matrix. An examination of the technique used to extract it reveals traces of a combination of several methods. The block of stone was first marked out by cutting parallel grooves to the desired depth, then detaching the block by means of wedges. The grooves reveal evenly spaced, vertical marks similar to those that would have been made by an enormous drill drilling closely spaced holes. The discovery near the quarry of spherical drill bits of bluish dolerite ranging in weight from 3,520 pounds to 9 pounds and measuring from 4½ inches to 6 inches in diameter has made it possible to explain the method used. It is believed that the workers actually bombarded the granite with spherical balls, thereby crushing the grain of the stone. In this way they slowly descended to the point where they could attack the stone horizontally. A groove was cut and wedges were driven into holes or slots ¾ inches x 3 inches deep and 5 inches apart; then the wedges

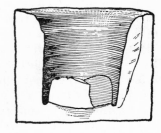

Fig. 27. Hieroglyph representing a drill used for hollowing out vases (M. Murray, *Saqqarah*, I, Fig. 65).

Fig. 28. Cylinder and "carrot" (Petrie, *Tools and Weapons*, pl. LII).

Fig. 29. Mold of a hole bored with a cylinder in a lintel of the granite temple of Gizeh (Petrie, *Tools and Weapons*, pl. LII).

Plate 11.
Obelisks of King Thutmosis I and Queen Hatshepsut, Karnak,
Eighteenth Dynasty. *Photo New York Public Library.*

were tapped one after the other. The small size of the holes rules out an alternative method the Egyptians were probably also acquainted with: wetting wooden wedges to make them swell, which caused the stone to be detached along the angle of separation.

Another method was used for drilling hard stone. The discovery of granite and diorite vases hollowed out by means of a cylindrical tube (Figures 28 and 29) indicate that this method was known in prehistoric times. Emery powder was used, samples of which (or more likely quartz powder, which is very abundant in Egypt) are believed to have been found. Petrie tells us that sawing and polishing by means of abrasives was undoubtedly a common process; it is, however, difficult to determine whether this abrasive was used in powder form or was mounted in the form of teeth in a metal tool.

Soft stone   In the area around Memphis the stone most commonly used for masonwork was extracted from quarries situated opposite the necropolis. These quarries yielded up the fine calcareous rock used in the construction of the façade of the Great Pyramid and the mastabas of Memphis. These quarries, which are still visible, took the form of large subterranean galleries supported by carefully squared pillars. To avoid encounters with fissures and hard surface areas in their work, the Egyptians did not hesitate to sometimes establish their quarries at quite a depth. As in the case of granite, the stone was extracted by means of two parallel channels dug frontally into the wall; work was begun from the top. When the block had been isolated from the rock mass, it was detached from the base as has been described above.

When it has a fine and homogeneous grain, calcareous rock can be sculpted, and lends itself marvelously well to the flexibility of the chisel. In this case it was equally suited to very beautiful masonry. Here, again, the Egyptians revealed their prodigious skill in the art of stoneworking. The stones are often laid without any kind of mortar. Beginning in the Middle Kingdom, very large stones were sometimes held together by means of bronze dowels or, as in the temple of Abydos, sycamore wood dowels marked with the king's seal.

The Egyptian architects went to great lengths to make the courses of stone uneven, with a view to strengthening the bonding. Thus one block of stone might be beveled, and the joint might be oblique instead of vertical, while another block might have notches along the joint of the course as well as the vertical joint.

In the case of very thick walls, only the interior and exterior facings were constructed of high-quality stone, often admirably polished and decorated. The interior, or filling, on the contrary, consisted of a coarse masonry of rubble, waste stone, and rock collected at random, all bound together with a mortar or clay mud.

The roofing of the structure was composed of large beams laid side by side; these were often cut from the same beautiful stone, and their span was sometimes as much as twenty feet. In buildings of the early dynasties, these beams, when seen from inside, look like palm-tree trunks because of their semicircular shape — a reminder of their plant origin.

Mortar   The Egyptians did not use lime mortar until it was introduced by the Greeks during the Ptolemaic period. The type of mortar they used most often was a mixture of plaster and sand, and sometimes crushed brick, combined in varying proportions. The builders of the stepped pyramid of Sakkara used a kind of natural cement which when analyzed proved to contain:

	CLAY	CRUSHED LIMESTONE (CALCIUM CARBONATE)	QUARTZ SAND
White layer	54%	11%	35%
Yellow layer	38.5%	53%	8%
Reddish layer	3%	14%	83%

The yellow layer is the most durable — so durable, in fact, that it can be removed only with hammer and chisel (Lauer).

Dried bricks were held together with clay, or rather with the same alluvium that had been used in their construction, made more fluid by diluting it with water and removing the stone particles. In one known case, however (the chapel of Huron, at Tanis), the dried bricks were bonded with a mortar of resin.

The vault   Although the Egyptians were familiar with the vault, they used it with the greatest timidity. During the First Dynasty they began to roof tombs with dried brick barrel vaults similar to those of houses still in use in Nubia. These vaults were built without the help of the arch or centering (Figure 30). First the load-bearing walls, rectangular in shape, were raised; then the wall was continued along the shortest side of the rectangle. Using this wall as a support, they began the vault by laying the bricks in oblique layers, gradually increasing until they reached the top. A second barrel was supported on the first one, and so on. In this way the vault was developed without the need of arches to support it. However, there exists an individual tomb, dating from the end of the reign of Pepi II (Sixth Dynasty), which has a semicircular stone vault apparently constructed with the help of a form. The official architects still continued to use the corbeled vault exclusively; an example is the vault of the hall of the Great Pyramid, where the seven courses progressively decrease, the last one at an interval of two feet.

In addition to its role in buildings, dried brick masonry was also used for scaffolding. In a country such as Egypt, which has never had forests, wood plays

FIG. 30.   Diagram of a three-layered vault in dried brick
(From G. Jéquier, *Manuel d'archéologie égyptienne,* p. 305).

a completely subordinate role, and the small quantity that was used — the large fir poles that preceded the pylons of the temples, for example, and perhaps also the door panels — was imported from Lebanon. Frameworks for projects requiring major operations were therefore out of the question. The Egyptians succeeded in meeting this need by means of dried brick ramps. A picture still in existence shows the construction of one of these ramps, which permitted the workers to carry construction materials to the desired height. In the section on major construction work we shall again have occasion to speak of these brick scaffolds.

Transportation of materials

The Egyptian land — oily and claylike in the Nile Valley, sandy on the edges of the desert — does not lend itself well to the use of the wheel, and so this instrument did not appear until quite late. On the other hand, the bas-reliefs of the Old Kingdom show that the scow or barge and the sledge were frequently used for the transportation of heavy materials. Statues and obelisks of sometimes considerable weight were in this way transported to every point in Egypt (Figure 31). At Tanis there are remains of at least three, and perhaps four, gigantic granite statues of Ramses II. One of these, called "Sun of the Princes," was famous in antiquity; it must have been at least sixty-five feet tall, with a pedestal and a dorsal pillar in keeping with its size. This statue, and at least two of the three others, to say nothing of numerous obelisks and monolithic columns of rose granite, came from the Aswan quarries. All these enormous blocks of stone were therefore carried for a distance of more than five hundred miles. They were obviously transported by water, and all that was needed was to let the

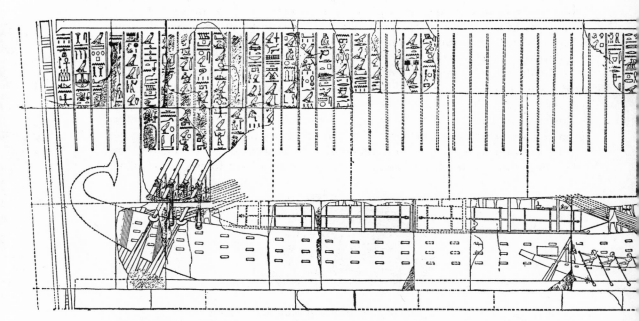

FIG. 31.   Transporting two obelisks on a large barge (from a bas-relief at Deïr-el-Bahari).

scows glide with the current. At the end of their trip, the monuments were loaded onto large sledges similar in shape to our ice sleighs. A large number of men were harnessed to the main towropes, and men perched on the front of the sledge poured water on the road in order to facilitate the passage of the sledge over the oily clay.

The motive power in every activity of any sort was always muscle power, human or animal, the latter being supplied by the ox and the donkey (the horse and the camel became known only much later).

Plate 12.
Colossal statue of Ramses II, Tanis, Nineteenth Dynasty.
Photo The Brooklyn Museum

Instruments for lifting   The Egyptians had no instruments for the multi-
                              plication of power, and the pulley was unknown.
They had only what we would call a "fixed" pulley, that is, a wooden beam fixed
at both ends, over which oiled cables were moved back and forth. An equally
rudimentary variant of this instrument was discovered in the neighborhood of
the Great Pyramids: a block of stone, semicircular in shape, in which a groove
was chiseled. The lower portion of the stone lengthens into a tenon that fitted
into a post or other support. The rope, one end of which was applied to the
force, the other to the resistance, slid through this groove. The result must have
been mediocre, since the normal reaction of the support increased by the force
of rubbing on the stone had to be taken into account.

The use of the modern type of pulley implies use of the wheel. Apart from
one isolated example, the wheel appeared only in the New Kingdom period. Even
then it was used only in war chariots, which were introduced into Egypt in the
same period as the horse, during the Syrian campaigns.

To the best of our knowledge, the only lifting devices known to the Egyptians were the lever, the cylindrical roller and several extremely crude jacks.
In moving large masses of stone, however, they compensated for their crude
tools by developing to perfection their knowledge of the laws of equilibrium.

For raising obelisks they used an ingenious system of dried brick ramps
(Figure 32). The obelisk needed only to be pulled on rollers, base first, to a
kind of gangplank built for this purpose, which took the place of scaffolding.
The base of the obelisk was shifted down onto the granite pedestal (Figure 33),
in which a groove had been chiseled to stop the slipping; at this point the
obelisk, its vertical center of gravity being at a 45-degree angle, was in balance.
Once past the 45-degree angle, the obelisk, by a simple law of physics, could be
raised by simple human muscle power.

The same method was used when sarcophagi weighing from 20 to 25 tons
had to be lowered sometimes as much as 100 feet into the ground. First, a shaft
for the descent of the sarcophagus was dug in the rock; a smaller shaft was dug
next to it, and the two were joined at the bottom by a passage. Both were then
completely filled to the top with sand. The heavy sarcophagus was placed on
the sand-filled opening of the large shaft, and the workers began emptying the
sand from the small shaft. When the bottom of the latter was reached, the passage joining the two shafts was emptied, and the sand in the larger one flowed
out. As the sand level dropped, the sarcophagus descended gently and arrived
unharmed at the desired level (Figure 34).

The Great Pyramids of   One of the most puzzling problems raised by Egyp-
the Old Kingdom   tian archaeology is the question of the method
                            used in the construction of those immense, gran-
diose monuments called the pyramids, the largest of which, the pyramid of
Cheops, still rises to a height of 453 feet. Astonishing as it may seem, no document that could tell us how they were built has yet been found. Although the
Egyptians liked to carve detailed scenes of their deeds and even scenes of their
private lives on their monuments, nothing exists to show the methods that
created the numerous pyramids that line the edge of the desert from Memphis

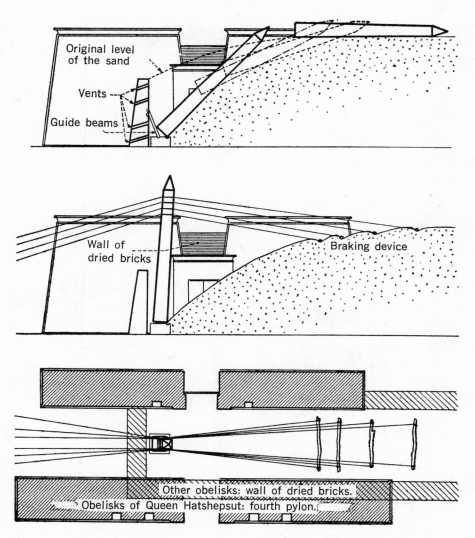

Original level
of the sand

Vents

Guide beams

Wall of
dried bricks

Braking device

Other obelisks: wall of dried bricks.

Obelisks of Queen Hatshepsut: fourth pylon.

FIG. 32.   Method of emplacement of obelisks during the reign of Queen Hatshepsut,
as reconstructed by H. Chevrier, architect in charge of the monuments
af Karnak (from *Bâtir*, 84 [April, 1959], p. 50).

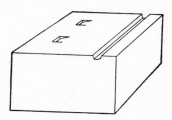

FIG. 33.   Pedestal of obelisk.

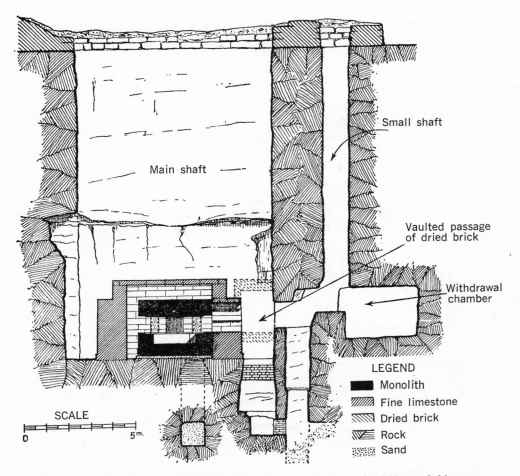

Main shaft

Small shaft

Vaulted passage
of dried brick

Withdrawal
chamber

SCALE

0   5ᵐ·

LEGEND
■ Monolith
▨ Fine limestone
▧ Dried brick
▨ Rock
⦂ Sand

Fig. 34. Vertical section showing the structure of the tomb of Hor at Sakkara
(from J. P. Lauer, *Annales du Service des Antiquités de l'Egypte,* Vol. LII, pl. 1).

to the Faiyum. It is very probable that the details of these undertakings were depicted in bas-reliefs that have now disappeared. The walls of the long covered passage that always connected the temple in the valley with the temple of the pyramid in which the funeral service of the king was performed were all decorated with the most varied scenes. Unfortunately, these ancient causeways are now almost completely destroyed.

The technique involved in the construction of the Old Kingdom pyramids belongs, by virtue of the tools it utilized, to the Aeneolithic or Chalcolithic period that marks the end of the Neolithic and Stone ages. This equipment was extremely rudimentary, for the men of this early period were not acquainted with bronze, which for all practical purposes appeared only in the Middle Kingdom.

The most experienced investigators, such as Borchardt and Lauer, agree on one point: The stones were brought to the site by means of dried brick and

Plate 13.
Pyramid of Chefren, Gizeh, Fourth Dynasty. *Photo Roger-Viollet*

earthen ramps, which were destroyed once they had served their purpose. There is no agreement, however, on the question of what these ramps were like. Lauer suggests the gradual raising of a ramp perpendicular to one side of the pyramid (Figures 35 and 36). (See also Lauer, *Le problème des pyramides d'Egypte* [Paris, 1948]; Borchardt, *Die Entstehung der Pyramide . . .* [Berlin, 1928].)

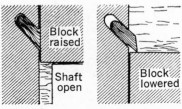

FIG. 35. Method of sealing a tomb in the Twelfth Dynasty (*Bulletin of the Metropolitan Museum*, 1933, Fig. 25).

FIG. 36. Relieving chamber above the sarcophagus chamber of the pryamid of Cheops (Perrot and Chipiez, *Histoire de l'art dans l'antiquité*, I, p. 227). →

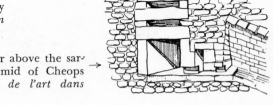

The great monuments of the New Kingdom

We should remember that the architect who constructed these monuments of sometimes considerable size did not have to solve the same problems as his modern counterpart. There were no machines for raising or moving the large blocks, the gigantic architraves, the heavy monolithic columns; there were no pulleys (these appeared with the arrival of the Greeks). The only equipment the Egyptians had were levers, rollers, ropes — and human muscle power. There was no wood for building scaffoldings; the system of embankments, inconceivable to us, had to be used. Traces of this method are still visible on several unfinished monuments.

In the case of large building blocks of stone, the blocks were brought from the quarry with the horizontal and vertical joints already prepared and their edges squared. (The facing, on the other hand, was left natural, in about the same state as our so-called "bossage" course.) The blocks were then carried to the designated place, where they were adjusted and fixed to the adjoining blocks, by means of dowels if necessary. When the course was completed, an embankment was made, and a dried brick ramp was constructed to permit the second course to be brought up on rollers. This procedure was repeated as often as necessary. All the parts of each course were brought up: walls, columns, and so on. The abacuses of the columns were reached, the architraves were brought up, and then the ceiling stones, which were laid on the architraves. The main body of the building was now complete. The embankments and ramps were taken away from the building, which appeared crude but in the main had been exactly calculated. There remained only to dress the façades, after which the decorators, painters, and even gilders (since entire sections of the monument were sometimes gilded) took over.

Plate 14.
Karnak, Nineteenth Dynasty, detail of the Hypostyle Hall.
Photo Roger-Viollet

This method of "blind" construction, which must often have required long periods of time, undoubtedly demanded perfect planning, precision-laid foundations, a methodical organization of the construction yards — in short, the total knowledge of a prodigious master builder. It justifies Loret's interpretation of the very name of the Temple of Amon at Karnak as "the most carefully planned of places."

Woodworking   The only native materials the Egyptians possessed for their carpentry and framework were the acacia and the sycamore. When it was possible they imported ebony from the Sudan, and various varieties of wood from Syria — pine, fir, cedar. The wood of the date palm and the doom palm, which were abundant in the Nile Valley, was too fibrous to be used for carpentry and frameworks. Despite the scarcity of wood, at the beginning of the historical period the Egyptian carpenters were already in possession of all the techniques of their art. They knew how to bind tenon-and-mortise pieces, to use dowels and tongue-and-groove joints, to glue, saw, drill, peg (they never used nails), and gild when they so desired. On the whole, their equipment bore a close resemblance to that of the modern carpenter, with the addition, as in the case of their descendants the modern Egyptian carpenters, of a noticeable preference for the use of the adze. With these tools they created furniture the perfection of which leaves nothing to be desired by our most skilled craftsmen. They made beds and headboards, chairs, chests, and other objects; the tomb furnishings still in existence help us to imagine what their work was like.

One of the strangest features of Egyptian carpentry is a technical procedure used in ship construction. The various parts of the hull were assembled, not by attaching them to the ribbing, but by literally sewing the planks together with thongs passed through closely spaced holes in the edges of the planks (Montet, *Scènes* . . ., 343). This method of construction was used even for large ships, such as the funeral ship of Cheops, more than ninety-eight feet long, which was fond intact in excavations at the foot of the pyramid.

TEXTILES

Spinning   Spinning and weaving are undoubtedly among the oldest crafts in the world, since woven objects have been discovered that date from Neolithic times (Caton-Thompson and Gardner, *The Desert Fayum*). Spinning differed very little from the method used by present-day fellahin: a spindle topped with a balance in the shape of a whorl, with a small hook at its end, to maintain the regularity of movement. After the bundle of fibers had been twisted and drawn out to a certain length, the spindle was turned, sometimes by resting its lower end on the thigh; when the thread was made, the worker stopped turning the spindle and wound the thread around it. Another method was to hang the bundle of fibers from the ceiling or any other support (Beni-Hassan, Thebes). In this case a larger spindle was used, and was rotated with both hands.

Weaving   The Egyptians learned very early in history to weave fabrics with a high standard of perfection. To give an idea of the weaving technique in use at the beginning of their history, we need only mention that the fabric in which the remains of King Djer (First Dynasty, approximately 3190 B.C.) were wrapped was woven with about 60 threads in the warp and 48 in the woof (our modern fine batiste has only 56

Plate 15
a and b Construction of a boat. Tomb of Ti, Sakkara.
Photo Georges Goyon

threads to the half inch) (Petrie). Fabrics dating from the Eighteenth Dynasty
have been discovered in which the ratio is 138×40 and 128×56. This difference between the woof and the warp, the former being less close than the latter,
was perhaps due not to the Egyptians' inability to weave a more even fabric,
but probably to progress, if we admit that the simplest and oldest of patterns is
the plain weave, in which the relationship is two to two.

What types of looms were used by the inhabitants of the Nile Valley? Mats
and coarse fabrics were made on crude horizontal frames equipped with a comb,
or cylindrical piece of wood. This comb had two grooves, one deep and one
shallow; its purpose was alternately to raise and lower the even and the odd
rows in order to permit the passage of the shutttle (Figures 37 and 38). Beginning in the Middle Kingdom, the artisans also had large vertical looms quite
similar to those used by the Persians for rugmaking.

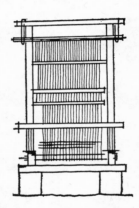

Fig. 37.   Loom
(Wilkinson, *Ancient Egypt,* II, 171).

Fig. 38.   Reconstruction of the loom
shown in Fig. 37.
(Johl, *Altäg. Webest.,* p. 54).

Fabrics were made chiefly of flax, because (according to Herodotus) the use
of wool was not permitted in the temples or in the tombs. Outside these places,
however, white woolen shawls were used. Pictures in the tombs of Thebes depict women dressed in transparent garments.

It is now an established fact that a variety of cotton was native to southern
Egypt, and Herodotus again declares (III, 47) that Pharaoh Amasis (died 525
B.C.) made a gift to Samos and the Minerva of Lindos of two corselets trimmed
with cotton.

It has been discovered that cotton felt was used in the dried brick joints
in the city walls of Tanis, probably to facilitate their drying.

Dyeing   The analysis of pieces of fabric found in the tomb
                  of Tutankhamen, which were woven in the Gobelin
method, has revealed that they were dyed (except for those that had their own
natural color) with madder. But fabrics found in other tombs indicate that the
Egyptians of the New Kingdom used indigo, henna, and safflower rose for dyeing linen fabrics.

POTTERY, GLASS, AND PAINT

Pottery The art of making earthenware vessels is also one of the oldest arts in the world. The Egyptians of the early dynasties naturally used terra-cotta vessels — in such quantities, in fact, that a historical site can be recognized from a distance by the abundance of the potsherds strewn about. Despite the presence of the raw material, however, the quality of Egyptian pottery did not approach the perfection of the products of other countries, particularly those of Greece. Ordinary pottery is quite coarse, and is yellow or red in color; its material is frequently mixed with straw or other impurities; it is often badly fired, and is rarely varnished or glazed.

Another type of pottery is Egyptian faïence, which was used particularly for the making of funerary figurines, scarabs, and beads for necklaces, amulets, and other ornaments. The inside of this faïence is grainy and often crumbly; it is composed of white quartz, slightly softened, without any visible addition of other substances. The exterior is covered with a glaze made of silica and sodium, with the addition of a copper coloring matter.

The figurines were made in terra-cotta molds, a great number of which have been found in the course of excavations. These figurines were molded in a single piece and then dried, carefully touched up, fired, and finally glazed as explained above, in colors ranging (depending on period and locality) from sky blue to bottle green.

Pliny's famous account of the invention of glass (XXXVI, 65) tells us: "A ship laden with natron (coming probably from Egypt) having run aground on a certain beach in Phoenicia, the merchants, wishing to prepare their meal, and finding no stones to hold their pot, used blocks of natron which they took from their ship. The heat, which caused the blocks to combine with the sand of the shore, produced glass."

Technically speaking, the process is correct; only the place and date are hypothetical (Lucas). Glass was known in the predynastic era in the form of glaze, which is simply vitrified matter. But to its makers this glass accidentally obtained by the beadmakers was only faïence.

Petrie has described the manufacture of vases in the time of Amenhotep IV. First a copper rod as thick as the neck of the proposed object was selected. A model in the size and shape of the intended vessel was made from clay probably mixed with sand, and was attached to the end of the rod by means of rags, traces of which are perfectly visible, as are the marks of clamps on the lower portions of certain vessels. The model was first placed in the kiln to give it a certain consistency. The glass was then spread over it in smooth layers, a process repeated as often as necessary to make the surface as smooth as possible. The object was decorated with varicolored threads of glass that the workman combined at will, being careful to roll the object occasionally over a well-polished surface so that the threads of color would penetrate the first layer of glass and the surface would be smooth and without rough spots. The handles and the foot were worked separately, and were applied with molten glass. The object was left to cool until the copper rod had contracted sufficiently to be removed without difficulty. The soft clay, its support now gone, crumbled easily and fell out when the vase was turned upside down. The object was never rubbed or polished. A similar method was used for the making of the decorated beads known to the Italian glassmakers as *millefiori*.

Colors and paints   The freshness of the colors in the Egyptian tomb paintings has often been remarked upon and described, particularly in contrast to our chemical paints. Actually, the durability of the Egyptian paintings is due chiefly to the fact that the paints were made of finely crushed natural mineral pigments; a secondary reason is that they were preserved in the darkness of the tombs and the exceptionally dry atmosphere of the Egyptian climate. The mineral substances used in their composition were ochers, either natural or modified by firing, copper oxides (malachite, chrysocolla), orpiment (natural sulfide of arsenic), chessylite (blue copper carbonate), charcoal or soot, and perhaps also colored stones crushed into powder, such as turquoise and lapis lazuli. The Egyptian blue, famous from the earliest days of antiquity, was a kind of calcined ore obtained by heating together a silicate, a compound of copper, generally malachite, lime carbonate, and natron, until partly vitrified.

The base was tempera, as in modern painting, prepared with a glue, gum, or albumin (egg white). The Egyptians also used encaustic painting, well known to the Romans, either by mixing beeswax into the pigment, which they applied with the brush, or by spreading it as a protective coating over the surface. They also used transparent varnishes made with the resins of the pine or the mastic tree, or sandarac.

METALWORKING

Metal tools   The problem of the metal tools used by the Egyptians is an enigma that has not yet been sufficiently explained. It is certain that the early inhabitants of the Nile Valley used cutting instruments for carving stone. It is also certain that iron, although in abundant supply as an ore in the Aswan region, was not generally used until the Saite period — that is, at a very late stage.

In contrast, numerous copper tools have been discovered at sites dating from the Old Kingdom. These implements — chisels, gravers, points — have exactly the same shape as their modern counterparts. Copper, a ductile and malleable metal, seems ill suited to the purpose for which it was used.

After analysis of a chisel for stone carving and a surgeon's lancet, both of copper, it was formerly believed that the chisel contained a small amount of beryllium or glycine, an alloy that makes it possible to obtain implements of a hardness comparable to tools made from the best modern steels (Ulivio Planta). Alternatively, it is possible that the copper implements used in the Ancient and New kingdoms were generally hardened by means of arsenic. The combination of arsenic and copper oxide, followed by hammering, hardened copper to a consistency comparable to that of our soft steels.

Still another explanation is that the copper in the samples examined had never been pure copper and that it is natural to find antimony, arsenic, bismuth, iron, manganese, nickel, and tin in alloys. These common materials could have accidentally entered into the composition of the ore, and thus the supposed secrets of a lost science may be only a myth.

Bronze The date of the appearance of bronze is quite uncertain, for the reason that it is very difficult for an archaeologist to distinguish at first sight between copper and bronze. However, despite the presence of several examples dating from the Ancient Kingdom, it can be said that bronze did not come into general use until the Middle Kingdom (around 2160 B.C.). It is probably of foreign origin, since it was known in other countries at a much earlier period (for example, at Ur in Chaldea toward 3500–3200 B.C.). Thus the use of bronze in Egypt represents a major technical achievement, since this metal is harder than copper. Its initial hardness, which in its natural state is 136, after hammering attains 257 on the Brinell test for a bronze with a 9.31 percent tin; for bronze with 10.34 percent tin it rises from 171 to 275 after hammering.* Bronze was preferred to copper for the making of *objets d'art* because the latter metal had a tendency, by absorbing gases during the molding operation, to become porous. However, the use of bronze did not entirely replace copper; the tomb of Tutankhamen, which dates from the New Kingdom (about 1344 B.C.), contained more copper than bronze objects.

Vernier, who made a special study of metal technology, estimated that bronze tools experienced a natural hardening when heated, due to the fact that the alloy exuded its tin. The material became harder and capable of taking a surface sharpening which gave it a keen edge. On the other hand, a bronze mortise chisel was found which had a high percentage (13.30%) of tin; the soft bronze band around the handle, 1.05 mm. thick, contained approximately 4.67% tin.

Gold There are no alluvial gold deposits presently being exploited in Egypt. A. H. Hooker found vestiges of placers in Wadi Korbiai, in the southeastern desert of Egypt. In contrast, there is a large number of auriferous quartz mines in the region that separates the Nile from the Red Sea, and some of them are still in production.

The Egyptian inscriptions distinguish two types of gold: Coptos gold, and that of Kush (Nubia). The gold extracted in the latter country, as well as that imported from Asia Minor, was used to make jewelry in the early dynasties, as can be recognized from their content. To the best of our knowledge the date of the oldest Egyptian mine has not yet been determined.

The eagerness with which the ancients sought out this precious metal was so great that we can safely say that there is not a single lode visible in the mountans of the Arabian chain that has not been explored by the Egyptians in the course of their long history. The gold mines in the environs of the Wadi Fawakir, between Coptos and Kosseir on the Red Sea, which are mentioned in the Turin "Papyrus of the Gold Mines," have galleries that descend about three hundred feet. These galleries, which spread out as they descend because they follow the vein, are supported on each side by quartz pillars deliberately

* C. H. Desch, "The Tempering of Copper," *Discovery*, VIII, 1927.

left standing to prevent the upper walls from collapsing (Figure 39). These passages are sometimes so narrow that only children or men reduced to the state of skeletons could work in them. This confirms the horrible description given by Diodorous of Sicily of the working conditions prevalent in these mines.

For this type of exploitation the Egyptians used the most primitive techniques, and the objects discovered in the mine at Fawakir, although they date from the Ptolemaic perod, indicate that Egyptian equipment was very limited: iron gravers, dolerite balls, scrapers, plumb lines with terra-cotta weights, minuscule baskets, torches made of palm-tree fibers. The presence of charcoal remains prove that the workers used fire to crack and break the quartz. The ore brought to the surface was ground to a powder by means of various types of mortars and grinders. This powder was washed in simple dishes into which the miners parsimoniously poured the water they drew from the neighboring wells. Basalt tables with inclined tops have been discovered. The miners spread either a layer of clay, or perhaps a sheepskin, over these tabletops; the running water carried off the waste matter, and the heavier gold remained caught in channels in the clay or in the hairs of the fleece.

To purify the gold, Diodorus tells us, the Egyptians placed a quantity of the metal in a vessel, together with a certain proportion of lead, a few grains of salt, a little tin, and a certain quantity of flour or the residue of barley steeped in water. The vessel was then placed in a glowing fire, where it was left for five days and five nights, after which the gold, by now almost completely pure, was removed.

FIG. 39. Cross section of a level in a gold-bearing quartz mine, Wadi Fawakir (sketch by the author).

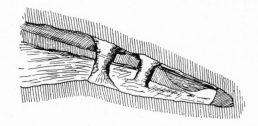

Despite the simplicity of these artisanal methods (Figure 40), as early as the Fourth Dynasty the Egyptians were capable of working quite a large quantity of gold at once, and later (Eighteenth Dynasty) of making the gold sarcophagus of Tutankhamen, which weighed fifty pounds and was engraved inside and out. The goldsmiths were able to make gold into thin sheets with which they covered wood, to plate copper by stamping, to do repoussé work with the hammer, cloisonné (which was more like a form of encrustation, since the small partitions were filled, not with enamel, but with a colored paste or colored stones), solid or hollow rings, and double or triple chains of all kinds. They were capable of making pieces of jewelry representing decorative bunches of gold by soldering grains together — a practice that indicates a degree of skill equal to that of our most skilled modern goldsmiths.

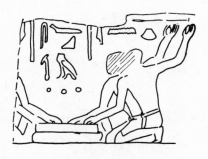

FIG. 40. Working precious metals.
Top, the founders (Sakkara).

Silver   Although silver objects dating from the predynastic period have been found, silver continued to be a rarity until the Eighteenth Dynasty. The tomb of Tutankhamen contained the first large silver objects to be found: a trumpet, and a vase in the shape of a pomegranate. The largest objects are the two massive silver sarcophagi of the pharaohs Psusennes and Sesac (Twenty-first and Twenty-second dynasties) discovered by the French Montet Mission to Tanis in 1939–1940.

THE GRAPHIC ARTS

Drawing and ground plans   In order to picture a building graphically, it is necessary to adopt the purely conventional procedure of giving a flat projection with its form and proportions reduced in equal quantities, without perspective. For this purpose ground plans, cross sections, and elevations, that is, three different drawings, are used. An exact idea of the whole building can be gained only by placing the drawings side by side and comparing them.

The Egyptians combined the three drawings in a unique way. The ground plan was first drawn in its relative proportions, apparently without showing the thickness of the walls. Then the façade was drawn in elevation along the principal side. Sometimes a side face was indicated — preferably at the left, the direction of Egyptian handwriting. Sometimes there was also a cross section, which was actually an elevation seen as if the wall were transparent. When space permitted, as in the case of the drawing of a villa with a garden, gates, kiosks, and trees, these were all drawn lying down, facing in the same direction (Figure 41), in accordance with what this writer calls the "law of levels." That is, the bottom of the first register indicates the foreground, the bottom of the second register indicates the middle plane, and so on, which permits us to compensate in imagination for the absence of perspective.

No trace of the scale indicating the relationship of the proportions is

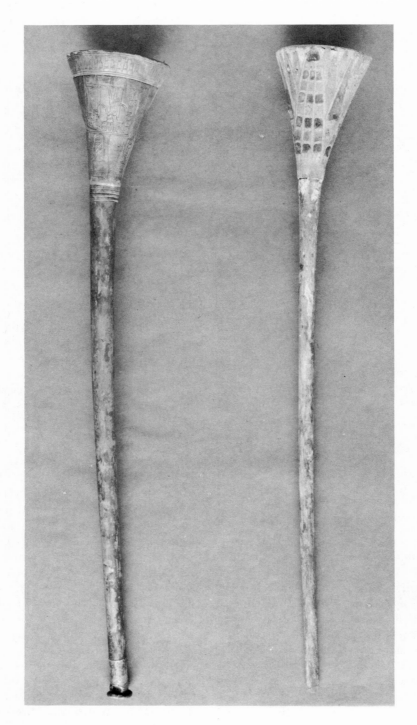

Plate 16.
Silver trumpet and case from the tomb of Tut-ankh-amen,
Valley of the Kings, Eighteenth Dynasty. Egyptian Museum,
Cairo. *Photo by Harry Burton, The Metropolitan Museum of
Art, with permission of the Ashmolean Museum*

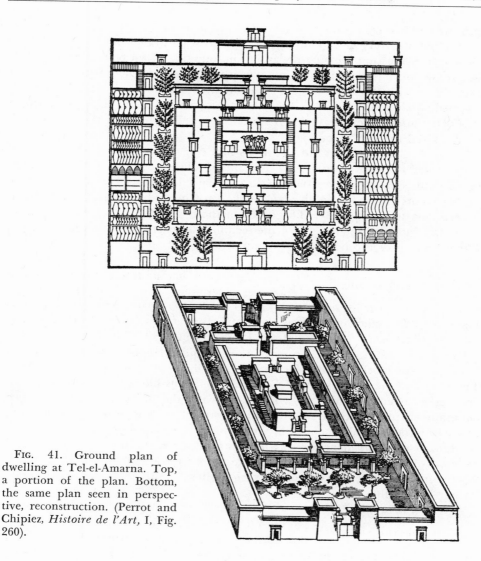

FIG. 41. Ground plan of dwelling at Tel-el-Amarna. Top, a portion of the plan. Bottom, the same plan seen in perspective, reconstruction. (Perrot and Chipiez, *Histoire de l'Art*, I, Fig. 260).

still in existence. In any event, the surviving flat projections show that the relative proportions were respected, but, depending on the requirements of the moment, their constant changed. The Egyptian ground plan could be defined as a ground plan with a variable scale. This is particularly evident in the pictures on Egyptian monuments: the principal figure that dominates the scenes is shown in heroic size, while the servants presenting the offerings are small figures.

From the technical point of view this method obviously lacks precision, and it is difficult to see how a workman could reproduce such a drawing without running the risk of falsely interpreting the architect's idea. Given the size of the Egyptian monuments and the extraordinary mastery the Egyptians had attained, it seem technically impossible that they could have been satisfied with such an "unmathematical" method. This problem would still be troubling us but for the discovery of a genuine elevation to scale drawn on papyrus (Figure 42) — a flat projection showing the front and one side of a wooden temple. Only the ground plan is missing; this obviously could have been drawn on another sheet that is now lost. The subject is drawn with black ink on a background of red-lined squares; the whole is treated in a modern fashion. The sides, which are not mentioned, can easily be calculated with the help of the "squaring," which played the role of our scale paper.

The conclusion is that the Egyptians had two methods for drawing ground plans. One was a figurative method used in bas-reliefs for the depiction of buildings. In this case they were obliged to use the tricks described above because the flat colors they used in the bas-reliefs were incapable of indicating depth or three-quarter perspective. The other method, similar to our modern technique, was intended for the use of the technicians, as is proved by the existence of the drawing of the temple just mentioned.

Relief Representation   Very few Egyptian relief maps have survived. There is, however, a ground plan on papyrus now preserved in the Turin Museum, and known as the "Papyrus of the Gold Mines" (Figure 43). This map, which was for a long time a mystery, has quite recently been identified as a representation of the region of Wadi Hammamat, which is in fact a gold-mining area. The papyrus is a kind of report on the transportation of a statue block intended for a king, from the *bekhen* stone quarry of Wadi Hammamat to the Ramesseum of Gurnah, where this block had been deposited. This report is accompanied by a sketch showing the main route between two mountain chains (which are shown in elevation), and a secondary route obstructed by rocks and bushes. Both, according to the map, lead toward the Nile. In addition the map mentions a cistern, a stele, a temple, and a "mountain of gold and silver," all perfectly recognizable on the terrain. By our standards the map is upside down, the top of the page being "south." Moreover, the scale has not been followed.

Papyrus   Papyrus (*Cyperus papyrus*) is a variety of cyperus that formerly grew in abundance in Egypt; it was particularly at home in the swamplands of the Nile Delta. Today it has completely disappeared from this region, and is found almost exclusively in the Sudan, around the lake regions.

The inhabitants of the Nile Valley used papyrus for numerous purposes, but the use for which it has remained famous is papermaking. Greek and Latin authors, as well as pictures on the Egyptian bas-reliefs, have given us information about the various operations used in producing this type of paper named for the plant from which it came.

The stem of the plant consists of several concentric, very light membranes, as thin as onion skins, which were detached while still thin and young. They were cut into strips approximately $13\frac{2}{3}$ inches long by $2\frac{1}{4}$ inches to $2\frac{3}{4}$ inches wide. These strips were laid one on top of the other at right angles, in the manner of our plywood; this combination constituted a sheet. The sheets were put

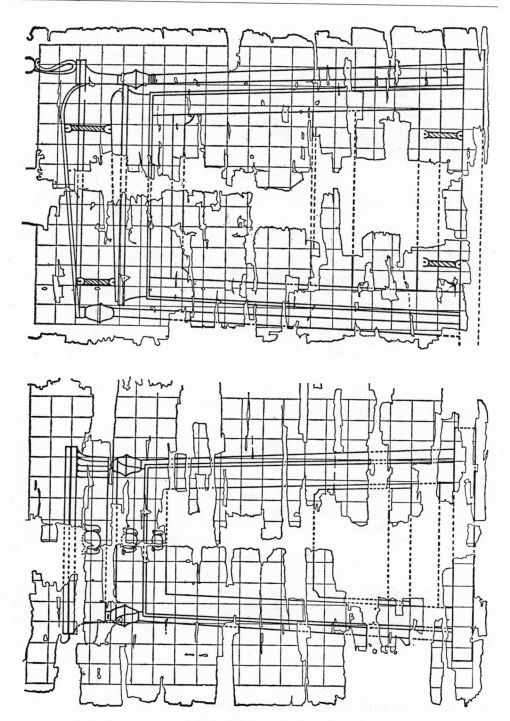

Fig. 42.   Papyrus from Gur'ab showing plan of a *naos*
(Petrie, *Ancient Egypt* [1926], pp. 25–27).

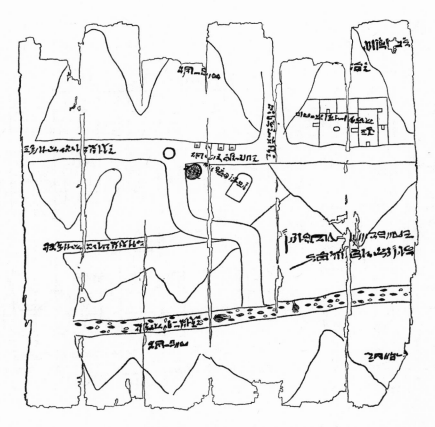

FIG. 43. The Turin "Papyrus of the Gold Mines"
(from Lepsius, Chabas, Lauth, etc.).

into a press, and dried; then they were beaten with a hammer or pounded to make them thinner (Desroches-Noblecourt). After being polished with an ivory tool, a shell, or a pumice stone, they were made into sheets usually 20 inches by 11⅓ inches in size, which were joined end to end to form a roll. The problems of the glue that held the sheets together and the sizing that prevented the papyrus from absorbing ink like a blotter are still somewhat obscure. Pliny, who described the method of making papyrus, seems to attribute to the Nile alluvium adhesive properties that it does not possess. We believe that the sheets were left in the water sufficiently long to become thoroughly steeped, just as was done with the chopped straw used in the composition of the plaster of the dried brick walls. This steeping operation covered the vegetable matter with a kind of mucilage sufficient to hold the strips together and to waterproof them to a certain degree.

MUMMIFICATION

Men in all ages have always had a deep horror of death, and have sought to find in the hope of a life after death a circumstance capable of softening in some way the definitive nature of their passing.

In Egypt this hope was given concrete form in the attempt to preserve the body. The Egyptians believed that the body was the shelter of the soul, and the corruption and disappearance of the one consequently resulted in the annihilation of the other. It was therefore necessary to find a way of protecting the corpse from decay.

In the prehistoric period, burying the corpse in the warm, dry sand of the desert sufficed for its preservation. Beginning in the First Dynasty, however, with the appearance of brick tombs, this method proved insufficient. A new method was invented that consisted of eviscerating the body, sprinkling it with natron, and wrapping it tightly in strips of cloth impregnated with resin. This extremely crude process, it was soon realized, was not very effective; it is a fact that the few surviving mummies from the Old Kingdom period were found in a poor state of preservation. Only in the New Kingdom was the technique of mummification brought to perfection.

The basic treatment for the protection of the corpse consists in evisceration and dehydration by burying the corpse under a heap of natron. In Egypt the latter, a natural compound of carbonate and sodium bicarbonate, also contains chloride and sodium sulfate. The action of the natron caused the dissolution of the tissue by saponifying them. Later improvements included stuffing the cavities of the body with cloths soaked in resin, while the outer areas of the corpse were coated with precious scented unguents whose importation had been made possible by the recent wars against Syria waged under the Thutmose and Amenhotep kings. The body was then coated with melted resin, and wrapped in strips of cloth that had also been soaked in gum or resin.

Inscriptions and chemical analyses indicate that the embalmers used the following drugs simultaneously: oil of cedar, the fluid oil of Gebty(?), oil of cummin, oil of Lebanon(?), wax, gum, fresh oil of concentrated terebinth, natron, and another mineral oil.

This practice continued until the end of the pharaonic civilization. Beginning in the Ptolemaic period and then in the period of Roman domination, with the democratization of the Egyptian religion and rites a definitely more economical method was introduced, which consisted of soaking the body in warm bitumen — a method that eventually completely replaced the older methods.

GENERAL LINES OF DEVELOPMENT

A civilization is better measured by the general level of its technology than by any philosophical discussions. Beginning from nothing, Egypt acquired with surprising rapidity all the elements of her technology. It is certain that her technology was complete, and in certain cases was even beginning to degenerate, as early as the Old Kingdom, that is, almost at the dawn of her history. Then Egyptian technology entered a period of stagnation during which the arts and crafts were modified and perfected but practically nothing new was invented. Social conditions and the structure of the Egyptian religion, which was extremely conservative, made it possible to profit from the existing impetus in order to preserve the same technical methods for three millennia.

BIBLIOGRAPHY

See also appropriate titles bearing on ancient technology in Bibliography for Part II, Chapter 1.

ALDRED, CYRIL, *The Egyptians* (New York, 1961).

BADAWY, ALEXANDER, *Architecture in Ancient Egypt and the Near East* (Cambridge, Mass., 1966).

———, *A History of Egyptian Architecture: The Empire (The New Kingdom)* (Berkeley, 1968).

BAIKIE, JAMES, *Egyptian Antiquities in the Nile Valley* (New York, 1932).

CHILDE, V. GORDON, *New Light on the Most Ancient East* (rev. ed., New York, 1969).

CLARK, C., and ENGELBACH, R., *Ancient Egyptian Masonry* (Oxford, 1930).

COTTRELL, LEONARD, *The Mountains of Pharaoh* (New York, 1956).

EDWARDS, I. E. S., *The Pyramids of Egypt* (rev. ed., Harmondsworth, Eng., 1961).

ENGELBACH, R., *The Problem of the Obelisks* (London, 1923).

FAKHRY, AHMED, *The Pyramids* (Chicago, 1961).

FORBES, R. J., *Metallurgy in Antiquity* (Leiden, 1950).

———, *Studies in Ancient Technology* (11 vols., Leiden, 1955).

GLANVILLE, S. R. K. (ed.), *The Legacy of Egypt* (Oxford, 1942).

HAYES, WILLIAM, *Most Ancient Egypt* (Chicago, 1965).

LANDSTROM, BJORN, *The Ship* (Garden City, N.Y., 1961).

LUCAS, A., *Ancient Egyptian Materials and Industries,* 4th ed. enlarged and revised by J. R. HARRIS (London, 1962).

MONTET, PIERRE, *Everyday Life in Egypt in the Days of Ramesses the Great* (New York, 1958).

PETRIE, FLINDERS, *Egyptian Architecture* (London, 1938).

SINGER, CHARLES, *et al., A History of Technology,* vols. I and II (London, 1954, 1956).

THOMSON, J. OLIVER, *History of Ancient Geography* (Cambridge, 1948).

CHAPTER 8

GREEK TECHNOLOGY

FROM THE technological point of view, the Creto-Mycenaean civilization of the second millennium hardly differed from the Mesopotamian and Egyptian cultures that have just been discussed. The decisive moment in the evolution of technology occurred toward the end of the second millennium (twelfth to eleventh centuries), when iron (or, more precisely, steel) metallurgy replaced that of bronze. Steel, which was probably invented in Anatolia or northern Syria around the fourteenth or thirteenth century B.C., brought about the radical modification of a certain number of basic principles of ancient technology, thanks to the improvements in equipment it made possible. In Greece these improvements did not become really apparent until the archaic period, more precisely in the sixth century B.C. Our study of Greek technology should therefore begin with this period and end with the Roman conquest of Greece in the second century B.C. Three groups of essential factors — geographic, historical, and psychological — must first be briefly summarized if we are to understand the technical level reached by the ancient Greeks, the considerable differences in the results obtained in each area of activity, and lastly the originality of their contribution.

In antiquity, as in modern times, Greece was above all a poor country, completely surrounded by the sea and possessed of a few plains (which were in many cases marshy) isolated among barren mountains, a few erratic rivers, and almost no mineral wealth except for the silver-bearing lead mines of Laurion, in Attica, a few gold mines, and a few marble quarries. From the historical point of view, Greece was divided into a multitude of rival city-states preoccupied mainly with their political quarrels and perpetual internecine wars. In the classical period, at least, there were few financial resources on which to base an industrial type of economy, especially in view of the fact that the abundance and cheapness of slave labor hardly encouraged the development of mechanization. Moreover, the contempt of the educated class for manual labor did not stimulate mathematicians and scientists to direct their efforts toward technical progress. Very often the machines they invented were interesting only from the theoretical point of view, and no thought was given to their actual construction and practical utility. On the other hand, the Greek feeling for plastic beauty permitted them to attain to a remarkable technological level in such areas as architecture and the arts.

Then, beginning in the second half of the fourth century, the Macedonian conquests caused the eruption of the narrow political framework within which

Greek political life had until then been constricted. Contacts were intensified with the rich civilizations of Egypt and the Orient. This brought about a radical change in the spiritual and economic life of the country. The technological consequences were, as we shall see, considerable.

CONSTRUCTION AND HOUSING

The architectural knowledge of the ancient Greeks was at its best in the domain of public life — in temples and official buildings. To the unsurpassed level reached in this area by Hellenic art — satisfying both the acuity of their esthetic sense and their desire to dazzle and surpass rival city-states — corresponds the perfection of its architectural technique.

Construction Primitive Greek architecture, like that of the ancient East, consisted of little more than dried brick walls usually supported on a stone foundation and sometimes reinforced by a wooden framework. This type of construction persisted throughout the classical period, particularly in domestic architecture, as is demonstrated by numerous houses that have been discovered, notably at Olynthus in Chalcis. Baked brick did not come into use until the fourth century, and even then its use never became general.

Beginning in the sixth century, however, an architecture of stone — limestone and marble — came into being, and here the Greek genius burst into flower.

Masonry The stone walls rested on crudely made, rather shallow foundations of tufa. In most cases the foundation was not a single continuous mass but rather a square frame, each side of which corresponded to the line of a wall. However, the terrain sometimes required the laying of a continuous foundation forming a kind of terrace. The Parthenon, for example, is built on an artificial bastion overhanging the southern slope of the Acropolis. Only the top or leveling course (the *euthynteria*) of the foundation had to be perfectly horizontal.

Plate 17.
Transverse section of the Athenian Acropolis, fifth century B.C.
  a. North Rampart (Wall of Themistocles)
  b. Summit of the acropolis and cistern
  c. Subbase of the Parthenon
  d. Retaining wall
  e. Mycenean Rampart
  f. South Rampart (Wall of Cimon)

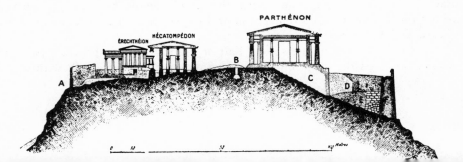

The walls themselves were constructed of evenly cut blocks of stone: orthostates (the bottom course), perpends (blocks of stone cut to the desired thickness of the finished wall), and headers (blocks of stone whose heads, or ends, faced out). To ensure the perfect contact of the joints that we still admire in the Greek monuments, the blocks were assembled, without the use of mortar, with their faces still undressed except for a narrow margin around the outer edge (wall). This process is known as *anathyrosis*. The vertical joints were roughened and slightly recessed in relation to this margin. The same method was used in the erection of the columns. Each drum was laid in place before being fluted. We shall later see, in our study of the beginnings of mechanization in Greece, by what means these blocks and drums were raised and lifted into place. The blocks were bonded together by means of wooden or metal dowels set in molten lead, the horizontal joints by means of iron gudgeons, the vertical joints by metal clamps of various shapes (dovetail, double "T," double "Γ"), or by simple double hooks driven vertically into the stone (Figure 44). The drums of the columns were bonded with cedarwood pegs set in molten lead that had been poured into square holes drilled in the center of the column (Figure 45). Only then were the blocks of stone polished and the columns fluted; in the latter operation, an intermediate polygonal stage preceded the carving out of the concavity of the fluting.

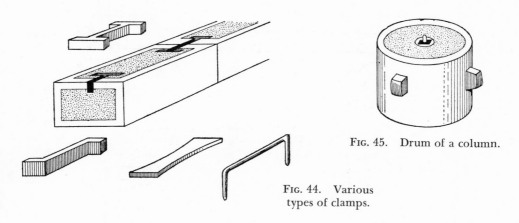

FIG. 45. Drum of a column.

FIG. 44. Various types of clamps.

"Reinforced" architecture Recent studies have revealed that the Greek architects, occasionally lacking confidence in the durability of the marble, believed they should strengthen certain parts of their edifices by reinforcing the stone with iron bars. Sometimes it was a question of strengthening foundations on a slope; in the Theban Treasury at Delphi, the leveling course was grooved to receive iron bars that formed a rigid framework. In other cases they attempted to reinforce an architrave or a lintel by cutting a groove into its lower face and inserting an iron bar; the ends of the bar rested either on the capitals supporting the architrave (as in the case of the Temple of Zeus at Agrigentum) or on the door jambs (for example, the Erechtheum at Athens). In the Propylaea, the end of one beam of the heavy ceiling (which was made of

heavy marble girders) rested on the center of the Ionic architrave. A groove cut in the underside of the architrave had a small projection at each end; the iron bar rested on these projections, and so received (at least in principle) the pressure of the beam, which was thus transmitted directly to the shafts of the columns and relieved the middle portion of the architrave (Figure 46). In the Temple of Apollo of Bassae (Arcadia), an iron bar was run through a parallelepipedal shaft drilled through the center of each beam of the portico. In still other cases, crossbars at right angles to the roof beams were used to support either the sculptures of the pediment, as in the case of the Parthenon (Figure 47), or the weight of the cornice, as in the Temple of Castor at Agrigentum.

Despite the technical skill they indicate, however, such methods of construction also reveal a complete misunderstanding of the possibilities of resistance of marble and, on the other hand, a considerable overestimation of the resistance of iron, on which they imposed stresses much greater than the actual tolerances.

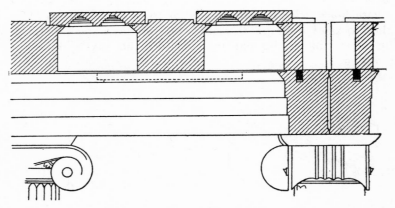

FIG. 46. Iron reinforcements of the Propylaea
(from *American Journal of Archaeology*, 26 [1922], p. 153, Fig. 3).

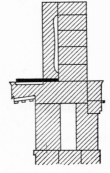

FIG. 47. Iron reinforcements in pediment
of Parthenon (from *American Journal
of Archaeology*, 26 [1922], p. 157, Fig. 5).

The roof   Greek roofing methods are a particularly characteristic example of the gap between the level of the Greeks' scientific knowledge and the level at which their practical achievements in some areas stagnated.

The roof of the Greek temple was nothing more than a simple piling up of crossed beams — a heavy, costly accumulation totally in defiance of the laws of the composition of forces. On these beams was laid a covering of marble or terra-cotta tiles, the numerous variations of which can be grouped into several principal types: Laconian (concave pantiles with joints protected by convex cover tiles), Corinthian (flat pantiles with raised rims and triangular or saddle-shaped cover tiles), and Sicilian (flat pantiles with raised rims and convex cover tiles) (Figure 48). Sometimes, as in the Temple of Apollo at Bassae, the cover tiles and the flat pantiles formed a single piece.

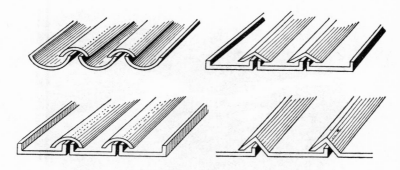

Fig. 48. Various types of Greek tiles.

Arches and vaults   Although several examples of brick vaulting are found in Eastern (particularly Assyrian [Nimrud]) architecture, the stone arch and vault began to come into general use especially toward the end of the Greek classical and during the Hellenistic periods. According to Diodorus of Sicily and Strabo, the vault may have been invented by Democritus of Abdera around 470 B.C. Until then the only method known had been corbeling, the simple superposition of blocks in successive overhangs. This method created doors or galleries the jambs or walls of which ultimately met at the top to form a more or less acute angle (Figure 49). The first examples of genuine arches and vaults known in Greece date from the fourth and perhaps even the fifth century B.C.; one of the gates in the ramparts of Oinidai (Acarnania) is already a true semicircular arch composed of trapezoidal blocks or voussoirs, each of which is held in place only through the reciprocal action of all the blocks. The baths of Gortys in Arcadia (third century) have two arches whose horizontal projection is curvilinear, while the first crossed barrel vaults appear in the terrace of Attalus I at Delphi. By the second century B.C. the Pergamenian architects were even able to cover stairways and sloping passages with semicircular vaults; examples can be seen in the amphitheater as well as the triple gymnasium of Pergamum.

The private dwelling   While the technical level achieved in Greek public architecture was remarkable, the same cannot be said of private dwellings, which were probably much inferior to the level of comfort in Cretan houses of the second millennium. Later we shall discuss the problem of water supply and waste removal. Heating was probably done by means of simple portable braziers. The doors of Greek houses probably pivoted on the ends of the jamb, which was embedded in sockets sometimes enriched with metal — a system then prevalent throughout the Eastern world. It is difficult to declare with certainty, on the other hand, that hinge pins were known in the Hellenic world.

The only major progress in domestic architecture was the increasingly widespread use of a genuine lock, which was perhaps of Egyptian origin. In the archaic epoch locks were still nothing but horizontal or vertical bolts that could

Fig. 49.   Cisterns with corbeled arch.
Nea Plevron, Acarnania. Third century B.C.

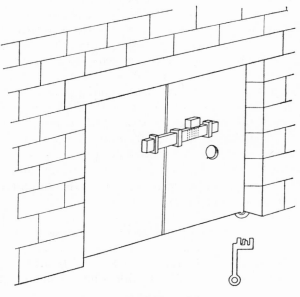

Fig. 50.   Key bolt and key.

be manipulated only from inside the house. The first improvement was the use of a kind of rudimentary key — a simple bent rod that, when thrust into a hole in the door, permitted the bar to be opened from outside; the door was closed by placing a strap across it. Obviously this system was not very secure. The next idea was to place wooden pegs (*balanci*) in an open-bottom box above the bar; when the latter was pulled closed, the pegs dropped into corresponding holes in the bar. To unlock the door, these pegs had to be raised by means of a special key inserted under the bar by passing one's arm through a slit in the door; at the same time the bar had to be pulled (Figure 50). These toothed keys of varying shapes, which probably came into use at the beginning of the fifth century B.C., provided much greater security than earlier systems.

TRANSPORTATION

Since geographical conditions play a decisive role in the history of transportation, the fact that roads were little used in ancient Greece, while navigation has always played a major role, is easily explained.

Overland Transportation

The poor condition of the Greek roads was equaled only by the archaic nature of the vehicles; both factors undoubtedly precluded transportation over long distances.

Roads   In the second millennium Crete had a fairly extensive network of well-maintained roads that linked the various Minoan cities. The typical Cretan road was laid on a foundation of rocks embedded in a clay mortar and covered with a layer of clay; it consisted of a center lane of basalt or sandstone paving stones bordered on each side by lanes paved with limestone. Vehicles probably traveled on these side lanes.

As for Greece itself, in contrast, all the early witnesses agree in condemning the bad state of the roads in the classical period. The Greek roads were generally simple paths sometimes interrupted by flights of steps; no maintenance was given to them. Bridges were few, and in almost all cases were constructed by corbeling. Obviously the only practicable form of transportation over such roads was the donkey or mule, as was still the case in Greece until a few years ago.

Only the sacred causeways for the processions bringing offerings from cities and rich private citizens to famous religious shrines were kept in repair. The major roads were those of Delphi, Eleusis, Olympia, Miletus, and Anaphe. These one-lane (two at certain crossing points) sacred roads appear to have been paved and to have had raised sidewalks. In rocky areas they were simply ruts cut into the rock to a depth of from $2\frac{1}{4}$ inches to $4\frac{1}{2}$ inches; they ranged from seven to ten inches in width, and the gauge varied from $3\frac{1}{3}$ feet to 6 feet; the wagons transporting the offerings moved over these "rails."

As for the cities, they were little better paved. The streets were often simply mud passages without any kind of drainage; at best they were covered with a layer of crushed stones (*lithostratos*) to which was sometimes added a layer of mortar. Only after the fifth century, and especially after the beginning of the Hellenistic period, were the streets of major cities paved with marble or granite paving blocks; in the same period urban planning came into existence, as is indicated by the plans of such cities as Miletus, Alexandria, Antioch, and Pergamum.

Vehicles and harnesses   The Greek vehicle was generally a two-wheeled wagon pulled by two draft animals attached one on either side of the pole. At first a military vehicle, it later became a chariot with a light frame, and front and side panels consisting of a latticework of leather or braided fibers; it was pulled by two (occasionally four) horses harnessed abreast. In the archaic period it began to be used for transporting people and merchandise. The carriage generally rested on two blocks of wood, one at each end of the axle. The latter appears in most cases to have turned with the wheels, if we can judge by the square or oblong head of the axle as it appears in pictures on the painted vases; in only a very few instances was the axle independent of the wheels.

The wheels were of various types. Simple wooden circles cut from tree trunks were probably still in use, but most often the wheels had four, six, seven, or eight spokes, or a crossbar with a hole in its center to admit the head of the axle; the ends of the bar supported the rim, which sometimes consisted of several pieces joined together. Lighter bars perpendicular to the crossbar were attached to the rim on each side of the axle (Figure 51). No trace of such a device prior to Greek antiquity is known to us, and undoubtedly it had almost completely disappeared by the fifth or fourth century B.C.

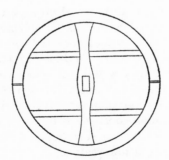

Fig. 51.   Wheel with crossbar.

Fig. 52.   Greek ship.
Painting on an Attic vase, around 500–475 B.C.

The harnessing of draft animals was still more rudimentary. As in all the ancient civilizations, the collar encircled the neck, strangling the horse or mule and obliging it to raise its head, so that the animal could not pull with its entire weight (see page 93). Moreover, it is unlikely that horseshoes were known in the ancient world, although the recent discovery of a horseshoe in a Hellenistic tomb at Ensérune (Languedoc, France) contributes a new element to this much-discussed problem. (We cannot, however, exclude the possibility that this is an intrusion of a much more recent object.) With unshod horses, it is possible that the carrying capacity of even the strongest vehicles did not exceed 225 pounds.

It is certain that the horse was utilized as a draft animal long before it was used for riding; no indications of the latter use are known prior to the eleventh century B.C., but after this the saddled horse gradually replaced the chariot for military use. It was probably ridden bareback or with a simple saddle blanket, although from certain remarks of Xenophon we can perhaps infer that the saddle came into existence at the beginning of the fourth century. Spurs and stirrups were still completely unknown.

Navigation

The Greek navy   In ancient Greece the navy was the principal method of transportation, and the fortunes of many private citizens depended on it. The merchant ships seem to have been small, with a capacity of approximately forty to sixty tons; they were round, sat low in the water, and had a not very prominent keel that permitted them to be hauled up on the beach. The bridge probably consisted of nothing more than shelters built in the front and back of the ship. These ships could be powered either by oars or by a quadrangular sail attached to a single mast by means of a yardarm perpendicular to the top of the mast (Figure 52).

It is not very likely that the Greek ships contained those numerous tiers of rowers that have for long been attributed to them. (The famous ship of Ptolemy Soter is supposed to have had up to twelve tiers of rowers!) The terms "trireme," "quadrireme," and so on, undoubtedly refer rather to the number of rowers used to move each oar. However, the existence of ships with two and perhaps three banks of rowers is not absolutely precluded.

The rudder seems to have consisted merely of one or two oars, without grommet, maneuvered by the helmsman, who had therefore to bear their weight himself. With such a lever, the smallest arm of which was on the power side and not, as is normal, on the resistance side, the possibilities for maneuvering were extremely limited. Thus navigation was reduced in most cases to a simple coastal traffic, and crossing large bodies of water directly must have been done only at favorable periods of the year.

The ports   While the Hellenic ships differed little from those of the Phoenician navy, port installations were radically improved under the Greeks. Numerous artificial ports were created by means of dikes linking small islands to the mainland; this was the case at Cnidus, where the construction of jetties permitted the establishment of two ports, the largest one being 1,968 feet long, and especially at Alexandria, where at the end of the fourth century B.C. the island of Pharos was linked to the mainland by a 4,264-foot-long dike with two openings by which the two ports communicated; in addition, a canal linked the sea with Lake Mareotis (Figure 53). At Seleucia, the port of Antioch, the entire harbor itself was excavated from a coastline that offered no natural shelter, and was linked to the sea by a canal. At the Piraeus, where three natural harbors were available for the reception of numerous ships, arsenals constructed toward the middle of the fifth century had dry docks where ships could be taken out of the water for repairs.

On the other hand, the project for piercing the Isthmus of Corinth, which was conceived around 600 B.C. by the tyrant Periander and then resumed by Demetrios Poliorcetes (end of the fourth to the beginning of the third century B.C.), was abandoned. The engineers had been under the impression that the waters of the Gulf of Corinth were higher than those of the Saronic Gulf.

Plate 18.
Sailing ship. Cup by Nicosthenes, sixth century B.C. Louvre Museum, Paris. *Photo Giraudon*

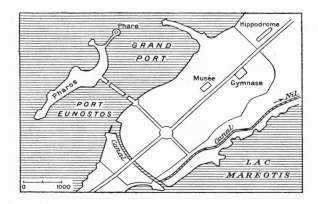

FIG. 53.   Map of the port of Alexandria
(from C. Merckel, *Die Ingēnieurtechnik in Alterthum,* Fig. 130).

The construction of one of the oldest lighthouses in the world was begun at Alexandria in 283, under the reign of Ptolemy Soter, at the northeast tip of the island of Pharos. The architect, Sostratos of Cnidus, built a stepped tower, at least 279 feet high, that narrowed toward the top. The light was produced by means of a large brazier and an immense mirror that projected the reflection more than twenty-five miles out to sea. It was destroyed in the fourteenth century by an earthquake.

THE BEGINNINGS OF MECHANIZATION

Although at the end of the archaic period mechanization was hardly better advanced in Greece than it was in the East, considerable progress was achieved in the classical and especially the Hellenistic periods, thanks to the work of great mathematicians such as Archytas of Tarentum (around 400 B.C.) and Archimedes of Syracuse (circa 287–212). While it is possible that we owe the invention of the screw, incorrectly attributed to Archimedes, to Archytas, it was Archimedes who increased its uses, applying it in particular to hydraulic machines. Probably it was also Archimedes who invented the toothed wheel and the gear train. As for the pulley (which seems to have been unknown in the Eastern empires) and its byproducts and complementary devices, the windlass and the crank, it is mentioned for the first time in the *Mechanica,* a work of the school of Aristotle (fourth century). Among the successors of Archimedes, we may mention Hipparchus of Nicaea (around 120 B.C., then Hero of Alexandria (between first and third centuries A.D.?); thanks to descriptions written both by Hero and by the Roman architect Vitruvius (first century B.C.), we are acquainted with most of the Greek inventions in the field of mechanization. It is, however, difficult to determine to what extent these are purely theoretical constructions or devices in common use; nevertheless, a few facts seem to be relatively certain.

The water mill   Although its origin has not yet been determined, the water mill does not seem to have been used in Mesopotamia or Egypt, and the only type known to the Greeks was quite rudimentary: a horizontal wheel with paddles pulled a vertical shaft that pierced the bedstone and turned the runner (Figure 54). The result was obviously poor, and required a rather rapid current. No real progress seems to have been made until the adoption, in the Roman period, of the mill with vertical wheel and lantern gear (that is, with small pegs on the surface of the wheel that acted as teeth). The windmill was still completely unknown.

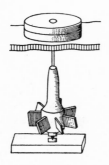

FIG. 54.   Reconstruction of a mill with horizontal wheel (from Charles Singer, *A History of Technology*, Vol. 2, p. 595, 540A).

FIG. 55.   Suggested reconstruction of a crane with single pole (from A. P. Usher, *A History of Mechanical Inventions*, 1st ed., Fig. 6).

Cranes   Not until the invention of the pulley was the problem of lifting large blocks of stone satisfactorily resolved. Unfortunately, attempts at reconstruction based on the descriptions of Vitruvius are arbitrary, and the dates at which the various types of cranes were invented are completely unknown to us. The most primitive type undoubtedly had only two pulleys placed at each end of a vertical pole, and a windlass that was used to maneuver the rope (Figure 55). Such a device did not permit lateral movements. The crane with three pulleys (Figure 56) (invented, it is claimed, by Archimedes) was an improvement; it had the advantage of reducing the friction on the rope. One of the pulleys was attached to a crossbar at the top of the crane, a second to the suspension hook; the end of the rope was attached to the crossbar or to one of the pulleys. The number of poles varied from one to four.

The blocks of stone were lifted by passing ropes under projections deliberately left along the principal sides (they were removed afterward) through U-shaped grooves in the faces of the stones, or through a channel in their undersides; hooks thrust into small lateral cavities in the stone were also used, and sometimes iron corners, trapezoidal in shape, which were inserted into holes

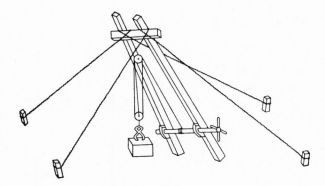

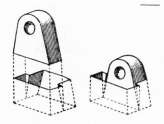

193

FIG. 56. Suggested reconstruction of a crane with two poles and two pulleys

FIG. 57. Method of lifting a block of stone by means of an iron wedge (from Charles Charles Singer, *A History of Technology,* Vol. 2, p. 661, Fig. 604).

of the same shape cut into the face of the block, and which were fixed firmly in place by a simple lateral movement (Figure 57).

The blocks of stone were not laid directly in their final positions. They were put in place with the help of pincers acting as levers, the end of which was supported on the underside of a block of the lower course, in a groove cut for that purpose.

The beam press   The simplest form of grape or olive press was the beam type, in which a very heavy weight was suspended at the end of a beam that acted as a lever. Obviously, it was difficult to lift the beam. Winches and pulleys to increase the pressure on the one hand and to help to raise the beam on the other undoubtedly came into use beginning in the Greek period (see below, p.211).

Automata   The Greeks' passion for the theatre, together with the lack of interest shown by the scientists in the practical utilization of their knowledge, explains why at this time the automata were the only mechanisms that made use of the ascending movements of air or hot water as well as the power of compressed air. At the same time changes of movement were obtained by the upsetting of balance produced by the sudden opening and closing of valves in the vessels through which the water flowed. This method was used to move statues of the gods and to cause doors to open by themselves; or to shoot artificial birds up into the air or cause them to sing. Needless to say, the machines that produced these effects were carefully hidden from view. The Alexandrian physician Hero devoted an entire treatise to the description of these automata.

Aside from this purely recreational use, we know of no other machines that were then operated on these physical principles.

HYDRAULICS

In a country as dry as Greece, the pressing problem of supplying water to the large cities and religious sanctuaries was solved with often remarkable methods personally supervised by the leading engineers. On the other hand, the poverty of an agriculture for which the winter rains sufficed, and the lack of financial possibilities and of cooperation among the city-states, explain the insufficiency of the solutions devised for the problem of irrigation, despite the example offered by the great Eastern civilizations.

Hydraulic machines   Water came either from wells or cisterns, which were generally constructed of masonry. The mechanism most frequently used for raising water was undoubtedly the shadoof (Figure 58), which is Egyptian and Mesopotamian in origin and is still frequently seen in the Orient: a balance beam with unequal arms suspended from the top of a post, with a heavy counterweight (generally of clay) which need only be pulled down in order to raise the vessel full of water (that is, the principle of the lever).

FIG. 58.   Shadoof. Painting on an Attic vase,
second half of the sixth century.

The *saqiyeh* (Figure 59) appeared in the second century B.C. Its construction was based on the principle of the toothed wheel invented by Archimedes, and consisted of a horizontal toothed wheel, turned by an animal, which pulled a vertical toothed wheel with buckets. The latter, as they plunged into a body of water, brought up a certain quantity of water that then fell into a canal or cistern. This machine obviously presupposes the existence of large bodies of water or large rivers, and for this reason its use appears to have been limited to Syria and Egypt, where it is found all along the Nile.

The invention of the worm screw, to which Archimedes contributed, was followed by that of the water screw (*cochlea*) (Figure 60). The screw was en-

FIG. 59.   Saqiyeh.

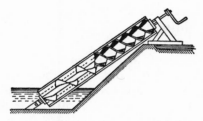

FIG. 60. Water screw (from A. P. Usher, *A History of Mechanical Inventions,* 1st ed., Fig. 13).

FIG. 61. Cross section of tunnel of Eupalinos at Samos (from C. Merckel, *Die Ingenieurtechnik im Alterthum,* Fig. 196). →

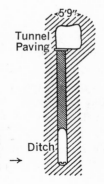

closed in a wooden cylinder, and in turning it drew the water up into the interior of the cylinder, which was sometimes inclined as much as 30 degrees; the screw was probably rotated by a crank. This device was used in the Hellenistic East for irrigation, and also in the mines, where it was used to remove water.

Water supply   The most widely used system for supplying water to the cities and sanctuaries was a simple application of the laws of gravity. The water flowed into conduits that followed the curves of the terrain. Sometimes these conduits were gutters faced with stone and plastered with hydraulic mortar; most often, however, and especially in distribution systems, terra-cotta or stone (occasionally bronze or lead) pipes were used, one end of a section of pipe being narrowed down so that it fitted into the next section. These pipes were often laid inside subterranean channels that could be examined through shafts located at regular intervals; this was the case, for example, with the channels that brought water from Mount Pentelicus to the city of Athens.

At Samos, the engineer Eupalinos of Megara (sixth century) attempted to compensate for the disadvantages of an excessively winding route by digging a tunnel through the mountain that separated the city of Tigeni from the inner lake that supplied it with water. (Such a method, it is true, had been used long before then by the engineers of Jerusalem.) The tunnel of Samos, approximately 3,600 feet long, was dug simultaneously from both ends, and as a result there was an angle at their meeting place. In addition, the levels at which the two tunnels had been dug did not coincide well at the joining point. The terra-cotta pipes were laid along the bottom of a ditch dug in the floor of the tunnel along the edge of one wall; in places this ditch was deliberately filled in with the help of excavated material and covered with paving stones, while in other places it was left uncovered (Figure 61). In certain areas the ditch was actually a second gallery dug independently of the first. The slope of the ditch is greater than that of the tunnel; at the opening at the city end there is a difference of level of about twenty-eight feet. Probably the slope of the tunnel was considered insufficient to ensure the flow of the water, and so the ditch was dug.

Risks of this type were eliminated when it was discovered how to utilize the principle of the siphon. Water passing through conduits constructed on the Venturi principle can, after crossing low areas under strong pressure, climb up a slope. An early example is the aqueduct of Patara in Lycia (of uncertain date), in which the water, flowing through stone conduits, descended the slope of a mountain, crossed a valley over a wall with two corbeled openings at its base, then climbed up the other side. But the most remarkable installation was that of Pergamum (second century B.C.), which carried the water from a cistern situated at an altitude of 1,230 feet to the top of the citadel (1,089 feet) after crossing two valleys 564 feet and 640 feet deep respectively. The pressure of the water at the bottom of the valleys was 20 and 17 atmospheres, respectively. The conduits, which have now disappeared, were undoubtedly cast in bronze or lead.

Sewers   In Minoan Crete, as also in certain Eastern civilizations, particularly that of the Indus River valley, around 2,000 B.C., the problem of waste disposal had been brilliantly solved. Such was not the case in classical, or even Hellenistic, Greece; the streets had no drainage system, and no sewer network seems to have existed. It is true that numerous vestiges of drains and sewers have been discovered in Athens, but they cannot be dated earlier than the Roman period; this is probably also true of the drains of Olympia.

Irrigation and drainage   Agriculture in Greece was always limited to dry cultures; no serious attempt at irrigation seems to have been made, except possibly in the region of Cyrenaica. As for the eastern portions of the Greek world, where irrigation was practiced on a large scale, this was a practice the tradition of which greatly predated the Greek colonization.

Only occasional attempts were made to drain the swamps that are so numerous in Greece. Around 450, Empedocles of Agrigentum drained the swamps of Selinus in Sicily by channeling the waters that fed them. In Greece itself several efforts were made to drain the unwholesome swamps of Lake Copais in Boeotia, the one underground outlet of which was constantly blocked. An early attempt, according to Strabo, was made by the Minyens in the second millennium; after that, nothing was done until Alexander had the subterranean canal dredged, and after him Crates of Chalcis began to dig artificial drains. All these attempts, however, were interrupted for political reasons.

MILITARY MACHINES

It was especially in the military domain, and particularly with the advent of the age of Alexander, that Greek science was translated into concrete achievements. In the Hellenistic period the rich and powerful states in which the Hellenic heritage was preserved and cultivated had to struggle constantly — no longer in petty quarrels between neighboring cities, but in violent clashes like Alexander's campaigns and the Carthaginian and later the Roman assaults on the Greeks of southern Italy and Sicily. This period witnessed the invention of most of the military machines that were to remain in use, basically unchanged, until the invention of gunpowder.

In the classical period the major tactic of besieging armies was still starving out the city's defenders or penetrating the city by means of treachery. Sometimes an attempt was made to set fire to the ramparts or to destroy them by means of battering rams and mines or to attack them at close range by building towers in front of the walls.

In the fourth century appeared the early machines that, at first utilized solely during sieges, were later (beginning with Alexander) mounted on wagons and used on the battlefields as well (for example in 335 during the Balkan campaign, then in 329 at the Battle of the Tanais). Our medieval hand-operated crossbows were unknown in antiquity.

The Greek war machines were based on two different principles of dynamics; some utilized the tension of an elastic body, others functioned by torsion. To the first category belong the *gastraphetes* and the *petrobolos*, designed for the hurling of arrows and stones respectively. The principle involved in both cases differed little from that of a simple bow, but the tension of the string, instead of being ensured by the hand, was determined by a sliding runner that could be fixed in the desired position by a rack (Figure 62). The use of cams for stopping or starting a movement seems to have been limited, in Greek antiquity, to various military machines.

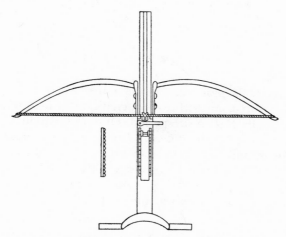

FIG. 62.   Gastraphetes (from J. Kromayer and G. Veith, *Heerwesen und Kriegsführung der Griechen und Römer*, pl. 16, Fig. 64).

A second group of ballistic engines, the *euthytonon* (Figure 63) and the *palintonon*, were activated by the liberation of forces built up by a torsion operation applied to a group of ropes. In these machines the "bow" was actually two independent arms, the pivot ends of which were attached to pivots around which a series of ropes were wound. The arms were pulled back by drawing the bowstring (to which the free ends of the arms were attached) back to the trigger; this created torsion in the pivot ropes. When the trigger was released, the torsion ropes unwound, pulling the bowstring forward and thus propelling the projectile, which had been placed in the runner. It seems that the first

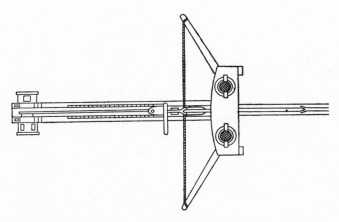

FIG. 63.   *Euthytonon (from J. Kromayer and G. Veith,*
Heerwesen und Kriegsführung der Griechen und Römer,
pl. 17, Fig. 66).

machines of this type made their appearance at Alexander's siege of Tyr (332 B.C.). These engines were probably capable of hurling arrows to a distance of approximately five hundred paces. The ropes utilized in their construction were generally made from the ligaments of deer or oxen, sometimes also of animal manes or hair.

INDUSTRY AND CRAFTS

The production methods practiced in the Greek world rarely surpassed the craft stage, and in comparison with earlier civilizations little progress was made in this area, even in the Hellenistic period. Only the exploitation of the mines of Laurion became a capitalistic enterprise, with the participation of the richest citizens of Athens, and only here were production methods appreciably improved; the other groups of crafts preserved a "cottage-industry" character. Only those techniques in which the Greek passion for the plastic arts played a role reached a level that in some cases was unequaled.

Mines and Metallurgy

Mines — Only the mines of Laurion, at the eastern end of Attica, show definite progress in methods of exploitation. In antiquity, as in modern times, they produced lead sulfide or galena with a large proportion of silver.

The ore was extracted through rectangular shafts 6¼ feet x 4¼ feet in size, which were very regular in shape and perfectly vertical. Every 33 feet the axis of the rectangle made a turn of from 8 degrees to 10 degrees, which was probably intended to facilitate access to the mine. The deepest shaft descended 386 feet to the level of a subterranean body of water. The descent was made by means of ladders, or tree trunks with steps cut into them, attached to the walls. The ore was removed by means of ropes pulled by winches and pulleys.

The various shafts were linked at the level of the deposits by parallel galleries; these were intersected by cross galleries that facilitated ventilation, while secondary access routes followed the slope of the veins (Figure 64). The roofs of the galleries were supported sometimes by stone pillars left in place when the vaults were being dug out, sometimes by artificial supports. Water screws were probably used to drain the galleries.

The ore was crushed in mills similar to the type used for the crushing of olives (see below, pp. 210–211 and Figure 72), and the fragments were then washed on large cement floors with streams of water that carried away the impurities.

FIG. 64. Level of a mine, as depicted on a Corinthian painted tablet of the sixth century (from Pfuhl, *Malerei und Zeichnung der Griechen,* 3, pl. 44, no. 186).

Metallurgy While mining did undergo a certain development in Greece, this cannot be said of metallurgy, which always remained extremely primitive and, in any case, never surpassed the technical level achieved by the great Eastern civilizations or the Hallstatt and La Tène cultures in Europe. It is even probable that a great portion of the iron used in Greece was imported in semifinished form.

The manufacture of brass by simultaneous fusion of copper and calamine or carbonate of zinc at a temperature of about 800 degrees was probably achieved for the first time in the Orient in the first millennium. Brass was used in Persia in the Achaemenid period (sixth to fourth centuries B.C.), but Greek metallurgy does not seem to have been acquainted with it.

On the other hand, it is possible that the vertical kiln with a chimney, the ancestor of our blast furnace, was already in use in the classical period. A kiln of this type that may date from the sixth century B.C. is believed to have been discovered at Agios Sosti, on the island of Siphnos. This kiln, built of brick with a clay facing, had an opening at its base to ventilate the kiln and permit the metal to flow out; the combustibles and ores were put in through the chimney.

Iron ore was supplied partly by small deposits such as those of Samothrace and Eubeus, which were probably exhausted long before the Roman conquest. It is also likely that European ores were imported into Greece.

Pottery

Pottery   The production of the Greek vases was not essentially different from, for example, that of pre-Hellenic pottery; in particular, the foot wheel seems to have been no better known in the classical period than in earlier ages. Certain pictures show the assistant to the master potter turning a hand wheel; the foot wheel does not appear to have been introduced until the third century B.C.

FIG. 65. Potter's kiln, as depicted on a Corinthian painted tablet of the sixth century (Louvre Museum, MNB 2858).

There were also few changes in the shape of the kiln as it appears in pictures on numerous painted Corinthian plaques (Figure 65) and in discoveries made on certain sites, for example at the Agora in Athens (eighth–seventh centuries B.C.) or at Megara Hyblaea in Sicily (sixth century). These were kilns with two stories; the top floor, which had a hemispherical cap, was used for the firing, while the bottom floor, where the fire was built, had a protruding mouth on one side and was covered with a barrel vault; a center shaft sometimes supported the intervening flooring. There was probably no ventilating chimney. The diameter of this kiln probably ranged from three to five feet. Only after the beginning of the fifth century were larger (14 feet 9 inches at the *kerameikos,* or potters' quarter, of Athens), sometimes rectangular kilns built; in the Hellenistic period perfected systems of piping that ensured the supply of hot air, a forerunner of Roman models, appeared for the first time. Kilns of this type dating from the Punic period have been discovered at Carthage.

The artistic value of Greek pottery results in part from the perfection of that lustrous black slip that has wrongly been called "Attic varnish." The beauty of this slip — that is, a clay solution — is the fruit of a gradually developed technique that reached its peak in the fifth century B.C. and that deserves to be studied carefully. In the Hellenistic era the technique of vase decoration was completely transformed by the introduction of molding and stamping, techniques which remained in fashion for several centuries.

The "Attic varnish"   The magnificent, lustrous black coat that until the last quarter of the sixth century was used for painting dark figures on a reddish clay background (the black-figure style), and then until the fourth century for the background of the vase while the decoration

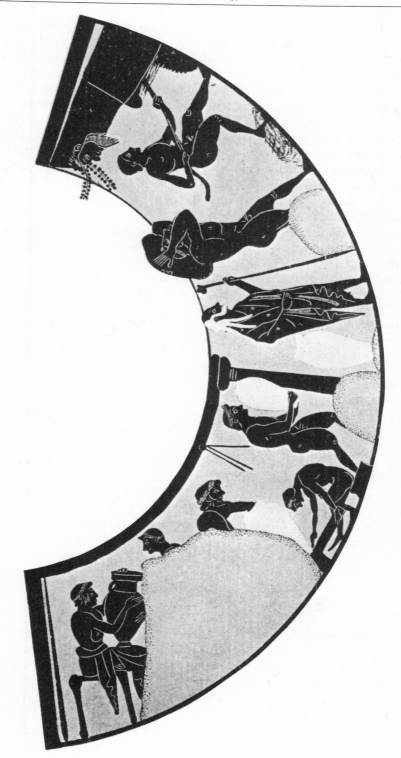

Plate 19.
Greek pottery workshop. Black-figured hydria, sixth century B.C. To the left, the assistant to the master potter is shown turning a handwheel. Museum Antiker Kleinkunst, Munich.

was reserved in red on the clay (red-figure pottery), has been the object of long controversies. It is now certain that this was neither an oil varnish obtained by firing nor a genuine glaze vitrified at a temperature of 900 to 1,000 degrees. Actually, it is simply a slip, and it is this same slip, sometimes red, sometimes black, that was used to decorate the molded vases of the Hellenistic period.

These color differences, as well as the contrast of the reddish clay of the Attic vases with the beautiful black color of the slip, were produced solely by the phenomena of oxidation and reduction, for the clay used in the making of the vase and that which forms the "varnish" are of the same composition. This clay contains an iron oxide that is normally red (ferric oxide Fe_2O_3), but acquires a black color by reduction (ferrous oxide FeO). As for the brilliance of the slip, it is the result of "peptization," that is, the beginning of vitrification under the action of an alkali such as wood ash. The painters of the Cretan and Mycenaean vases were already acquainted with the conditions in which it occurs: the solution of the slip must contain only very fine particles of clay; the addition of certain materials such as gall facilitated the process considerably; certain grades of clay (for example, illite, which contains a large proportion of alkaline elements) are more susceptible to this transformation.

The partially dried vase was first completely covered with a very thin layer of the slip in diluted form. This gave a brilliant luster to those areas of the vase in which the red clay background was to be visible. The design was then drawn with a solution of the same clay thickened by the addition of wood ash, and the black "varnish" background was added in the same manner. The incisions used in black-figure pottery for the details of the decorations were added before firing.

Firing was done in several phases. In the first phase the temperature reached 600 degrees, and a very oxidizing atmosphere was maintained that gave a beautiful red color to the clay and a more even tone to the decoration. Then the supply of air was cut off and the atmosphere became reducing, while the temperature was pushed up to around 950 degrees. The ferric oxide was now transformed into ferrous oxide, an operation that may have been facilitated by causing steam to form inside the kiln; the decoration turned a beautiful black color, while the clay acquired a gray tone. The temperature was now lowered, and hot air was again admitted. The very thin slip covering the clay background did not prevent the latter from reoxidizing, while the thicker layer that formed sometimes the figured decoration, sometimes the background of this decoration, resisted oxidation for a longer period and retained its black color; the objects had to be quickly removed from the kiln at the right moment, before the black areas became oxidized as well. If the vase was to be completely black, as in certain Hellenistic relief vases, the process was terminated after the second phase of firing.

White highlights were sometimes obtained with the help of a nonferruginous clay, while violet touches were produced by the addition of ocher, which contains up to 80 percent iron and whose physical structure permits easy reoxidation.

Molded and stamped pottery As early as the fourth century, certain Greek vases were being decorated with a medallion motif or with relief decorations usually inspired by toreutics or goldsmithing. Ordinarily the decoration was molded directly from a metal

Plate 20.
Forge. Athenian amphora, sixth century B.C.
Photo Guillaume-Budé

Plate 21.
Women pounding grain. Krater, sixth century B.C.
Photo Guillaume-Budé

object and was reproduced with the help of plaster matrices. Beginning in the Hellenistic period, however, the entire vase was very often covered with a decoration in molded relief, the latter being sometimes composed of a series of isolated motifs produced separately and then applied with barbotine (a thin clay paste) to the surface of the vase before firing. These are the so-called "Pergamenian" vases, which appear toward the middle of the third or during the second century. In another type the relief was molded directly with the vase; these are the bowls incorrectly termed "Megarian," whose origin is still obscure but the oldest of which undoubtedly date from the third century B.C. The production of these Megarian bowls combined molding and throwing. First a hollow mold was made, and the motifs that were to constitute the decoration of the bowl were impressed on the inside of the mold with stamps. The mold was fired, after which it was placed on a wheel and clay was poured into it. The vase was shaped on the wheel; then the wheel was stopped and the clay was pressed into the hollows of the decorative impressions on the mold. The inside of the vase was smoothed on the wheel, so that no finger marks remained visible (as happened in certain Etruscan vases made according to a similar technique).

Sculptured vases and statuettes   Sometimes it is not the decoration of the vase that is in relief, but the vase itself that is in the round; in this case its production is similar to that of terra-cotta statuettes. At first, double molds were used. Then the assembly-line production of statuettes in certain centers such as Tanagra in Boeotia, or Myrina near Smyrna, led to the perfecting of a new technique, in which as many molds were used as there were parts in the figure to be obtained. This made possible a considerable number of combinations with a very limited quantity of molds. Airholes were drilled in the mold to avoid distortions in drying. Before the pieces were completely dry, they were assembled by means of a slip; a very fine slip was applied over the entire work, which was then fired over a very low fire. The decoration of the statuettes was painted after the firing.

Glass and Enamel

Glass   Until the Hellenistic period, molding was the only method used in the production of glass. Of Egyption origin, this method was also practiced in Greece, especially after the Archaic period, with typically Greek forms. The glass was molded around a compressed sand core attached to the end of a rod. Decorative bands colored by means of metallic oxides were then applied to the surface of this first layer of glass, and the outside of the vase was polished by turning it on a very soft marble slab, while the interior had a grainy appearance after the sand core was removed. This technique was practiced until the first century B.C. in its principal centers of Phoenicia and the region around Alexandria. However, it could not compete with glassblowing, whose origin remains obscure; undoubtedly it appeared for the first time in the Orient, perhaps in the second century, and did not attain its full development until the Roman period. The Hellenistic necropolis of

Myrina (third to first centuries B.C.) is interesting in that it contains examples of both methods.

Blowing presupposes a much more highly developed knowledge of the properties of glass. Thus if a perfectly transparent product is desired, the heating and especially the cooling must be done very slowly. On the other hand, a temperature of at least 1,100 degrees is required in order to avoid the formation of air bubbles. As for blowing, it can be done with or without a mold.

Enamels   The technique of enamelwork, which is older than that of blown glass, seems more definitely Greek in origin. The earliest surviving enamels — the gold rings from Kouklia in Cyprus — are isolated examples that date back to the Mycenaean Age. Then there is a gap until we come to the examples from the classical period, after which specimens become quite numerous.

Enamel is a glass consisting of sand or flint (50 percent), lead (35 percent), and sodium or potassium (15 percent). The latter ingredients were obtained from sodium carbonate imported from Egypt (especially from Wadi Natrun) or from the treatment of various vegetable ashes. The material produced by the fusion of this mixture is colored by the addition of various metallic oxides and solidified, after which it is crushed into a powder so that it can be purified and used to fill cavities left for this purpose in the metal to be decorated. After a new fusion and solidification, the enamel is polished. The cloisonné and champlevé techniques seem to have been practiced beginning in the Greek period.

Spinning and Weaving

Spinning and weaving   Wool was the principal textile used by the ancient Greeks; geographical conditions were favorable for sheep raising. Flax, in contrast, was little cultivated, and its fibers were imported from Egypt. Silk appears to have been unknown, while cotton was undoubtedly introduced into Greece after Alexander's campaign in the Indus Valley (327–325 B.C.)

The textile industry was still in the artisanal stage. The spindle was the only instrument used by the spinners: the Greek loom, however, did show a slight improvement over the looms of the Eastern civilizations, although it was still a crude affair (Figure 67) with a vertical warp suspended to a crossbar supported on two posts. The warp was stretched by weights attached to the end of each thread or group of threads. A movable bar separated the odd and even threads or groups of threads; a second movable bar to which the odd threads were attached permitted them to be passed in front of or behind the even threads (Figure 66). The shuttle, which was still only a long, flat wooden needle, wove the woof in the opening thus made between the two series of threads that composed the warp. The weaving was done from top to bottom; a comb squeezed the woof as it was woven. This was at first a simple toothed comb, later a wooden frame with metal threads between which the warp passed. The woven piece could be only as long as the total height of the loom.

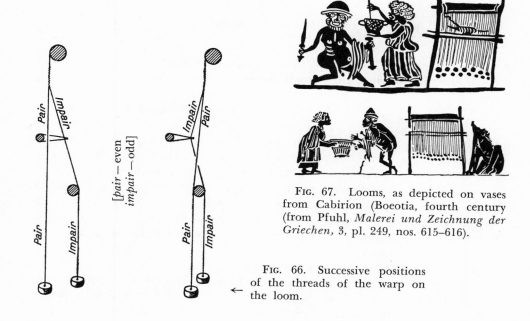

[pair — even
impair — odd]

Pair
Impair
Pair
Impair

Pair
Impair
Pair
Impair

FIG. 67. Looms, as depicted on vases from Cabirion (Boeotia, fourth century (from Pfuhl, *Malerei und Zeichnung der Griechen,* 3, pl. 249, nos. 615–616).

FIG. 66. Successive positions of the threads of the warp on ← the loom.

Woodworking and Painting

The saw and the lathe Two major innovations are probably of Greek origin. One was the bowsaw, which was completely unknown in the Bronze Age; it appears on an Attic red-figure vase dating from the end of the sixth century, in a decorative scene showing the goddess Athena constructing the famous Trojan horse (Figure 68). The other major invention was the wooden lathe, a quite primitive machine composed of a certain number of fixed parts to hold the headstock that turned the piece of wood being worked; the headstock was rotated by means of a bow, the cord of which was wrapped

FIG. 68. Athena modeling the Trojan Horse, Decoration on an Attic vase from the end of the sixth century (from A. Furtwängler and K. Reichhold, *Griechiesche Vasenmalerei,* pl. 162, 3).

around the headstock, while a chisel or gouge held against the piece of wood worked the material into the desired shape.

Encaustic painting   The Egyptians used encaustic only to protect an underlying layer of paint; the Greeks seem to have been the first people to use encaustic painting for itself. Its use varied from coating hulls of ships to decorating marble statues and reliefs. The Metropolitan Museum of New York possesses a krater of Italian provenance, dating from the beginning of the fourth century, which depicts a statue painter surrounded by his equipment (Figure 69). The pigment was first mixed with wax, and the paste thus obtained was applied with a spatula; in the case of a work of art the end of a red-hot iron rod was passed over this layer, which was naturally uneven, to soften and smooth it.

The fresco technique — paint dissolved in limewater and spread over a freshly plastered wall — does not appear to have been practiced before the end of the Hellenistic period; it was widely used particularly in the Roman period. Apart from encaustic painting, the Greeks were familiar only with distemper painting, in which the color was dissolved in a substance that served as a binding, and was then spread over a surface prepared with the same substance, generally a gum. Oil painting was completely unknown.

Fig. 69.   Encaustic painting of a statue. Decoration on an Attic knater of the early sixth century (from D. von Bothmer, in *Metropolitan Museum Bulletin,* February, 1951, Fig. p. 157).

AGRICULTURAL TECHNIQUES

By comparison with the agricultural practices of great Eastern empires, Greek agriculture presented no new features. We have already seen that irrigation was scarcely practiced in Greece and that drainage of swampy lands was exceptional. Greece was and still is a country of dry cultures for which the winter and spring rains provide sufficient water and which require only a very superficial plowing. The Hellenic plow of this period was still very primitive: it was a simple *aratrum* without wheels, with a single handle (unlike the Oriental plows) and a plowshare that apparently was not always equipped with a metal blade. It was attached to a beam — first vertical, then lengthening into a horizontal curve — which led up to the yoke. The instrument was pulled by oxen harnessed under a horn or neck yoke on each side of the draft beam (Figure 70). The animals were probably unshod.

FIG. 70. Peasant with his plow. Terracotta figurine from Tanagra, seventh century (from *Bulletin de Correspondance Hellénique*, 17, 1893, pl. 1).

The principal products cultivated were the grape and the olive, together with the cereal grains; thus the role played by mills and presses in the economy of ancient Greece is easily understood. We have already seen that agriculture, together with war and architecture, was one of the few activities in which the Greek scientific genius could be expressed in concrete machines, for example, the water mill and the beam press activated by a windlass. But these are late achievements that were not completely developed until the Roman period. In the classical period the mills and presses still used man or animal power. There

Plate 22.
Plowing and sowing. Cup by Nicosthenes, sixth century B.C.
Photo Guillaume-Budé

were two types of machines, depending on whether the vertical pressure was accompanied by a lateral back-and-forth movement or by a rotating movement. The former were grain mills and olive crushers; the latter were grape or oil presses.

Millstones and crushers   In the most primitive type the runner, instead of turning around a vertical axle, operated in a back-and-forth direction supplied by the movement of a shaft that was attached to a pivot fixed in the bed stone outside the runner (Figure 71). The millstone was a box-shaped, hollow stone with a slit in the bottom through which the grain fell onto the bedstone. This method seems to have been practiced until the Hellenistic period.

In cases where the rotating movement was centered on a median shaft, the problem of motive power was not fully solved until the invention of the water mill. Until then animal or human labor was used to turn the runner by means of arms protruding from its edge.

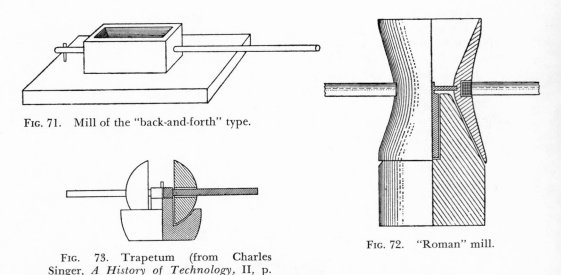

FIG. 71.   Mill of the "back-and-forth" type.

FIG.   73.   Trapetum   (from   Charles Singer, *A History of Technology*, II, p. 112, Fig. 80).

FIG. 72.   "Roman" mill.

Two types of rotating mills are particularly worthy of note. One type, although often called the "Roman" mill (Figure 72), already existed in Greece in the classical period, where it was used to crush lead ore in the mines at Laurion. Its bed stone was conical, while the runner looked like an hourglass; the lower half of the runner fitted like a hat over the conical peak of the bed stone, while its upper portion acted as a funnel. To prevent the rubbing of the two stones on each other from producing a harmful stone grit, a narrow space was left between them.

The second type, the Roman *trapetum* (Figure 73), principally intended for the crushing of olives, consisted of two vertical planoconvex runners attached to each end of a horizontal axle that turned on a vertical pivot. The bed stone was a hemispherical trough, the walls of which were shaped to fit the runners; in this case, again, a narrow space was left between the bed stone and the runners. The excavations at Olynthus have uncovered examples of the *trapetum* that date from the fifth century B.C.

Presses — Apparently the only oil or grape press known to Greek antiquity was the beam press, which functioned on the principle of the second type of lever, in which force is exerted by vertical draft from top to bottom. In the most primitive and simple type this was done by means of weights suspended from the balance beam. When the crank was discovered, a rope attached to the end of the beam and maneuvered by a windlass replaced this primitive system. The next development was the two-pulley press described by Hero (Figure 74), in which the reduction of force increased the power of the apparatus; a second set of pulleys perhaps made it possible to raise the beam when desired. The screw press was apparently unknown until the Roman period.

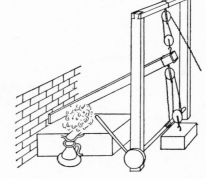

FIG. 74. Beam press with double set of pulleys (from A. P. Usher, *A History of Mechanical Inventions*, 1st ed., Fig. 9).

WEIGHTS AND MEASURES; BOOKS

The Greek scale differed very little from its pre-Hellenic predecessors, which had two platforms and an arm without a pointer. Two major inventions, however, must be credited to the Greek genius: coinage and parchment.

Coinage

The use of coin became possible only with the knowledge of how to refine gold, and especially how to test its purity by means of a touchstone. The latter, which was undoubtedly discovered in Asia Minor, utilized the property possessed by gold of leaving on certain polished stones a mark whose color was determined by the purity of the metal.

Coinage was probably invented by the kings of Lydia in the seventh century B.C., when they conceived the idea of stamping ingots of known weight. As for silver money, tradition claims that it is Argive in origin and that it too dates from the seventh century. Until then the medium of exchange in archaic Greece, as in its predecessors the civilizations of the Bronze Age, had been objects of a certain type of metal, for example, rods in the form of spits (*obeloi*).

At first only one side of the coin was struck; the motif was sunk in a mass of bronze or iron that served as an anvil, and the blade of metal to be made into coins was forced into this mold with the help of a press. Striking of the reverse side of the coin came into practice in the middle of the sixth century; the motif was sunk in the head of the press, which was held in the hand and pressed into the metal with hammer blows. Obviously the stamping of the motif on the reverse side of the coin was less even and less well centered than on the obverse; in addition, the ancient Greeks attached much less importance than we do to the perfect identity of all the pieces in the same series.

The Greek monetary system was essentially based on the weight of the coin. The same word was used to designate both a measure of weight and its monetary equivalent; the drachma represented a given quantity of metal of any sort, whether gold, silver, or bronze, and therefore had a different value depending on the individual case.

In the beginning there were many systems with various standard units (the mina, the drachma, the obolus) that corresponded to a different weight in each region of Greece. The system of weights was based on the talent. Each talent represented 60 minas (one mina = 60 shekels), and weighed sometimes 60.552 kilograms and sometimes 30.276; consequently, the shekel varied from 16.82 grams to 8.41 grams.

As for the monetary system, the principal unit was the drachma, or one-hundredth part of the mina; its principal multiples and submultiples were the tetradrachm (four drachmas), the didrachm or stater (two drachmas), the hemidrachma or triobolus (a half-drachma). The obolus, or sixth part of a drachma, could also serve as a standard unit, and like the drachma had numerous multiples.

In the so-called "Aeginetan" system, which was used particularly in the Peloponnesus (Sparta and Argos), the drachma weighed 6.28 grams and the obolus 1.04 grams; in the Attic system, which was also used in Euboea, the drachma represented 4.36 grams of metal and the obolus 0.73 grams. Rhodes, Corinth, and other major cities each had a different system.

Papyrus and Parchment

Papyrus   In the Archaic period the basic method of communication of ideas was the wooden tablet covered with wax (*pinakes*), on which the signs were drawn with a stiletto. In the fifth century papyrus, imported from Alexandria, began to be used throughout Greece.

After the message had been written on them, the sheets of papyrus were pasted together end to end to form a roll, or *tomos*. The length of the roll was variable, until in the Hellenistic period the standard roll was set at twenty sheets; the arrangement of the columns and lines on each sheet was also more or less standardized. The roll was held in the right hand and unrolled with the left.

Parchment   The invention of parchment was a major achievement. Animal skins, especially those of sheep and goats, had undoubtedly long been in use as "writing paper." The discovery of a new method of preparation that required no tanning or dressing made possible the widespread use of parchment. After careful washing the skins were soaked for twenty-four hours in pure water, and then were soaked for a period of from eight to fifteen days in a viscous liquid containing approximately 30 percent freshly slaked lime. The coarsest of the hair was removed from the skins, which were then given a second lime treatment. Lastly came the delicate operation of scraping and rubbing, after which the skins were left to dry.

It is difficult to determine the truth of ancient stories that parchment may have been invented at Pergamum (whence its name) in the time of King Eumenus II (197–159 B.C.), in order to compensate for the cessation of shipments of Egyptian papyrus owing to the rivalries opposing the Ptolemies to the Attalids of Pergamum. It seems likely, however, that its discovery does in fact date from the second century, although it was not until the Roman period that it completely supplanted papyrus. In the same period appeared the first "codices" (the forerunner of our modern book), a form for which the papyrus was not suitable.

The achievements of the Hellenistic period in the history of Greek technology were apparent only in certain areas of activity; agricultural and craft techniques, on the whole, remained primitive. But the enlargement of the historical framework, the appearance of rich, centralized states, the impetus given to material progress by certain rulers, and the encouragement and support given to philosophers and scientists, particularly at Alexandria and Pergamum (to say nothing of Syracuse, where this tradition antedated even the Hellenistic period) — all these factors contributed greatly to technological development in the Greek world. Not until the coming of the *Pax romana,* however, did the many inventions of the Hellenic genius find their full expression in concrete achievements.

BIBLIOGRAPHY

I. SOURCES

HERON OF ALEXANDRIA, *Pneumatics* (transl. and edited by Bennett Woodcraft [London, 1951]).

HESIOD, *Works and Days* (Loeb Classical Library, Cambridge, Mass., 1954).

II. GENERAL WORKS

Cambridge Ancient History, ed. by J. B. BURY, S. A. COOK, F. E. ADCOCK, and M. P. CHARLESWORTH (Cambridge, 1923–39), vols. 3–8.

FARRINGTON, BENJAMIN, *Greek Science,* 2 vols. (Harmondsworth, England, 1944).

FORBES, R. J., *Studies in Ancient Technology* (11 vols., Leiden, 1955–).

GLOTZ, GUSTAVE, *Ancient Greece at Work* (London, 1927; paperbound ed., New York, 1967).

MCDONALD, WILLIAM A., *Progress into the Past: The Rediscovery of Mycenaean Civilization* (New York, 1967).

MICHELL, H., *The Economics of Ancient Greece* (Cambridge, 1940).

NEUBERGER, ALBERT, *The Technical Arts and Sciences of the Ancients* (New York, 1930).

PENDLEBURG, J. D. S., *The Archaeology of Crete* (paperbound ed., New York, 1965).

ROSTOVTZEFF, M., *Social and Economic History of the Hellenistic World,* 3 vols. (Oxford, 1941).

SINGER, CHARLES, *et al., A History of Technology,* Vol. II (London, 1956).

TOUTAIN, JULES, *Economic Life of the Ancient World* (London, 1930).

USHER, ABBOT PAYSON, *A History of Mechanical Inventions* (2nd ed., Cambridge, Mass., 1954).

VAUGHAN, AGNES CARR, *The House of the Double Axe: The Palace at Knossos* (New York, 1959).

WACÉ, A. J., *Mycenae* (Princeton, 1949).

III. BUILDING

BLEGEN, CARL W., and RAWSON, MARION, *The Palace of Nestor at Pylos in Western Messenia,* vol. I, Parts I and II (Princeton, 1967); vols. II–IV to follow.

DINSMOOR, WILLIAM B., *The Architecture of Ancient Greece* (2nd ed., London, 1950).

IV. TRANSPORTATION

CASSON, LIONEL, *The Ancient Mariners* (New York, 1959).

FORBES, R. J., *Notes on the History of Ancient Roads and Their Construction* (Amsterdam, 1934).

GIBSON, CHARLES E., *The Story of the Ship* (New York, 1948).

V. MECHANISMS

BRUMBAUGH, R. S., *Ancient Greek Gadgets and Machines* (New York, 1966).

DRACHMANN, A. G., *Ancient Oil Mills and Presses* (Copenhagen, 1932).

———, *Ktsebios, Philon and Heron* (Copenhagen, 1948).

———, *The Mechanical Technology of Greek and Roman Antiquity* (Madison, Wis., 1963).

ROTH, H. LING, *Ancient Egyptian and Greek Looms* (Halifax, 1913).

VI. MINING AND METALLURGY

CALHOUN, G. M., "Ancient Athenian Mining," *Journal of Economic and Business History*, XXXV (1931), 333 ff.

CHARBONNEAUX, JEAN, *Greek Bronzes* (New York, 1962).

FORBES, R. J., *Notes on the History of Ancient Roads and Their Construction* (Amsterdam, 1934).

———, *Metallurgy in Antiquity* (Leiden, 1950).

VII. WEAPONS AND WARFARE

ADCOCK, F. E., *The Greek and Macedonian Art of War* (Berkeley, Calif., 1957).

HACKER, BARTON C., "Greek Catapults and Catapult Technology," *Technology and Culture*, 9 (1968), 34–50.

OAKESHOTT, R. EWART, *The Archaeology of Weapons: Arms and Armour from Prehistory to the Age of Chivalry* (New York, 1960).

SNODGRASS, ANTHONY, *Early Greek Armour and Weapons* (Edinburgh, 1964).

SPAULDING, OLIVER LYMAN, *Pen and Sword in Greece and Rome* (Princeton, 1937).

TARN, W. W., *Hellenistic Military and Naval Developments* (Cambridge, 1930).

VIII. AGRICULTURE

FUSSELL, G. E., "Farming Systems of the Classical Era," *Technology and Culture*, 8 (1967), 16–44.

HEITLAND, W. E., *Agricola: A Study of Agriculture and Rustic Life in the Graeco-Roman World* (Cambridge, 1921).

IX. ARTS AND CRAFTS

DAVIDSON, JEAN M., *Attic Geometric Workshops* (New Haven, 1961).

ÉTIENNE, H., *The Chisel in Greek Sculpture* (Leiden, 1968).

NOBLE, JOSEPH VEACH, *The Techniques of Painted Attic Pottery* (New York, 1965).

PHILIPPAKI, BARBARA, *The Attic Stamnos* (Oxford, 1967).

VERMEULE, CORNELIUS C., *Some Notes on Ancient Dies and Coining Methods* (London, 1954).

WATERER, JOHN W., *Leather in Life, Art and Industry* (London, 1946).

CHAPTER 9

THE ROMAN CONTRIBUTION TO TECHNOLOGY

OUR PURPOSE in this chapter is to describe, not so much a period at the end of the Roman Empire in which various techniques inherited from earlier societies were in use, but rather the Roman contribution to the technological progress of humanity. Having gradually extended their empire to include all the territories bordering on the Mediterranean Sea, the Roman conquerors brought home to Italy a number of the techniques they had observed in Greece and their Eastern possessions. Having become the masters of the "barbarians" — the Europeans, Asians, and Africans who had remained untouched by Hellenic civilization — the Romans learned from these peoples (especially in the West) a number of techniques that then became part of the Greco-Latin patrimony. In addition, they were themselves the authors of several inventions that they succeeded in popularizing.

The rulers of Rome and Italy had to gradually absorb and perfect an enormous mass of diverse techniques. In Italy itself, and in Sicily, this people had begun before the second century B.C. to absorb the heritage of the Greeks of southern Italy and mainland Greece, the Etruscans of central Italy, Campania, Tuscany, and the Celts in the north and in Cisalpine Gaul. In the course of the second century the Romans achieved a firm foothold in the mine-rich Iberian Peninsula, Greece, and the eastern portion of North Africa, as well as in the Hellenized areas of Asia Minor and southern Gaul. In the first century B.C. the principal additions to their empire were two countries, at opposite ends of the Mediterranean, which were particularly rich in natural resources and craft traditions: free Gaul and Alexandrian Egypt, including Syria. The first century of the empire and the reign of Trajan saw the acquisition of still-uncivilized areas: the remainder of Berber North Africa, the island of Britain, that portion of Germania on the right bank of the Rhine, Thrace, all the countries of the Danube, and eastern countries of ancient civilization: a portion of Arabia, the heartlands of Asia Minor, Armenia, Mesopotamia. Here, on the banks of the Euphrates, the Roman armies stopped. Here, too, they came into contact with the Asian caravans — but the techniques of the Far East did not travel the Silk Route. The list of technical innovations that appeared between the end of the Hellenistic period and the end of the Roman Empire is a genuine measure of Rome's contribution to technology. These were made by innovation, by development through constant use of techniques that had existed only in embryonic state in the Hellenistic world, and by incorporation into the Roman civilization of techniques that had already been developed among the "barbarians."

216

The Roman civilization, even more than the Hellenistic civilization produced by the fusion of the Hellenic and Eastern worlds, was an urban one, and it spread by the path of conquest over immense and often far-flung territories. The technical progress accomplished during the *Pax Romana* was therefore concerned particularly with the construction of cities, their means of subsistence, and the roads that linked them. The fact that they put within reach of the numerous peoples of a far-flung empire practical and economic methods and instructions for mass production that could improve their material comfort, if not their food supply and general standard of living, is an important aspect of this progress.

The cities   The typical Roman city was the military colony, which was simply a permanent camp constructed in permanent materials, with its *pomerium* (a cleared space delimiting the city), its quadrangular city wall, its orientation to the northeast or northwest, its two principal avenues bisecting each other at right angles, its four gates, and its secondary streets laid out in a checkerboard pattern. It is, if you will, a Roman creation, but it is a composite of borrowed elements, some taken from the Italic peoples of the age of the lake dwellers, others from the Greeks of southern Italy, still others from the Etruscans. The checkerboard idea was taken from the Italic peoples and the Greeks, the orientation and the intersecting principal arteries from the Etruscans — a ritual element, preserved along with the religious character of the foundation of cities. The systematization of these elements, and particularly the invariably rectangular wall of the camp and its offshoot, the city, are the product of the Roman military genius and the requirements of simplification for installations in open country, which were built for an unvarying unit. All over the world the Romans built many nonmilitary cities with irregular boundaries — but the streets were invariably laid out in checkerboard pattern. In any event, the foundation of cities and the creation of methods of rapid intercommunication quickly brought the techniques of construction and major public works to the forefront, and it is in these fields that the fundamental technical achievement of the Roman is to be found.

CONSTRUCTION AND PUBLIC WORKS

The originality of the Roman contribution to architecture lies neither in its town planning nor in its styles (which are derived from Greek types), but in certain types of edifices it invented, the study of which belongs to the history of architecture (the triumphal arch, the amphitheatre, the aqueduct, the bridge, the victory memorial monument known as the *trophaeum,* the pantheon, the votive column or decorated gravestone, the villa), and in procedures or materials of construction (mortar, brick lacing, window glass), in certain specifically technical equipment (roads, hydraulics, heating by hypocaust), and in certain techniques characteristic of monumental art (for example, mosaics).

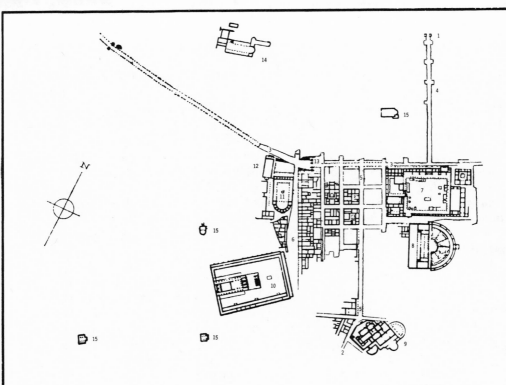

Ministry of Public Education

and Fine Arts

———

TIMGAD

(THAMUGADI)

Condition of the Excavations in 1896

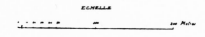

ECHELLE

200 Metres

1. North Gate
2. South Gate
3. Decumanus Maximus
4. North Cardo
5. South Cardo
6. Capitol Way
7. Forum
8. Theatre
9. Baths
10. Capitol
11. Marketplace
12. Annex of the Marketplace
13. Arch of Trajan
14. Cathedral
15. Christian basilicas
16. Basilica of Gregory the Patrician
17. Byzantine fort

Drawn by the Chief Architect of Historic Monuments of Algeria

Plate 23.
Plan of the Roman city of Timgad, Algeria. Founded in A.D. 100 by Trajan.

Cement masonry   If the Romans were able to cover a large part of the Mediterranean world with edifices that are often colossal and were generally built in record time, the credit must go to their use of cement, which they probably learned from the palaces of the Hellenistic kings of Asia Minor (particularly Pergamum), to which its use had until then been limited. The Romans popularized and perfected cement to such a point that we may rightfully call them, if not its inventors, at least its re-inventors.

The practice of assembling large blocks of stone without mortar continued in use throughout antiquity and down to the Middle Ages. However, masonry built with small stones required mortar, and the smaller and more irregular the stones, the more solid the mortar had to be. Thus mortar formed the major part of the wall. A mixture of lime, sand, and clay was well suited to this role: it was a wet, pliable paste into which the stone rubble for the future wall could easily be compressed, and it hardened in drying and became even harder with the passage of time, so that ultimately it combined with the lumps of stone it enveloped into a kind of conglomerate that was harder than stone. Because it could easily be made in large quantities, it permitted the Romans to build quickly and in quantity the large public works and colossal edifices with which they covered the world they conquered; thanks to this abundant and unusually solid material, they hesitated before no outlay of money, no daring undertaking.

The only secret of this Roman cement was the quality of its component materials and the care taken in combining them and then in pressing the stones into the paste thus formed until the smallest air bubble or drop of water disappeared. (In later ages this professional conscientiousness was lost, and medieval cement is distinguished from its predecessor by its lack of solidity, its poor quality, and the cavities left in the cement and between the stones due to poor mixing and insufficient compression.) The concrete with its filling of rough stones was compressed with a pounder into the space between vertical planks, inside which the facings had been placed. In the case of vaults, compression was accomplished partly through the pressure of their weight against wooden arches made of rough planks; since the cement quickly formed a kind of monolith with the stones or bricks laid flat on the vault, the pressure of the vault on the walls was lessened, and it could often be supported, even without a keystone, on thin walls.

Numerous analyses of Roman cement have showed that it contained only a more or less fine grade of sand, lime containing charcoal dust, clay, and sometimes gravel and crushed brick, in proportions close to those mentioned, for example, by Vitruvius: one part of lime to three parts of sand (or two to five). The sand and water were mixed with the help of a binding made from lime and clay. Samples of cement recently analyzed in Paris had been made with binding agents which, although they contained less lime (20 to 36 percent) than modern cements (45 to 65 percent), nevertheless permitted a remarkable hardness. The great difference between the Roman bindings and modern cements is that the former are the result of a simple, although careful, *mixture* of lime and clay, while the cements are the product of a *combination* produced by firing these two elements.

Brick facings and lacing courses The principal problems involved in this type of masonry were how to prevent the wet wall from buckling while it was being constructed, how to ensure the perfect cohesion of the facing, and how to prevent cracks, which under the action of heat or extreme cold could spread through the entire structure. To guard against these difficulties, the Roman builders left the regularly spaced supporting timbers of the scaffolding in place in the wall. They also ran brick courses along the entire wall, approximately 3½ feet apart in the period of the Late Empire (beginning in the second century) and at closer intervals thereafter; these courses acted as a constant check on the horizontality of the wall, and ensured the cohesion of the facing along its entire surface. This technique, which was particularly well developed in Gaul and northern Italy, was perhaps a survival of the famous Gallic practice of using frameworks of intersecting wooden timbers in fortification walls; Caesar described it in speaking of Avaricum (modern Bourges), and archaeologists have discovered numerous examples, for example at Vertault (Côte-d'Or) and Murcens (Lot). In any case it was based on the same principal, and had other advantages as well, the principal one being that it prevented dampness and cracks from spreading in either direction.

As for the facing, various solutions to the problem of cohesion were invented: small square or rectangular stones with elongated bonders at the angles; diagonal laying of courses in a netlike (*opus reticulatum*) or herringbone pattern; alternation of stone courses and brick lacing courses; proliferation of relieving arches over openings and even in the facing itself. Despite the sometimes great depth and often enormous thickness of the foundations, every possible method of lessening the weight of the walls was utilized, including the use of semicircular recesses in the walls, progressive decrease in their thickness, and the proliferation of bays and windows in the upper portions. The use of brick, a prefabricated and easily cut element, made it possible to correct all the irregularities of detail in the delicate joinings of the facings — for example, at the junction of the rubble wall and the stone blocks used for the door and window frames, pilasters, reinforcements and buttresses, and at the springing of the arches. These bricks, according to analyses recently made in Paris, were very similar in composition and quality to modern bricks. Lastly, the adherence of the coats of stucco that covered the entire masonry and even the blocks of stone, inside as well as out, was ensured by deep grooves that imitated the joints of building stones and were cut into the wall at right angles to the mortar joints, and by geometrical incisions made on the faces of the small stones of the facing (Figure 75).

Roofing and windows In roofing, the Romans generalized and systematized the use of the Greek tile to such a point that we still speak of the "Roman tile" — a curved tile (*imbrex*) covering the jointed edges of two flat tiles (*tegulae*). The use of slabs of stone or marble to cover important (particularly religious) edifices had been known to the Greeks, but the gilded bronze tile used for example on the Pantheon in Rome seems to be a purely Roman variation.

The window openings were closed with wooden or canvas shutters, oiled

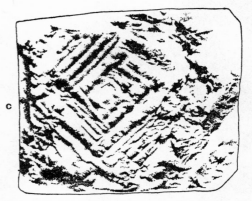

Fig. 75. Stones used in small-stone masonry work. The geometrical incisions served to retain the facing. Baths of the Cluny Museum, Paris (from findings by Henri Poupée).

skins left to hang free or attached to a frame, or slabs of mica. Around the third century the Romans added another material to the list: window glass, at first opaque and still very similar to mica, later transparent. Examples of it have been found in Belgium and Germany, under the window openings of large buildings. Lactantius (fourth century) mentions window glass, and Prudentius even mentions stained-glass windows.

Vaults and framework; bridges The incredible development of the vault in the Roman era was simply the exploitation of an earlier innovation invented by the Greeks and perfected by the Etruscans, which made possible the colossal style, the immense naves with semicircular barrel and groined vaults, the use of corridors placed one above the other to support tiers rising into space, the multiplication of stories and staircases, the arches of bridges, the labyrinths of the sewers. The vault and the cupola, supported by walls that were masses of extremely hard cement, became the customary roofing method for a large number of public buildings. The vaulted cupola over a square space, supported by squinches and pendentives, later successfully replaced the corbeled cupola; it became especially popular in the Middle Ages.

In similar fashion, the perfecting of the art of the wooden framework made possible not only the erecting of vertiginous scaffoldings but also vast, mighty roofs over the superstructures of such buildings as temples, porticoes, and theater stages (thanks to the improvement called the "truss" and the triangular framework, which would not buckle under stress), as well as scientific constructions such as bridges thrown across major rivers. Caesar's description of the Rhine bridge built by his troops in ten days (55 B.C.) permits us to judge the importance of these works of art, and also their novelty. This was not a pontoon

Plate 24.
Vault of the Frigidarium of the Baths of Paris. Cluny Museum.
Photo Jean Suquet, Paris

bridge of the type that Xerxes had thrown over the Bosphorus and Hannibal over the Po, but a bridge built on piles. It may have been the prototype for the numerous stone bridges which, thanks to the vault, became the inevitable complement of the road network throughout the Roman Empire. The bridge of Alcántara (Spain) has arches 88 feet high; those of the "Augustinian bridge" at Narni, north of Rome, are 105 feet high. In Rome itself, the piles of the Ponte Sant' Angelo (known to the Romans as the Aelius) plunge 16 feet into the bed of the Tiber.

Stairways and multistoried buildings   The facility with which both small and large vaults could be used also made possible the proliferation of stairways of masonry, and consequently the construction of multistoried buildings. Very often each story had its own individual stairway. The average stairway in the houses of Rome and Ostia ran for three stories; houses with five and six floors existed, and one Roman building, the *insula* of Felicula, was definitely taller still — a genuine "skyscraper." This "aerial" Rome, denigrated by one poet, presupposed a great skill in the construction of stairwells in masonry and wood. In this domain the Romans even

Plate 25.
Façade of the Roman amphitheatre at Nîmes, France, first century B.C. *Photo New York Public Library*

performed feats of strength, as for example the spiral staircase and the "screw" of the Trajan column. The role of vaulted access halls and stairways is of primary importance in a colossal edifice such as the amphitheatre, which was built to hold tens of thousands of spectators; the façade of the amphitheatre of Nîmes, the best preserved but by no means the largest Roman amphitheatre (its major axis is 436 feet, as against 613 feet for the Colosseum), had 60 Doric arcades in 2 stories, 3 flights of 10 tiers, 5 vaulted access galleries, 124 *vomitoria* (openings for entrance and exit), and 162 stairways; the building was crowned by 120 corbels that held the poles of the canvas "roof."

Roads   In the field of transportation the Romans developed a construction technique that for many centuries supplied civilized Europe with her roads. For the few "royal roads" of the Orient, the bad Greek roads, and the simple paths of the barbarian countries of East and West, they substituted a roadbed, preferably straight and along a ridge, with a deep foundation, set in a vast trench and flanked by drainage ditches and shallow embankments. Their real innovation was the superposition, in varying order, of layers of large stones, gravel, rocks laid in obliquely, and (in exceptional cases) mortar (Figure 76); these gave the road, which was slightly convex and dominated the surrounding territory, a kind of elasticity that contributed to its solidity and compensated for the deficiencies of the harness and carriage systems then in use. The network of roads that linked all areas of the Roman Empire has been estimated at 55,890 miles, exclusive of the 125,000 miles of secondary roads. Pliny recommended a width of 18 feet; the main roads were sometimes twice this width.

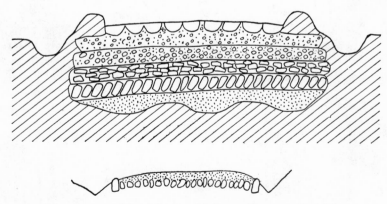

FIG. 76.   Cross section of Roman highway. Top, a main highway; bottom, a secondary route, shallow and unpaved.

"The foundations of the Roman roads are not constructed of masonry; they are not like large walls buried in the ground. Basically they consist of layers of rock, which ensured proper drainage, and an inert, porous core that has become solid only through the passage of time and the utilization of the road" (Grenier). City streets and approach routes to cities had a surface layer of paving stones; in this layer, and in the rock of mountain roads, ruts were cut as guides for chariots and carriages.

The cohesion of these foundations to the ground was ensured by digging three ditches with two intermediate slopes in the bottom of the trench (which was at least three feet deep (Figure 76); this "toothed" surface permitted a genuine imbrication of the mass of the road. A similar system was sometimes used for the foundations of fortifications.

Another great convenience for public transportation and traffic throughout the empire was the marking of the main roads at regular intervals by milestones, separated by secondary, more closely spaced, landmarks that sometimes indicated the distance in relation to the "golden" or metropolitan milestone at the center of Rome.

The city wall

Unlike the highways, fortifications do not appear to have been the object of significant innovations after the Hellenistic period. One feature, however, is that posterns became fewer as sorties were replaced by combat at the top of the ramparts. This type of combat required long, spacious curtains. The principal gates became powerful bastions, often with interior courtyards.

In civil architecture two technological innovations appeared that were to transform the material life of the individual and become powerful instruments of Romanization: the improvement of aqueducts by the discovery of an easier method for crossing valleys, and the heating "from underneath" of public buildings and well-to-do homes.

The aqueduct

The aqueduct had existed in the Eastern and Greek civilizations, and the Romans simply developed the principle and applied it on a grand scale, with an audacity of location equal to their prodigality of construction. Rubble masonry and the massive production of lead made it possible to use the "bridge aqueduct" (a new development) and the "inverted siphon" for crossing valleys. The Romans could now lay an aqueduct, like a road, across any site. Once they had chosen a site for a city for reasons of strategy, commerce, or agreeable location, they did not have to depend on the local water resources. They were ready to dig tunnels and construct ducts fifteen, twenty-five, or fifty miles long and, consequently, to make any number of detours in order to give the gradient the slight incline required for the resistance of the channel and the regularity of the flow. If the natural slope was too great, they had recourse to waterfalls, like the twenty-four falls between the Morvan region and the city of Autun, which broke the flow of the mountain waters in order to bring them gently into the city.

For most of its route the duct ran underground, for reasons of health and security. It was built in the form of a trench, often along the side of a slope in order to follow the topography.

The drilling of a tunnel was generally done by two teams starting from the opposite ends of the line and advancing toward each other. An African inscription in the neighborhood of Lambaesis tells of the error of direction made in the drilling of the tunnel for the aqueduct of Bougie around the middle of the second century: one team having dug too far to the right, the two groups missed each other by half the total proposed length! The assistance of the navy and auxiliaries was required to correct the error. The methods of orientation used by these engineers are not known to us. Fortunately, other tunnels, drilled for roads, were more successful: the Furlo pass of the Via Flaminia (131 feet long), the tunnel under the Apennines from Naples to Pozzuoli (984 feet long), the Hagneck tunnel in Switzerland, near the lake of Biel (2,624 feet), and at Briord (Ain) in France (754 feet).

Generally the greatest skill was displayed in the harnessing of springs and subterranean bodies of water, the construction of ascending drains and filtering basins, the installation of reservoir-barrages, inspection shafts, and waterworks. The gradient was never less than 12½ inches per mile. Valleys of average size were crossed in a straight line, thanks to the bridges with their rows of arcades placed one above the other; the most beautiful of these is the Pont du Gard (the aqueduct of Nîmes), 157 feet high. Very deep valleys could be crossed only by the "inverted siphon," a series of lead pipes laid parallel (to avoid excessive pressure) that descended from a storage tank and then rose, thanks to the pressure thus acquired, to a receiving tank on the slope opposite. This system, which was not new but had been greatly perfected, has been studied in one of the aqueducts at Lyon, where it must have required the use of over sixteen miles of lead pipes, representing approximately two thousand tons of metal. It was also used for the crossing of certain rivers (for example, the Rhône at Arles) and, on a smaller scale, for the distribution of water to cities by piping from waterworks. Its use may have been even more widespread.

As for sewers, they were well developed in the cities, where they ran under the streets; they carried off waste water, and thus did not become blocked up by the calcareous deposits that gradually obstructed the aqueducts. The treatise of Frontinus, administrator of the water supply at the end of the first century, gives us considerable information about the nine aqueducts of Rome; he describes the system of sluices regulating the flow of water, the valves, dishonest "punctures" in the pipes, the use of clandestine taps, and repairs. Perhaps the only structure more constantly in need of frequent repairs and reinforcement than the aqueducts was the widespread system of heating known as the "hypocaust."

Canals and irrigation   In contrast, the digging of canals, in which the Eastern and Greek civilizations had set an example, does not appear to have made much progress, apart from several beautiful achievements which are, however, few in number (the canal built by Corbulo to connect the Rhine and the Meuse, the canal of Drusus paralleling the Rhine delta, the canal built by Marius from the environs of Arles to Fos [the "Fossa Marina"], which continued in use until the end of antiquity). The canal of Corinth was begun in the reign of Nero; routes were established across the isthmus of Suez, two of which are known to us. The lakes of Albano, Nemi, and Fucino were supplied with outlets. Irrigation does not appear to have been practiced on a large scale in the countryside; it was limited to the outlying areas of cities, private estates, and farms, without being systematically organized over large areas.

Plate 26.
Pont du Gard. Roman aqueduct at Nîmes, France, first century
B.C. *Photo Pan American Airways*

Heating   The problem of heating was solved in a manner that integrated it closely with architecture itself. It became a matter of capital importance for two reasons: first, the Romans, inhabitants of a rather warm country, conquered cold countries in which they developed means of comfort for themselves and for the conquered populations who adopted their material civilization; second, in both warm and cold countries they made systematic use of private and particularly public baths with numerous large steam rooms. Particularly in the public establishments they employed on a large scale the system of the hypocaust (Figure 77). This was an ancient idea — it had been used as early as the third millennium B.C., in the Indus Valley civilization, for heating bathrooms. The floor of the room to be heated was raised on piles; around these piles circulated hot air, which was supplied by a fire placed against an outer wall of the area so that it could be tended from an adjoining room. The air was drawn up by flues in the walls (harmful gases were removed by the same method). The draft passed directly under the fire, an arrangement that permitted maximum heating of the air directed into the hypocaust. The floor of the room to be heated was constructed of large tiles, which were excellent conductors of heat; it varied from ten to twelve inches in thickness, and was covered with paving stones or mosaics. This compact mass prevented cracking and the passage of gases into the room.

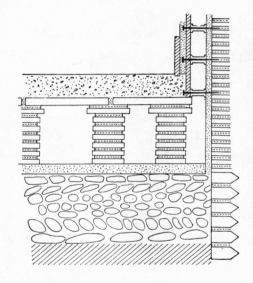

FIG. 77. Cross section of a hypocaust with wall pipes. Baths, Collège de France, Paris (from de Pachtère, *Paris à l'époque gallo-romaine*).

Much discussion has centered upon the question of whether the terra-cotta tubes placed in the walls behind a stucco or other facing were actually used for heating or were simply an escape hatch for gases or a means of drying the walls, particularly in the case of the baths. It seems difficult to accept the idea that they did not at least help to retain the heat in the walls and the facings that covered them. Charcoal was the principal fuel burned in the furnaces, which were numerous and small (although a single fire could be used to heat two rooms at once simply by using communicating hypocausts). The heat obtained cannot have been great, except in the very small steam rooms, and every possible method was used to increase it: very thick walls that, once heated, retained the heat; limiting the number of opening; continuous maintenance of the fire, which caused a buildup of the heat (the fires in the public baths were kept going twenty-four hours a day, and were used in the morning and early afternoon by the women, the rest of the day and the evening by the men); lastly, the room requiring the greatest heat was generally given a southwest exposure, to profit as long as possible from the heat of the sun.

The fragile hypocaust, which in most cases had small brick pillars bonded with fireproof clay to support the heavy floor of the room and its occupants, was exposed to a corrosive heat, and excavations indicate that frequent repairs and reconstruction were necessary. It was undoubtedly for this reason that the Romans perfected (at least in the West, and particularly in Gaul) the system of constructing the flooring on a masonry base in which they installed a network of pipes communicating with each other and with the adjacent furnace. Toward the end of antiquity the hypocaust began to be replaced by the large fireplace, the principle of which had long been known but which did not come into use in the West until around the fourth century A.D.

ART

The mosaic   Among the techniques related to art, the one most fully developed by the Romans was the floor mosaic, a technique closely related to architecture. The Romans made no innovations in wall painting (which was, however, widely used); the mosaic, in contrast, became in their hands a kind of industrial decorative art. Born in the Orient, known to the Egyptians and the Aegean peoples, used by the Greeks, it had such prodigious success in the Roman Empire that but for its even greater flowering in the Byzantine age we might have been able to regard it as a Roman art par excellence. Particularly in North Africa, the number of mosaic vestiges that have been discovered point to a kind of passion for this type of decoration. The Romans perfected the technique to the point of using minuscule cubes that permitted great finesse in design and color. In addition, they executed the center motif (*emblema*) separately, like a genuine picture, in a very wide range of colors (thirty, on the average), inventing numerous methods of arranging the cubes: small cubes for the figures, larger cubes for the background, in concentric circles for the medallions, diagonally for the borders, and so on. They even went to the trouble of creating — in pictures on which people walked! — effects of relief and modeling. The care with which the foundation (the *opus signinum*) was laid contributed to the solidity of the work; it was composed of ordinary cement reinforced with crushed brick, pebbles, stones, gravel, and small bits of marble.

Wall mosaics (and sometimes pavements) were usually made with pieces of flagstone that could be either regular or irregular in shape; this material, called *opus sectile,* was probably imported from the Orient in the time of Claudius. The classic mosaic material, however, was the *opus tessellatum,* made of small cubes of stone; the increasingly small size of these cubes permitted mosaic artists to achieve great virtuosity in this art, which closely imitated painting while re-

Plate 27.
Longshoreman. Paving mosaic, Ostia. *Photo Roger-Viollet*

taining a style very different from that of painting and suited to the limitations of mosaic technique. The *opus vermiculatum* required not only stones of various types but also hard and precious stones, marble, and (in rare cases) colored glass. Artists working in this medium were capable of doing repairs and even of constructing movable pavements (for example, for use in generals' tents). Around the fourth century the use of gilded cubes became common. Mosaic was also used for walls, although painted walls were more common; sometimes, too, walls were encrusted with various materials, especially shells.

Pottery   The principal developments in the field of pottery had all taken place prior to the development of Roman civilization, which never valued this material very highly; to the Romans, a piece of pottery was not a work of art but an article for everyday use. The wheel, the relief mold, and firing to red or black color, depending on the degree of heat, the slip, painting, incision, and stamping — none of these techniques were new. The Roman contribution to the potter's art was the systematic use of the mold, smooth or with incised decoration, for assembly-line production of enormous quantities of objects often decorated with reliefs.

Italy and the Roman world were the principal producers of pottery with relief decoration, either molded, appliquéd, or applied with barbotine. In Italy, the workshops of Arezzo became the leaders in the field shortly before the turn of the first century A.D. They were later supplanted by those of Gaul, while Italy created factories for the production of "light," thin *sigillata* ware, which kept the Italian market supplied until the Late Empire. Gaul specialized in molded relief vases (the *"terra sigillata"* of the archaeologists), with which they inundated the entire Roman Empire in the first and second centuries A.D. The very simple method used to produce the semitranslucent red varnish with the silken sheen which particularly characterized this Gallic pottery, known in the Roman era, has recently been reconstructed. Made at La Graufesenque (Aveyron) and Lezoux (Puy-de-Dôme), and the only mass-produced "industrial art," it was the same as the Greek method already described on pages 203, 205. In the third and fourth centuries objects were decorated by holding a small patterned wheel against the object as the latter was turned, thus repeating the same motif indefinitely; this very simplified equipment, which a potter could carry in his pocket from one shop to another and from one region to another, ensured, in a period of growing insecurity, a very large production of wares at a low price. The fifth century saw the development of the gray or red so-called "Visigothic" pottery (which is actually of late Gallo-Roman origin), which was decorated by stamping isolated motifs directly onto the object.

The two-storied potter's kiln consisted of three parts. At the bottom, sometimes buried in the ground, was the main fire and an adjacent hearth divided into two parts by a wall, which helped to support the perforated floor of the third portion — the firing chamber above the second hearth, which was elongated to prevent the objects from being burned. Ventilation was ensured by the upper opening in the domed firing chamber, which sometimes rose to a height of several yards. Only one such firing chamber, near Cadiz, has been discovered intact.

Plate 28.
Gallo-Roman bowl with marbled decoration found at Bordighera.
La Graufesenque, middle of first century A.D. British Museum,
London. *Photo British Museum*

Plate 29.
Gallo-Roman ceramic jar with applied molded and slip decoration
found at Felixstowe, England. Probably Lezoux, late second
century A.D. British Museum, London. *Photo British Museum*

Figures incised on rejected potsherds, some fifty of which were found at La Graufesenque, indicate that an average load of 30,000 objects could be fired at a single time.

Glassware The Romans made more original contributions in the field of glass. Jewelry of poured glass and pressed glass vases had long been known, but not until around the middle of the first century B.C. do we find (in a text of Cicero, *Pro Rabirio Posthumo*, XIV, 40, written in the year 54) the first mention of glass in Rome, and blown glass did not appear until around 20 A.D. (It had been invented shortly before by the Syrians.) Glass, being a much more flexible material than pottery, could be used to make vases of all shapes (Figure 78): long, slender, animal-shaped; it could also be molded in the shape of metal vases. The Romans did not value glass for itself; they wanted it to be an imitation of other materials, notably the hard and marbled stones. They nevertheless acquired a great technical virtuosity in its production, and succeeded at genuine feats of skill, for example, vases made of a

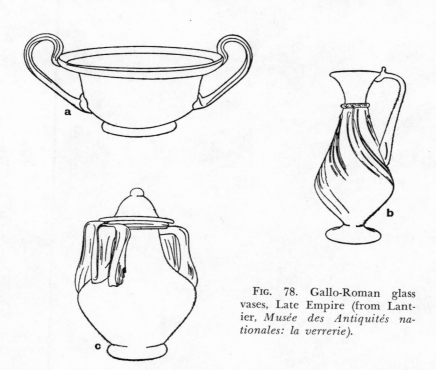

FIG. 78. Gallo-Roman glass vases, Late Empire (from Lantier, *Musée des Antiquités nationales: la verrerie*).

kind of netlike glass lace. In the second century they were already making a translucent white glass that resembled crystal, as well as window glass, the metallic mirror, and a magnifying glass — a lump of glass filled with water. However, metal vases continued to be preferred; the period of the Flavians saw the birth of the molded silver vase with decorative frieze in relief.

Plate 30.
Glass vase. Landesmuseum, Trier. *Photo Landesmuseum, Trier*

Enamelwork The technique of enamel working was closely linked with that of glass. It consisted of brushing a paste, made with glass powder colored with metallic oxides and blended with water, on a piece of metal (bronze or iron) in which hollows formed a decorative motif; the piece was then fired to ensure the adherence by fusion of this paste to the champlevé metal. This method was known not only in Egypt but also in the Caucasus, Persia, Greece, and Etruria, where the enamel thus obtained was generally thin. Beginning in the fifth century B.C., it came into vogue among the Celtic peoples; they used a red enamel (colored by a cuprous oxide), and at least in the beginning seem to have regarded it as a substitute for coral. The technique was perfected in Gaul and the British Isles in the Roman period by the introduction of polychromy. Philostratus of Lemnos, writing in Greek in the third century A.D., describes it as follows:

"The Barbarians who inhabit the Ocean know how to lay colors on red-hot copper, where they are glazed and transformed into an enamel that is as hard as stone, maintaining the outlines of the figure painted on it" (*Paintings,* I, 28). This polychromy was the result of the meeting, probably in Gaul around the first century, of these workers in red enamel and the glassmakers. The glass powder was then combined with pieces of glass arranged in a mosaic, so that metal partitions were no longer necessary. This technique flourished between the first and third centuries, and produced pieces of jewelry and harness, fibulae, and even vases.

METHODS OF TRANSPORTATION

Techniques involved in methods of transportation made some progress in the Roman period, especially through the adoption of ideas from the Western peoples and particularly from the Celts. The latter, former nomads of European plains rich in wood and iron, were remarkable chariot and carriage makers, and the Romans adopted most of their two- and four-wheeled vehicles.

Vehicles and harness The Latin names of most vehicles are loan words from the Celtic: *carpentum* (a wagon built on a "frame" [French *charpente*]), *carrus* ("chariot"), *benna* (a kind of basket on wheels; whence *bagnole,* the French slang word for a "jalopy"), the four-wheeled *reda* and *petorritum,* and many others. A theory was recently propounded that the front wheels of the state chariot found in the royal tomb of Vix (Côte-d'Or) were independent.

The Roman vehicles permitted fairly rapid travel. The customary day's journey was twenty-eight miles, with relays every six to seven miles. On the other hand, the imperial messenger was expected to cover an average of forty-seven miles a day, and there were forced marches that were remarkable performances: In a trip from Pavia to Germania, Tiberius covered nearly 186 miles in twenty-four hours; in 69 A.D. a standard-bearer covered the ninety-nine miles between Mainz and Cologne in twelve hours; the year before that, an imperial envoy had gone from Rome to Clunia in Spain in seven days, traveling between 155 and 185 miles per day.

The harness apparently was not greatly improved in this period. The shoulder collar was still unknown; horses, being yoked around the neck, could not pull heavy loads, and the strongest vehicles of the army had a maximum carrying capacity of a half-ton for two horses (today a single horse can pull three times this weight). The animals were always harnessed abreast, harnessing in single file being unknown. The saddle was modified and perfected until it resembled the modern saddle (Figure 79), but the stirrup was unknown and the horseshoe was not in general use. No word for "horseshoe" exists in Latin. It does not appear in works of art; very few examples have been discovered; and, since a horseshoe is a heavy object, we are never sure in these cases that it did not sink from a higher stratum. On the other hand there was what the archaeologists call a "hipposandal," a kind of metal sole raised at the sides and back so as to envelop the foot (Figure 80); it was undoubtedly used to protect an injured hoof. Some author-

FIG. 79. Roman saddle (profile view) from a Germanic funerary stele).

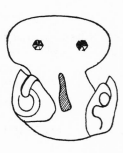

FIG. 80. Metal hipposandal. The leather thongs used to attach it are missing.

ities argue for the existence of shoeing in the West in the Roman era, and even as early as the Iron Age. It seems astonishing, however, that the Celts and the Iberians should have known how to shoe their horses several centuries before the Christian era, while the Greeks, then enjoying the full flowering of a refined civilization, seem to have been ignorant of this practice. The horseshoe seems to have been born, either during or after the Roman Empire, as a result of the hardness of the roads in contrast to the natural terrain. Small spurs existed in the La Tène Age, and were adopted by the Roman Army, but the large spur did not appear until the end of antiquity.

Land transportation was thus very limited. Water transportation for heavy goods was all the more important for this reason, and towing by human draft was commonly practiced.

Ships

Navigation made a certain amount of progress in the Roman era. The Gauls of the Morbihan region, whom Caesar defeated in 56 B.C., used iron rigging, oak hulls held together with long nails, and semirigid sails of skins sewn together, which were far superior to canvas sails. The Romans may have borrowed their system of sails, for Lucius

Plate 31.
Blacksmith. Ostia. *Photo Giraudon-Alinari*

(second century A.D.) describes the ship that transported wheat from Egypt as having sails made of skins. In any event the Romans were familiar only with the "square" sail, that is, a square or rectangular sail perpendicular to the ship's principal axis. (The so-called "lateen" sail, trapezoid or triangular in shape, and set obliquely in relation to the axis, is Arabic in origin and owes its name to its appearance, with the spread of Islam, in the Latin waters of the Mediterranean.) The Roman sail, however, was an improvement over the Greek version. Rings evenly distributed over its surface permitted the sail to be gathered up at the top of the mast before being dropped and attached to the bulwark (this practice was already followed by the Egyptians), instead of dropping directly off the yard. A second mast in front of the principal mast projected out over the bow (Figure 81).

Fig. 81.   Roman ship with ram, sails, and oars
(from a mosaic at Soussee, Tunisia).

On the basis of the quantities of foodstuffs they could hold, the tonnage of a certain number of ancient ships has been calculated. The usual capacity in the Hellenistic period was 130 tons, but in the Roman period the imperial transports attained a capacity of 340 tons, and the largest ships of the fleet more than 1,300 tons. Pliny the Elder mentions the case of the ship specially constructed to transport Egyptian wheat to Ostia: it could be loaded with an obelisk and a pedestal weighing 500 tons, and then with 800 tons of lentils as ballast. The speed of these ships was greatly dependent on the favorable or unfavorable direction of the wind. A merchant sailing vessel could do a maximum of four to six knots on the open sea in a favorable wind, two to two and a half knots in an unfavorable wind. The trip from Marseille to Rome took from two to three days, the return trip four and a half to six days. Marseille to Alexandria was a voyage of twenty to thirty days, the return trip fifty to seventy-eight days. Vitruvius describes (in *De Architectura*, X, 9) the Roman method of reckoning distance traveled, which he considered "one of the most ingenious things we have inherited from our "ancestors." Paddle wheels were suspended at each end of an axle which passed through the hull, and by means of a system of geared, perforated drums, pebbles were dropped into a brass receptacle to announce each mile. A similar system may have been fitted to the wheels of overland transport.

The numerous and varied types of ships, often adopted from the "Sea Peo-

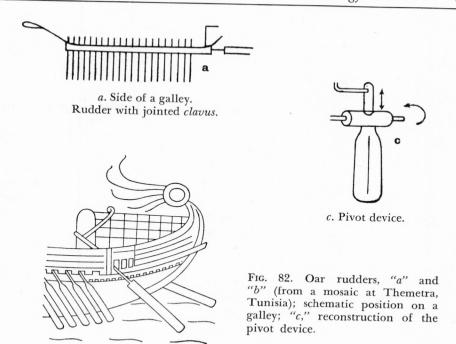

a. Side of a galley.
Rudder with jointed *clavus.*

c. Pivot device.

FIG. 82. Oar rudders, *"a"* and *"b"* (from a mosaic at Themetra, Tunisia); schematic position on a galley; *"c,"* reconstruction of the pivot device.

ples," and particularly from the Gauls, had no very original features. For his voyage to Britain, Caesar seems to have invented a compromise between the galley and the sailing vessel by placing rowers in light transport sailing ships; the war galley was little developed, and in fact degenerated during the Roman period, for the complete Roman mastery of the seas that resulted from the Battle of Actium (31 B.C.) did not encourage technical progress. The rudder (Figure 82) continued to be made of two large oars, one on each side of the stern. The efficiency of the oars was not determined, as has been claimed, by human muscle power. Instead it was controlled with the help of a device similar to the present-day oarlock, the maneuvering of which required no major physical effort. The top part of the oar passed through the opening in the "oar box," and was rested in an oarlock device (view "c," Figure 82) which acted as a pivot, so that the blade of the oar could be slid upward in a vertical direction and also rotated on its axis. To the projecting top portion of the oar (the *clavus*) was attached a small jointed handle that was maneuvered like the handle of a crank (Figure 82). The fact that the blade was bisected by the axis of rotation, as in the modern balanced rudder, decreased the physical effort required to operate the rudder. However, this instrument was still very inferior to the sternpost rudder, which was the major contribution of the medieval period. We do not know by what means the Roman navigators determined position, or how they noted it on a map. In any event, thanks to these ships the Romans were able to reach the North Sea, the waters of western Africa, and (via the Red Sea) the Indian Ocean and the China Sea.

Ports   Whereas the Greek coasts offered numerous natural ports to ships, the western Mediterranean was rather poor in such natural harbors, and in most cases the Romans, once they had become a seagoing people, had to create artificial ports. They brought to this task the same technical qualities they had brought to the construction of roads and bridges. Thus they constructed a series of harbors that formed the port of Ostia at the mouth of the Tiber, particularly the large harbor of Trajan (now silted up), the construction of which was described by Pliny the Younger. Port facilities were also created at Civitavecchia, Terracina, Pozzuoli, and Messina in Italy, Cherchel and Leptis Magna in Africa, Trebizond, Alexandria in the Troad and Pompeiopolis in the East, and Fréjus in Gaul. The latter port (Figure 83) was dug out of the mainland to shelter the fleet that defeated Cleopatra at Actium; not until the seventeenth century did it become completely silted up and return to the state of arable land. The Romans did not know how to protect artificial harbors from the alluviums created by nearby torrents. On the other hand, their construction of wharfs was quite scientific; excavations in Vaison (Vaucluse) have uncovered the paving stones of the wharf on the Ouvèze River, which were embedded in oak piles driven into the bank and filled in with very fine, hard earth fill.

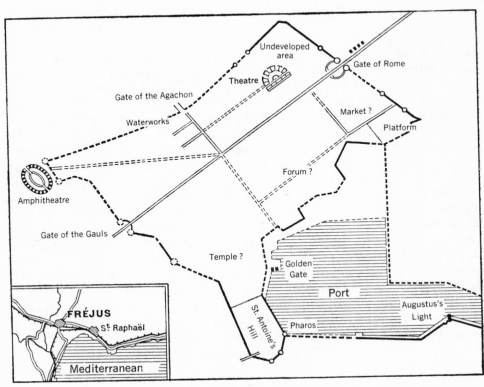

FIG. 83   The ancient port of Fréjus, now several miles inland.

Plate 32.
A ship in the port of Ostia. Bas-relief. Villa Torlonia, Rome.
Photo P.-M. Duval

Signaling and communication of news

These methods of transportation were accompanied by well-developed methods of signaling. Most ports had lighthouses, sometimes constructed (as at Ostia) on artificial islands formed by stone-filled platforms. The lighthouse at Boulogne in Gaul, constructed by Caligula in the first century, was a twelve-story octagonal tower corresponding to the lighthouse Dover in Britain; it was not destroyed until the seventeenth century. The lighthouse of La Coruña in Spain was equally famous. The light was supplied by burning resinous varieties of wood. Pliny the Elder states that the continuous light from these lighthouses might cause navigators to mistake them for stars; from this statement we may infer that the flashing light was not in use in the Roman period.

News was relayed overland by the system of a succession of fires built on hilltops, or from tower to tower by means of signal flags; we have no detailed information on the latter system. News circulated quite rapidly. During the Gallic Wars, for example, the news of the uprising of the Carnutes in 52 B.C. was transmitted by criers to Gergovia, 149 miles from Orléans, in a little over twelve hours. These vocal messages had thus traveled almost twelve-and-a-half miles an hour. Caesar describes how it was done (*De bello Gallico*, VII, 3):

"When an event of some importance occurs, the Gauls call the news from one field to the next and from farm to farm; step by step the message is received and passed on. So they did, and what happened at Cenabum [Orléans] at daybreak was known before the end of the first evening by the Arvernes, approximately 160 miles away."

As for the ordinary messenger traveling at maximum speed, in the year 238 the messenger who departed from Aquilea, in northeastern Italy, in order to announce to Rome the death of the Emperor Maximinus arrived in the capital on the fourth day.

TOOLS

Most of the tools used by the Romans had already been developed in the course of earlier ages. The Romans did nevertheless make several additions (Figure 84) to this stock of basic equipment.

The craftsman's equipment

While pivoted tongs were already known in the Second Iron Age, pivoted scissors made their appearance in the Roman period. These were far superior to the usual hairpin-shaped "shears" made of a curved blade of metal. (Improvements made in these shears, however, for specimens with a detachable blade have been discovered in Egypt.) Other innovations were the plane, the saw in a wooden frame, and the leather punch (known only from a funerary stele at Autun that dates from the second or third century). A painting now in the Naples Museum depicts a brace and bit. The drill gradually came into general use for sculpture in the second century, while the use of the trowel was already common. The appearance of the milling cutter and the helicoidal drill is generally dated from the Roman period. Credit for the invention of the three-pointed drill (French

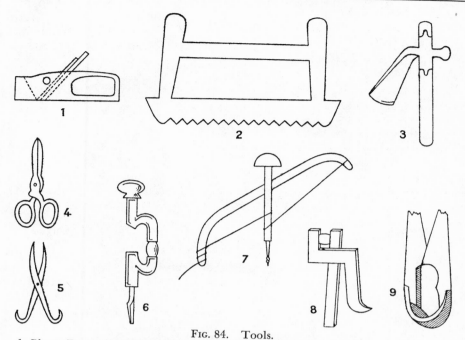

FIG. 84. Tools.
1. Plane (Pompeii). 2. Saw (Rome). 3. Adze *(ascia)* (Gaul). 4. Scissors (Rondsen, first century A.D.). 5. Scissors, (Priène). 6. Drill (from a painting at Naples). 7. Drill (Thebes, Egypt). 8. Punch (from a bas-relief at Autun). 9. Shears with interchangeable blade (Pompeii).

tarière) is generally given to the Gauls, because its Latin name, *taratrum*, appears to be of Celtic origin, and one classical author speaks of the *"tarière gallique."* If the screw and the nut actually existed, their use must have been limited, for archaeological investigation has uncovered only wooden pegs and nails, of all sizes and in large quantities. Before the screw and nut can be used on a large scale, the technique of mass-producing calibrated screw threads must first be learned. The screw is difficult to produce; it does not appear to have been used in carpentry, where the peg predominated. The rotating key (Figure 85) made its appearance together with the modern type of lock, replacing the bent, toothed (the so-called "Laconian") key of the preceding period. Its use was widespread; numerous museums have beautiful collections of keys, and some of them — keys for chests and caskets, for example — are lovely works of art. A specimen of a movable padlock is still in existence.

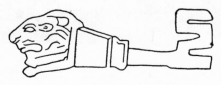

FIG. 85. Key decorated with lion head.

Plate 33.
Cutler's shop. Altar, first century. Vatican Museum.
Photo Giraudon-Alinari

Agricultural tools   Agricultural tools were also evolving in this period. The use of the spade became more widespread, although the hoe continued to be the principal tool. The plow underwent a major transformation in the moist plains of Celtic Europe with their heavy soil, grassy knolls, and long fields: the Mediterranean *aratrum* was replaced by a heavy wagon (this is the meaning of the Gaulish word *carruca,* whence the French word for plow, *charrue*) with a moldboard on which two wheels supported the front and the beam of the plow. From then on, the form of the plow was definitively established; it had a handle, a pole to which the animals were harnessed, an iron plowshare, sometimes two ears that served as a moldboard, and wheels. However, even in Gaul the plow did not supplant the *aratrum* until Merovingian times. Pliny the Elder, who mentioned the existence of the wheeled plow in Bavaria, attributes to the Gauls the invention of a kind of harvester that could do the work of a team of men armed with scythes (*Natural History,* XVIII, 296):

"On the vast estates of the Gauls, an enormous box with teeth, supported on two wheels, is moved across the crop by an ox that pushes the device; the heads of grain, torn off by the teeth, fall into the box."

This harvester is pictured on two funerary bas-reliefs discovered at Arlon and (in 1958) at Buzenol in Belgium (Figure 86) and on a caisson of the arch in Rheims known as the Porte de Mars. For threshing, the workers had a choice of the sledge, the roller, or the flail, all rudimentary forms of progress over simple treading.

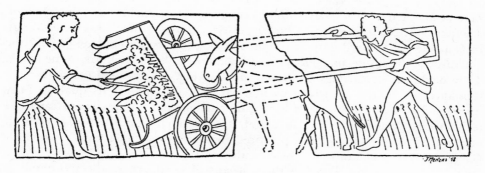

FIG. 86. Gallic harvester (from bas-reliefs at Buzenol and the Arlon museum, *Le Pays gaumais,* [1958], p. 129).

The various types of curved sickles constituted a complete series of precision instruments needed for harvesting grapes. A wooden vat was used to hold the wine — an invention, according to Pliny, of the Celts of subalpine Italy, who were fond of beer. These vats, hooped like modern barrels, were carefully constructed with the help of the adze (*ascia*), in a variety of sizes. In capacity and solidity they represented a great advance over the terra-cotta amphora, and enormously facilitated the transportation of wine, especially in the Rhineland. The amphora continued to be used, most often in Italy.

The direct-action screw press, which was an improvement over and soon replaced the beam press, also appeared in the Roman period. In the beam press, one end of a beam was hinged to the wall, while the other end was raised and lowered with the help of a windlass, or even a screw, fixed to a drum in the ground by means of an enormous, very heavy counterweight — a cumbersome and costly monolith. The crushed grapes or olives were placed in bags and pressed by the beam. More modest (particularly domestic) establishments used a small press without a beam but with crossbars fixed with wedges. After the appearance of the screw press, the typical large press was no longer the beam type with counterweights, but a direct-action screw press that bore directly down on the platform crushing the olives; the screws were activated by turning a bar.

AGRICULTURAL TECHNOLOGY

Agricultural technology made remarkable progress in the Roman era. Italy is a country of several rich regions, and the Romans were essentially a rural people that conquered the most fertile areas of the ancient world: Egypt, North Africa, Gaul. They were greatly concerned with agricultural techniques, and important treatises on this subject, which constitute a major portion of the Latin technical literature (see Bibliography), are still extant.

The Romans were the first people to breed cattle for the sake of fertilizer. They attempted to eliminate fallowing by rotation of crops, but did not completely replace the triennial system, which they may have practiced, by the biennial plan. They made systematic use of natural fertilizer — that is, manure — and were acquainted with niter fertilizer; liming was practiced in Gaul and Britain. The Romans, excellent arboriculturers, perfected grafting and the use of seedlings. In Gaul they developed grape varieties suited to the ocean climate, created the vineyards of the Bordeaux region, and crossed the banks of the Loire in the north. They made use of cuttings, grafting, and layering. They often added resinous substances to their wine, which by modern standards spoiled its quality. They stimulated agriculture by the draining of land, notably subterranean canals, a technique perfected in Italy. The manufacture of pitch by distillation of resin in urns buried in the ground was practiced in various Roman provinces, notably Gaul and Africa. The raising of oysters, already known to the Greeks, was remarkably developed, especially along the seacoasts of Gaul, where the technique of bedding them in the "open sea" was used. The preserves, pressed meats and cheeses of Gaul held no secrets for the Romans.

SURVEYING

These empire builders developed to perfection methods for measuring land. As the heirs of the surveyors' traditions of Alexandrian Egypt, in the days of the republic they already knew how to determine the principles of a rational and equal division of the land by means of the *groma*, a sighting instrument. They were thus able, for example, to divide, almost in a single operation, a major portion of Tunisia into equal sections. These have been discovered in the last few years, thanks to aerial photography. The systematic use of surveying to fix tax rates throughout the empire led to the perfection and widespread use of the surveying method, which consisted basically of dividing the land into rectangular or square plots along each side of two main perpendicular axes — the *cardo*, oriented by and large toward the north, and the *decumanus*. While surveying is of Eastern origin, the intersection of two axes was probably a Roman idea. The results were engraved on marble tablets and posted in public places in the cities. Hundreds of fragments of several surveys, the most important of which was made in 77 A.D., were discovered in 1949 in the city of Orange in southern France; they indicate the land division in the environs of the colony of Arausio. Not until the advent of the Roman civilization was surveying developed to such a peak of perfection and used over such wide areas of land.

MEASURING

The principal Roman instrument of weight measure was apparently the balance with one platform and one weight suspended from the unequal arms of a graduated index. This is the *statera*, or "Roman scale," as contrasted with the much older, two-platform *libra*. (The word "Roman" is in this case, however, a derivation via Provençal and Spanish of the Arabic name for the balance and weight, and this instrument is also the typical Chinese scale.) It was made in all sizes, for the goldsmith or minter as well as for the seller of wood; some of them — for example, the balance from Antium (Porto d'Anzio) preserved in the Cabinet de France at the National Library in Paris — had different systems of markings on various faces of the index and several suspension hooks, so that the weighing capacity could be varied; thus objects varying from one to three pounds, four to ten pounds, ten, twenty, or thirty pounds, could be weighed on the same instrument. These devices have been in existence for centuries, right down to modern times.

The units used by the Romans did not permit them to perform minute calculations; the lightest weight they possessed was the "scrupule" of 1,137 grams. However, this did not prevent minters from making even smaller calculations, probably by means of double weighing. Similarly, on instruments for measuring length (for example, the "folding foot rules" of the masons and carpenters, more than one specimen of which has been found), the smallest unit was the "finger" (1.85 centimeters), but this did not prevent them from making fine joinings on pieces of furniture or marble facings. These "foot rules" were often graduated in Roman units and (in Gaul, for example) in local units of measure on the reverse side.

Measures of weight were based on the duodecimal system of the pound (*libra, pondus*) of 327.45 grams, of which the "ounce" (*uncia*) was the one-twelfth part and the *scrupulum* the 288th part. The unit of length was the foot of 29.57 centimeters, of which the *uncia* was the one-twelfth part, the "finger" one-sixteenth. Five thousand feet were a "mile" (1,478 meters), corresponding to one thousand perches of five feet (1,478 meters). The monetary unit was the *as*, of which the *uncia* was one-twelfth; the denier was equal to 16 *as*. Thus the units of 12 and 16 were commonly used. Local units of measure were also used.

Plate 34.
A pharmacist's shop. Altar. Épinal Museum
Photo Giraudon-Alinari

MACHINES

a. Lifting machines   As far as machinery, the principal area of technology, is concerned, Rome appears to have lived off her Hellenistic heritage, a portion of which can be studied in the treatise of Vitruvius. In the field of lifting and hydraulic machines, at the very least, the machines described in his *De architectura* are combinations of five elements distinguished by the Greek mechanicians: the wedge, the lever, the block with several pulleys, the windlass, and the screw. The lifting machines described by Vitruvius were probably very recent models; they were based on the windlass or the axle, the block, and the large wheel turned from inside by human muscle power; this wheel was also used in hydraulic machines.

The use of the lifting wheel (noria) for raising water was probably known on the Mediterranean coast before the beginning of the Roman expansion. The water pump, however, seems to have been invented by the Romans. Philo's treatise contains a description and drawing of a two-cylinder force pump (Figure 87), and a few remains of pumps found in excavations bear a strong resemblance to the one he describes. The barrel of the pump was cut from a block of wood, and contained several vertical, cylindrical chambers. Two of these were the actual cylinders of the pump, and probably contained pistons. Openings with valves permitted the water to be drawn in and then forced into the other chambers, which were the conduits; the latter opened into a kind of distribution chamber with an opening to which a lead pipe was probably fitted. These pumps were used in fire fighting, and undoubtedly also for drawing water from wells into reservoirs.

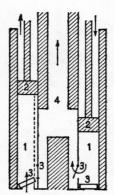

FIG. 87. Two-cylinder force pump for pumping water.
1. Cylinders. 2. Pistons. 3. Valves. 4. Discharge outlet.

b. Motive machines   Few improvements had been made in the animal-powered mill; its results were not particularly spectacular, but it was nevertheless an improvement over the hand mill, the only type known in classical Greece. The water mill, which produced energy, was already known to Mithridates, and is described by Vitruvius. The Romans improved on it by replacing the horizontal wheel on a vertical shaft by a vertical wheel with a gear train. The result was a major improvement over the old system, and the new mill, used for flour and oil, spread through the provinces during the empire; in Rome, it was operated by water supplied by the aqueducts. This hydraulic mill

Plate 35.
A lifting machine. Bas-relief. Lateran Museum, Rome.
Photo Giraudon-Alinari

was a valuable acquisition. In Gaul its invention was often attributed to local saints — Orens of Auch, who created the mill at the lake of Isaby (fourth century), and Césaire of Arles, who utilized the waters of the Durançole at Saint-Gabriel (sixth century). However, the oldest example discovered in the Western world is the Barbegal mill near Fontvieille, in the region of Arles, in which the waterfall from an aqueduct turned vertical paddle wheels that transmitted power through a horizontal axle to a small vertical wheel geared into a toothed wheel; the shaft of the latter turned the millstone. Two series of eight mills built in receding tiers spread over an area of about sixty yards were in operation at Barbegal (which was probably a state monopoly), the ruins of which can be dated from the fourth century. An anonymous author writing at the end of this century describes a paddle boat that must have been inspired by the one described by Vitruvius for measuring distance (see above) and that probably did not get beyond the papers of this "idea-man of the Late Empire," as Salomon Reinach calls him.

c. Military machines Military machines were also a heritage from the Hellenistic past, but the Romans made such distinctive use of them that, here again, their perfections acquired the nature of genuine inventions. In the days of the Late Empire the name *mechanici* was applied only to the manufacturers of these machines (the two other categories of manufacturers being, in the opinion of the fourth-century writer Pappos of Alexandria, the manufacturers of lifting machines and the manufacturers of pumps, screws, and other machines). The propulsion machines, which still operated on the principle of the torsion of a set of ropes, are listed by Vitruvius as: the ballista, transported by wagon, and capable of hurling weights of 250 pounds to a distance of approximately 150 yards; the catapult, a kind of bow for hurling arrows, transformed, however, into a machine and operated by means of a tourniquet; and the scorpion or onager, of which Ammianus Marcellinus (fourth century) gives a good description. Based on the principle of the sling, the scorpion was basically a piece of wood that, when pulled down in front, imitated the propelling action of the human arm, and hurled balls of stone. Still on the subject of propulsion weapons, around the second or third century a hunting bow with a blocking device was invented; two bas-reliefs of this period from Le Puy (Figure 88) are the first known pictures of this device. Its power is equivalent to that of the crossbow, but it lacks the crank that makes it possible to increase the power of the human arm; however, Vegetius, writing in the fourth century, uses the word *arcuballista* for it.

As for battering weapons, in addition to the classic ram Vitruvius and Vegetius mention the beam with a scythe on one end, used to tear stones out of rampart walls, and the pointed beam, used to punch holes in the walls.

METALLURGY

It is now an established fact that, beginning in the second and third centuries, the Romans became acquainted, through their auxiliary forces of barbarian cavalrymen, with the long, damascened, wrought sword with very hard blade and soft core (the chief qualities of the Merovingian sword) manufactured in

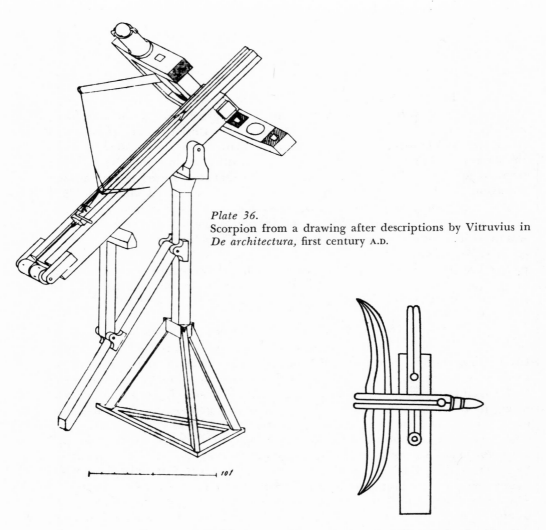

Plate 36.
Scorpion from a drawing after descriptions by Vitruvius in *De architectura,* first century A.D.

FIG. 88.   Bow with locking device and quiver.
Bas-relief from Le Puy.

workshops on the Rheno-Danubian frontiers. The method of partially steeling iron had been known since the second century B.C. In the barbarian swords of the Roman age, the case-hardened iron cutting edges were soldered onto a soft iron damask core, which was made of several strips twisted and soldered together (Figure 89). The steel was probably obtained, not by alloying iron and carbon, but by the exploitation of an ore of exceptional quality, such as the variety found in Noricum and Carinthia; Pliny mentions the so-called "seric" and "Parthic" grades of iron as being the only ones that contain "only pure steel, for all the others contain a mixture of softer iron. Here and there in the Roman world can be found a vein that supplies iron of this quality, for example in Noricum" (*Natural History,* XXXIV, 145). This is probably natural steel obtained from ores rich in manganese, not iron prepared by case-hardening to produce steel.

On the whole, the metallurgical industry did not make very remarkable technological progress during the Roman Empire. It continued to be limited by the combustible used — charcoal. The Romans nevertheless succeeded in separating

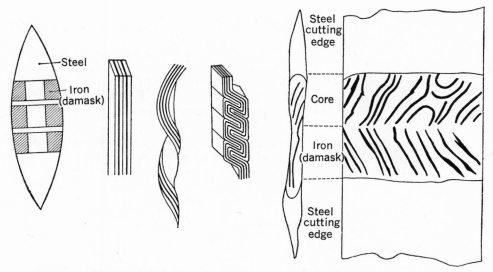

FIG. 89. Cross sections and diagrams of damascened sword, with stages in production of the damask (in the center) (from E. Salin, *La Civilisation mérovingienne*, Part III).

gold from copper ore and silver from lead, in perfecting the technique of gold leaf, and in obtaining mercury (particularly in Spain in the first century). Some progress was also made in the exploitation of mines, thanks to the improvement of methods of wheel pumping, and it was probably in the Roman Age that the bellows was perfected, the effects of which instrument were to become apparent in the early Middle Ages (Figure 90); a bellows was added to the lower portion of the furnace for the reduction and fusion of the ore, by the addition of nozzles.

A very special technique was the tin-plating of copper and silver, described by Pliny the Elder in his remarks about "white lead," that is, tin (*Natural History*, XXXIV, 167):

"White lead, if it is not alloyed, is good for nothing, but copper is plated with it in such a way that only with the greatest difficulty can it be distinguished from silver. This is an invention of the Gauls, and its end products are called *incoctilia*, that is, tinned metals. In time silver also came to be tin-plated, for use in embellishments for horses, draft harnesses, and still later for carriages themselves. This process was used for the first time at the *oppidum* of Alesia; the Biturigi shared the credit for it."

As for fibulae, which were often enameled (another specialty of the Gauls), continuing improvements were brought about by the substitution of the hinged fibula for the spring type made with a single brass wire, and by the creation of

FIG. 90 A fourth-century forge, as pictured on the sarcophagus of Saint-Aignan (Aveyron).

the buckle, the hinged clasp, and (after the third century A.D.) the cruciform fibula, which continued to be made until the fifth century (Figure 91). The fibulae were mass-produced, in the first century, in northern Gaul, especially in the workshops of Master Aucissa.

Aside from these novelties, forge equipment — anvil, hammer, the bellows — was still primitive, and was to remain so for a long time to come.

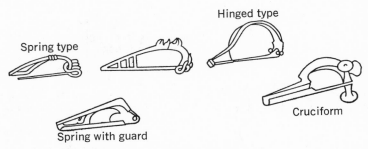

FIG. 91.   Gallo-Roman fibulae.

TEXTILES

In the area of textiles, no innovations were made in spinning, weaving, or milling, except for the use of the weaver's comb for separating the threads, which appeared in the Late Empire. Wool, canvas, and cotton continued to be the principal fabrics in use; felt was known. But the style of garments gradually changed, as the Romans installed themselves in the West in countries colder than their own. In Gaul and Germania they learned the use of furs and (particularly in Gaul) the principle of making garments from individual assembled pieces; breeches that later became trousers, short- and long-sleeved shirts and tunics, long and short hooded mantles, mufflers, woolen and cotton stockings and socks, the caped cloak, and laced boots, adopted by the Romans and soon by the entire western portion of the empire, inaugurated the reign of "European" dress. The garments, cut out with shears and sewn together with bone or metal needles, were closed by fibulae (buttons, usually of the shank variety, seem to have been used only as an ornament, not as the customary method of fastening). It is not known whether knitting existed.

MISCELLANEOUS INVENTIONS

Several novelties made their appearance in the area of daily life. There was, for example, the wax candle with a wick of vegetable fiber, papyrus, tow, or rush, which was found to be better than the tallow candle. "Soap" (its name, *sapo*, is Gaulish) appears to have been originally a hair dye, and was introduced into Rome as a cleaning agent only in the fourth century. It was at first a mixture of tallow, ashes, and probably a liquid made from red herbs, and had healing virtues (for example, for scars on the face and to prevent itching); it later came to be used for cleaning the skin and washing clothes.

The Gallic wicker armchair with high, rounded back is the ancestor of our rattan chairs; wickerwork had by then achieved a high degree of perfection. In Roman Gaul, again, the four-legged dining table, covered with a fringed table-cloth around which the guests sat, replaced the half-moon table and the dining couch of the Italians; the latter, on the other hand, created the "work-couch" for intellectual work, and, for sleeping, the bed with a flexible spring made of a metal latticework. The furniture industry in general made great progress in this period.

Hunting, the favorite occupation of the upper class, continued to develop without any striking innovations, but the Romans adopted the Gallic dogs, especially the greyhound (*vertragus*), for hunting.

An apparent innovation in the theatre was the curtain, at least that type of curtain of which indications have been found in the ruins: it was a piece of fabric, attached to posts, which was raised from a ditch in front of the stage to form a screen. The stretching of canvases (*vela*) over theatres, amphitheatres, and even circuses to protect the public from the rain or the hot sun was also a Roman idea. These edifices were crowned with perforated brackets for the poles that supported the *vela*; in Rome ship's rigging was used to raise these *vela*, but exactly how they were raised and supported at such a dizzying height with the help of ropes fixed at the bottom remains a mystery.

COMMUNICATION OF IDEAS

The book, or *codex*, composed of individual sheets joined together was the successor to the excessively long and not very practical papyrus *volumen*. For writing on the papyrus (parchment after the end of the third century) pages of this *codex*, ink made by diluting a solid stick of lampblack was used. The date of the oldest extant specimen is uncertain; some authorities date it from the first century. In any case the new form, like parchment (which was already in use in the third century B.C. at Pergamum, whence its name), was used at first for works consisting of numerous volumes; the Christians quickly came to prefer it for the Holy Scriptures, and thus appear to have greatly contributed to its popularization. By about the end of the third century, it had completely supplanted all earlier methods. In addition to being smaller and easier to handle than the roll, the fact that the parchment had a smoother surface and much greater strength made its contribution to the progress of writing all the greater.

Another, less important, innovation also encouraged the communication of ideas: the creation of a Latin-language stenography. This was definitely already in existence at the time of Cicero (who used his slave Tiro as his stenographer), and improvements were later made to the system, particularly in the imperial administration, where the official stenographers (*notarii*) became important functionaries during the Late Empire. The system is not known to us; its principle probably dates back to the Greece of the fourth century B.C., but it did not come into general and official use until the Roman Empire. The poet Ausonius (fourth century A.D., preceptor of the future Emperor Gratian) addressed his private stenographer in the following terms (*Ephemerides,* 7): "Slave, skilled minister of rapid writing, come here quickly. Prepare the two sides of your tablets, where

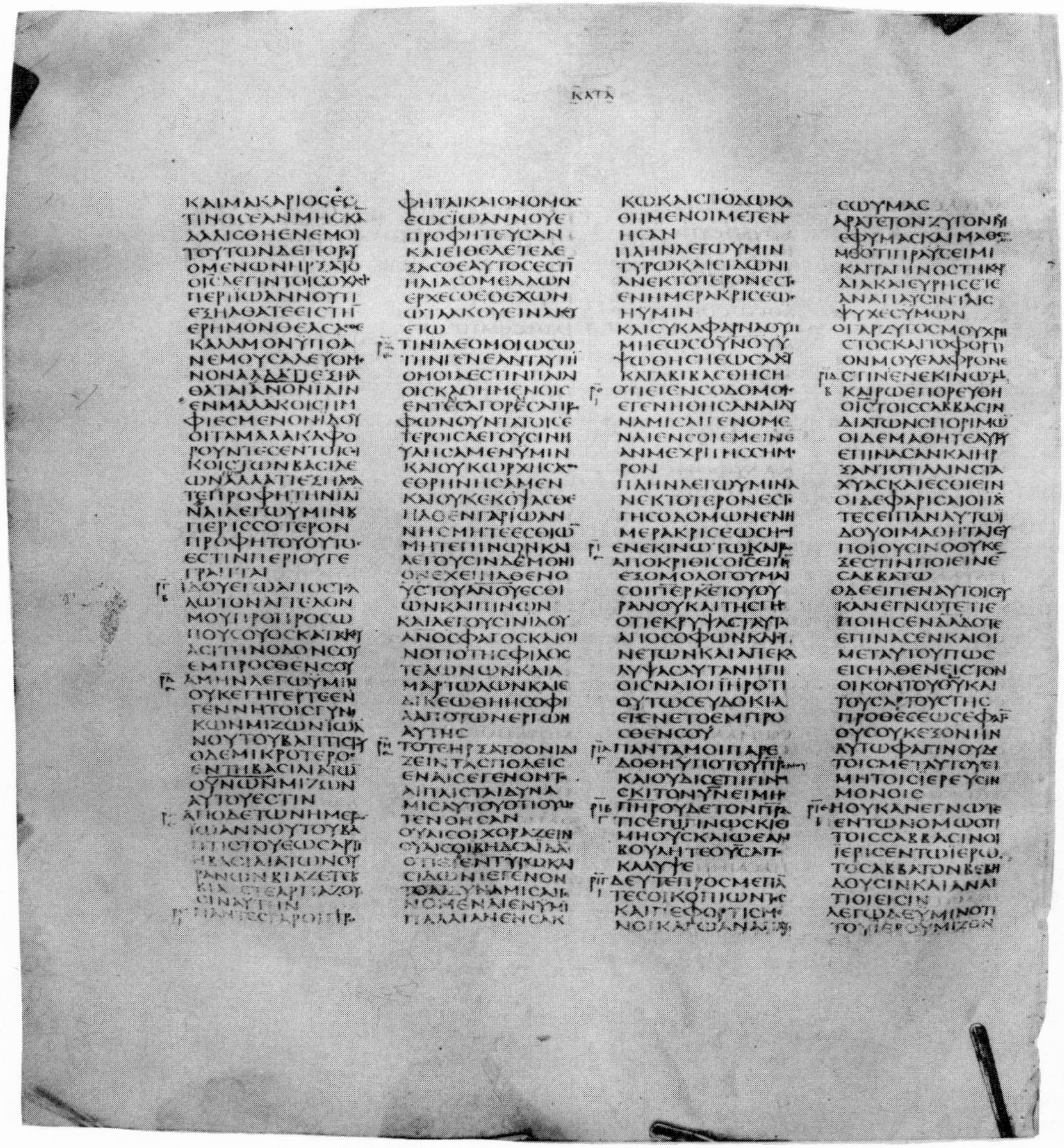

Plate 37.
Manuscript page from the Codex Sinaiticus, A.D. 400
British Museum, London. *Photo British Museum*

a great number of words, expressed in a few signs, combine as if they formed a single word. . . . Torrents of words roll off my tongue, but your ears do not err, and your hand, sparing in its movements, flies over the surface of the wax."

SUMMARY OF THE ROMAN CONTRIBUTION

Although in the area of science Rome lived off an already ancient Hellenic heritage (for little scientific progress was made after the second century B.C.), the least we can say is that in the technological domain she demonstrated a rare power of accomplishment and a remarkable power of dissemination. Rome's genius is that she perfected still-rudimentary techniques for the day-to-day use of an immense empire and that she did not hesitate at any collective effort; the expression "a Roman project" is still used to designate an almost superhuman accomplishment, a technological success far beyond ordinary achievements.

As regards its effective popularization, the role of invention was a weak one, and can be summarized in a few pages, while a treatise on the techniques of the same period would easily fill several volumes that would form a summary of the techniques of antiquity. It is nevertheless a fact that the practical application of earlier inventions and their perfecting for general use, which resulted in maximum efficiency and rapid and generally economical utilization, are more important in the history of technological development than an isolated series of inventions. The essential role played by Rome is that she put at the disposition of numerous peoples techniques that until then had been but poorly developed.

Three groups of techniques may be regarded as the contribution of the Roman era: those the Romans inherited from the Hellenistic civilization and perfected by increasing the number of uses to which they could be put; those they borrowed from the barbarian peoples and introduced unchanged into Latin, Western civilization; and those whose actual invention can be attributed to them.

Romans perfected: cement masonry, which permitted them to make innovations in architecture by creating various types of edifices and by developing the vault, the cupola, the staircase, the bridge, the aqueduct (with siphon), and the sewer; tiling, roofing, and the hypocaust; roads, ports, surveying, the rudder; mosaics, molded pottery, blown glass, fibulae; the water mill, military machines; arboriculture, particularly viticulture; oyster cultivation; the use of natural fertilizers and, to a limited extent, fallowing.

They borrowed: most types of vehicles; possibly the semirigid sail; the plow and the harvester with wheels; the barrel; garments made by assembling individual pieces of cloth; soap; polychrome enamel work; the damascene sword; the wicker armchair.

Most of these items were borrowed from the West — from Celtic Europe, rich in metals and in industrial and craft traditions, and especially from Gaul. On the other hand, they acclimated in the West the chestnut, peach, and apricot trees.

They invented: the vaulted cupola; the windowpane; the milestone; the bridge aqueduct; the balance with weights(?); the direct-action screw press; the bow with a stop device; a certain number of tools: the scissors, the plane, the frame saw, the punch, the borer, the brace and bit, the door key, the weaver's comb; the wheel pump for mines; the bellows with nozzles; the use of chemical fertilizers; the wax candle; the metal bedspring; the theatre curtain and the *vela;* the *codex;* Latin stenography.

This summary would have to be fleshed out with a study of the degree of development achieved in each technique, and especially by an assessment of the technological civilization of the Roman Empire. Did the Romans make full use of the methods at their disposal? Were their technical inventions developed to the fullest extent possible? The powerful and complex Roman administration apparently did not include any office charged with the mission of promoting and coordinating research and development for purposes of technical progress, and encouraging the advancement of science. The existence of the craft guilds, which were, moreover, very powerful, ensured at most the handing down of traditions and procedures. Given the Romans' ignorance of statistics and the absence of any idea of demography, technical progress was achieved empirically rather than intellectually. Rome was remarkable in construction and major public works, thanks to the vast opportunities for action offered to their activity, inventive spirit, and the emulation of the engineers and architects. She was notable in the crafts, thanks to the spread of tools, skill, and the contribution of the provincial experience and traditions. The Romans were weak, however, in the utilization of mechanization for industry and the application of science to the development of machines and of scientific knowledge to agricultural production.

A major portion of the knowledge accumulated by the Greeks relative to machines remained buried in theoretical treatises. The existence of slave labor partly explains this early lag in mechanization. Owing to the lack of major drainage and irrigation projects, and the lack of an agricultural policy based on technical experiments and research and applied over large areas, agricultural progress was concentrated, not on large-scale production of the cereal grains, but especially on achieving splendid successes in arboriculture. The dissemination of the olive tree and especially the grape are the most spectacular. In industry no efforts were made systematically to improve machinery or to create factories. Only a handful of large landowners established isolated craft enterprises that consisted of individual workshops or (in the Late Empire) were able to group together on one estate all the industries necessary to its management. This was the complement of industrial decentralization, which was encouraged by the existence of a multitude of "working" provinces, but the extreme skill demonstrated by the engineers in the field of public works was never channeled into the problem of increasing production.

Thus the Roman Empire was no better able than the ancient empires of the East, the Greek democracies, or the Hellenistic monarchies, to encourage an agricultural revolution that would have made it permanently self-sufficient or an industrial revolution that would have led to the birth of mechanization. Classical antiquity witnessed the birth of machines, but mechanization had to await the coming of another era.

BIBLIOGRAPHY

I. GENERAL WORKS

CARCOPINO, JEROME, *Daily Life in Ancient Rome* (New Haven, 1940).
DE CAMP, L. SPRAGUE, *The Ancient Engineers* (Garden City, N.Y., 1963).
DILL, SAMUEL, *Roman Society from Nero to Marcus Aurelius* (New York, 1959).
FELDHAUS, FRANZ M., *Die Technik der Antike und des Mittelalters* (Postdam, 1931).
FORBES, R. J., *Studies in Ancient Technology* (11 vols., Leiden, 1955–).
FRANK, TENNEY, *Economic History of Rome to the End of the Republic* (Baltimore, 1920).
———, *An Economic Survey of Ancient Rome* (Baltimore, 1940).
NEUBERGER, ALBERT, *The Technical Arts and Sciences of the Ancients* (New York, 1930).
ROSTOVTZEFF, M., *Social and Economic History of the Roman Empire* (Oxford, 1926).
SINGER, CHARLES, et al., *A History of Technology*, Vol. II (London, 1956).

II. SOURCE MATERIALS

Over 450 classical Greek and Latin works have been translated into English in the Loeb Classical Library (Cambridge, Mass., Harvard University Press). In addition, the following volumes are of special interest in the history of technology:

FRONTINUS: *The Two Books on the Water-Supply of the City of Rome by Sextus Julius Frontinus* transl. by Clemens Herschel (London, 1913); *The Stratagems and the Two Books on the Water Supply of Rome,* transl. by Charles E. Bennett (Cambridge, Mass., 1950).
VITRUVIUS, *The Ten Books on Architecture,* transl. by Morris Hicky Morgan (New York, 1960).

III. BUILDINGS AND PUBLIC WORKS

ASHBY, T., *The Aqueducts of Ancient Rome* (Oxford, 1935).
BLAKE, MARION ELIZABETH, *Ancient Roman Construction in Italy from the Prehistoric Period to Augustus* (Washington, 1947).
———, *Roman Construction in Italy from Tiberius Through the Flavians* (Washington, 1959).
BROWN, F. E., *Roman Architecture* (New York, 1961).
BURR, WILLIAM, *Ancient and Modern Engineering and the Isthmian Canal* (New York, 1902).
FORBES, R. J., *Notes on the History of Ancient Roads and their Construction* (2nd ed., Amsterdam, 1964).
MACDONALD, WILLIAM L., *The Architecture of the Roman Empire* (New Haven, Conn., 1965).
PACKER, JAMES E., "Structure and Design in Ancient Ostia," *Technology and Culture,* 9 (1968), 357–388.
STRAUB, HANS, *A History of Civil Engineering* (Cambridge, Mass., 1964).
VAN DEMAN, E. B., *The Building of Roman Aqueducts* (Washington, 1943).
WHEELER, MORTIMER, *Roman Art and Architecture* (New York, 1964).

IV. INDUSTRIAL ARTS

CHARLESTON, R. J., *Roman Pottery* (London, 1955).
GOODMAN, W. L., *The History of Woodworking Tools* (London, 1964).
MERCER, HENRY C., *Ancient Carpenters' Tools* (Doylestown, 1950).
NEUBURG, F., *Glass in Antiquity* (London, 1949).
RICHTER, G. M. A., *The Furniture of the Greeks, Etruscans, and Romans* (London, 1967).
SINGER, CHARLES, *The Earliest Chemical Industry* (London, 1948).
WHITE, K. D., "Technology and Industry in the Roman Empire," *Acta Classica: Proceedings of the South African Classical Association,* II (1959), 78–89.

V. FOOD AND AGRICULTURE

FUSSELL, G. E., "Farming Systems of the Classical Era," *Technology and Culture,* 8 (1967), 16–44.
GILFILLAN, S. C., "Roman Culture and Dysgenic Lead Poisoning," *The Mankind Quarterly,* V, 3 (Jan.–March 1965), 123–48.
HAWKES, CHRISTOPHER, "The Roman Villa and the Heavy Plough," *Antiquity,* IX (1935).
HEITLAND, W. E., *Agricola: A Study of Agriculture and Rustic Life in the Graeco-Roman World* (Cambridge, 1921).
MORITZ, L. A., *Grain-Mills and Flour in Classical Antiquity* (Oxford, 1958).
WHITE, K. D., *Agricultural Implements of the Roman World* (New York, 1967).

VI. ARMAMENTS AND WARFARE

See, in addition, appropriate listings in Bibliography for Part II, Chapter 2, "Greek Technology."
RODGERS, WILLIAM LEDYARD, *Greek and Roman Naval Warfare* (Annapolis, 1937).

VII. MINING AND METALLURGY

DAVIES, OLIVIER, *Roman Mines in Europe* (Oxford, 1935).
RICKARD, THOMAS A., *Man and Metals* (2 vols., New York, 1932).
TYLECOTE, R. F., *Metallurgy in Archaeology: A Prehistory of Metallurgy in the British Isles* (London, 1962).
WHITTICK, G. C., "The Casting Technique of Romano-British Lead Ingots," *Journal of Roman Studies,* LI (1961), 105–111.

PART THREE

SOUTHERN AND FAR
EASTERN ASIA

CHAPTER 10

THE TECHNIQUES OF THE ANCIENT FAR EAST

THE HISTORY of the development of technology in the Far East is not well known, and a full discussion of it would require space and illustrations that are not available to us. We have therefore limited our discussion to China, and to the most basic elements within this already narrow limitation. However, a few references are made to Japan, whose technology will be discussed separately in the second part of this work. Though like all the premechanistic civilizations, even in its most technical activities the Chinese civilization made use of subjective, magic, or moral concepts, our intention is simply to summarize its technological development.

In very early times the geographical area of China was the setting for numerous cultures — Maritime, Northern, Southern, Southwestern, Northwestern — which differed little from the neighboring cultures (Indonesia, Tungus, Tibetan, and Turkish). Suddenly, in the Upper Neolithic period, a superior Chinese civilization appeared in the great northern plain, and has continued to exist down to modern times, without, however, eliminating the less highly developed neighboring cultures. We have not discussed the technology of the non-Chinese peoples on Chinese territory, and have limited our work to a very rapid examination of specifically Chinese techniques, in the following order:

General development of Chinese techniques from early times to the six-
teenth century
Metallurgy and mining
Chemistry
Agriculture
Textiles
Techniques for the expression of ideas
Housing
Time and space techniques
Military techniques

GENERAL DEVELOPMENT OF CHINESE TECHNOLOGY
FROM EARLY AGES TO THE SIXTEENTH CENTURY

The beginnings of Chinese technology Three legendary heroes symbolize the early stages of Chinese technology. Fu Hsi, the hunter, is supposed to have invented cooking and the eight trigrams (*pa-kua*), into which all the elements of the cosmos can be classified and related to the primordial forces of nature. He represents a preagricultural stage of the Chinese civilization, of which he is regarded as the founder. His successor, Shen Nung, the "divine worker," was in ancient China the god of fire who transforms the jungle into arable land by means of fires. He taught the people to trace furrows in the soil with a wooden plow, and to sell the produce of the fields in markets. He also discovered the healing plants, and for this reason was honored as the god of medicine; the writing of the first *Pên-ts'ao (Materia medica)* is attributed to him.

Huang-ti (the "Yellow Emperor") ruled over the one hundred clans of the Hsia (the proto-Chinese) people. To him is attributed the invention of cloth weaving, writing, ox and horse harness, river navigation, the breeding of silkworms, and surveying. He ordered the construction of houses, palaces, and temples, and coined money; he also divided the country into provinces and districts. From his dialogues with his councilors (the chief among these being C'hi Po) resulted the *Canon of Medicine (Nei-ching)*.

These three mythical rulers were followed by the almost historical dynasties of the Hsia (twenty-first to sixteenth centuries B.C.), Shang (sixteenth to eleventh centuries B.C.), western Chou (eleventh century to 771 B.C.) and eastern Chou (720–249 B.C.).

    The Shang period   We know very little about technology during the Hsia period. The Shang period was characterized by a mastery of bronze that produced bronze objects unequaled anywhere else in the world. During this period, also, the cultivation of millet and wheat encouraged cattle raising. Chariots were pulled by horses and oxen; elephants were used in warfare. Artisan-slaves supplied the labor for an industry of luxury items oriented toward silk and jewels. State supervision, already well developed, stimulated important technical and scientific activity. The government employed and supervised geomancers, soothsayers, alchemists, mathematicians, doctors, and astrologers. The latter observed eclipses of the moon (1361 B.C.) and the sun (1261 B.C.). Pottery had already reached a highly skilled stage of development.

    Chou dynasty
 (eleventh century B.C. –
     249 B.C.*).* Under the Chou rulers agriculture made great progress. The number of species of plants under cultivation was increased. The utilization of iron and founding made it easy to cast a large number of tools and instruments for agricultural and artisanal purposes, and to create coinage. Large cities with more than 70,000 inhabitants existed, and roads facilitated interurban traffic at all hours of day and night. Hydraulics engineers harnessed rivers by means of major projects, some of which are still in existence. The Chou potters invented stoneware, an extremely durable, completely vitrified clay.

The first Chinese astronomical instruments, the style (gnomon), the template for finding the Pole Star, and the water clock (clepsydra), all date from this period. Human labor was made easier with the help of machines, several of which were invented by Lu Pan, a famous carpenter and possibly a contemporary of Mo Ti (circa 480–397 B.C.). Tsui Liang (circa 386–334 B.C.) built water mills whose yield was ten times greater than that of hand-operated machines. Toward the end of the western Chou period, the use of lands cleared by brush fires was gradually replaced by that of permanent fields, some of which remained the property of the village collectives, while the others were given to the nobility; Chinese agriculture thus developed from the practice of temporary clearing to that of cultivation by fallowing at regular intervals (H. Maspéro).

The first imperial dynasties (Ch'in to South-and-North dynasties, third century B.C. to sixth century A.D.)

Ch'in Shih Huang (221–210 B.C.) unified the systems of weights and measures, the monetary system, and the sizes of chariot axles, and laid out the network of "imperial roads."

The Han period surpassed the level of its predecessors in volume of exploitation as well as in the methods used. By the beginning of the Han Dynasty (211 B.C.), a powerful industrial and commercial class was amassing enormous fortunes through the manufacturing of armaments, gold objects, cloth and embroideries, dyes, and lacquered objects. It exploited iron and bronze foundries, saltworks (both sea salt and rock salt), and mines. Deep drilling was done with piston bellows. In this period the toothed wheel appeared as a device for the direct or indirect transmission of energy. The oldest known gear train appears to be the one represented by a terra-cotta mold used for casting bronze ratchets which had sixteen teeth and a square arm with a hole for the shaft (J. Needham). Achievements in the field of astronomy included the sighting tube, the azimuthal sundial, first the equatorial and then the ecliptic armilla, and the armillary sphere.

During the Eastern Han period (25–220 A.D.) protoporcelain was made, and hydraulic energy was used to power the bellows in iron foundries. Paper was invented in 105 A.D. by Ts'ai Lun (died in 121), and made possible an embossed reproduction of texts engraved on stone steles; in this way reverse images with white characters on a black ground were obtained. Later (in the sixth century), the reverse engraving of texts on blocks of wood made it possible to obtain a normal image by xylography (Paul David).

Rhinocerous-hide armor and harness gradually disappeared from the various provincial armies that had contact with the Barbarians; they were replaced by leather or metal scale armor and a more mobile and more efficient cavalry that made use of the stirrups, the saddle, and the bit. General Chu-ko Liang (181–234) invented the twenty-arrow crossbow, and a light vehicle (the mechanical wooden ox and horse), drawn by human beings, for the transportation of food.

Ma Chün (Three Kingdoms period, 220–265) constructed the first south-pointing carriage, a mechanical, nonmagnetic device in which a hand always pointed to south no matter what the direction of the chariot (Li Shu-hua). This is the first example of a differential gear train. Ma Chün also perfected the catapult and the weaving loom. Sun Chüan, King of Wu, may have had vessels capable of reaching the South Seas.

Coal began to be employed in northern China, natural gas in Szechuan.

The Sui and T'ang dynasties (581–907)

During the Sui Dynasty vast public works were begun. Two million workers built the city of Lo-yang, and a million others were assigned to the repairing of the Great Wall and the digging of the Grand Canal. A system of messengers and postal relays was established. The imperial administration established a large number of plants for the manufacturing of brocades, felts, carpets, pottery, and metal objects. Iron chain bridges and single-arched stone bridges (sixth century) were built. (The bridge of Chao-Chou in Hopei Province is still in existence.) White porcelain with feldspar glaze was an important achievement of this period (D. Lion-Goldschmidt).

The T'ang period is partly contemporary with the Nara period in Japan, whose greatest emperor was Shômu-tennô (724–741). To contain her collections of artistic and scientific objects the Dowager Empress Komyô ordered construction (in 756) of the famous wooden building called the Shôsoin, or Imperial Treasury of Nara, which houses millions of objects — furnishing, fabrics, musical instruments, *objets d'art,* medicines, and so on — from China and other Asian countries.

The Sung Dynasty (960–1279)

Metalworking in gold, silver, copper, tin, lead, and particularly iron flourished during the Sung period, thanks to the widespread use of coal. A new textile, kapok, became popular. Huang Tao-p'o perfected cotton weaving in Kwangtung (end of the thirteenth century). The perfect mastery of silk weaving led to the popularization of paintings woven in silk. Xylographic printing was perfected, and printing with movable clay type was invented. For the first time there appeared libraries with more than 100,000 volumes.

The porcelain industry achieved a high degree of perfection, owing to the high temperature of its kilns and the use of glazes and special types of clay. Long-distance navigation became possible, thanks to the use of sea charts and celestial maps, ships with watertight bulkheads, and the magnetic needle. Numerous technical treatises were written on agriculture, legal medicine, strategy, architecture, and other topics.

The Yüan Dynasty (1260–1367)

Technology continued to be perfected during the Yüan period. Great progress was made in the crafts through the creation of craft centers in which Mohammedan and Christian prisoners of war worked for the state; in this way the technique of enamelwork was introduced into the Far East. Egyptian experts arrived to perfect the sugar industry. Conversely, Chinese artisans were sent to central Asia to work in war industries. In scientific technology Western engineers, astronomers, and doctors introduced new ideas.

Kuo Shou-ching (1231–circa 1316), the greatest hydraulics engineer of the time, made improvements in the Grand Canal, and completed the calendar known as the Shou-shih-li, which contains the best calculation of the year made before the arrival of the Jesuits. This led him to invent several methods of chronometry and to perfect the existing astronomical instruments.

The Ming Dynasty (1368–1644)

The progress made in naval construction and navigation (Fukien and Kwangtung) permitted Cheng Ho to reach East Africa. Progress was also made in cotton weaving, the cultivation of which was made compulsory by Emperor

Ming T'ai Tsu. The famous porcelain factory of Kingtechen (Kiangsi) turned out 159,000 objects in 1591, and a "porcelain" tower almost 125 feet high was raised in Nanking in 1591. It remained standing until 1863.

Emperor Cheng Tsu (1403–1424) constructed in the suburbs of Peking the Monastery of the Great Bell (Ta-chung-sze), which owed its name to a bell 23 feet high and 12 feet in diameter, covered with 220,000 characters and weighing between 42 and 52 tons. The temple was constructed *after* the bell had been put in place. The existence of movable lead characters (*Pi-ling, Wu-chin*) and the popularization of wood engraving and color lithography indicate a high degree of perfection in the book industry. Sung Li, a talented engineer, completed work on the Hai River system, which included the rebuilding of the Grand Canal (circa 1375). The Great Wall was also rebuilt.

A panoramic view of Chinese technology in the Ming Dynasty can be gained from three works. The most important is the *T'ien-kung k'ai-wu (The Exploitation of Natural Resources)* of Sung Ying-hsing (English translation by E-tu Zen Sun and Shiou-Chuan Sun, Pennsylvania State University Press, 1966), a general handbook of technology consisting of eighteen sections illustrated with splendid woodcuts. This work gives an excellent summary of

FIG. 1.   Founding brass for the making of bells
(from the *T'ien-kung k'ai-wu,* [Exploitation of Natural Resources], by Sung Ying-sing, 1637); published by the Commercial Press; reprinted in Shanghai in 1954.

Chinese techniques of plowing, weaving, boring, saltworks, pottery, oil, paper, firearms, dyes, alcohol, hydraulics, mines, coinage, pearl oysters, jade working, and other subjects. The book was destroyed, probably because of its author's antidynastic attitude; it was reprinted in Japan in 1771 by Eda Masuhide, and a second Japanese edition was issued in 1825. On the basis of the Japanese edition of 1771, T'ao Hsiang (1871–1940) prepared a new Chinese edition in 1929. The latest edition, based on the original woodcuts from the tenth year of the reign of Ch'ung Chên (1628–1644) was published in Peking in 1959. An important study of this work was begun in 1956 by Yabuuti (Yabuuchi) and several collaborators. Among the writings left by Li yu (1611–1680), a poet and architect who ordered the building, near Nanking, of the famous house in the garden "as large as a mustard seed" (an allusion to the small size of the valley in which the house was built), particular mention should be made of the *Hsien-ch'ing ngou-ki*. This monograph was republished separately in 1921 by the Association of Chinese Architects (*Chung-kuo ying-tsao-hsueh shih*) because of its importance.

Knowledge of Western technology was increased by the writings of Wang Cheng (1571–1644). He was particularly interested in the applied sciences, and in collaboration with Father Terrenz translated European writings on mechanics, doing the illustrations himself. His work (*Yüan-si ch'i-ch'i t'u-shuo-lu-tsui*) was printed in Peking in 1627, and reprinted in 1830 and 1844; through it certain mechanical terms became part of the Chinese language.

FIGS. 2 & 3. Iron metallurgy.
Left, puddling; *right,* a blast furnace.

METALLURGICAL AND MINING TECHNIQUES

Metallurgical Techniques

Bronze (t'ong)　　　From the fourteenth to the third centuries B.C., the social and political structures of northern China were characterized by bronze, which was utilized for the manufacture of weapons, tools, and pots. In the most delicate as in the largest objects (for example, a large pot weighing 1,540 pounds), the bronzeworkers of the Shang period appear as unsurpassed masters.

These techniques achieved their greatest perfection in the Ming Dynasty, when a bell weighing 114,400 pounds was cast for the Monastery of the Great Bell (*Ta-tchung-sze*) in the suburbs of Peking. Bronze mirrors were in use until the Ch'ing Dynasty; concave mirrors were used to make fire through the combustion of artemisia leaves by means of the sun's rays, while flat mirrors were used for reflecting images.

The Japanese also cast gigantic statues that are found nowhere else in the

Fig. 4. Casting bronze tripod vessels
(from the *T'ien-kung k'ai-wu,* op. cit.).

world. The 700-year-old *daibutsu* (Great Buddha) of Kamakura is more than 43 feet tall. The Nara *daibutsu* (recast in 1180, after the original statue of 743–745 was destroyed by a fire) is more than 52½ feet tall; 1,071 tons of bronze were used in its construction.

Copper   Copper began to be worked sometime around 3500 B.C. in China, and in the seventh century B.C. in Japan. Hardening in water and gall were known, as in the West. There were white, red, yellow, and green varieties. The possibility of producing copper deposits that would adhere to iron is an ancient idea that long predates European electroplating.

Iron (t'ie)   Both the word "iron" and the object appeared during the Spring-and-Autumn period (722–481 B.C.). However, the Chinese iron metallurgists, being heirs of the bronzeworkers of the preceding period, did not so much forge iron as cast it, and they very soon became masters in the art of founding, which they used in the making of steles carved with characters. The Imperial Code of the State of China is reproduced on the walls of an iron tripod cooking vessel of 513 B.C. The casting of iron came into practice beginning in the first century B.C. It did not become common in the West until the fourteenth century.

In the Era of the Warring States (fifth to third centuries B.C.), casting molds were in general use for the casting of hoes and axes. Small blast furnaces existed after the first century B.C. Thanks to the double-acting piston, the continuous blowing of the combustible (coal was used for this purpose beginning in the fourth century A.D.) made it possible to obtain higher temperatures than were possible in other Asian centers of iron metallurgy (first century A.D.). Blowing, initially done by several hundred workers, was later (beginning in the first century) performed by hydraulic power. The drop-hammer forge, which was operated by hydraulic power and was similar to the pounder used for husking rice, was probably known in China before it spread through eastern Europe (Styria, twelfth century), and it may therefore be of Asian origin (Haudricourt, 1948).

Steel, which was known before the founding of the Chinese Empire, was obtained by empirical methods: reduction of the oxide by rubbing with oil or melting of the reduced iron and its decarbonization by a blast of cold air. Soldering of hard steel to soft steel (third century A.D.) became a Japanese speciality, and made possible the manufacture of the famous sabers, beginning in the eleventh century. This technique was utilized by the Merovingian arms manufacturers; its origin in Europe is unknown. Modern analyses have revealed that the medieval Chinese sabers were made of high-speed tungsten steels that did not become known in Europe until 1900 and that required a highly perfected technique that was ultimately lost.

The metallurgy of sidearms, introduced into Japan by Chinese and Korean smiths, made great progress. Emperor Mommu Teno (697–707) ordered that every weapon be signed by its maker. Beginning in the eighth century, the sword was replaced by the saber, and during the Heian period the Japanese technique was perfected to such a high degree that the Japanese saber became the best in the world, capable of cutting a blade of straw as well as a stack of coins. The ore used was titanic iron or magnetic oxidized iron. The blade was usually heterogeneous (iron and steel) rather than homogeneous. The hardening process, too, differed for the cutting edge and the body of the blade. The peak period of saber production occurred between 900 and 1450; the work was done by hundreds of shops, and there were numerous technical variations. Arai Hakuseki (1657–1725) devoted an entire work, entitled *Honchō gunkikō* (the 1913 English-language edition, *The Sword Book in Honchō Gunkikō,* by Inada Hogitaro and Henri L. Joly, was reprinted by C. E. Tuttle, New York, in 1963), to side arms; it has two volumes of text and two volumes of illustrations.

Other metals   Antimony was discovered in 698. Zinc, discovered in India in the fourteenth century and in the West in 1509 (by Erasmus Ebener of Nuremberg) may have been known at the beginning of the Christian era (J. Needham); Rhazes speaks of a Chinese metal, *Khar sini,* that appears to be zinc. This metal was sometimes obtained by direct reduction; the zinc ore was treated in porous vessels placed in a bed of burning coals. Apparently the metal thus obtained was not pure but unrefined zinc; it was imported from the Far East by the Dutch and the Portuguese in the seventeenth and eighteenth centuries.

Silver was often extracted from silver-bearing lead ore. Brass ("yellow copper," *huang-t'ung,* or, as the alchemists called it, *yü-shih* is mentioned around the fifth century of our era. In the T'ang Dynasty its basic ore was smithsonite ($ZnCO_3$) (*lu-kan-shih*) with the addition of copper. This technique is described in the treatise *T'ien-kung k'ai-wu* (The Exploitation of Natural Resources), by Sung Ying-Hsing (1637).

"White copper" (*po-t'ung*), mentioned by a fourth-century author, appears to have been a nickel silver. It contained zinc, nickel, and copper, with traces

FIG. 5.   Making zinc by cupellation.
The metal is poured into pots,
which are then sealed, chilled, and stacked up.

FIG. 6.   Smelting tin.

FIG. 7.   Preparing mercury from cinnabar
(from the *Tien-kung k'ai-wu,* op. cit.).

of iron; smelted in Yunnan, it was refined at Canton, and whitened by the addition of unrefined zinc. Known in Europe as "petong" (Father Halde, 1736) or "pactong," it was imported from China in the nineteenth century and was soon (1824) imitated in Berlin under the name of "white metal."

Lead (*ch'ien*) was used early in history as a cosmetic; white lead (*po-ch'ien-fên*) was first produced in Japan in 692. Gold (*kuang-hsin, man*) was the first metal known by the Chinese. A rarity in the second millennium B.C., by the first millennium it was somewhat better known; it came from Siberia. The existence of gold encrustations in the sixth century B.C. leads us to believe that by this period a small stock of the metal existed. By the first century of the Christian era, the Chinese were exploiting a gold mine. The Greeks and Romans, who were gold experts, were able to recognize the metal by fusion, density, or touchstone; the latter may not have become known in China until the fifteenth century.

The amalgam now used for filling teeth may be the material mentioned (under the name of "silver paste") in the *Materia medica* of Su kung (659). Li Shih-chên (1590) gives the following formula for this substance: 900 parts of tin, 100 parts of mercury, and 45 parts of silver.

In the years between 86 and 81 B.C., a famous debate was carried on concerning the monopoly of salt and iron. The government institutions were vigorously criticized by the intellectuals, and the chief censor, Sang Hung-yang, defended the government's practices. This lively controversy is described in the *Yen-t'ieh lüen* (Speech on Salt and Iron), a literary work composed by Huan K'uan (circa 73–49 B.C.), recently (1957) the subject of a study by Professor Kuo Mo-jo, president of the Chinese Academy of Sciences.

Mining Techniques

The technique of boring shafts permitted the Chinese to dig mine shafts in order to reach metal deposits, brine, and rock salt, and to utilize petroleum and natural gases produced by coal as heat sources. The gas was piped through terracotta pipes.

The salt industry dates from the first century B.C., and is known to us through the works of Ch'en Ch'uen (1334) and Li Yung (nineteenth century). The shafts were sometimes very deep; one such shaft, dug in 1080, was 3,000 feet deep (Su Shên). It is possible that the concept of the artesian well, the European version of which dates from 1126 (at the monastery of Lillers in the Artois region) and the theory of which dates from 1010 (Al Biruni), is a Chinese idea transmitted to the West by the Arabs (J. Needham).

The drilling of a well was followed by the installation of a pumping system that required windlasses, pumps, pipes, and canals. The same technique has been followed for almost 2,000 years: a well is drilled, and interlocking bamboo

FIG. 8. Removing toxic gases from a coal mine.

FIG. 9. A prospector of precious stones descending into a shaft (from the *T'ien-kung k'ai-wu*).

tubes draw out the brine from a level ranging from 1,640 feet to 3,280 feet. Buffalo-driven waterwheels or windmills lead it to boilers operated by coal or natural gas, where it is evaporated. A brick recently found in the environs of the provincial capital of Szechuan, in a tomb dating from the East Han period (25 B.C.–220 A.D.), shows the extraction of the brine (*yen-shui*) from a salt shaft (*yen-ching*), and its treatment by evaporation (Wen Yü and R. C. Rudolph).

CHEMISTRY

It is often difficult to distinguish between Chinese chemistry and alchemy, since the Chinese authors rarely differentiate between their procedures. Writings generally mention the raw materials without giving their exact nomenclature, using different chemical or alchemical names for the same product. Moreover, the system of explanation of chemical transformations is for the most part common to both sciences, and in turn belongs within the general system of interpretation of nature. Except in cases where the presence of an alchemical process is revealed by the search for its supreme goal, "the liberation of the cadaver," a study of the vocabulary is required in order to distinguish between chemistry and alchemy. We may simply say that alchemy began to prosper around the fourth century B.C., side by side with a more "realistic" chemistry.

Glass   It was long believed that glass, which became known in the eastern Mediterranean area around 1500 B.C. and was imported into China, began to be manufactured by the Chinese themselves only much later. The technique of glassmaking is supposed to have been imported from Bactria into China during the reign of the Wei Emperor T'ai-wu (424–452), when Indo-Scythian craftsmen opened a factory in Shensi, and a Western glassmaker in Nanking was able to "change stone into crystal" (G. Sarton).

The discovery of glass in prehistoric sites in Japan, China, and Korea proves that glassmakers were much more numerous in ancient China than was formerly believed. This is the only possible explanation for the presence of glass beads, disks, and other small objects found by C. W. Bishop in Han tombs near Loyang (Honan), in the same strata as bronze objects dating from the third century B.C. The hypothesis that they were imported is negated by a physicochemical study of these objects, which shows a high index of refraction and a characteristic content of lead and barium; such glass was not made in Europe until 1884, by Schott of Jena, who created a glass with a high index of refraction for scientific use. The high lead and barium content thus militates in favor of a Chinese origin, predating at least the Han period. However, glass beads did reach China via the steppe routes across Asia. Glass from the Roman East dating from the first centuries of the Christian era has been found in Silla-period sites in Korea, and Sir Aurel Stein made similar discoveries along the routes of central Asia. It must therefore be recognized that glass may have been manufactured, as well as imported, at a very early period in China. Because of their transparency, hard stones imported from the West may have given the Chinese esthetic satisfaction and thereby delayed the development of their own glass industry (Leroi-Gourhan).

In any event, imported glass existed in China beginning in the Han period; it was manufactured in Shensi and Nanking by foreign craftsmen. Glass spectacles may have reached China either by way of Champa (seventh century) or central Asia (thirteenth century), or by the sea route (fifteenth century). They seem to have been imported into China in the fourteenth century; the words for them (*ai-na, ai-ti*) are of Arabic origin. They were very costly items. Prior to their importation rock-crystal glasses may have been in use (Sung Dynasty).

Pottery   The Neolithic Chinese potters made painted pottery (*ts'ai-t'ao*) that surpassed that of other Eurasian centers. The black pottery of Lungshan, with its extraordinary eggshell-thin walls, and surfaces as smooth and shining as lacquer, indicates that their successors developed to perfection the techniques of modeling and firing purified clays. By the Shang period (fifteenth–fourteenth centuries B.C.) thin-walled glazed vessels were already in existence.

Silicious glazes, when combined with a melting agent, resemble glass; they never become part of the pottery itself, and are thus differentiated from the glazes of very fine feldspar, glazes like porcelain itself, which combine with the pottery. This is one of the fundamental discoveries of the Chinese ceramics makers at the end of the Chou period. It presupposes very powerful kilns that permit the obtaining of stoneware, a completely vitrified matter; by the East Han period, the temperature of the Chinese pottery kilns could be brought up to 1,300 degrees, while in the rest of Eurasia 800 degrees to 900 degrees was the maximum. So the way was paved for the appearance of protoporcelain and, in the period of the Six Dynasties (222–589), jade-green glazed ceramicware.

Genuine porcelain, as contrasted with opaque pottery (*t'ao*), appeared during the T'ang period; in 851 the merchant Suleiman was greatly impressed by its comparative translucency. This porcelain was a superior grade of stoneware

FIG. 10.   Making pottery.

FIG. 11.   A brick kiln; modeling the clay.

vitrified until it became translucent, a process made possible by a clay that acquired a white color at around 1,450 degrees. This feldspathic, nonfusible clay (*kao-ling* [kaolin]), the "bone" of the porcelain, was combined with white stone (*po tuen-tsu*) (the "flesh"), the fusion of which formed a cement surrounding the particles of kaolin. Even at this early period porcelain factories existed at Kingtehchen (modern Fowliang, Kiangsi), a site that later became famous for its porcelain.

Sung-period ceramics are very varied: there are stoneware, porcelaneous pottery, celadons, and incomparable porcelains as clear as a mirror, as thin as paper, and as sonorous as jade (R. Grousset). The discovery of new enamel colors made possible polychrome decorations in which overglaze or low-temperature ("*petit-feu*") glaze combined with an underglaze or high-temperature ("*grand-feu*") glaze, a technique attributed to Tz'ü Chou. We can assume that by the end of the southern Sung period (1127–1279 A.D.) the decoration was being executed either on the damp clay before the applicaton of the glaze (underglaze enamels) or, on the other hand, on the fired but still-soft glaze. The enamel colors immediately adhered, and were then fixed by a second firing. The best pieces were fired in clay cases that were either refractory or, in cases where air was required to oxidize the colors, porous. By playing on the different coefficient of expansion of the pottery and the specially treated glaze, a fine crackle was obtained (D. Lion-Goldschmidt). At the peak of perfection of the porcelain technique, its products possessed the three classic characteristics of hardness, sonority, and translucency.

During the Yüan period foreign markets were developed requiring the industrialization of the techniques of production. This period saw the introduction of cobalt black ("Mohammedan Blue"), which becomes pure blue under certain conditions of firing.

The great age of Chinese pottery was the Ming period. Three techniques were used for painted decoration: "*grand-feu*" colors, enamels painted on a fired glazed surface and then refired in a muffle (a clay oven for firing pottery without direct exposure to the flame), and enamels applied on unvarnished, already fired material (biscuit). Combinations of under- and over-glaze enamels required several firings; numerous combinations were possible. Small oval ("rice-seed") holes in the clay, which were filled with transparent glaze, increased the dangers of the firing, and presupposed great skill (D. Lion-Goldschmidt).

The Ming period was also characterized by the decision of Emperor Hung Wu (Chu Yüan-chang) to use mass-produced glazed bricks as a construction material, for example in the famous "Porcelain Tower" of Nanking, which was 325 feet high. The emperor also ordered the construction of an imperial porcelain factory at Kingtehchen (Kiangsi), whose tremendous production (159,000 pieces in 1591) was exported to every corner of the world.

Much more attention has been devoted to the esthetic aspect of Chinese porcelains than to their technical aspect. We still lack precise technical information, although numerous kiln sites have been found by archaeologists, and not until a general study is made of the remains discovered will we be able to classify the pottery of this period under its technical, archaeological, and artistic aspects.

The technique of Kingtehchen was revealed to Europe by Father d'Entre-colles (1712–1722) and by Stanislas Julien, who made a partial translation (in 1856) of Lan Pu's *Ching-të-Chën t'ao-lu* (The History of the Porcelains of Kingtehchen; English translation *The Potteries of China*, by G. R. Sayer [London: Routledge and Kegan Paul, 1951]). Professor Yüan Han-Ch'ing, in *Studies in the History of Science* (Peking University, 1956), supplies the following table for the composition of the porcelains of this period:

	Kiangsi Varieties	Kingtehchen Varieties	Porcelains of Japanese origin	Porcelains of English origin
SiO_2	51.64%	65.75%	79.27%	73.93%
Al_2O_3	32.08	21.73	16.06	20.62
Fe_2O_3	1.54	1.03	0.25	0.57
CaO	0.84	0.10	0.21	1.25
MgO	0.31	0.27	0.15	0.21
K_2O and Na_2O	2.03	4.98	3.94	2.08

The Ming potters utilized numerous colors — vermilion, green, red, brown, yellow, dark blue — the composition of which included iron and copper oxides. Many mineral colors were also used: mud-colored ocher, a blue made from calcareous lapis lazuli, an emerald green derived from copper protoxide, vermilion prepared with sulfide of mercury, red obtained from lead oxide, white lead, or calcinated oyster shells. For a long while the Chinese vermilion and red surpassed the European colors, thanks to better utilization of the mechanical procedures of porphyrization and division, and to the repeated distillation of the cinnabar. Fuchsine rose was known in China long before it came into use in Europe (de Mely). It is mentioned by Father Grossot in 1840, whereas aniline colors were unknown industrially in Europe before 1856, when they were discovered by J. Perkins (1838–1907).

Lacquering   The technique of lacquering with a milky liquid obtained by making incisions in the bark of the *Rhus vernicifera* (sumac — *ch'i-shu*) was known as early as the Shang Dynasty. Born in central and southern Asia, it arrived in Japan (via Korea) in the sixth century, and in Vietnam in the fifteenth century.

The sap of the lacquer tree is collected in waterproof baskets; here, through the action of distillation, a film is formed which when skimmed off and separated forms the true lacquer. It is strained through a cloth and is then colored, occasionally by vegetable dyes (for example, indigo) but more often by mineral dyes; in this way black, brown, and red lacquers are obtained. Through oxidation lacquer turns most of the mineral salts, except vermilion, oxide of bismuth, orpiment, and certain ochers, to black.

Before lacquering a wooden, preferably porous, object, a series of operations (very careful planing, sandpapering, polishing, and cementing) is required to give the surface the smooth appearance of a ceramic piece and a metal. In certain cases a paper coating or a canvas or silk backing is necessary. The object is then given several coats of lacquer with a brush and is left to dry (which it does more rapidly in damp than in dry surroundings). The last stage of production is the decoration, in which drawings are executed on the object in the desired colors with a hard or soft brush, depending on the consistency of the varnish. In some cases metal leaf (gold, silver, beaten tin) is applied three-quarters of the way through the drying process. The encrustations may be covered with a glaze wash; in the case of gold, this can be done only after several months have passed. Encrustations of ivory, hard stones, and decorative beads are frequent in cabinet-work. Leather, metal, silk, paper, and canvas, as well as furniture, can be lacquered. One method of lacquering invented in the Six Dynasties period (222–589) consisted of covering a hemp core with as many as thirty-six layers of lacquer, which were then sculpted. The material to be sculpted could also be formed of a combination of tow, paper, eggshell, and oil of camellia, to which was added a coral red varnish (*t'i-hung*). Technical improvements made it possible to obtain gilt lacquer (in the Yüan Dynasty) and sculptured lacquers with bronze, tin, or engraved wood backgrounds (Ming Dynasty). The lacquer surface was perforated so that the colors would adhere.

The *Ch'i-ching* (Laquer Classic), written by Chu Tsuen-tu (of unknown date), has been lost. The only basic technical writing is the *Hsui-shih-lu* of Huang Ch'êng (circa 1567–1572).

Sugar

Honey was the principal sweetener used in ancient Eurasia. However, it could be gathered only in small quantities, and therefore substitutes were quickly found: date honey, fig honey, evaporated grape juice, manna produced by the parasitic action of insects on certain plants (*Najacoccus serpentinus, Tralutina mannipara*, the exudation of *Fraxinus ornus*). In Northern India the possibility of extracting a sweet substance from the sugarcane had long been known, but refined sugar was not produced until quite·late (around 300, at the time when its use for medical purposes was becoming common). From its Indian homeland the sugarcane traveled to Iran, and from there the Arabs introduced it into Egypt (in 750), Sicily, and the western Mediterranean areas. Since Alexander's expedition this part of the world had known of the existence in the Orient of a reed containing a honey (*sackharon*) that owed nothing to the bees. The writings of Strabo, Aristotle, Dioscorides, Pliny, Paul of Aegina, and Isidore of Seville confirm this idea; however, they confused tabasheer with sugarcane. In 627 the Basileus Heraclius found a rather large stock of cane sugar, together with silk, pepper, and cotton, in a castle of the Sassanid King Chosroes. The sugarcane was not mentioned in modern European literature until the seventh century; the first large imported shipment of it, in Venice, is recorded only in 996.

In the East the sugarcane traveled from India to Indochina and China. *The History of the Sui Dynasty* (581–617) mentions "stone honey" (*sheh-mi*), a favorite with the Chinese, as a Persian product; this may have been a kind of sugar in white powder form. The sugarcane (*tche*) may have been introduced into China around 200 B.C., from the south rather than by way of the commercial routes of the north. But the method for obtaining sugar from the cane was not known in China before the T'ang Dynasty. *The History of the T'ang Dynasty* (618–970) tells of a mission sent by Emperor T'ai-Tsung for this purpose to Maghrada in India in 647. The *T'ang-shuang-p'u* (Treatise on Sugar), written by Wang Shao (Northern Sung, 960–1126), attributes the invention of the procedure for preparing crystallized sugarcane (known to the ancient Chinese as *shuang-t'ang*, to their modern descendants as *ping-t'ang*) to a bonze of Szechuan named Chou who may have revealed the secret of his art around 775 A.D. The same work gives a detailed description of the procedure, which consisted mainly of crystallizing the liquid obtained by the mechanical treatment of the sugarcane, after it was reduced by prolonged boiling. The *T'ien-chüng k'ai wu* describes the preparation of sugar candy from the same raw material. The sugarcane industry, which began to flourish during the T'ang Dynasty, must have given competition to the older techniques that utilized honey, grapes, or rice. The *Ch'i-min yao-shu* (Arts and Sciences Concerning Agriculture, and Necessary to the People), a work on political economy by Chia Sze-hsieh, a prefect during the Wei Dynasty (386–534), explains the preparation of maltose and the making of the sugar by the action of malt on rice.

The numerous synonyms and homophones for "sugar" that are found in the earliest writings of the Chinese civilization are sufficient evidence for the variety of its origins and techniques of preparation. The character *t'ang*, by which it has been represented since the fifth century A.D., was preceded by *fu*, *san yi*, *t'ang* — to mention only a few. The various types of sugarcane (*Saccharum officinarum*) are cultivated in the southern provinces of Fukien and Kwangtung, which until a recent period accounted for 90 percent of Chinese sugar production.

During the Mongol period Egyptian experts came to China and trained the inhabitans of Unguen in their methods. Sugarcane became for the Chinese a supplement to the limited production of sugar from honey and sweet substances.

Alcoholic fermentation   Alcoholic fermentation of grains was known beginning in the Chou Dynasty, and special treatises in the literature of the early centuries of the Chinese Empire mention a large number of fermenting agents (*ch'ü*).

The sixth-century work known as the *Ts'i-min yao-chou* discusses some ten methods of fermentation. It is possible to follow the evolution of this technique by comparing the instructions given in this work with those in two treatises of the Sung period, the *T'ung-po chiu-ching* (Wine Classic), by Su Shen (circa 1036–1101), and the *Pei-shan chiu-ching* (Wine Classic) (1117) of Chu Yi-chung. These writings reveal the changes in the choice of the basic grain (millet, glutinous rice, and so on) and the utilization of techniques of refermentation.

Grape wines (*p'u-t'ao-chiu*) appeared shortly before the Christian era, and are mentioned in the early dynastic histories. The variety and quality of these beverages is attested to by the great number of regional wines that are mentioned in, for example, the *Chiu hsiao-shih* (Little History of Wine) of Sung Po-jen (Yüan Dynasty), who mentions the "Fen Wine" of Shansi, the "Wine of Great Ferment" of Szechuan, and the "Yellow Wine" of Chekiang, whose alcohol content is often quite high. The Chinese poets have often sung of the pleasures of wine, among friends or "alone, under the moon. . . ."

Wine thus played an important economic, social, and cultural role in ancient China. But was it really wine? If we restrict this term to the product that results from fermentation of the juice of the grape, it cannot be applied to China, where the seeds of various grains are used much more widely than the fruit of the vine; other equally incorrect expressions have further complicated the terminology. Although certain writings distinguish the alcohols of fermentation (*hua-t'ao*) from the products obtained after distillation (*fen-chiu*), the word *chiu* or "wine," widely used in both senses, is a source of constant confusion. The alcohols of fermentation are beers rather than wines, with a fairly high weight of extraction, often "composed" with fruit skins, leaves, and flowers which were macerated before filtration. The alcohols of distillation may also be "composed," other substances being added before or after distillation.

The identification of aqua vitae (*shao-chiu, po-kan-chiu, cheng-liu-chiu*) is doubtful, for the reasons already mentioned and also because grain alcohols (to which grapes are added during production) can lead to confusion. The distillation of grain alcohols may not have become known until the Yüan period (1260–1367), and was probably borrowed by the Chinese from the Irano-Arab culture (E. von Lippmann). Without entering into a detailed discussion of the terminology of the vine, wine, and alcohol, we may however mention that the term *A-la-ku*, which in the Mongol period was used to designate fermented beverages, may be derived from the *arrack* of Asia Minor. This may have been an aqua vitae of Western origin, made from grain and introduced into China during this period. The Mongols introduced hydromel and kummis, leading to a major development of the alcohol industry. (In 1254 the Parisian goldsmith Guillaume Boucher was ordered to construct a "magic fountain," of the type which had existed in Baghdad and Byzantium [829], to slake the thirst of the imperial courtesans.) Special treatises like the two treatises on wine mentioned above describe medicinal breads (*ping*) made of glutinous (*no-mi*) and nonglutinous rice (*k'ang*), ginger, and other drugs; a preparation very similar to that of *ping* produced a leaven. The *Pei-shan chiun-ching* gives detailed descriptions of how these products were used and how they acted (which have been explained in terms of modern chemistry by P.-L. Deniel, Plagnol, and Richard).

Salts and various chemical substances
The *Pen-ts'ao* contains important lists of salts and various chemical substances; their Chinese names, which are for the most part alchemical in origin, are too numerous to be mentioned here.

Metallic salts. — These include silver chloride and sulfide; copper acetate, sulfate, carbonate, and tannate; iron sulfate, acetate, tannate, oxide, and sulfide; tin oxide; zinc sulfate; lead carbonate, sulfate, acetate, tannate, and oxide; and mercury calomel, sublimate, sulfide, and oxide.

Cinnabar was known in the second millennium b.c., mercury around the end of the first millennium. The corrosive sublimate was used for medical purposes around 629 a.d. The antiseptic, insecticide, and anticryptogamic properties of copper sulfate were used as a preservative for wooden posts and in the preparation of the soil.

Nonmetallic salts. — Alum was used for purifying water; magnesium salts were remedies for diseases of the bone marrow, the bones, and the blood vessels. Hydrated silicate of magnesia was prescribed for cases of gravel, selenite for bone diseases. Glauber salt was used for its laxative properties; sodium borate, in lotions; potassium nitrate was utilized as a diuretic.

The alchemists became acquainted with saltpeter and sulfur in the first century B.C. The combination of these two bodies in definite proportions became common in the third century, and by the beginning of the seventh century it led to the production of fireworks and firecrackers.

In the tenth century Chinese military technology began to utilize explosive mixtures, under the name of *huo-yo* (sulfur, saltpeter, carbon, asphalt). One of these mixtures is mentioned for the first time in an official publication, the *Wu-ching chüng-yao* of Tseng Chüng-liang (Sung Dynasty), the latest edition of which dates from 1959.

FIG. 12.   Preparation of sea salt by boiling brine.

FIG. 13.   Making sulfur (from the *T'ien-kung k'ai-wu*).

Metalloids and various products. — Sulfur was used as a depurative, vermifuge, emmenagogue, an antidysenteric, and a cure for leprosy and skin diseases. A mixture of sulfur and alum was used to make medicinal cups for drinking a wine that strengthened the eyesight and the nervous system. Arsenic was known in three principal forms: flowers of arsenic and the two sulfurs already known to Aristotle — orpiment and realgar. For the Chinese the yellow-colored orpiment had a *yin* character, while the red realgar was *yang* — an example of classification, and the beginning of a system that permitted the Chinese to explain the reactions of the two sulfurs. Kaolin, which was used medically and industrially, did not become known in Europe until the eighteenth century.

The magnet is mentioned by the famous alchemist Ku Hung (circa 281–349), the author of the *Pao-p'u-chü*; he also mentions mica (*yün-mu*), which was used in preimperial China for making palace windows, embalming material, and the magic food of the alchemists.

AGRICULTURAL TECHNIQUES

A large body of literature devoted to agronomy, of very early date, indicates that agriculture was the object of constant attention on the part of the Chinese and that it developed in a progressive fashion.

The early Neolithic cultivators probably used a technique that was suitable only for small, semisedentary human groups with vast spaces at their disposal, and who did not seek a high yield from their work: the technique of forest "strips," which is still used in the upper regions of southeastern Asia (Castagnol).

After felling and burning the trees, the cultivators used simple planting sticks to create temporary cultures that lasted two or three years, or until the fertility of the soil was exhausted. While the soil was being regenerated (a process that required between twenty-five and fifty years), another part of the forest was destroyed and its soil cultivated. This is the method that must have been used in the then well-wooded lands of Hopei, Shansi, Honan, and Shantung provinces, on the borders of which regions archaeological sites have been dis-covered.

This technique was not suitable for the sedentary peoples of the great plain, since the problem of manuring had not been solved, except in the suburban areas that could make large-scale use of human fertilizer. Long fallow periods were required for the regeneration of fertility, whence the necessity of food-gathering to supplement a limited agricultural production. Nevertheless, the essential traits of Chinese agriculture were already recognizable, and an agricul-tural economy was coming into existence that included the preservation of large stocks of grain for long periods of time.

The peoples of the great plain were organized into already very hierarchical societies in which even the Shang and Yin peasants were required not only to feed themselves but also to feed that portion of the population that specialized in other activities. The increase in the area exploited and in the yield from the cultivated area was achieved by clearing the wooded alluvial lands with improved equipment (for example, the animal-drawn plow). The cereal grains (paniculate millets, glutinous and nonglutinous rice, and barley) could now be cultivated, and the cultures became permanent ones.

During the Chou Dynasty agriculture was "nationalized." Various govern-ment bureaus were in charge of the ground, its preparation, seeding, drainage of low-lying areas, and the irrigation of dry lands, in accordance with a calendar. Uncultivated ground was plowed with the *lei-ssü,* a metal tool that loosened the subsoil and increased the supply of water available to the plant. The arable por-tion was turned over with a plow (*li*); cross-plowing and hoeing permitted a rational working of the soil.

New cultures — spikelet millet, rice, hemp, sesame, wheat, and others — be-came known. Rice had already acquired sufficient importance to warrant the ap-pointment of special government officials to direct its cultivation. Horticulture was quite well developed; gourds, peach trees, plum trees, quinces, and jujube trees were also cultivated. The protection of the plants from the numerous wild ani-mals was ensured by hunting and brush fires.

During the Han and Sung dynasties Chinese agriculture learned new techniques and acclimated new plants through contacts with tropical and steppe vegetation. Improved plowing tools and the practice of manuring stabilized agriculture practices, the technology of which was codified in numerous treatises. Six Han treatises, now lost, are known to have existed. One of these, attributed to Fan Sheng-cheh (first century B.C.) described a technique, utilized in dry areas, known as "digging a small piece of land," which resulted in a harvest of 497 bushels per 100 square meters. Major progress was made both in volume of exploitation and in the methods employed.

Chia Szu-hsieh (circa 533–546) wrote the first complete treatise, *Agriculture, or Important Facts for the Well-Being of the People (Ch'i- min yao-shu)* (repub-

FIG. 14.   Wheat seeder.

FIG. 15.   Harrowing.

FIG. 16.   Winnowing mill.

FIG. 17.   A pedal noria.
(All except Fig. 18 are from the
T'ien-kung k'ai-wu, op. cit.)

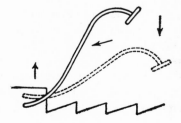

FIG. 18.   The *lei-ssü,* a metal tool
for loosening up the subsoil.

lished in 1957). He improved on the work of Fan Sheng-cheh on many points that did not become known in Europe until much later. One of these techniques, which may have been practiced much earlier, but had never before been described, concerns the rotation of crops, the selection of seeds, and the grafting of fruit trees. Chia, who was prefect of Kao-yang, realized the disadvantages of cultivating the same plant in the same ground for several years; he knew that it was advisable to alternate millet and haricots and that leguminous plants had to be cultivated before nonleguminous plants. The selection of seeds is described in detail, and the fact that certain kinds of millet and peas do not grow in certain types of soil is mentioned. Grafting was a common practice. Peonies and wild monochrome and numerous polychrome chrysanthemums were extracted. Grafting of the pear tree (*Pirus betulaefolia*) was common (it did not spread to the West until the eighteenth century). Treatises on veterinary medicine appeared in the third century B.C.; the annals of the Sui Dynasty (581–617) mention eight such treatises.

During the Sung Dynasty numerous treatises on agriculture were written. Chu Fu describes (circa 1101–1103) the cultivation of aquatic rice, an arrival from southern China at the end of the T'ang period. Skilled cultivation produced many varieties of vegetables and fruits: eighteen types of beans, seven of rice, eleven of apricots, eight of peas. Grape cultivation, known in the Sui Dynasty, was intensified in the Yüan period, and imported Egyptian experts improved the production methods of cane sugar.

The numerous treatises on agriculture written during this period include the *Nung-sang chi-yao* (circa 1273–1274) and the *Nung-shu* of Wang Chen, which formed the basis of the Ming treatises, the most famous of which is the *Nung-cheng ch'iuan-shu* of Hsu Kuang-ch'i (1562–1633).

The Chinese had a good knowledge of soils and fertilizers. Lacking the abundant manure available to the Europeans, they compensated for this lack with green nitrate fertilizer that they plowed into the soil, the preparation of compost, lime, the mud of old walls, and mud from ditches (Castagnol).

Horticulture China has been called the mother of gardens (E. H. Wilson). Having escaped the effects of the glaciation that destroyed so many European plant species, the country succeeded in retaining a very abundant flora ranging from tropical to temperate to Alpine, uninterrupted by desert areas.

Chinese gardens were in existence at the beginning of the Christian era, but treatises or monographs on horticulture, the earliest known in world literature, did not appear until the Sung period (960–1279). Later, treatises on painting, such as the *Mustard Seed Garden,* proved an artistic iconograph of ornamental plants, the drawing of which was often the monopoly of artists who specialized in this subject. Treatises on the art of making bouquets and floral arrangements also contained discussions on horticulture.

Beginning in the T'ang period (618–907), the horticulturists (who were sometimes poets as well) succeeded, by using various single-color wildflowers, in obtaining numerous multicolored varieties which were sold at high prices

(H. L. Li). The preferred flowers became the peony (*Paeonia suffruticosa*), chrysanthemum (*Chrysanthemum*), the "Japanese" apricot (*Prunus mume*), the peach tree (*Prunus persica*), lotus (*Nelumbo crucifera*), orchids of the *Cymbidium* variety, camellia (*Camellia japonica*), narcissus (*Narcissus tazetta*), roses (*R. multiflora, R. rugosa, R. chinensis*), *Hydrangea, Viburnum, Pittosporum, Gardenia*, various types of lilies, jasmin (*Jasminum sambac*), *Hibiscus, Magnolia, Azalea, Gardenia*, and *Begonia* (H. L. Li). Some of these plants (for example, jasmin and narcissus) were imported from the Arabic countries around the fifth century, and Chinese flowers in turn began to change the faces of European gardens in the eighteenth century (H. L. Li).

Miniature gardens were a special type of garden that included herbs (*Acorus calamus, Raphus humilis*), heather (*Adiantum lycopodium*), flowers (*Narcissus poeticus*) and dwarf trees (*Cedrus odorata, Citrus reticulata, Cycas revoluta, Juniperus sinensis, Picea asperata, Pinus tabulaeformis, Pinus thunbergii, Tamarix chinensis, Tachycardus excelsa, Ulnus parvifolia*, and *Paeonia suffruticosa*).

TEXTILES

Wool and cotton   Wool was known to the Neolithic civilizations in the Iranian area, which domesticated the sheep. It was never of great importance in China, except in the northwest. Felt, born in central Asia, became known to the Chinese in the fourth century B.C. Cotton, well known to the Mohenjo-Daro civilization, did not travel quickly to the East; the Chinese bought it in India or Java in 340, and found it in Turkestan in 706. Chinese paper did not contain cotton until the thirteenth century, when Huang Tao-p'o perfected the weaving methods of Kwangtung.

Hemp and flax   Hemp and flax were imported from Turkestan around the beginning of the Christian era. Grass cloth was also known at a very early period. *Chamaerops excelsa* (the low-growing fan palm) was used in the making of rope for junks and for leaf garments worn in South China and North Vietnam.

Silk   The Eurasian paleo-industries obtained multicolored silk from various Lepidoptera: *Bombyx mori* (the true Chinese silkworm), *Antheraea mylitta* (the Indian tussah moth), *Antheraea pernyi* (Mongolian), *Philosamia cynthia* (Mediterranean), and others, some wild, other raised domestically. Silk weaving first appeared and underwent considerable development in China. Damask silks, already decorated with remarkable designs, existed as early as the Yin period. Weaving looms with four or more heddles were known in China in the second century B.C.; they made it possible to produce silk brocades (*kin*) and damask silks (but not genuine satin) — marvelous fabrics that aroused the admiration of the Greco-Latin world. Pliny seems to make a careful distinction between Asian silk (*serica*) and local

wild silks (*bombyx*), the existence of which is known in Cos in the third century B.C. The Median dress which Alexander the Great admired, and which is
mentioned by Herodotus and Xenophon, was made with local Irano-Indian
silks, or with Chinese silks alone or blended with the former. Coptic and Syrian
silks made with bobbins of silk imported from China or with local or Indian
wild silks seem to have formed the major part of their stock. Beginning in the
first century B.C. the workshops of the Near East were weaving silk thread or
decorating fabrics with Chinese silk (Forbes).

Fig. 19.   Raising silkworms.

Fig. 20.   Throwing (twisting) silk.

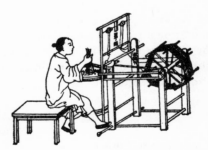

Fig. 22.   Unraveling a piece of silk.
(from the *T'ien-kung k'ai wu*, op. cit.).

← Fig. 21.   Weaving.

The historians, however, were confusing China with the cases of central
Asia — a pardonable error, for beginning in 419 Khotan and Sogdiana were
raising silkworms, as did Korea and Japan beginning in 300. Another, more
serious, error claimed that silk was a vegetable product, a kind of cotton (*byssus*)
obtained from the bark of trees. For the Byzantine Empire silk was a very necessary product; for the most part it was worked in imperial workshops by carefully chosen workers. Forbes tells of Byzantium's difficulties in importing silk,
the exodus of the independent Byzantine and Syrian silk manufacturers, and the
fruitless deputations sent to Iran, Turkey, and China. The solution appeared

with silkworm eggs, brought from Serindia in 553 by two monks; thanks to these eggs, the silk industry was firmly established in the Byzantine Empire.

Weaving The silk-weaving technique (*cheh-ch'eng*) was constantly being perfected in China. Woven painting on silk was already quite well developed in the T'ang Dynasty, and has remained popular until modern times.

The remarkable textiles obtained in China demonstrate the superiority of the Chinese looms. There were two types of looms, both of which left the operator's arms free: the draw loom, which could be found both in China and in Europe, and the pedal loom, which appears to be unique (and probably Chinese) in origin (Montandon). Pictures of Han looms can be seen in twelfth- and thirteenth-century writings; in many ways they resemble those of medieval Europe. Double-thread (each thread of the warp and woof is double) and single-thread (woof threads are single, warp threads double) silks could be made on these looms. Methods of producing fabrics from the cocoon are clearly described in the *Keng-cheh-t'u* (Album of Agriculture and Weaving), published in 1210 and republished in 1462, 1696, 1739, and 1879.

Dyeing Printed fabrics were known as early as 140 B.C.; lacquered fabrics were also known at a very early period.

THE EXPRESSION OF IDEAS

The use of characters for writing was known in China at a very early date. In Asia Minor the medium for writing was the papyrus, the Mesopotamian clay tablet, or the sheepskin (parchment) or calfskin (vellum) roll. The ancient Khmers used a kind of palm-tree paper; the Hovas, Malays who emigrated to Madagascar, used bark paper (*havoha*, fourteenth century). This may suggest comparison with the Oceanic *tapa*, to which paper is comparable (Leroi-Gourhan).

The evolution of writing followed a different path in China. Neolithic libraries contained collections of flat animal bones and turtle shells. Legislative or canonical texts were later cast in bronze or iron caldrons, or were engraved on stone steles; the characters were also "written" on wooden slats (*fang, pan*) or on pieces of bamboo (*chien, ts'ü*) with a bamboo stylet coated with a kind of lacquer (*ch'i*). Writing on silk (*po, chien*) came into use in the third century B.C., during the reign of Ch'in-shih-huang-ti.

The *pi*, a brush for drawing characters, was introduced by Mung T'ien (died 210 B.C.). Red ink made from sulfate of mercury, reserved for the emperor's use, appeared in the Han Dynasty.

Paper The wooden slats strung on two leather cords were replaced by rolls (*chiuan*) of silk or paper kept in lacquered tubes. Much about the discovery of paper (*cheh*) in China still re-

Fig. 23.   Selecting vegetable substances
(soaking and cleaning bamboo).

Fig. 24.   Preparing and reducing →
bamboo to a pulp.

Fig. 25.   Spreading the pulp on a rectangular screen as large as the future piece
of paper.

The making of paper (from the *T'ien-kung k-ai wu,* op. cit).

mains obscure. By the third century B.C. various types of paper were being made
in Asia, on a very small scale and with a great variety of materials. Ts'ai Lüen's
use of the bark (*shu-fu*) of the mulberry tree (*Broussonetia papyrifera*) in 105
A.D. appears to be a definite fact. This discovery, however, did not eliminate
the manufacturing of paper from various products: silk, bamboo, flax, rice, or
wheat straw. A combination of straw and nettle-tree wood produced *hsiuan-cheh,*
a special paper with long fibers, a smooth surface, and great strength, which can
last for a hundred years. This is the preferred material of painters and calligraphers; of better quality than machine-made paper, it is still being made on the
craft level in Ngan-huei. In any event, China quickly eliminated the non-

Chinese centers of paper production, and gained the monopoly of its production in Asia. Its technique seems to have been brought from China to the Middle East by Chinese workers taken prisoner by the Iranian Arabs at the Battle of Talas in 751.

From Samarkand the "Paper Route" (which has been carefully studied by A. Blum) passed through Baghdad (793), Cairo (900), Fez (1100), Palermo (1109), Játiva (1150), Fabriano (1276), and Nuremberg (1390); another route traveled along the coast of North Africa, through Spain (beginning of the thirteenth century), and into France.

In addition to serving as the medium for writing, paper was used in the making of fans, screen, lanterns, napkins, bandages, and house partitions. When oiled, it was used in umbrellas, raincoats, air cushions, as a strong wrapping, and for imitation windows.

Ink

Ink (*mo*) obtained from gummed pinewood soot (*sung-mei*) dates from the fourth or fifth century; lampblack ink was not known until the tenth century. Ink manufacturing, like that of paper, is basically dependent on materials that can be exploited locally; Shensi pine was used in the Han period, Kiangsi pine under the Ch'in, Shansi and Ngan-huei pines by the T'ang emperors, and so on. According to the *T'ien-kung k'ai-wu*, tung oil (*t'ung-yu*), light oil (*ch'ing-yu*), and pork-fat smoke were other ingredients used in the making of ink. The use of smoke black and cinnabar in ink production was common in Egypt long before the Chinese knew of it (Sarton).

Early Forms of Printing

The Buddhists, however, did not completely abandon stone as a method of expression. Steles on which canonical texts are engraved are preserved in nine caves in the Hill of Inscriptions and Steles, in the Fangshan district near Peking. These caves have been the subject of work and investigation since 1956.

While Asia showed a preference for paper, Europe defended parchment, which can be scratched (palimpsest), and which as late as the fifteenth century was Jean de Gerson's preferred material. Printing appears to have been derived from the use of the seal (an extremely important object in the official circles of the Far East) and stele engraving.

Xylography

Xylography, which replaced stone with a wooden plank carved in reverse, was invented in 770, and made possible the reproduction of Buddist texts on printed rolls of paper. In 868 the Diamond Sutra was printed for the first time on a roll (rather than folded and sewn sheets) of paper. This is the oldest known book; it was discovered in 1907 by Sir Aurel Stein in the caves of Tunhuang.

In the ninth century the popular Buddhist and Taoist works began to travel west toward the Middle East, and east toward the Orient. Until 932 there were only two printing centers in China. Fung Tan (882–954) is regarded as the inventor of the art of printing because in 932 he ordered the printing of a critical edition of the classics, which until then had been carved on steles. The printing of 130 volumes was completed in 953. The industrial stage of printing was reached with the translation in 982 of the *Tripitaka*: the printing of the Buddhist canon required 130,000 plates. The imperial libraries of the northern and southern Sung contained, respectively, 70,000 and more than 100,000 volumes.

Typography   The next invention was typography. Pi Sheng invented (1041–1048) the first movable characters, which were individually engraved in viscous clay, then hardened in the fire, and assembled on a framework made from a mixture of resin, wax, and paper ash. Thereafter abandoned, this method was reinvented and improved around 1314 by Wang Cheng, magistrate of Ching-tü (district of Hsiuan-chou, in Ngan-hui Province. Wang made the handling of the characters easier by storing them in compartments on a table that turned on a vertical axle. During the Ming Dynasty (1368–1644), the movable wooden characters were utilized largely for the publishing of the official court gazette (*ti-pao*).

Wang composed a treatise on agriculture that stands today as one of the great classics in the history of Chinese and Far-Eastern technology. The author was not satisfied to describe the development of the agricultural sciences in China; he told also of the invention of printing and its cultural results. Wang probably wrote his treatise between 1295 and 1301; the preface dates from 1313. Wang compiled such a large number of works that he was forced to abandon the customary xylographed blocks. He composed his own printing characters, and persuaded an artisan to carve them on movable blocks (*huo-tsou*). After two years of work he possessed more than 60,000 characters, which he used to publish a local gazette, printing more than a hundred copies in one month. This innovation was destined to have very great repercussions in the province. Two years after his discovery, however, Wang was transferred to the district of Yung-fung (Kiangsi Province), and took his collection of letters with him. He then decided to publish the *Nung-shu,* and to include in it a report on his experiments with the movable characters, in order to leave his experiment to posterity. In addition he created very beautiful illustrations of the history of Chinese agricultural techniques.

The Korean   In Korea, metal (lead and copper) typography was
metal characters   regarded as a major invention (1403), and numerous
books were printed during the fifteenth century. The Korean characters spread to China (end of the fifteenth century) and Japan (1596), but were abandoned, and when the Europeans arrived in the Far East the traditional inking of wooden blocks was the customary procedure. This is due to the fact that the movable character, which is essential in languages with a phonetic alphabet, is of less interest for the languages that have ideographic symbols. Moreover, the fragile mulberry-tree paper did not permit printing on both sides of the page with metal letters. The composition of the ink (on which Shen Chi-sun wrote a treatise, the *Mo-fa chi-yao,* 1398) was another reason for the abandonment of the metal characters. G. Sarton remarks that at this period

Oriental typography spread west and east and achieved phonetic transcriptions: *ouïgur* Mongol, derived from the Aramaic, and *onmum* Korean, derived from Sanskrit (1419–1450), like the Japanese syllabary.

Spread of Chinese printing   This expansion toward central Asia explains the transfer of printing to Egypt by the Mameluks during the period 900–1350. Egypt, however, is a remarkable exception in the Arab world. Islam permitted the use of paper to write the word of Allah, but not printing. (The first Arabic-language book was not printed in Cairo until 1825.) The spread of printing through Eurasia was thus blocked by the Moslem barrier. Europe had to reinvent, after and independently of the Chinese, xylographic printing and then printing with movable characters. This required an alloy that would permit the making of suitable metal characters, a method for the precise assembling of the characters, and an oil ink that would produce a varnish. These three discoveries, accomplished in Mainz around 1440, permitted the rapid spread of the "wonderful secret," despite the protests of the Italian Humanists (this subject will be discussed in the second volume of this work).

European printing appeared in the Far East with the romanization of Japanese (*romaji*, 1548). For a long period the missionaries employed both the new printing and the traditional xylographic printing (China, 1584). The earliest known books printed with metal characters are Aesop's fables, printed in romanized Japanese in 1593, and the *Contemptus Mundi* of Antonius Harada (1610). The Jesuits sent xylographed books to Rome; they were greatly admired by the Europeans. In 1546 the Venetian Paul Giovo saw a Chinese book presented by Don Manuel, King of Portugal, to Pope Leo X. He was astonished that the Chinese printed in the European manner, and wondered if printing had not come to Europe from China by way of "the Muscovites and the Scythians." In his essays (1588), Montaigne, after a visit to the Vatican Library, noted that while European printing was regarded as a miracle, "other men at the other end of the world, in China, were using it a thousand years ago."

Xylographic printing was perfected, and permitted the printing of illustrated books, playing cards, and bank notes. Xylography and color lithography, also common in the Yüan period, made possible the printing (in the Ming Dynasty) of remarkably illustrated books and prints for popular consumption. The first religious manuscripts, written on paper by the Manichaeans of Chinese Turkestan, were richly illuminated; they started a fashion that was imitated during the Arabic, Jewish, and Christian Middle Ages and that required the work of numerous specialists, scribes, calligraphers, miniaturists, illuminators, and, later, bookbinders. China appears to have almost completely ignored this fashion.

Imperial and private Chinese libraries are characterized less by beauty of bindings and plates than by the abundance and utility of the books. The National Library of Peking now contains 4,200,000 volumes dating from the Sung, Ming, and Ch'ing periods. The encyclopedias *Szu-K'u* (of the four deposits) and *Ku-chin* (old and new) were preserved in the *Wen-yüan-ku* (1782).

DWELLINGS

We possess no documents describing the manner in which the domestic dwelling place was conceived, in early ages, in a space as vast as China. Only in the case of the official (that is, civil, military, and religious) architecture do archaeological discoveries, historical narratives, and treatises on architecture permit us to reconstruct a developmental diagram that begins in the sixteenth century B.C.

The rural house — Huts made of clay or branches, and houses on piles, were exceptional in the Chinese civilization. The typical Chinese building was rectangular, and was roofed by means of a horizontal superposition of binders "working by flexion," and capable, when necessary, of covering large areas.

The rural house does not seem to have changed greatly in the course of the centuries. It was essentially a rectangle, 10 feet by 6½ feet, built on a layer of packed earth; several of these rectangles could be placed end to end or at right angles to each other in an L or U shape around a courtyard. The framework consisted solely of corner posts that were not inserted into the ground but simply rested on stone bases; they were connected at the top by crossbeams. This framework exerted a downward rather than outward thrust. The walls, which did not support the roof, served only to fill the empty spaces and to isolate the house from the outside; they were of mud or packed earth (in northern China) or brick (in southern). The roofs, which were sloping (except in northwestern China, where terraces were used as in central Asia), were of straw (wheat, kaoliang, rice). Tile, which was common in central and southern China, was an indication of a certain amount of wealth. The subterranean dwelling existed in Honan, Shansi, and Shensi. As in Spain, France, and eastern Europe, there were subterranean villages and cattle sheds, abundantly supplied with water, in the loess and rocky regions. In the southeast, genuine floating villages were formed by hundreds of riverbeoats.

The city dwelling — The city house was built on the same principles as its country cousin, with a sculptured roof, a paved floor, and usually no ceiling. It consisted of several groups of intersecting buildings separated by courtyards and gardens. Screens (*yin-pei*) warded off evil influences. These houses were squeezed within ramparts that were sometimes very high (forty-nine feet) and sufficiently wide to allow two horsemen to pass side by side on their rounds. Inside, walls and secondary gates often isolated the various quarters from each other.

Official architecture — The official architecture first appeared during the protohistoric dynasties. Ch'in shih-huang-ti, who ordered the completion of the Great Wall, also ordered numerous palaces built, and in an area slightly more than 621 miles around the capital city of Hsien-yang there were more than 270 imperial residences with parks, like those of Suchow, famous since the Ch'in Dynasty, and miniature gardens, which were decribed for the first time by Wang Wei (699–759).

Ming imperial architecture was characterized by two types of construction. There were light buildings built of wood and bricks and roofed with tiles; they were colonnaded, an ancient style that had hardly been modified. There were also heavy buildings often roofed with cemented stones. The former were long, rectangular parallelograms (*t'ing*) with numerous individual columns without capitals, and a framework that was sturdy but also abundantly decorated and sculpted. The walls did not play any supporting role; the whole structure was open, airy, and supplied with verandas. The building was constructed on a terrace connected by three stairways (two side stairs and one in the center) to a vast esplanade. For a long period flat roofs were used. (The roof with raised corner rafters, characteristic of Chinese architecture, did not appear until around the eighth century.) The roof was a two- or three-story affair, and thus might occupy half the total height of the dwelling. The importance of each roof in relation to the others varied. They were covered with varnished round tiles, the color of which was determined by the purpose for which the building was intended.

Palaces and cities   Astronomical observatories were built for the official astrologers, and watchtowers for security purposes. The capital was a cluster of palaces built along a north-south axis, enclosed within parallel walls. This technique of "squares" (*fang*) oriented toward the cardinal points was inherited from model imperial cities such as Changnan (the capital during the Sui and T'ang dynasties). At Peking, for example, it was necessary to cross first the Chinese city and then the Tatar city in order to reach the imperial city, in the center of which was located the forbidden Purple City, the residence of the Son of Heaven. The walls of the Tatar city, rebuilt in 1406–1427, were 46 feet high, 62 feet wide at the base and 55¾ feet wide at the top, and 14¼ miles long.

The tomb of the sovereign was often built in the grandoise proportions of a palace, as for example that of the Ming Emperor Shen-tsung (1573–1619), which was discovered in 1959.

Religious architecture   There were various varieties of religious architecture. For example, there was the minuscule courtyard pagoda consecrated to a domestic divinity, the national or regional temple (*tsung-miao, wu-miao, uen-miao*) built in memory of a king, a genius, or a famous person, the temple honoring Confucius (551–579 B.C.) or his disciples (*k-ung-miao, wen-miao*), and the Taoist (*kuan*) and Buddhist temples(*szu*), or "pagodas." The first pagoda, that of the Monastery of the White Horse (at Lo-yang in Honan Province), was built between 58 and 75. Since then Buddhist temples have appeared in many different aspects. There are, for example, underground temples or caves that, when suitably prepared and decorated with statues and frescoes, served from the fourth to the seventeenth centuries as shelters for the greatest Buddhist sanctuaries. The site of Tunhuang in Kansu included, in an area of 1¼ miles, 480 caves with 2,400 statues sculpted between the years 386 and 1368. The caves of Mai-tsi-chan in Kansu contained more than a million statues ranging in date from the fourth to the seventeenth centuries. In both

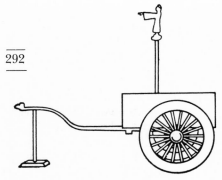

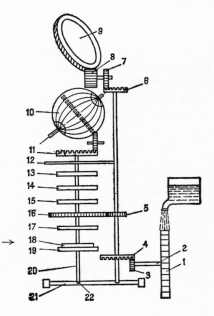

Fig. 26. The "south-pointing chariot" from the *Chung-kuo ku-tai k'u-hsiu-chia* (The Scientists of Ancient China), published by the Academy of Sciences, Department of the History of Science, Peking, 1959.

Fig. 27. Astronomical clock of Su Sung and Han Kung-lien (1086–1089), *from Chung Kuo tsai chi-shih-ch'i fang–mien ti fa-chan* (Developments in the area of Chinese Devices for Measuring Time); Professor Lieu Hsien-chou of Ch'ing-hua University, Academy of Sciences, August, 1956.

The drawing shows only part of the mechanism: (1) A wheel turned by hydraulic power; (2, 3, 4) mechanism by which power is transmitted to a toothed wheel that turns an axle. By means of the two gear trains (5, 8) the axle turns an armillary sphere (9), a celestial globe (10), and a clock, in which the hours are indicated by the appearance of various figures.

sites the statues were accompanied by numerous frescoes.

The most typical temples were, however, the traditional low wooden buildings with single or multiple roofs. An example of this type is the Fukuang Monastery on Mount Wu-t'ai (857), the frameworks of which are the oldest in China. They are, however, later than those of the Temple of Horyuji, in the suburbs of Nara, which is one of the oldest wooden buildings in the world (607). Japan also contains the largest known wooden building, the Todai-ji, which shelters the *daibutsu* (Great Buddha) of Nara (747–751), rebuilt after a fire. It is more than 117 feet high; its east-west dimensions are more than 187 feet, its north-south dimensions 164 feet. The Golden Pavilion of Kyoto is characterized by 33 intercolumnar spaces (*Sanjusanjendo*, 1164). Rebuilt in 1266, it is 390 feet long, and is considered one of the longest wooden buildings in the world.

Multistoried stone or marble temples required the partial or total abandonment of the Chinese techniques. Examples include the white pagoda of the Miao-ying Monastery in Peking, built by the Nepalese architect Anika circa 1348, and the Temple of the Four Pagodas built in Peking by Pancha Darma around 1403. The first multistoried temples date from the time of the Empress Wu (688).

There were also Buddhist towers (stupas), the earliest of which are square, and date from the fifth century; they later evolved into hexagonal (Sung Dynasty) and octagonal and round (Ming Dynasty) forms. The oldest of these towers still in existence is that of Sung-yu Monastery, on Mount Sung-shan in Honan, built in 523; it has fifteen stories.

The architecture of the late eleventh century is known to us through the *Ying-tsao-fa shih* (analyzed by P. Demiéville in the *Buletin de l'École française d'Extrême-Orient,* 1925), that of the sixteenth through Li Yu.

Furnishings   The rural home was very sparsely furnished. In the north, the bench known as the *kang* acted as a bed, table, and heating unit, thanks to a system of interior piping that brought warm air from a fire outside the house. While the Japanese and the Koreans remained faithful to the practice of squatting on mats, the appearance of the chair, arm-chair, and couch (*kang-chuang*) around the Sung period marked the transition from the squatting to the seated position in China. Chinese furniture is some-times richly decorated and very often lacquered in black or red with the resin of *Rhus vernicifera.* This techniques dates from the Chou Dynasty (sixth to third centuries B.C.). Ming-period painted and sculptered lacquer objects are known, as well as lacquered objects colored by projection of pigmented powders on a thick, still-soft varnish.

TECHNIQUES OF TIME AND SPACE

Measuring and determining time and space

Clepsydras and water clocks   Rudimentary clepsydras for the use of the army and the astronomers came into existence in the Chou Dynasty (eleventh to third centuries B.C.). The hy-draulic clock appeared in the Han period; it measured the duration of a whole day, but could not show the seasonal inequality of the days. The same period saw the appearance of bronze ratchets with a square opening so that they could be attached to a transmission shaft, which had a gear train with sixteen teeth. This device, which was utilized as an agent for the transmission of energy, does not appear to have led to the making of clockworks. Yi-hsing (725) was the first to invent escapements, the starting point for mechanical clockworks. The clep-sydras were perfected in the Sung Dynasty by Shen Kua (1032–1096) and Su Sung (1020–1101), leading to the making of astronomical clocks, which will be dis-cussed later.

Chariots that measured distance   In 1027 Lu Tao-lung presented to Emperor Jen Chüng a chariot that could measure the distances it covered (*ki-li-ku-ch'ih*). By means of a mechanism with eight wheels, totaling 285 teeth, it moved two arms, one of which struck a drum each time a *li* had been covered, while the other rang a small bell after every ten *li*. This model was reconstructed by Giles and Hopkinson in 1909.

Knowledge of magnetism and discovery of the compass   Discussions of the problem of the compass have been made more difficult by the confusion between the "south-pointing chariot" (*cheh-nan-chi' ih*), a purely mechanical device that has no relation to magnetism (which was discovered in the third century), and the "south-pointing needle" (*cheh-nan-chen*), mentioned for the first time by Shen Kua (eleventh century). The "south-pointing chariot" is frequently mentioned in annals written between the third and the thirteenth centuries; their functioning was explained by Yen Su (1027) and Wu Tu-jen (1107). It was a very complicated mechanism composed of twenty-four toothed and two plain wheels, and two pulleys. It has been studied by Giles (1909), Moule (1924), and Wang Chen-to (1937), but none of them has been able to reproduce the model described by Yen Su.

In his 1954 study of the origins of the compass, Li Shu-hsua notes that:

1. The natural magnet and its property of attracting iron were known in the third century B.C. The jurist Han Fei-tsou (died 233 B.C.) mentions a needle that points south (*szü-nan*).

2. The attraction of an iron needle by a natural magnet is a notion that was acquired in the first century A.D. In the *Luen Heng*, Wang Chung (born 27 A.D.) describes a spoon pivoted on a square bronze platform whose handle points south when it stops turning.

3. The first mention in world literature of a magnetic needle is from the hand of Shen Kua (1032–1096), in his *Mung-ch'i pi't'an* (Essays Written at the Villa "Torrent of the Dream," the author's home). He describes its preparation by rubbing on the lodestone, the possible setting — floating, suspended, or pivoting — and its deviation toward the southeast. The pivot setting was perfected (in the south-pointing fish and the south-pointing turtle, described by Ch'en Yüan-ching) around the end of the Sung period. Tseng Kung-liang, in his *Wu-ching tsung-yao* (Principles of Military Science), 1040–1044, also speaks of the fish that points south. The second reference to the magnetic needle is found in a medical work, the *Pen-ts'ao yen-yi*, written by Ku Tsung-shih around 1116. Because of the importance of the idea of "south" in Chinese cutlure, it is possible that these devices were at first utilized for ceremonial and divinatory purposes.

4. The first mention of the use of the artificial magnet for navigation is made by Chu Yü (*P'ing-chou k'u-t'an*, 1119), but we can safely say that the compass was known before then. Hsiu Ching, who was sent on a mission to Korea (1123), describes the compass with floating needle (*feu-chen*). After his mission to Cambodia (1295), Chou Ta-kuan mentions that the ship's pilot steered according to the indications of a magnetic needle.

The date of the invention of the compass card, with twenty-four points (the azimutal) for the Chinese and thirty-two points (or sidereal) for the Irano-Arabs, is unknown. Hashimoto believes that it was combined with the compass around the end of the thirteenth century.

Astronomical instruments　　The oldest astronomical instrument is the gnomon (*pei*), which was already known in the eighth century B.C. It was a simple rod (*piao*) planted vertically in the ground, and sometimes accompanied by a template (*t'u-kuei*) that made it possible to determine certain characteristic shadows, particularly the shortest shadow, which corresponds to the summer solstice. These templates, the size of which was originally adjusted to correspond exactly to the shortest or longest length of the shadow

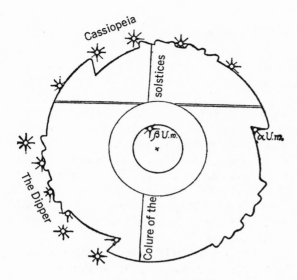

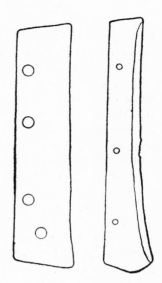

FIG. 28. The *hsiuan-chi* is an improved version of the *pi*, an instrument three thousand or four thousand years old, conceived between the eighth and sixth centuries B.C. Its center corresponds to the celestial pole as it was in the sixth century B.C. The circumpolar stars are placed in the notches on its edge. By marking on the three large teeth the twelve angles corresponding to the twelve hours of the Chinese day, this "star-dial" can be used as a chronometer, for to mark the location of the stars on their daily round is to determine what time it is. It also makes it possible to know the date of the solar solstices, and also the period for which the instrument was constructed (from Henri Michel, "Méthodes astronomiques des hautes époques chinoises," *Conférences du Palais de la Découverte*, 1959.

FIG. 29. Templates for measuring the angle of gnomon (*t'u-kuei*). In actual use they are in horizontal position. The shortest are 14 inches long; the longest, 17 inches. They were made for gnomons of 8 and 10 feet, at the latitude of Lo-yang, 600 years before Christ (from Michel, op. cit.).

in a given locality and for a given height, were reference points rather than measuring instruments. They were later graduated, as is indicated by the tablet that accompanies the bronze gnomon erected in 1049 at K'ai-fung, the Sung Dynasty capital.

A contemporary of these instruments was the *hsiuan-chi*, mentioned in the early writings of Chinese literature, whose function has long been a subject of discussion (Figure 28). It is a flat crown with notches and crenellations on the periphery of its outer circumference. It was the examination of this instrument that revealed its astronomical utilization, which until then had appeared obscurely in brief references in the *Shu-ching* and the *Chou-li*. In brief, this gauge made it possible to determine the position of the axis of the world by sighting Ursus Major, at a period when it was not in the immediate vicinity of a brilliant star.

By the third century B.C. the Chinese astronomers were also using the sighting tube (*wang-tung*), which was replaced by the telescope in the seventeenth century.

At Lo-yang, the Han capital, there was an azimuthal sundial, a square tablet of jade, 11 inches long; the diameter of the circle was 9½ inches. The circumference was graduated in accordance with the course of the sun for the longest day of the year.

In 52 B.C., Keng Shou-ch'ang presented a bronze armillary (*yüan-yi*), to the emperor, "which made it possible to measure the movements of the Sun and the Moon, and to verify the form and movements of the sky."

In 85 A.D., Chia K'uei invented an ecliptic armillary (*huang-tao-yi*), which was basically a circle with the ecliptic inclination. Around 130, Chang Heng added meridian and horizontal circles and a sighting tube to this instrument. The device operated automatically by hydraulic power. This was the equatorial armillary sphere (*hung-t'ien yi*) that was to become the classic instrument of the Chinese observatories; descriptions of it still exist.

The celestial globe (*hung-t'ien hsiang*) was a solid; it showed the equator, the twenty-eight lunar houses, the Milky Way, and the major constellations. The globe of Ch'ien Lu-cheh (South and North dynasties) was bronze; the stars were indicated by pearls. The Sun, the Moon, and the planets moved along the ecliptic; the instrument was turned by a hydraulic clock.

An improved system of armillary sphere, worked by a hydraulic clock at the of a tower, was built by Su Sung, who gives a description of it in his *Hsin-yi-hsiang-fa-yao* (1090), a book illustrated with very precise drawings.

At the beginning of the Mongol period, the Emperor Kubilai ordered a certain number of Arabic instruments from the Persian observatory at Maragheh. The history of the Yüan Dynasty describes these instruments: an armillary sphere, an alidade with two viewfinders, a dial with equal hours, celestial and terrestrial globes, and a *torquetum*, an ancient astronomical instrument that represented the movement of the equator in relation to the horizon. The astronomer Kuo Shou-ching was commanded to establish the Peking Observatory, where he installed copies of these instruments. Later innovations were made by the Jesuit missionary-astronomers.

The flat projection of the Chinese constellations is probably very old; in 672 a planisphere carved on stone was lost in the river Tai-tung. A celestial map engraved on stone in 1247 is still in existence; it shows approximately 1,440 stars.

Mapmaking   We have no exact information about Chinese maps of the pre-Christian period. Tradition claims that in the Yin and Chou dynasties special officials were appointed for the purpose of establishing maps, probably for military and political reasons, and of taking the bearings of land or sea routes. The oldest mention of what could be called a map dates from 227 B.C. In any event, cartography was sufficiently developed by the third century A.D. to permit P'ei Hsiu, Minister of Public Works under the first Chin Dynasty emperor, to formulate the principles of mapmaking with regard to the rectilinear divisions (*fen-shuai*), orientation (*wang*), distances (*tao-li*), top and bottom (*kao-hsia*), angles (*fang-hsieh*), and curves (*yü-cheh*). P'ei Hsiu's map was scaled to 2 inches = 100 *li*, and was 10 feet long. In 801 Chia Tan made a map 30 yards long, in which 1 inch = 100 *li*. It was called *Hai-nei hua-yi-t'u*, or "Map of China and the Barbarian Countries Within the Seas." Judging by its proportions, it must have included almost all of Asia. By 1093 Shen Kua had made relief maps in wood and wax.

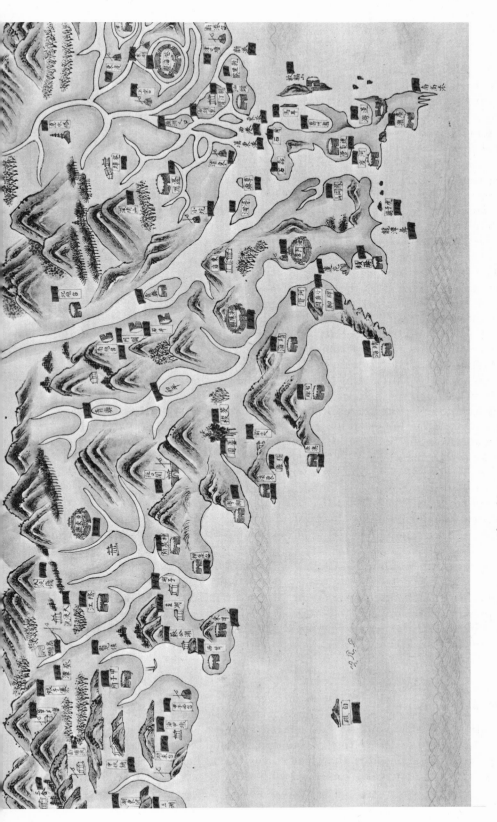

Plate 38.
Kien-Lung map (detail). Bibliothèque Nationale, Paris.
Photo Bibliothèque Nationale

The oldest extant maps are carved on a stone stele dated 1137, located in Shansi. Each face of the stele represents a map of the Chinese Empire covering approximately 7,000 *li*, with a scale of ½ inch = 100 *li*. One map may have been carved a century earlier; the other, divided into squares of 100 *li*, reveals older sources. Another map of China, also stone, carved in 1247, is a copy of an original that dates from 1193. North is at the top.

In the Mongol period a large map was placed in the encyclopedia of the administration. Made in 1320 by Chu Szih-pen (1273–1337), it measured 7 feet, and showed Asia, Europe, and Africa. The original has been lost, but several copies are preserved. A manuscript copy on silk, dated 1402, is preserved in the Buddhist University of Kyoto; it measures 5¼ feet by 4½ feet. Fuchs has shown that it is a synthesis of two earlier maps, both of the fourteenth century. The triangular form of Africa was already indicated on it.

The large nautical map of Mao K'uen has been the object of numerous studies. It was prepared as the result of a series of naval expeditions sent by Emperor Yung-lu between 1405 and 1421 to the Indian Ocean, the Persian Gulf, and the Red Sea. The original map of 1421 has been lost. The total length of the 1621 edition is 18 feet long by 8 inches wide. Routes are shown by dotted lines on which are indicated compass directions, the distance traveled during the daylight hours, and the results of soundings. The coast is seen from a distance; its profile is flattened in the interior, and all the information about the coast from Amoy to the Red Sea is given. The map has been copied many times, and its technique inspired descriptions of the coasts of Asia until the eighteenth century. One of the longest rolls of this type, the *Ngao Tung Yang lien ti t'u,* which dates from the Ming period, is preserved at the National Library in Paris. It is a large-scale map (at least 1/500,000).

The Mongol map of Chu Szih-pen was revised by Lu Hung-hsien (1504–1564), who cut it into pages and published it in 1555 in the form of an atlas, the *Kuang-yü-t'u,* in four volumes. It is known to us only through reprints. In 1655 one Father Martini published in Amsterdam a map of China taken from this atlas; in 1946 Fuchs published forty-eight of its maps under the title *The Mongol Atlas of China.*

A great number of original Chinese maps have been preserved; the National Library at Peking contains more than two hundred, ranging in date from 1600 to 1750. Neither the inventory nor the history of these maps, nor that of the maps preserved in European and American collections (the Hummel collection in Washington is rich in documents of this type) has as yet been undertaken in systematic fashion.

The Mastery of Terrestrial Space

Roads and trails   Chinese carriage roads are very old, but the imperial roads, which are standardized in relation to the length of the axles, and are two and one-half feet wide, date from the Han period. As straight as their Roman counterparts, and, like the latter, built especially for strategic purposes, they avoided curves and had watchtowers, shelters for travelers, and post relays connected with a rapid service of official messen-

Plate 39.
Double row of buried wagons (detail). Found at Liu Li Ko Hui
Hsien, Honan Province. Fourth–third centuries B.C.
Photo Britain-China Friendship Association, London

gers. This service was improved particularly under the Yüan emperors, and sometimes supplemented by carrier pigeons. The network of carriage roads served northern China in particular. Outside these roads, however, there were no good routes. On the loess plateaus the trails gradually sank into the ground, and for dozens of yards could support only heavy, solid, two-wheeled carts. In southern China the roads were frequently nothing more than paved paths, twisting between rice paddies or following along ridges with extremely steep slopes. In the mountain regions simple paths permitted transportation only by animal or man.

Land transportation   The Chinese appear to have borrowed the wheel at a very early date from Western cultures, and they made great use of it, either as a method for transmitting power or as a means of transportation. In a semicontinent as varied as China, numerous forms of transportation came into existence: the sledge, horse cart or oxcart, donkey, camel, yak, packhorse or saddle horse, and transportation of passengers, whether by sedan chair with two rods, a hammock with one rod, or in a wheelbarrow with center wheel, sometimes provided with an auxiliary sail.

The saddle, the stirrup, and the bit, which made possible the cavalry charge (third century B.C.), were invented in an indefinite area extending from central Asia to western Siberia. This invention gave the horsemen of the steppes (who also used metal armor) an obvious superiority over the sedentary peoples both Western (for example, the defeats of the Emperor Valerian in 260 and of the Emperor Valens in 378) and Chinese. Haudricourt has demonstrated that a line stretching from central Vietnam to the Baltic creates two zones: to the south was the area of the Indo-Greek chariot with pole and with two animals harnessed abreast, while in the northern zone this type was replaced by a chariot with shafts instead of a pole, and drawn by a single animal. The earliest depictions date from the Han Dynasty, but its existence is perhaps older (Figure 30). In any event, a mechanism with a rational method of draft well adapted to the conformation of the horse was discovered by the Chinese 1,400 years before the Europeans discovered it (Lefebvre des Nouettes). It consisted of two shafts prolonged into a forked yoke, for support rather than draft purposes, attached at the ridge of the neck very close to the ears; the system was completed by the reins, a breast strap, and a breech band, the two latter items being attached to the shafts and not to the chariot itself. The yoke gradually disappeared and the shafts were shortened; by the eighth and ninth centuries they were straight and were supported by a ridge strap, while the breast strap continued to be attached to the shafts. The shoulder collar attached by traces to the vehicle, the whiffletree, and the horseshoe were inventions after the sixteenth century.

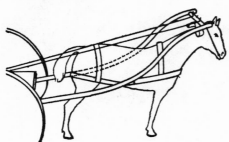

Fig. 30. The Chinese horse harness at the time of the Han Dynasty (from Lefebvre des Nouettes, *La conquête de la force motrice animale: Chine et Japon*).

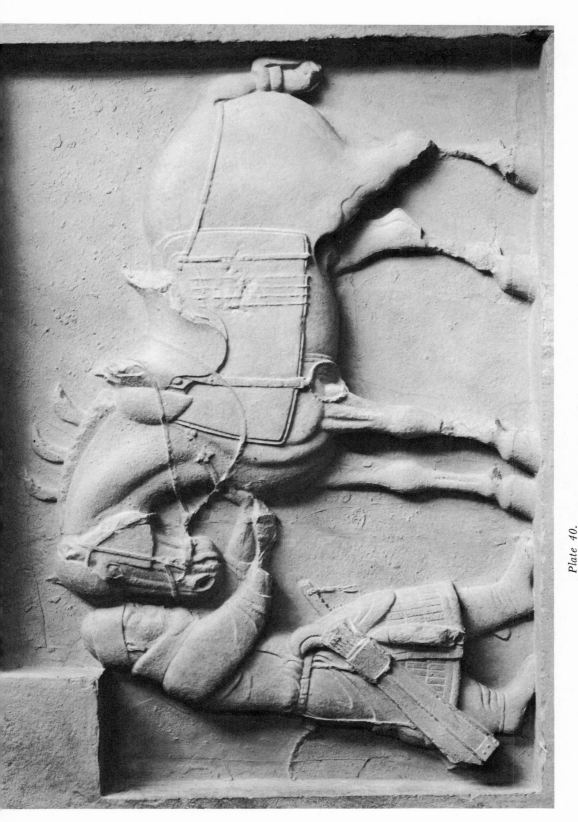

Plate 40.
A harnessed horse. Bas-relief on the tomb of the Emperor
Tai Tsung, seventh century. *Photo Giraudon*

Plate 41.
Head of a hub. Warring-Kingdoms period. Musée Cernuschi, Paris.
Photo Giraudon

Mastering Water

River hydraulics　　The Chinese civilization has always had to contain the inundations of the major rivers and drain the swamplands. Since Emperor Yü the Great the millions of river dwellers in the lower basin of the Yellow River, the scourge of the sons of Han, have had to build and rebuild enormous dikes under the direction of special officials, in order to "open the door of the dragon" and "clean the nine rivers."

In the Warring-States period Si Men-po (Wei Kingdom, 424–287 B.C.) ordered the digging of twelve canals along the Shang River, a branch of the Yellow River, to absorb the surplus water resulting from the floods. Li Ping (Ch'in State, circa 300 B.C.), with the help of his son Eul Lang, finally conquered the Min River, a branch of the Blue River, by the *Tu-chiang-yen* system, in which the Min was divided into internal (canal) and external (the natural bed of the river) channels. The canal irrigates the entire plain of Ch'eng-tu, and stone and bamboo dikes slow down the dispersion of the canalized waters into 520 secondary canals. This system has been in existence for 2,200 years.

The State of Ch'in achieved superiority vis-à-vis her neighbors with Chen Kuo's numerous canals along the three rivers Wei, King, and Lu — one of the reasons why its royal family became the first imperial dynasty. In the fourth century B.C. the Blue and Hai rivers were united by canals, forming the embryo of what would become in the Sui Dynasty the Grand Canal. The Kin Canal, along the Ning-hsia plain (a distance of forty-seven miles) turned this plain into an oasis. It has been in existence for twenty-two centuries.

Chia Jang (seventh century B.C.) decided to build a stone dike for miles along the north bank of the Yellow River at Hsiun Hsien (Honan), with lower sluice gates to irrigate the plain in time of low water, and upper gates to permit the draining off of the excess waters in floodtime. The project was not accepted. Later, Emperor Yang-ti (605–618), using gigantic forced labor crews, ordered the digging of the Grand Canal, for geomantic as well as economic reasons; starting from Hangchow in the south, it crossed Lo-yang, the capital, and ended at Chu-chün in the north. This waterway, a combination of several earlier canals, included a main section, the *T'ung chi ch'u*, and two secondary branches, the *Chiang-nan hu* and the *Jung chi ch'u*. A dike built in the seventh century between Wu-sung and Hangchow (a distance of 124 miles) created a polder that stabilized the branches of the delta of the Yang-tsou chiang.

Kuo Shou-ching (1231–1316), a great mathematician and engineer, lengthened the Grand Canal by the Tung-huei-hu Canal, which linked Peking with Tungchow. Sun Li (circa 1375, a hydraulics engineer, completed the Hai River system, which included the Grand Canal, by major projects that required 300,000 forced laborers for a period of seven months. In the Ming period P'an Chi-hsiun (second half of the sixteenth century) devoted his life to mastering the Yellow River. His order — to construct dikes to retain the water and let the water carry away the sand — summarizes the ideas of the old Chinese hydraulics engineers.

Bridges　　In the case of bridges, achievements varied greatly. Flexible rattan or bamboo bridges suspended over mountain torrents were the only bridges known to the non-Chinese peoples in

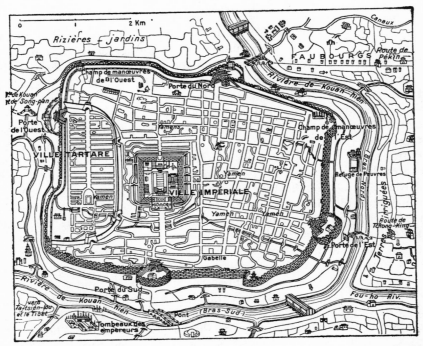

Fig. 31. Map of Ch'eng-tu, from a Chinese map *(taken from Variétés sinologiques)* (from J. Sion, *L'Asie des Moussons* [Armand Colin, 1928]. Note the various types of bridges.

the Chinese territory, but the superior Chinese civilization had various types. The wooden variety were either plain or surmounted by wooden houses with tile roofs (covered bridges). Iron bridges and bridges suspended by means of iron chains (eleventh century) (Needham) were built. Stone bridges were straight, with roadbeds paved with flat stones; they were used to cross small streams. Sometimes they were vaulted and had arches. The first known single-span bridge is that of Chao-chun (Hopei), which dates from the beginning of the seventh century. The bridge of Lu ku chiao (1189–1194), the so-called "bridge of Marco Polo," was 350 paces wide, and had 11 arches.

The bridge of the Myriads of Longevity (*Wan-shui-chia*), constructed in 1303, crossed the Min River at Fuchow; it was 3,100 feet long, and had 39 arches. Another bridge, 8,266 feet long with 362 arches, was constructed at the same period at Ch'iuan-Chu. Brick bridges were used to cross the moats in front of citadels.

Navigational techniques   The boats used in the Chinese territory were of various types. The rafts on waterskins used on the Yellow River were completely different from the circular wicker boats of the south. Paddle boats were constructed at various periods, especially in the Sung Dynasty.

The sampan (etymologically speaking, a light boat made of three planks) was sometimes very carefully constructed; however, only special ships which

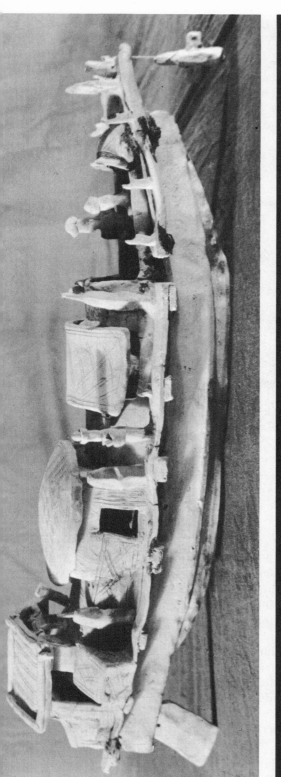

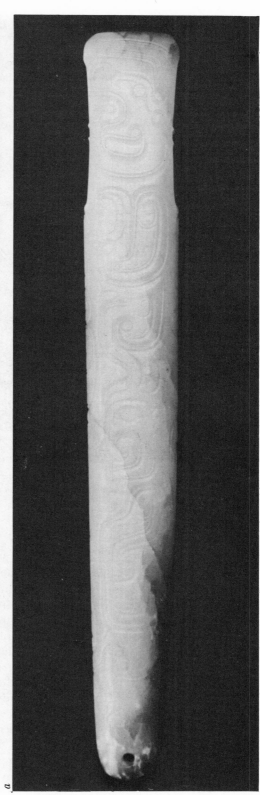

Plate 42.

a. Model ship found in a tomb to the east of Canton. Late Han Dynasty. *Photo Britain-China Friendship Association, London*
b. Chisel decorated with dragons. Warring-Kingdoms period. Private collection, Kansas City. *Photo Giraudon*

a

b

served as a model for Nordic racing yawls, were seagoing. We shall limit our study to these, based on the study by P. Paris. Beginning in 219 B.C., the fleet of Ch'in shen-huang-ti, utilized for the conquest of Yü, had multidecked boats. These were probably riverboats that may have been capable of short coastal voyages. The large seagoing junk may have existed in the ninth century; it definitely existed by the eleventh century. In the form in which it is depicted in the Bayon at Angkor, it was probably born between 817 and 1200 (G. Groslier and P. Paris).

FIG. 32.   Seagoing junk
(from the *T'ien-kung k'ai-wu*, op. cit.).

After this period the ships of Fukien and Kwangtung were able to reach Timor, the Spice Islands, and Singapore, and later the Indian Ocean. At the end of the twelfth century an army of 100,000 men set sail in 4,000 ships bound for Japan, while another fleet sailed for Java. At the end of the thirteenth century the geographer Chou Ta-kuan, visiting Angkor, remarks "that here it is the Chinese who become sailors." The Chinese Navy reached its peak with the fleet of Admiral Sheng Hu, whose ships, 440 feet long and 180 feet wide, reached East Africa in 1431–1433. They had bulkheads (Niccolò de Conti, fourteenth–fifteenth century) and numerous cabins (Marco Polo, 1275–1292); the navigators had compasses and maps, and followed precise routes.

Also characteristic of the junks was their flat receding bow, their midship frame placed slightly to the rear of the middle of the ship, their sterncastle and their lack of keel.

The tiller oar permitted perfect mastery of the oar in forward sculling (a practice that contrasts the Chinese with the Indians, Arabs, and Europeans, who use backward sculling). The stern oar with automatic angle of attack, constantly maintained at the same height by a rope joining its handle vertically to

Fig. 33. A Sung ship with sails, oars, and paddlewheels (from the *Fang-hai chi-yao,* twelfth century).

the bridge "is one of the most ingenious of those Chinese inventions which always tend to increase the productivity of the human effort" (P. Paris).

The sternpost rudder with vertical axis was known to the Europeans and the Arabs in the thirteenth century, with the Europeans perhaps slightly in advance of the Arabs (Guilleux La Roerre). It is seen in Sung paintings, where it appears to be suspended rather than fixed to the stern by metal hinges.

The use of battens to spread the sails made it possible to tack ahead and to change tack with a single movement of the bar, without getting off course. This rigging was known at a very early period in China; it has been successfully adapted for European and American racers and schooners.

Movable keels, utilized especially by the *tangway* of Kuang-chou-wan, were discovered independently of (and probably prior to) the movable European keels.

Chinese naval construction served as a model for that of Korea, Japan, and Vietnam.

Mastery of the Air

Kites had been known in the Three Kingdoms period, (220–265 A.D.) and were utilized as games, musical instruments, ritual objects, transmitters of messages, measurers of distance, weapons, and parachutes. The latter idea was probably borrowed from Europe (Duhem). The combination of wings with a screw propeller is pictured for the first time in the illustrations of a Chinese novel (1819). These are not sufficient to prove, however, that the Far East invented the glider (Montandon, Giles, B. Laufer).

MILITARY TECHNIQUES

War in China was usually ceremonial in character, its goal being less to annihilate a force than to limit the size of the battle, while attempting to give style to the instinct of destruction, like the medieval horsemen, the Renaissance condotierri, and the generals of the baroque period. In addition, geomancy continued to be of considerable importance in the construction of fortresses and strategic and tactical ideas. The same cosmological diagrams were used to explain the structure of an organ, a therapeutic prescription, an alchemical transmutation, or the destiny of a dynasty. Moreover, in ancient China sidearms had a social and symbolic value that was more important than their efficiency.

The first treatise on military art reveals, around 500 B.C., a rather heavy army equipped with rhinoceros-hide breastplates and bronze weapons, with chariots playing the role of cavalry. In the Han period the cavalry was organized along the lines of the barbarian model. From central Asia came the automatic magazine crossbow, which was stretched with the thumb protected by a stone or metal ring. These bows came in various sizes, and hurled light and heavy arrows, flaming arrows, and even clay or stone balls. Near the royal coffin was placed an automatic bow device, the purpose of which was to kill possible violators of the tomb with its crossfire.

Wang Ngan-shih (Sung Dynasty) instituted a major military, administrative, and technical reform. A bureau of studies was created to experiment on and perfect military matériel, and rewards were established for new inventions.

Fortifications

The largest Chinese fortification is the Great Wall — a grandoise creation, not to be compared with the Roman, Rhenish, or Danubian limestones. Around 220 B.C., ancient individual fortifications were combined to form an earthen rampart thirty feet high, the sides of which were paved with stones or bricks. There were two watchtowers per *li*, stationed at points of strategic interest (by the Ming period there were 20,000 of them). Runoff of water was ensured by a double canal running along each side of the wall. Every emperor after Shih Huang-ti maintained the wall, which kept out barbarians and wild beasts and served as a symbol of the grandeur of the empire. It had to be rebuilt several times during the T'ang and Ming periods, and the structure we see today dates for the most part from this period. The labor force recruited for the initial construction and for repairs was on the scale of the wall itself; it consisted of forced-labor battalions of more than 100,000 men, whose mortality and misery have inspired much literature.

Mo-ti (Warring-States period) describes the penetration of strongholds with saps and tunnels. He also mentions rolling towers that carried armored soldiers.

Artillery

Mo-ti also describes propulsion machines that could hurl projectiles to a great distance. However, we lack precise information on this subject.

The use of fire as a weapon evolved in three stages. There was first the construction of "fire dragons" moved by hot air; they projected smoke and flames that were intended to terrify the enemy. Such devices were used by the Carolingians, the Byzantines, and the Mongols; perfected by the German engineers of the fourteenth and fifteenth centuries, they were still being used by the Russians when Napoleon invaded that country in 1812.

Pyrogenous and toxic projectiles (*huo-ch'iu*) were made from an explosive powder (*huo-yu*) and petroleum, sulfur, or lime. They were not self-propelled, but were hurled with a bow (*huo-chien*) or carried on flaming arrows (970) or catapulted by a purely mechanical gun (*huo-p'ao, p'ao-ch'eh*). Incendiary grenades, reminiscent of the Greek fire Callinicos, 678) and the flaming sticks of Marcus the Greek (eighth century), are mentioned in the year 1000. Did the Chinese, who learned of treacle and cloisonné enamels (eighth century) from

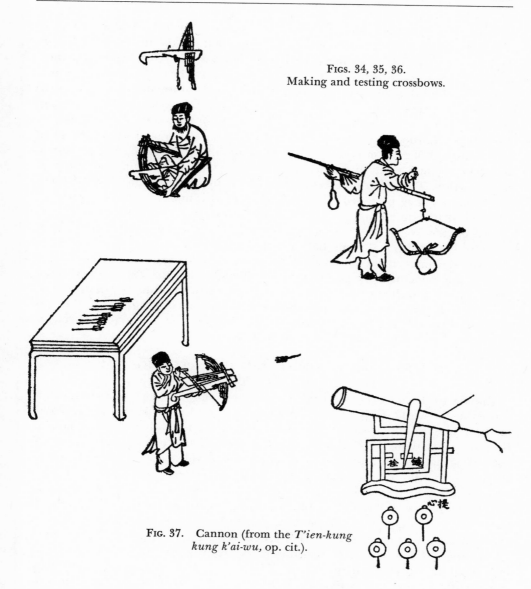

FIGS. 34, 35, 36.
Making and testing crossbows.

FIG. 37.   Cannon (from the *T'ien-kung
kung k'ai-wu,* op. cit.).

Byzantium, perhaps also learn the formula for explosive preparations, or were similar inventions created simultaneously in the Near and Far East? It is difficult to answer this question.

In the third stage, the Chinese Army used projectiles that were self-propelled by means of an explosive powder made from saltpeter and sulfur and directed by means of a tube; these were portable cannons (*t'ieh-huo-p'ao*) constructed at first of bamboo (1257), then in metal (1275). This stage corresponds to the invasion of China by the Mongols, some of whom were victims in the explosion of several Chinese arsenals, and who found themselves being opposed by major artillery.

Cannon powder became known to the Persians under the name of "Chinese salt," and to the Arabs as "Chinese snow" (*talga sin*), in the thirteenth century. Both peoples utilized it in different types of firearms. To what extent did they modify the Chinese discovery? We know, moreover, that the Polos gave the Mongols technical advice on catapults and cannons — information that was used to good advantage in the taking of Hsiang-yang (1273), one of the last Sung strongholds. To answer the question, we would have to establish a comparison between the European and Chinese arsenals at contemporary periods. As soon as they learned of cannon powder (Roger Bacon) (1248), the Europeans neglected its incendiary and toxic properties in order to reinforce its explosive and propulsive action. Thanks to progress in metallurgy, by the second half of the fourteenth century they possessed heavy artillery. Thus cannons seem to have been invented in the same century at both ends of Eurasia.

When the Portuguese arrived in China (1517), the Ming army had heavy artillery, but traditional arms had lost none of their prestige. In Europe, in contrast, the importance of firearms was considerable, and the European artillery were undeniably superior to the Chinese cannons. The Chinese purchased pieces of ordnance at Macao. In the seventeenth century a priest, Father Schall, created a cannon foundry (1636), which made it possible for the imperial armies to defeat the Russians and the Mongols and to conquer Formosa.

THE STABILIZATION OF CHINESE TECHNOLOGY

Beginning in the fifth century B.C., Chinese technology was one of the most advanced in the world. China continued to advance under the Han emperors, and retained her brilliant place until the beginning of the Ming Dynasty (1368). González de Clavijo, the Ambassador of Castille to Tamerlane (1402), tells of an old saying that the Chinese have two eyes and the Franks one, while the Moslems are blind. It can be said that at this moment Western technology and Far-Eastern technology were on equal footing, more advanced craftsmanship and knowledge having reached the "Vinci-an stage" of knowledge at both ends of Eurasia, with perhaps a few elements of the Galilean method (J. Needham).

However, when the Ch'ing Dynasty succeeded the Ming 276 years later (1644), the Far East was still a medieval society, whereas modern science and technology had been born in Europe. This gap widened during the following three centuries, for reasons that are undoubtedly numerous and difficult to isolate. At least some of these reasons, however, were as follows.

Opposition to progress   The major inventions (printing, cannon powder, the compass) had only slight repercussions in the Chinese world. They were not always discussed in special treatises (Shen Kua's description of the compass in the eleventh century appeared in a literary anthology), and, moreover, met with considerable opposition in ruling circles. Despite its supply of artillery, the strategy of the Chinese army remained ceremonial, astrological, and geomantic in character. Despite the tremendous development of the techniques of navigation, and the astonishing trips of Cheng Hu, long trips were forbidden.

Neo-Confucianism The principal factor responsible for this behavior was Neo-Confucianism. Beginning in the Han period, it possibly betrayed the naturalist thought of Confucius, and definitely stifled the development of the later schools of thought — the jurists, and the followers of Mo-ti, circa 480–397 B.C. — which had glimpsed the possibility offered by logic and a scientific and technical system of thought (Hu Sheh). Neo-Confucianism, which subordinated all other cultural factors to ethics as the solution even of economic and political problems, sought to train, not scientists or technicians, but sages. A product of a bureaucratic feudal structure that was rural in character, it conditioned a static system of thought in a static society. It smothered all possibility of the development of law by opposing its paternalistic concept of justice to a juridical solution of conflicts. There is a certain relationship between the appearance of the concept of juridical law (or elaboration of a general precept valid for a large number of cases) and that of natural law. The Neo-Confucian attitude was systematically hostile to the development of commerce, navigation, and industry. The men engaged in these activities — and there were many such men in southern and central China — were extremely favorable to the development of Western technology; their progressive mentality was diametrically opposed to that of the Neo-Confucianists, for whom the idea of permanence and a return to the Golden Age was the only acceptable one. As a result of a strange confusion between the bureaucrat and the intellectual (S. de Beauvoir), the megalomanic intellectuals turned their sociological system into a theocracy that prevented the manifestation of all heterodoxical opinion and activity.

Geographical isolation The development of Western technology was the fruit of constant communication among national and foreign technological centers, which stimulated the activity of the inventors. These conditions did not always exist in China, where there was a succession of empires separated by long periods of anarchy during which a great part of the acquired scientific knowledge may have been lost. China was from the very beginning of her existence a federation of independent states, rather than a unified country, in which numerous obstacles were opposed to free exchanges among the various intellectual centers. Despite the spread of printing, the great Sung mathematicians used different methods, and did not exchange information about their work. These methods were forgotten during the Ming period and, like certain methods of porcelain production, were not rediscovered until the eighteenth century. On the international level, the cessation of the development of technology was felt throughout Asia. China ceased to receive, either by sea or over the Silk Route, those contributions that had effectively stimulated the Chinese scientific milieus during the T'ang, Sung, and Yüan dynasties.

The language barrier A final factor that militated against technical development was the absence of adequate linguistic equipment. Because of the unlimited singularization of its monosyllabic words and the lack of a syntax and grammatical forms in which subject, verb, and adjective develop, the Chinese language was unable to create the fundamental abstract words that promote mental operations (M. Granet). It never experienced the development of a logic similar to that of Greece or India, which permitted the construction of an abstract Euclidian geometry or a Buddhist metaphysics. The inadequacy of ancient Chinese mathematical thinking results, in the opinion of Mikami and Fu Szu-nien, from the absence of the idea of rigorous proof, which is linked to the inadequate development of formal logic and associative thinking (J. Needham).

We might add that the classic written language, which was rich in clichés and literary metaphors, did not favor the exposition of technical inventions. The transposition of such inventions from the spoken to the written language met with great difficulties. During the Sung Dynasty efforts were made to introduce the spoken elements into the written language, which resulted in a great development of technology.

Static nature of philosophical concepts Additional factors that distinguished Europe from the Far East were the way of life and the psychology that were born of the Reformation and the Renaissance. In Europe the "Gothic" Age was succeeded by the "Promethean" Age, an introverted philosophy by an extroverted philosophy. That is, instead of persisting in the negative Sino-Hindu and proto-Christian attitude toward the world, Europe was to attempt to reach an understanding with reality and to achieve, through comfort and social reforms, a "Kingdom of God" that could be "of this world." Action became superior to contemplation; the cosmological concept of culture failed. Religion, now detached from the world, was henceforth restricted to the realm of the soul (Luther). Universalist concepts were replaced by individualistic, nationalistic concepts. Politics became a technique that ensured the supremacy of the state over culture (Machiavelli); man ceased to be the servant of the world in order to become its master.

These changes were made possible, as Needham has shown, by the emergence of a capitalist, manufacturing, and mercantile economy favorable to the spirit of enterprise, which destroyed the mythic, qualitative concept of time, space, and cosmography. These factors were henceforth expressed in uniform figures valid for the entire world. From this resulted the diffusion of calculation and mathematical factors together with acquired biological knowledge. Under these socioeconomic circumstances the practical applications of science underwent a considerable development, and suggested to man that he could acquire a power increasingly superior to that of the natural forces that threatened him.

BIBLIOGRAPHY

The most comprehensive work in Far Eastern technology is still in the process of publication. It is JOSEPH NEEDHAM, *Science and Civilization in China* (Cambridge University Press, 1954–), of which seven volumes are projected, some comprising two or three parts, each of volume size, and making a total of eleven volumes. By 1969, seven volumes (through Vol. IV, Part 3) had been published. In addition, Needham's monumental scholarly project has spawned a number of other volumes, by Needham and his collaborators, some of which are mentioned below:

BAILLIE, GRANVILLE HICKS, *Clocks and Watches, an Historical Bibliography* (London, 1951).

BARNARD, NOEL, *Bronze Casting and Bronze Alloys in Ancient China* (Tokyo, 1961).

BEDINI, SILVIO A., *The Scent of Time*, American Philosophical Society Transactions, n.s. 52, Part 5 (Philadelphia, 1963).

BLUM, ANDRÉ, *On the Origin of Paper* (New York, 1934).

———, *The Origins of Printing and Engraving* (New York, 1940).

BOYD, ANDREW, *Chinese Architecture and Town Planning, 1500* B.C.–A.D. *1911* (Chicago, 1962).

BURTON, W., and R. L. HOBSON, *Handbook of Marks on Pottery and Porcelain* (London, 1928).

BUSHELL, S. W., *Description of Chinese Pottery and Porcelain* (Oxford, 1910).

CARTER, THOMAS FRANCIS, *The Invention of Printing in China* (New York, 1925); 2nd ed. revised by L. CARRINGTON GOODRICH (New York, 1955).

CHANG, KWANG-CHIH, *Archeology of Ancient China* (rev. ed., New Haven, Conn., 1968).

DAWSON, RAYMOND (ed.), *The Legacy of China* (Oxford, 1964).

FORBES, R. J., *Studies in Ancient Technology* (11 vols., Leiden, 1955–).

FRANKEL, J. P., "The Origin of Indonesian 'Pamor,'" *Technology and Culture*, 4 (1963), 14–21.

GOODRICH, L. CARRINGTON, and CHIA-SHENG, FENG, "The Early Development of Firearms in China," *Isis*, 36 (January 1946), 114–123, 250–251.

GROOT, GERARD J., *The Prehistory of Japan* (New York, 1951).

HARTWELL, ROBERT, "Markets, Technology, and the Structure of Enterprise in the Development of the Eleventh-Century Chinese Iron and Steel Industry," *Journal of Economic History*, XXVI, No. 1 (March 1966), 29–58.

HEEKEREN, H. R. VAN, *The Stone Age of Indonesia* (The Hague, 1957).

———, *The Bronze-Iron Age of Indonesia* (The Hague, 1958).

HETHERINGTON, A. L., *Chinese Ceramic Glazes* (London, 1937).

HOMMEL, RUDOLF, *China at Work* (New York, 1937).

HUARD, P., and M. DURAND, *Connaissance du Vietnam* (Imprimerie Nationale, 1954).

HUARD, P., and M. WONG, *Chine d'Hier et d'Aujourd'hui* (Horizons de France, 1960).

HUNTER, DARD, *Papermaking* (2nd ed., New York, 1947).

KARLGREN, BERNHARD, "New Studies on Chinese Bronzes," *Bulletin of the Museum of Far Eastern Antiquities*, 9 (1937), 1–117.

KATES, G., *Chinese Household Furniture* (New York, 1948).

KIRBY, E. STUART, *Introduction to the Economic History of China* (London, 1954).

KO HUNG, *Alchemy, Medicine, and Religion in the China of* A.D. *320; The "Nei P'ien" of Ko Hung*, transl. and ed. by J. R. Ware (Cambridge, Mass., 1966).

LATOURETTE, KENNETH S., *The Chinese, Their History and Culture* (3rd rev. ed., New York, 1946).

LI, CH'IAO-P'ING, *The Chemical Arts of Old China* (Easton, Pa., 1948).

LI CHI, *The Beginnings of Chinese Civilization* (Seattle, 1957).

LIN, HSIEN-CHOU, "Wang Cheng and the First Book on Mechanical Engineering in China," *International Journal of Mechanical Science* (Oxford), 2 (Oct. 1960), 30–39.

LOEHR, MAX, *Chinese Bronze Age Weapons* (Ann Arbor, 1956).

MITUKUNI, YOSHIDA, and KIHEI, KOYANA, *Western Asia at Work* (Kyoto, 1964).

NEEDHAM, JOSEPH, "The Pre-Natal History of the Steam-Engine," Newcomen Society *Transactions*, 35 (1962–63), 3–58.

———, "Chinese Priorities in Case Iron Metallurgy," *Technology and Culture*, 5 (1964), 398–403.

———, *The Development of Iron and Steel Technology in China* (Cambridge, 1964).

————, "Science and Society in East and West," *Centaurus,* 10 (1964), 174–197.

NEEDHAM, JOSEPH, LING, WANG, and PRICE, DEREK, *Heavenly Clockwork: The Great Astronomical Clocks of Medieval China* (Cambridge, 1960).

READ, THOMAS T., "Chinese Iron: A Puzzle," *Harvard Journal of Asiatic Studies,* 2 (December 1937), 398–407.

SAKAKIBARA, KOZAN, *The Manufacture of Armour and Helmets in Sixteenth Century Japan,* ed. by H. Russell Robinson (Rutland, Vt., and Tokyo, 1963).

SAMOLIN, WILLIAM, "Technical Studies of Chinese and Eurasian Archeological Objects," *Technology and Culture,* 6 (1965), 249–255.

SUN, E-TU ZEN, "Wu Ch'i-chun: Profile of a Chinese Scholar-Technologist," *Technology and Culture,* 6 (1965), 394–406.

WANG, LING, "On the Invention and Use of Gunpowder and Firearms in China," *Isis,* 37 (July 1947), 160–178.

WERTIME, THEODORE A., *The Coming of the Age of Steel* (Chicago, 1962).

————, "Asian Influences on European Metallurgy," *Technology and Culture,* 5 (1964), 391–397.

WITTFOGEL, KARL, *Oriental Despotism* (New Haven, 1957).

WU, CHIN-TING, *Prehistoric Pottery in China* (London, 1938).

YABUUTI, KIYOSI, "Sciences in China from the Fourth to the End of the Twelfth Century," *Journal of World History,* IV, 2 (reprinted in Guy Metraux and François Crouzet [eds.], *The Evolution of Science* [Mentor Book MQ505, New York, 1963], 108–127).

YING-HSING, SUNG, *T'ien-Kung K'ai-Wu, The Exploitation of Natural Resources,* transl. by E-Tu Zen Sun and Shiou-Chuan Sun (University Park, 1966).

CHAPTER 11

TECHNOLOGY IN INDIA

The history of technology in India has at its disposal a large but still unexploited, an even uncataloged body of material. This chapter cannot claim either to fill this gap or to measure its exact extent, but must confine itself to revealing its existence and to indicating the changes of perspective that will be indispensable once the study of Indian technology is further advanced.

Present state of our knowledge — On the map of industries and crafts, India is represented by a white area sparsely sprinkled with black dots. This sparsity is due to the immensity of the subject, which has discouraged researchers. Eighteenth- and early nineteenth-century observers, however, were already drawing attention to certain Indian techniques, such as the dyeing and printing of fabrics, and to the Indian looms, which they described and drew and brought to public attention through their private collections of popular figurines.

India herself has never lost interest in her own technology. Bas-reliefs and painting have always depicted scenes of contemporary life in which craftsmen are busy at their work, outside or in the streets, as they are still to be seen today. In addition collectors have always been interested in the earthen, wood pulp, or colored wooden figurines commonly sold at least as early as the nineteenth century; an example is the Lamare-Picquot collection, which dates from the first half of the nineteenth century and which is now preserved in the Musée de l'Homme in Paris.

Indian technology did not benefit to the same degree as that of China from the research of the Jesuits in the eighteenth century, nor did it benefit from translations of technical treatises, because the Indian presentation of information is much less didactic and clear than that of the Chinese. There are innumerable Indian treatises, but they are devoted solely to preferred subjects, and completely ignore the others. Moreover, they lack drawings for the description of practical procedures, which are transmitted directly from teacher to pupil without written notes. Chinese techniques were partially published, while those of India have remained more inaccessible to the public.

However, the Indian achievements described in foreign works (not to mention Indian texts), and the evidence supplied by archaeology and modern observers, reveal the wealth and importance of Indian technology. The diffusion of the ideas and scientific knowledge of India, which accompanied her commerce to upper Asia and China on the continent, and to the Indochinese Peninsula and the Indonesian Archipelago across the eastern and southern seas, brought about tremendous exchanges of a practical type between India and all eastern Asia, as well as western Oceania. We may therefore expect excellent results from a comparison of the techniques of India and eastern Asia once they have been sufficiently described and classified.

The nature of the India techniques
The Indian techniques can be divided into several major groups:

Techniques of sustenance: agriculture, cattle raising, fishing, hunting, preparation of food.

Techniques of settlement: the dwelling place, construction of cities, regional organization.

Techniques of equipment: crafts and industries.

Esthetic techniques: personal ornamentation, dance, music, fine arts.

Physical techniques: care of the body, hygiene, exercises, notably, the psychophysiological discipline of yoga.

These techniques are found in very different degrees among the various peoples of India. The most isolated and most poorly equipped forms of the human community exist side by side with a highly civilized society that itself is subdivided, in accordance with a relatively ancient and still common principle, into classes of varying levels and functional specializations. These classes are in turn divided into social units ("castes") that are often specialized along professional lines. Caste distinctions have been officially abolished, and industrialization and the civil service are tending to replace the caste idea by categories of employment. The castes have nevertheless dominated every aspect of technical activity throughout history, and their structure is only gradually and incompletely weakening, as the spirit of professional pride replaces the notion of innate possession of profession by caste.

Information about techniques
The information available to us about Indian technology is furnished directly, by existing material achievements, and indirectly, by pictures, descriptions, and witnesses. It is fairly well known, in a fragmented way, but on the whole much work remains to be done in order to complete our knowledge. Archaeological finds, which are few in number in relation to the multiplicity of known sites awaiting systematic excavation, are already providing valuable information about material equipment. From an examination of such finds we infer the state of the ancient techniques. Aerial observation and the examination of aerial photographs promise the discovery of information about regional technical organization, with systems of irrigation, communication, and urban or monumental sites.

The indirect information furnished by pictures on monuments often offers a complete and very useful picture of achievements that have completely disappeared or that exist only in a vestigial or ruined condition because of their perishable material, particularly in the case of wooden buildings. It also provides information about the use and even the existence of craft techniques that could not leave material remains.

Literary documentation is abundant but difficult to interpret. In most cases the texts merely allude to technical facts, in a perfectly adequate, obviously rich vocabulary whose precision can be guessed, while the subject to which it refers cannot be determined. However, the preservation of this vocabulary in the modern language can make it possible to determine at least in part what it signifies. The work of comparing modern meanings with their ancient contexts where corresponding designations already appear has unfortunately made little progress.

The information supplied by the richest texts relating to techiques concerns the plans and general rules of construction (not the art of building as such) contained in the *Çulvasûtra* for the ritual buildings of ancient Brahmanism, and the *Çilpaçâstra* and *Âgama* for the buildings of classical Brahmanism. They also deal with pharmacy, describing the preparations made from plants and the chemical operations required in handling mineral drugs. Texts relating to the production of utensils and instruments are rudimentary and few in number, even as regards astronomical instruments. Bronze techniques, very frequently used in the casting of statues, are not described, although they are attested to by the statues themselves and by detailed descriptions of the forms to be achieved. The instruction in the techniques of painting is somewhat better.

As for physical culture techniques, they are the subject of an abundant literature, but they are described in terms of the results obtained rather than in terms of how they are practiced.

Generally speaking, works on Indian technology are classifications of achievements rather than operational manuals. Even instructions for the practice of magic, which constitute a special body of literature and which are given in detail, are often obscure or incomplete — a fact that is explained, however, by their secret nature.

An entire category of allusions to miraculous weapons and vehicles, allusions found especially in the epic poems of the Mahabharata and the Ramayana, has lent credit in certain Indian quarters to the belief that India already possessed artillery and aviation in antiquity. A mechanical flying peacock is mentioned in a poem from southern India. All these cases obviously refer to imaginary objects used by gods and heroes, of which no mention is made in genuine technical works.

Some foreign witnesses supply valuable evidence for Indian technology. In Northwest India Alexander's Greeks were struck by the industrious skill of the natives, who produced imitation sponges which they sold to the Macedonian soldiers to replace the worn-out genuine sponges the soldiers had brought from the Mediterranean. Long before then, in the time of Herodotus, India was already known for fabrics woven from a "wool grown on trees" — that is, cotton.

Modern investigation has ample material on which to work, thanks to the variety of regions and races, and consequently of technical products; this is particularly important in the case of the transition from individual to collective craftsmanship using unified techniques, and in the accomplishments of industrialization. Where archaeology has revealed the identity of modern and ancient instruments, observation of current practices in India permits us to reconstruct, by analogy, the ancient methods of their utilization.

THE TECHNIQUES OF SUSTENANCE

Agriculture   Except for a few tribes who practiced (and sometimes still practice) cultivation on patches of burned land, the Indians since antiquity have been cultivating barley and numerous other types of grain, such as raggee (*ragi,* or African millet — *Eleusine coracana*), all kinds of peas and beans in the upper regions, and rice in the lowlands. Plowing has always been done with very light wooden plows with pointed wooden (frequently iron, in modern times) plowshares (Figure 38) that lightly scratch the muddy earth in the rice paddy. The plow is pulled by oxen or buffalo. It lacks any device that would permit it to be rolled along the roads, but it is sufficiently light to be carried on the workman's shoulder. Plowing, or working the land with instruments that turn the soil, buries the stubble left by the preceding harvest (or, in the case of fields that have been burned prior to this superficial turning of the earth, the ashes left by the fire).

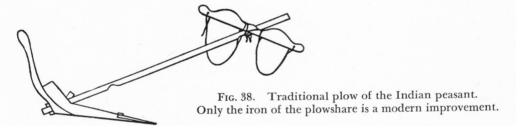

Fig. 38.   Traditional plow of the Indian peasant.
Only the iron of the plowshare is a modern improvement.

The flooded rice paddies, which are separated by earthen dikes, are spread with fertilizers, cow dung (if the total available supply is not used as fuel for cooking), and especially the green foliage of certain trees (nowadays it is the foliage of the "bendy tree" or Portia tree — *Thespesia populnea*), which is left to rot in the water. Once the rice seed has sprung up in the starting beds, the shoots are pulled up and replanted in the rice paddies. The mature, harvested rice is thrown on open spaces and trampled by oxen and buffaloes to break open the ears. It is then gathered up and beaten on the ground to force out the grains. Lastly it is winnowed and spread out to dry in the sunshine. The husking and pounding of the grains is done in a narrow mortar with a long wooden pounder, which is grasped by the middle and used in vertical position. Pounding is also done with a piece of wood attached to the long end of a balanced beam; the beam is operated from the opposite (short) end.

The principal crops, in ancient as in modern times, are spices, oil-yielding crops, bananas, mangoes, jackfruit, sugarcane, coconut palms, Palmyra palms, Areca palms, betel, the kapok tree, cotton, and jute. In addition to the edible elements of its fruit, the coconut palm has always supplied fibrous elements for the making of rope and (like the Palmyra palm and several other species) the sap of its spathe for the production of a palm wine called "toddy" in the north, *kallu* in the south. Betel, a climbing plant cultivated for its leaves, which are chewed with lime and the betel nut, has always been grown on stakes in carefully enclosed gardens. Sugarcane, like cotton, has always been essentially an Indian product. The Greek name for sugar, *sakcharon,* is derived from a popular form, *sakkarâ,* from the Sanskrit name *çârkarâ,* which was specifically applied to coarse sugar. In addition to this product and sugar candy, the most rudimentary techniques of extraction supply brown molasses, which continues to be prepared and used. The device used for crushing oil-yielding fruits is a mill, which is operated by an animal attached to the end of a piece of wood perpendicular to the axle of the millstones.

Cattle raising   The raising of bovine animals in India dates from a very remote period and continues to be of the greatest importance. Its purpose is to supply draft animals for agriculture and transportation and to supplement the food supply with dairy produce. Protected by a custom that is utilitarian as well as religious, the bovines, especially the cows and the zebus, were never killed to supply food for the ruling social classes, and their general protection for the purposes of dairy products and the requirements of agriculture is confirmed in the modern Constitution of India. Elimination of the practice of keeping cows in urban agglomerations will come about only gradually, through the introduction of modern methods of storing the milk, so that it can be transported; otherwise it must continue to be collected right at the door of the house.

No special techniques are practiced in cattle raising; the animals are simply led in large herds to uncultivated lands. But the techniques of preparing dairy products have developed along lines that are very characteristic of the entire Indian world. Given the precarious conditions of their nutrition, the yield of the cows is very low — whence the necessity for large numbers of animals. In most cases the milk collected is not used in its original form, but is turned into curds, a kind of yogurt, whey, and butter. The preparation of butter involves a special technique. The curdled milk is churned in a vertical churn rotated in a vessel by means of a rope wound around its middle and pulled first in one direction and then in the other. The butter collected is immediately melted, then chilled and decanted. It now has the consistency of light oil, and is called "ghee."

The raising of sheep, pigs, chickens, and ducks has not given rise to any special techniques. Sheep raising supplies wool for weaving, and leather.

Fishing and hunting   Seagoing fishermen use hollow boats made of planks sewn together with coconut fibers — a construction that gives the hull the flexibility necessary for the passage of sandbars. Rafts made from trunks of trees lashed together are also used.

Their fishing techniques include many varieties of pots and nets. One technique of collective fishing, in use along the seacoasts, consists of stretching a huge net between several rafts and drawing it slowly in to shore.

Techniques of hunting utilize the bow, sling, and the pike, depending on the animals being hunted; they are complemented by techniques of trapping, especially among the hunters of fowl. Animals, birds, and reptiles must be transported live to the markets, since dead creatures would immediately begin to decay in the tropical and subtropical climate.

Preparation of food The techniques of food preparation utilize a mortar for grinding grain. The mortar has a hemispherical cup in which is fixed a hemispherical stone capped with a short cone; the cone serves for the rotation (by hand) of the stone. Spices are crushed by rolling a stone cylinder over a stone platform; this device has frequently been found in archaeological excavations. Vegetables are cut with a concave knife set vertically in a piece of wood resting on the ground and held with the foot; the vegetables are pushed against the blade of the knife. Cooking is done on crude stoves over a fire of wood, charcoal, leaves, or dried cow dung. The most commonly used utensils are of pottery, and consist mainly of jars and bowls (the jars are often copper). For pots, large leaves are sewn together and arranged in the form of plates; large pieces of banana leaves, or metal (mainly copper) platters, are also used for this purpose. Spoons are made from wood or the crown of a coconut. After the hands and mouth have been washed, the meal is eaten with the fingers of the right hand.

SETTLEMENTS

Buildings and cities The earliest archaeological sites studied, at Harappa and Mohenjo-Daro in the Indus Valley, reveal the existence of an extraordinary technique of construction and layout of cities during the period from the middle of the third to the middle of the second millenniums B.C.

The cities of the so-called "Indus" civilization (which, however, is now known to have extended into Gujarat), reveal a symmetrical arrangement of streets bordered by brick houses. These cities are remarkable not only for their high ramparts, as at Harappa, but still more for a system of sewers placed under the streets and connected to pipes leading from the bathrooms of the houses.

In the later Vedic civilization, in contrast, information supplied by texts leads us to envisage a single, very crude and fragile, type of house consisting of a framework of posts topped by a thatched roof supported by rafters, with mats for walls and partitions. Brick was used only for the construction of altars to the God of Fire and secondary hearths; stone had a very limited use as a subsidiary material. The information in the texts deals only with religious construction, and leaves us in ignorance of what may have been the true human dwelling; the inadequacy of these texts, however, agrees with the absence of identifiable archaeological remains from the Vedic period.

In the still later "Brahman" period, the remains of buildings and monuments are built of bricks that are generally large (modern Indian bricks are smaller, being comparable to modern European bricks). No traces remain of the wooden houses and buildings, but their appearance is preserved in pictures on bas-reliefs and in the stone architecture of many monuments, which imitates wooden architecture. In addition, in the northern regions, in Cambay and Kashmir, as well as in Nepal, wooden architecture still exists, with its complex frameworks, covered balconies, and pavilions with elaborate sculptures.

The art of cutting and building in stone has always been remarkably well developed, in contrast to the art of wood and brick construction. Remains of large stone walls mark the sites of fortresses or old palaces, as at Patna, the site of the former capital of the Maurya kings of the fourth and third centuries B.C. Many ancient sanctuaries of the west (regions of Kathiyawar, Bombay), eastern India (Orissa), and the south (the Tamil country) are artificial grottoes cut out of the rock, complete with columns and interior decoration.

The technique used in cutting these grottoes is clearly revealed by those that were left unfinished. A series of squares, marked out with a chisel on the rock walls, determined the direction of attack. The crudely hewn elements were gradually refined and polished, and finally elaborately sculptured. Certain temples, like those of Mamallapuram (Mahabalipuram) south of Madras, or Ellora, northeast of Bombay, are monoliths carved out of enormous blocks of rock, like gigantic statues.

Buildings and monuments — Temples, fortresses, large palaces, and Mogul tombs were built of stone — rose or red sandstone in the north, in the regions of Delhi, Mathura (Muttra), Agra, and Jaipur, or white marble, like the famous Taj Mahal of Agra, ocher sandstone in Orissa, granite in the south, with superstructures of brick in the case of very tall temples. The technique of construction used in building stone temples began with the complete construction, on the ground, of the upper floors and top. Once the stones had been exactly fitted together, they were marked. Then the structure was dismantled and carried up piece by piece, by means of wooden ramps, to its proper place, where the stones were reassembled in accordance with the markings. This method of construction, which made it possible to do the greatest portion of the work conveniently on the ground, prefigured the modern method (called "anastylosis") that is used to reconstruct ruined monuments whose elements can be found and identified.

Stonecutting, like stone sculpture (Figure 39), remains common in regions that have quarries. Simple fences have often been built of stone, notably with enormous granite slabs crudely but evenly cut. The cutting of the blocks was (and still is) done by drilling small, closely spaced, cubical holes along the line selected for breaking, thus decreasing the resistance along this line (Figure 40). Dry wooden cubes are forced into these holes and then wetted; their swelling causes the stone to break.

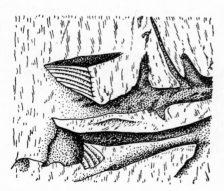

FIG. 39. The beginning of a sculpture in a monolithic stone temple of the eighth century, showing the line of attack of the chisel.

FIG. 40. Preparation for cutting into a ridge of rock.

Plate 43.
a. Unfinished monolithic temple. Mavalipuran. *Photo Filliozat*
b. Monolithic temple. Mavalipuran. *Photo Filliozat*

In the oldest constructions, like the sandstone walls around the monuments of Bharhut (Satna) and Sanchi and the marble walls of Amaravati, the stones were cut and fitted in imitation of wooden elements, vertical posts being dug laterally from mortises in which the horizontal stone crossbars were fixed.

Important buildings or parts of important buildings were stuccoed and painted. Stuccowork was frequently used to cover stone sculptures, either as a base for colored decorations or to increase the size of the original monument or to modify its architectural design. Copper facings or even gilded tiles are frequently found above main sanctuaries of temples.

In marble monuments of the Mogul period, the encrustation of semiprecious stones or small mirrors decorates wall surfaces and makes them gleam. Part of these encrustations was the work of Italian and French artists working at the Mogul courts in the seventeenth century. Finely carved screens of sandstone or marble were often used to cover the openings in these monuments.

Homes

Ordinary houses were built of brick, and had an outer veranda and a central courtyard surrounded by a cloister with wooden columns. They were roofed with tile, and frequently had a roof of palm-tree leaves over the courtyard. The terraces and flooring of well-to-do homes were made of a hard, smooth cement. The flooring of poorer houses consisted of packed earth that was given a daily coating of cow dung blended with water; this formed a relatively smooth coating that eliminated dust and the slight surface irregularities propitious to the development of insects.

The house of the Indian farmer has always been built of low mud walls, with a roof of palm-tree leaves that extends quite close to the ground. There are also low round huts with walls of bamboo stalks encircled with fiber ropes, and square dwellings with roofs supported on posts stuck into an earthen footing, the walls being made of bamboo screens or braided palm leaves.

Water was obtained with modern installations from pools, cisterns, and wells; major buildings had pipes. The removal of waste water was ensured by deep gutters and open sewers.

Public works

The very numerous pools required for the continual baths of the population (the frequency of the use of the baths in India astonished medieval Chinese travelers) were dug, in most cases by human labor, to hold the water during the rainy season. The largest ones were bordered with stone steps.

Irrigation was one of the most highly developed techniques in India, beginning with the protohistoric Indus civilization. Major hydraulics projects are frequently mentioned in ancient royal inscriptions; they consisted chiefly of dikes built to create reservoirs, and irrigation canals.

The irrigation of the fields and rice paddies, when it was not done by means of canals or ditches, was accomplished by various devices. The water was drawn from the well in a waterskin on the end of a rod, which balanced around an axle located at about one-third of its length. A man standing on the shortest end of this arm alternately lowered and raised the skin; an assistant emptied it on the side that was to be watered. Animals were also used for this

Plate 44.
Two devices for raising water. Pillaiyarkupom. *Photo Filliozat*

a

b

Plate 45.
a. Device for raising water. Karasamir. *Photo Filliozat*
b. Oil mill. Periya-Kalapet. *Photo Filliozat*

chore, by means of a sloping embankment of earth against the well. Two oxen, descending the slope, pulled the waterskin out of the well by means of a rope that was attached to their yoke and passed over a pulley. When the animals walked backward up the bank, they caused the waterskin to plunge back into the well.

Another technique, used on the edge of a pond or a river, utilizes a rod that balances on a post planted on the bank. The long end of the rod extends out over the water; the other is weighted with a stone. A wooden trough is placed beneath the end of the rod that is resting on the riverbank; it is attached at the other end to the long part of the rod (Figure 41). A man standing in the water lowers this end, thereby plunging the gutter into the water, then lets it rise by the effect of the stone counterweight, which raises it and causes it to pour its water into a ditch on the bank.

Fig. 41. Modern drawing (early nineteenth century) of a water-raising device composed of a beam and an elongated trough.

Main roads were established at a very early period; they were planted with trees and had inns for travelers, as is indicated by the inscriptions of Asoka in the third century B.C. Cisterns built on piles and, more frequently, stone porches were located at regular intervals. These porches, as tall as a man, permitted travelers to put down and pick up the burdens they carried on their heads.

The public works employed a considerable labor force armed with picks and shallow baskets no more than nineteen to twenty-five inches in diameter; these baskets were used for carrying soil and materials in small quantities.

Methods of transportation The traditional methods of transportation were the pack animals — elephants, donkeys, buffaloes, camels (in the northwest), and especially chariots and light carriages pulled by oxen, and palanquins carried by men. The vehicles generally had wheels with very large rims. The large chariots of gods used in processions in southern India and Orissa had large and sometimes numerous high wheels. These chariots were of carved wood, were shaped like the bases of tower sanctuaries, and had a light wooden frame to hold the statue of the god; the frame was covered with canvases, hanging curtains, and garlands of flowers. These chariots were pulled (by means of ropes) by the crowd, not by animals.

The rudimentary palanquin was a net bed suspended from a long wooden rod supported in front and in back by porters.

Small burdens were carried on the head, often on the hip (jugs and children) and, in the northeast, perhaps on both ends of a rod the center of which was rested on the shoulder. This method of carrying burdens is widely used in the Indochinese Peninsula and throughout the Far East.

EQUIPMENT

At a very early period a very well-developed craft tradition was already supplying all the tools needed by a society that was refined but little encumbered with furnishings.

Furniture The traditional furniture of the Indian house consists especially of chests and beds. The bed is a wooden frame supported on four feet, on which a rope net is stretched. The net, which is made of parallel, closely woven strands whose tension can be varied, is more tautly stretched at the head of the bed; it is covered with a mat. Interwoven straps are sometimes used instead of a net.

Swings made of heavy wooden platforms suspended on chains permit those who sit in them to enjoy a breeze by means of a slight push from a foot left to rest on the ground.

The traditional lighting is the oil lamp with a cotton wick, either alone or in groups that form a candelabra.

Crafts The various crafts that supply equipment have always been divided among closed communities that avoid intermarriage and even contact in daily life, especially during meals. These communities, which have come to be called "castes" (although the word is ambiguous, and also designates the major social classes into which society is theoretically divided), very often specialize in crafts which they have turned into local monopolies: carpentry, smithing, pottery making, leatherworking, and so on. Techniques are handed down from father to son, orally and by example, within these groups, and thus are rarely set down in detailed manuals. Technical books are hardly more than books of formulas, especially in the cases of medicine and chemistry, and even these are only notes that presuppose a knowledge of the fundamental procedures.

Metallurgy Metallurgical techniques, of which we have no descriptions, nevertheless produced masterpieces at a very early date, such as very large columns of almost pure iron cast in a single block. The most remarkable specimen is the so-called "Delhi" Iron Pillar, which now stands near Qutb Minar in a suburb of New Delhi. It bears a Sanskrit inscription dating from the fourth century A.D.; it stands more than 23 feet high,

Plate 46.
Making mirrors. Miniature, Mogul School. Indian Museum,
Calcutta. *Photo Indian Museum*

is 16 inches in diameter at its base, 12 inches at its capital-crowned top, and weighs more than six tons. It was cast in a single piece, including the capital, without joints.

The techniques demonstrated by such monuments were not exceptional. Even before the fourth century they were being widely practiced, along with those of bronze and tin, gold and silver, for Indian iron is mentioned as an item of trade in the first century by a Greek text, the *Periplus of the Erythrean Sea*. In the twelfth century these techniques were still producing remarkable results, as is indicated by the iron beams used in the construction of several temples in Orissa.

Until the beginning of the nineteenth century steel was obtained by case-hardening. Iron extracted from magnetic iron ore was heated in crucibles for four or five hours, together with dry wood shavings and the leaves of various plants, over charcoal fires fanned by bellows; it was then reheated to reduce the excess carbon, and sprinkled from time to time with water, whey, or liquid butter.

Copper, brass, and bronze were employed in very large quantities. Bronze statues three feet tall and taller are very numerous, especially in the south. Small statues were produced in great quantities. Metal encrusted on metal was a specialty of artists of various regions such as Hyderabad and Tanjore.

A metal cement described by a sixth-century text, the *Brhatsamhita*, consisted of eight parts of lead, two parts of bell metal, and one part of brass, cast together.

Chemical techniques Artisanal chemistry, which was very well developed, not only permitted numerous alloy preparations, but was also very advanced in the treatment or use of various minerals, particularly mercury salts, together with organic, vegetable, and animal products. All sorts of preparations resulted, from the most varied drugs to the stucco and cements used in construction and the production of various objects.

Included in the specialties of Indian technology are several very hard cements (*vajralepa,* "adamantine paste") obtained with lime and resins first extracted and blended for decoction, then reduced to liquid form by prolonged periods of distillation. It was chiefly on dry surfaces obtained by procedures of this type that mural painting was done. A glue prepared by prolonged boiling of buffalo skin, the product of which was reduced and preserved in dry sticks and diluted when needed, served the painter as an adhesive.

The lacquer technique, including its name (*laksha* in Sanskrit, *lakkha* in the vernaculars), originated in India.

Textiles The most remarkable Indian techniques are undoubtedly those concerned with clothing. Others, such as faïence, have been completely neglected. Indian dishes, unlike their Chinese counterparts, were made of terra-cotta or metal, and were very simple affairs. The development of highly perfected techniques of weaving and dyeing, especially for cotton, and the preparation of indigo dates from ancient times. As we mentioned earlier, the use of fabrics made from cotton was already known in India in antiquity.

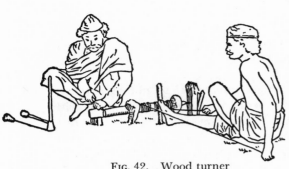

FIG. 42.　Wood turner
and his assistant.

FIG. 43.　Dyeing
hanks of silk.

FIG. 44.　Woman winding
dyed silk.

FIG. 45.　Woman
spinning cotton.

FIG. 46.　Woman winding
cotton thread.

FIG. 47.　Woman weaving cotton.
The man carries a bow for cleaning the cotton.

FIG. 48.　Beating canvas.

FIG. 49.　Making a loincloth
on a vertical loom. The man is knotting
the threads.

The weaving of wool, especially in Kashmir, and of silk in Kaci (Benares) is a traditional industry famous for the fine quality of its fabrics.

The typically Indian garment for women is the sari (a word that is of feminine gender in most of the languages of India, but neuter in the south), formed of a piece of cotton or silk cloth about six yards long, woven in a single piece, with a decorative pattern along the edges and one of the ends. This garment, which has neither seams, buttons, nor hooks (as is also true of men's garments), is rolled around the body, and its decorative end is thrown over the shoulder as a scarf or is used to cover the head.

The Indian textile craft has been well developed in recent centuries, especially in the production of pieces of cotton fabric, generally checkered, which are used by the Moslems and are often exported to Malaysia, Indochina, and Indonesia.

ESTHETIC TECHNIQUES

Esthetic techniques include, in addition to those of the fine arts (sculpture, painting, metalworking, and various decorative techniques), goldsmithing and the working of precious stones on the one hand, music and dancing on the other.

Precious stones and jewels The finding, cutting, and setting of precious and semiprecious stones is a very ancient Indian specialty. Pearl fishing, practiced off the southern part of the eastern coast, is known to have been in existence since the end of the fourth century B.C., from the evidence given by Megasthenes, the ambassador of Seleucus Nicator to the court of King Chandragupta. Indian lapidaries, who unfortunately described the characters of the stones rather than techniques for their use, were numerous.

Semiprecious stones were principally used as inserts in marble, to form decorative floral motifs; this use can be seen in the Mogul monuments of Delhi, Agra, Fatehpur Sikri, and elsewhere. These decorations are found side by side with beautifully carved marble or sandstone screens; a great number of these screens have been preserved.

The making of jewels, bracelets, necklaces, ankle rings, pectorals, earrings, nose jewels, diadems, and forehead pendants has always been a very busy industry. Personal ornaments are made not only for women but also for statues of the gods and human or zoomorphic figures. Breastplates of embossed silver have often been made for stone statues of the gods.

Musical instruments There are many Indian musical instruments, some of which are very simple. There is, for example, the ektar, used by beggars as an accompaniment for their song; it is made of a calabash with a bamboo handle, at the end of which a key holds the single metal string which is passed over a wooden bridge in the center of the calabash. The most important of the other stringed instruments, in contrast, have a very elaborate technique. The most famous of these is the vina, with its two sounding boxes formed of two large calabashes, one at each end of the body of the instrument. Another instrument, the sitar, is a guitar, the sounding box of which is formed of a large half-calabash closed with a piece of wood. It has six strings the player causes to vibrate by means of a finger device; the resonance of these strings in turn causes the vibration of an underlying series of strings that are not touched. There are many types of drums: large and small kettledrums, a double hourglass-shaped drum, a double drum that is cylindrical or swollen in the center, a type called the *tabla* (used in pairs), and so on.

Percussion instruments include various types of cymbals: small cymbals used to beat time, bells, earthen urns struck with rods, and xylophones. Wind instruments include straight and transverse flutes, and the oboe.

The applied arts   Music is allied with dancing, the technique of which involves the use of many costumes and much jewelry.

Much less developed is the luxury industry of profane *objets d'art*. These are few in number, consisting of small caskets needed for betel, spittoons, cosmetic boxes divided into numerous compartments, plates, ewers, lamps, and (in the Mogul civilization) parade arms.

These objects involve the use of those very remarkable production techniques known (incorrectly) as damascening. One method consists of inserting gold, silver, or copper wires in grooves cut in the metal surface of the object. These grooves are narrow at the surface and grow wider as they penetrate below the surface. Once the wire is inserted, the edges of the groove are tightened over it by hammering, fixing it firmly in place. Another method consists of applying thin gold or silver leaf on surfaces roughened with a graver and then hammering so that the slight lips raised by the graver fold over this gold or silver leaf. The applying of embossed silver appliqués on copper plates is a specialty of southern India (region of Tanjore).

PHYSICAL CULTURE TECHNIQUES

Although it is not customary to include techniques of the body in the history of the technology of various countries, for two reasons it is necessary to make note of their unusual role in India. The simplicity of household and personal equipment in daily life is not solely the consequence of a very low general standard of living and a resignation to misery encouraged by a philosophy of renunciation. It has always characterized, even at the level of the most refined luxury and among social groups that are by no means inclined to follow doctrines of renunciation, the Indian way of life. In India the sensation of physical and psychological comfort has always been sought less from artificial external means than from the exercise of personal discipline. Walking barefoot, sleeping on a hard surface, eating frugal meals are not misfortunes but skills. Thus habits acquired in childhood free the individual from his dependence on objects. Possible deprivation is less keenly felt, and techniques of luxury are less important.

In addition, the preparation of the craftsman or artists for technical accomplishments very often rests upon a preparatory psychophysiological training determined by precise rules. This training conditions and dominates the operational processes, and therefore a consideration of it is indispensable in order to complement, by an understanding of the traditional Indian technician, our knowledge of his methods and achievements.

Hygiene and care of the body The ability to dispense with complex technical equipment in daily life has never implied simplicity of individual habits, still less a neglect of personal care. On the contrary, it is acquired only by the use of complicated physical techniques. The rules of hygiene and diet are extremely numerous and elaborate; they had already caught the attention of the Greeks in Alexander's army. Even at that time they were already being applied to domesticated animals, particularly to elephants. They give close consideration to conditions of temperament, climate, seasons, and circumstances. Manuscript collections of these rules and medical instructions are still among the most universal family possessions, and when printed are sold in the marketplaces along with religious books, legends, books on the interpretation of omens, and astrology and moral precepts, the latter being instructions for spiritual comfort that complement the technical instructions for material comfort.

One of the most indispensable elementary requirements is bathing, which has always astonished most foreign visitors, particularly Chinese Buddhist pilgrims like the seventh-century Yi-tsing. Hydrotherapy has had recourse to numerous techniques described since the early centuries of the Christian era in medical treatises: steam bath, spraying, inhalations, fumigations, sweating, and so on.

The washing of garments, done during bathing, goes hand in hand with the latter, although the general washing is left to the launderers' caste (*dhobi*), who beat the pieces of wet linen on stones with their arms, instead of rubbing them or beating them with a paddle.

Exercises for mind and body Massage, done by barbers and specialists, is another highly developed physical technique. Gymnastic exercises for the development of the muscles and general toning have been very much practiced in the Indian armies, and even more so in training for competitions and exploits of endurance, which have been highly regarded since early antiquity. The principal, specifically Indian technique of the body is yoga which aims at the most complete possible mastery of body and mind.

In their preliminary stages, and because of their provisions for diet, hygiene, and training in postures that are usually difficult to maintain, the yoga techniques all resemble the basic ascetic exercises. Very soon, however, they depart radically from the latter. Physiologically speaking, yoga is based on a fundamental regulation of breathing (*pranayama*) and on the assumption of stable postures (*asana*). When developed further, it may include the prolonged suspension of respiration, and achieve deliberate restriction of the movements of the heart, registered by the electrocardiogram and clinically demonstrated by the cessation of the pulse and of the sounds of auscultation. In most cases it includes the assumption of complex postures, training in the isolated use of muscles that normally function as a group, and an action on the rectal and vesical cavities, which can then be contracted and expanded at will. This is called *hatha-yoga*.

Psychologically it consists of successive stages that include the neutralization of sensory stimulations (*pratyahara*), fixing of attention (*dharana*), meditation (*dhyana*), total fixation of psychic activity on a single idea (*samadhi*), and, in its ultimate stage, the maintenance of the psychic being through the extinction of this thought in a stable position where it remains itself, liberated from all sense reaction, whether affective or intellectual. This is the "royal yoga," or *raja-yoga*.

Since the physiological and psychological conditions of the exercise of yoga complement and benefit each other, it is always (though in varying proportions) both a psychotechnical and a corporeal gymnastics. When practiced in any degree for a control of the physical and psychical individuality, it naturally serves as a preparatory discipline for the artist who wishes to utilize his skills to the maximum and who derives from it at least the benefit of an automatic concentration on his activity.

BIBLIOGRAPHY

ACHARYA, P. K., *An Encyclopedia of Hindu Architecture* (Oxford, 1946).

BASHAM, A. L., *The Wonder That Was India* (New York, 1954).

BROWN, PERCY, *Indian Architecture (Buddhist and Hindu Periods)* (Bombay, 1942).

Cambridge History of India, ed. by Edward J. Rapson (Cambridge, 1922).

Central Road Research Institute, *History of Road Development in India* (New Delhi, 1964).

CHAKRAVARTY, TOPONATH, *Food and Drink in Ancient Bengal* (Calcutta, 1959).

GARRATT, GEOFFREY T. (ed.), *The Legacy of India* (Oxford, 1937).

GUPTA, H. C. DAS, "Bibliography of Prehistoric Indian Antiquities," *Journal of the Asiatic Society of Bengal,* n.s. 27 (1931), 5–96.

LAL, B. B., "Further Copper Hoards from the Gangetic Basin and a Review of the Problem," *Ancient India,* 7 (1951), 20–39.

MacGEORGE, G. W., *Ways and Works in India, Being an Account of the Public Works in That Country from the Earliest Times up to the Present Day* (London, 1894).

MAJUNDAR, R. C., "Scientific Spirit in Ancient India," in *The Evolution of Science,* ed. by Guy S. Metraux and François Crouzet (New York, 1963); reprinted from *Journal of World History,* VI, 2.

MARSHALL, JOHN H., *Mohenjo-daro and the Indus Civilization; being an official account of archaeological excavations at Mohenjo-daro carried out by the government of India between the years 1922 and 1927* (3 vols., London, 1931).

PIGGOTT, STUART, *Prehistoric India* (Harmondsworth, England, 1950).

RAY, P. C., *History of Chemistry in Ancient and Medieval India* (new ed., Calcutta, 1956).

SUBBARAO, BENDAPUDI, *The Personality of India* (Baroda, 1956).

SWARUP, SHANTI, *The Arts and Crafts of India and Pakistan* (Bombay, 1957).

TOY, SIDNEY, *The Fortified Cities of India* (London, 1965).

WHEELER, MORTIMER, *Civilizations of the Indus Valley and Beyond* (London, 1966).

WHITE, LYNN, JR., "Tibet, India, and Malaya as Sources of Western Medieval Technology," *American Historical Review,* 45 (1960), 515–526.

PART FOUR

ISLAM AND BYZANTIUM

CHAPTER 12

THE MOSLEM WORLD
(SEVENTH TO THIRTEENTH CENTURIES)

I N 636 the Arabic conquerors began to extend their rule over a vast territory stretching from the Garonne River in France to the Oxus River in central Asia — a process that required less than a century. It was at first a centralized empire ruled in the beginning from Medina and later from Damascus. When the power of the caliphs became centralized in Baghdad, it turned its eyes to the East, and by force of circumstances neglected the Mediterranean basin; the reasons for the secession of Spain and North Africa were probably geographical rather than political. Then other disturbances broke out in the East with the birth of local dynasties whose ambitions, which generated conflicts, clashed violently. For the man in the street, however, for the farmer and the traveler, daily life did not change; the same religion, the same language, a uniform kind of existence, were found everywhere in the empire. New groups came into existence; brilliant Moroccan dynasties and a glorious Syriac-Egyptian Empire sprang up, then, in the East, a disturbing Mongol thrust that later diminished while the Ottoman power gained a foothold in eastern Europe, as if to compensate for the Christian conquest of Spain.

It is extremely difficult to paint a picture of the customs of the Islamic world, to learn the details not only of daily life but also of dress, furnishings, and all the objects with which individuals like to surround themselves. We have little information about Islam's technical and economic structures, which remained at the archaic level. It is nevertheless interesting to attempt an investigation, keeping in mind that the Islamic civilization gave a great impetus to commerce and to certain industries, for it had able craftsmen at its disposal.

AGRICULTURE AND RURAL LIFE

The continuation of the organization of large rural estates in the regions conquered by the Arabs, with the fellah kept in servitude, contributed to economic prosperity. The new masters found that practices and techniques to ensure continuous vegetation or to defend the land against the invasion of the waters were already in existence; the ancient civilizations, not the Moslems, had covered Mesopotamia with a network of canals. In Persia, Afghanistan, and Turkestan, the Moslems benefited from a system of artificial irrigation that harnessed springs in the mountains and brought the water to the fields by means of underground conduits. This method was common in Arabia, Armenia, and the oases of the Sahara; the gardens surrounding the city of Marrakesh owed their splendor to the water supplied by subterranean galleries.

Machines for raising water In Egypt and Mesopotamia, in order to raise water, the farmer was sometimes satisfied merely to scoop it up with a rope and a bucket. When the height was not too great, he used the drum with Archimedean screw, or the beam elevator, known to the Egyptians as the shadoof (see the chapters on ancient Mesopotamia and Egypt).

The animal-powered noria existed almost everywhere; this was a machine pulled by oxen or camels wearing blinders so that they would not be made dizzy by the circular movement. They turned a horizontal wheel, which in turn moved a vertical wheel with a chain of scoops, jars, or buckets, or (as is seen in one twelfth-century miniature) encircled by a hollow rim.

In places where the riverbanks were very steep — on the Euphrates, the Orontes, at Hama on the Yesil Irmak, at Amasya, or at Toledo in Spain — the hydraulic wheels were turned by the current of the river, acting as the motive power for flour or paper mills. The wheel of Toledo was gigantic; it measured 90 cubits (about 145 feet) in diameter. From Spain the technique spread to Morocco.

In Transoxiana (Sogdiana), Susiana (Elam), and Yemen, irrigation was carried out by means of barrages with sprinkling sluice gates. Seistan was the country of the wind, and its inhabitants made use of its power to turn millstones and at the same time to draw from wells the water needed for watering their lands.

"A millstone is attached to the end of a wooden cylinder, 1⅔ feet wide and from 11½ to 13 feet high. This cylinder is arranged vertically in a tower open to the Northeast to catch the wind blowing in this direction. The cylinder has wings made of bunches of rushes or palm-tree leaves, which are arranged along the movable shaft around its axle. The wind, penetrating the tower, exercises strong pressure on the wings, thereby turning the shaft and the millstone" (Khanikoff, *Mémoire sur la partie méridionale de l'Asie centrale*, page 166).

Elsewhere, in Persia as in North Africa, the artesian well was used:

"A very deep shaft is dug, and its walls are carefully shored up. The digging is continued until a very hard layer of rock is reached. This rock is attacked with picks and pickaxes to make it thinner. The workers then return aboveground, and throw a mass of iron into the shaft. The layer of rock breaks, permitting the water beneath it to rise. The shaft fills with water, which overflows and forms a stream on the ground. Sometimes the water rises so quickly that nothing escapes it" (Ibn Khaldun, *History of the Berbers*, III, 299).

Irrigation and drainage Moslem law contains explicit provisions regarding water. The dredging of the canals and the strengthening of their banks could become a collective enterprise imposed on the inhabitants of the riverbank or done at their expense. Methodical organization of irrigation was necessary for certain crops, such as sugarcane or rice. The rice

الاول البذاور ربح الثاني الذائم وما وقف اذا اردف بالبذور نقص صاحبه
في العيد و قوم بالبذور وخرج من البذور و بعض للبذور فهذه عشرة مثله

دفن عادكم وزنه لذلكم ولو زدتم زدنا وان عدتم عدنا قال المخبر
هذه الحكاية نورد عليب امر لحاجيه اللاتي هالت لما انها لت ما حارث

Plate 47.
A noria with hollow rims. Manuscript. Bibliothèque Nationale,
Paris. *Photo Bibliothèque Nationale*

Plate 48.
Noria and aqueduct. Hama. *Photo Musée de l'Homme—Bénézech*

paddies had to be kept constantly flooded, and the water had to be changed every three days. In Persia, as in Egypt, the distribution of the water was precisely regulated; hourglasses were used to measure the length of time the farmers utilized the water, and systems of sluices made it possible to gauge the volume of the water. Leo the African speaks of measuring clocks that operated by water: "When they are empty, the watering period is over." Mention is also made of skilled personnel and even teams of divers.

In Lower Mesopotamia, at the beginning of the Arabic period, an excessive flooding of the Tigris and Euphrates inundated the region to the point where the water formed permanent lakes. During the reign of the Ommiad Caliph Muawiyah (661–680), drainage projects were begun; the reeds that covered these lakes were cut down, and the water was pushed back with the help of dikes. This extremely difficult work was done with Negro slaves brought from the east coast of Africa; the slaves eventually revolted in a slave war that continued for many years.

Stability of equipment Agriculture had rudimentary tools at its disposition that the "somnolent" agronomists felt no need to modify; in certain regions (Egypt, for example), agriculture was hardly more than gardening. The *aratrum* remained very primitive, with neither a colter nor a moldboard; it did not have to turn the earth over, and did not penetrate deeply into the soil. It was so light that it could easily be carried on a man's shoulder; Joinville commented with astonishment on this plow "without wheels." A variety of animals were harnessed to it — the camel, the ox, the donkey — with crude harnesses made in most cases of braided or rolled alfa grass, rarely of leather (Figure 1). Other tools can be seen in miniatures dating from the end of the thirteenth century: a spade with triangular or trapezoid blade (Figure 2); the shovel for digging; the harvester's sickle, either smooth or toothed, with a rather long handle, and a blunt-tipped blade; the pitchfork, the rake, the hoe, and the omni-

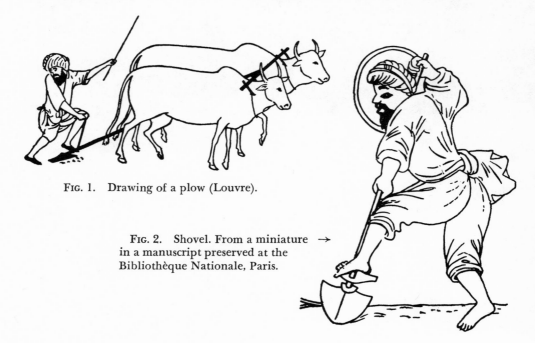

FIG. 1.   Drawing of a plow (Louvre).

FIG. 2.   Shovel. From a miniature →
in a manuscript preserved at the
Bibliothèque Nationale, Paris.

present basket. The peasants husked the cereal grains by beating them with a club. Treading of sorts was done by the simple trampling of animals walking in a circle, or else with the *noradj*, a kind of chopper utilized in Syria and Egypt; it had a pick and a winnowing device, the latter equipped with a six-tined fork. Rice was husked with a mallet. On small estates the grinding of grain was done with a hand mill, a very simple type of crushing device; this task was assigned to the women. Some cities had large mills with water-driven millstones; elsewhere, floating mills were located on rivers for the use of the inhabitants of the riverbanks. These mills, which were installed on rafts, were found in Spain, and were also sufficiently numerous on the Tigris River to warrant the existence of special laws to regulate their use.

Agricultural writings   Needless to say, agricultural work could not be adapted to the Islamic lunar calendar, and local almanacs utilized the ancient calendars to ensure the efficient progress of the work. It may be questioned whether the Arabic technical writings were simply observations of an already existing situation or had an influence on farm life. In any event we cannot ignore the abundant agronomic literature of Andalusia, and particularly the famous Calendar of Córdoba, drawn up in 961: "These treatises," according to Lévi-Provençal (*Histoire de l'Espagne musulmane*, III, 264), "are almost the only ones that describe the varying characteristics of the Andalusian land and give detailed descriptions of the care which must constantly be given to it: manuring, irrigation, plowing, pulling up weeds, struggling against parasites, birds, and locusts."

The rural dwelling   In the two continents ruled by Islam, the external structure of the villages varied, depending on the region and the available materials, the nature of the soil, and ancient customs. The small Egyptian market town had a special characteristic: the huts were built close together to conserve all possible cultivable land. Its appearance has certainly remained the same for centuries: a cluster of low huts huddled against each other. Constructed on a hillock that the Nile floodwaters did not reach, by its color and the uniformity of its houses the village often resembles a gigantic molehill. The floors are always made of beaten earth. In addition to these Egyptian huts built of dried bricks and mud, there are buildings of rammed earth or daub, sometimes with a lath framework. There are sugarloaf villages in the environs of Aleppo, and Kabyle communities in North Africa perched on peaks that can be reached only with difficulty. In southern Arabia there are multistoried houses built of sun-dried bricks decorated with geometrical designs, with small towers and machicolations. Then there are the much-praised bastioned walls built by the sheep-raising population of central Morocco, with their granaries and court-yards for the livestock, and their "dungeon-like silhouette and appearance of a Rhenish stronghold."

Except in regions where stone and basalt are available, the principal building material is dried brick; the interior of the buildings is not very comfortable, and is certainly less well kept than a nomad's tent. There is hardly any rural "furniture" aside from the indispensable utensils for cooking food or washing clothes. On the whole, Islam has paid little attention to the peasant class, whose sole *raison d'être* is to obtain a return from the soil sufficient to permit the land-owners to live comfortably.

In certain rural centers temporary markets were held at regular intervals for the sale of poultry, animals, and fodder; food for the cities was supplied by the wholesale vegetable market. The Egyptians continued the ancient method of "chicken ovens," an artificial method of incubation in which the eggs were hatched without hens. The chicks were born in such quantities that they were sold in bottomless measuring devices which were placed in the purchaser's basket.

TECHNIQUES OF URBAN LIFE

Urban organization   In a penetrating study (*L'islamisme et la vie urbaine*), William Marçais has demonstrated that in order to achieve its social and religious ideal Islam required an urban life. The Moslem world found, preserved, or founded a great variety of cities. There were, for example, "Fez, the highland city, Tunis, the seaport, and Kairouan, the steppe city." In addition Kufa, Basra, Wasit, Mosul, Baghdad, El Fustat, Cairo, and Chiraz were established in the East, and, in the Maghreb, Kairouan, Mahdia, Algiers, Oran, Tiaret, Tlemcen, Fez, Marrakesh, Bougie, and Rabat. About fifteen of these cities are still in existence. The result of this urbanization was to settle the nomads, though the process developed less in response to efficient agriculture than out of a desire for security.

Mosques and markets   The Moslems took as their model the arrangement of the ancient city, divided into two equal sections by a major avenue bisected perpendicularly by another avenue. Like many of the world's cities, the Moslem cities were focused on the religious center, the mosque, which was surrounded by markets. Each street in the market had its specialty, as in the markets of antiquity and medieval Europe; the Spanish traveler ibn-

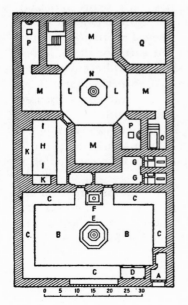

FIG. 3. Ground plan of a public bath. (W. Lane, . . . *Manners and Customs of the Modern Egyptians*).

Jubayr (twelfth century) tells us that "upon leaving the area of one trade you find yourself in that or another, and so on until you have seen every type of urban craft." The shops resembled "large boxes, one side of which has been removed to expose the inside." In his book on Aleppo, Jean Sauvaget describes the method of closing these shops with a wooden double shutter, the upper portion of which forms a porch roof when raised, while the lower, hinged portion serves as a counter and a seat.

The proximity of the principal mosque and the markets should be emphasized, as well as the fact that the doors of the sanctuary of Mecca bore names of merchants. At Seville, the moving of the main mosque caused the transferal of economic activity. Each city had one or several mosques that in the early days served as a court and a religious and political meeting place. Thus the main marketplace corresponded in general to the ancient Forum.

Inns During the period of conquest, and sometimes continuing after the conquest, the original city was a kind of camp for the occupation services, the army, and the administrative offices; Kairouan was originally a parade ground. The short-lived capital of Prince ibn-Tulun in Egypt was demolished before it had progressed beyond the stage of a palatine city. It was apparently customary for cities that began as military and administrative centers to become places of business and centers of intellectual life. Most of the large cities of Persia and Mesopotamia had solid city walls, not to mention the splendid walls of Cairo.

The cities were organized for commerce, in the sense that special buildings were set aside for the storage of merchandise and for lodging businessmen. Depending on the period or the use to which the building was put, these caravanserais were called either by the Persian name *khan,* the Greek names *funduk* or *kaisariya,* or the Arabic word *wakala* (which passed into medieval French as *okelle*). These buildings were uniformly constructed in the form of a square around a large paved courtyard with a portico that supported a winding gallery. The ground floor consisted of large warehouses; the second floor contained apartments — genuinely monkish, unfurnished cells with bare walls, which their users were obliged to furnish, in addition to doing their own cooking. A single door similar to that of a fortress provided access to the building; this protected the residents from harm in case of riots. Everything was oriented toward the easy exchange of merchandise — a guarantee of economic expansion.

The "hamman" The Moslem city increased the number of baths (*hamman*) it had inherited from antiquity, without making any changes in the structure. The bath consisted of a dressing and rest room, a steam room, and sometimes a room of intermediate temperature. The bath played a social and hygienic role of primary importance in all the Moslem countries. A doctor from Baghdad, traveling through Egypt in the twelfth century, declared that "Nowhere have I seen better built and furnished baths, whether from the esthetic or technical viewpoint." In certain regions (Egypt, for example) little progress had been made in solving the problem of fuel; buffalo dung had been in use since antiquity. Elsewhere scraps of wood, underbrush, and manure (in the forests of the Atlas Mountains, charcoal) were used.

Supplying drinking water   One Arabic author outlines three conditions for the successful development of a city: a source of water in the vicinity, pastureland, and wood for heating. Certain cities benefited by the supply routes of drinking water built in antiquity, or built basins to ensure the supply. In Persia, subterranean aqueducts supplied the cities with water. Several fortunate cities had wells; in others, like Samarkand, lead pipes had been in use for a long time. Still others (Antioch, for example) had an abundant supply of running water for the use of the inhabitants and for watering its gardens; it was piped through small channels to feed the fountains in the homes of the rich. In Egypt, at El Fustat, and later in Cairo, water was brought to the houses by camels and donkeys, and by teams of water carriers. Heat and dust were the enemies to be fought in the streets of Cairo, and this battle required the mobilization of a large number of camels. Eight thousand animals, each carrying two leather water bottles that contained the equivalent of two small kegs, distributed Nile water. The number of water carriers has been estimated at 100,000; around their necks the men carried goatskin water bottles with canvas pipes from which they dispensed drinking water, in silver and copper cups, to all comers. In Seville the Almohades of Morocco put an end to this practice by constructing an aqueduct.

THE CONSTRUCTION OF BUILDINGS

Building materials   For large official buildings the builders did not have to go far to look for their materials: they used stones taken from the monuments of antiquity in Mesopotamia, Persia, and Egypt. For the sake of solidity, the cautious master masons sometimes used drums from columns as perpends in their masonry walls. Beginning in the twelfth century, the walls of ramparts and citadels often had bossages.

Persia continued to be the land of brick, and its building industry therefore required brickmaking factories (which by law had to be located outside the cities) to supply dried and baked bricks. Brick factories prospered everywhere in the East. Standards were set for their manufacture; as an example, the bricks of the Egyptian mosque of ibn-Tulun (ninth century) were 7 inches long, 3 inches wide, and $1\frac{1}{2}$ inches deep. The brick courses, alternately upright and in herringbone pattern, were often strengthened by wooden lacing courses. Beams and rafters came in standard sizes that we are unable to determine; dialect words indicate the diameters of the pipes in each region.

The city of Aleppo was particularly remarkable; its quarriers and masons carefully cut and polished blocks of stone: "Neither mud walls with scaling surfaces, nor unstable wooden planks, nor masonry with crumbling mortar: this is the city of stone" (Sauvaget). Everywhere else the houses were built of small-stone masonry.

Mortar   Ibn-Khaldun tells us: "Two pieces of wood, the height and width of which vary from one region to another (generally, however, they measure four cubits by two) are raised on foundations, the space between them being equal to the width of the foundation

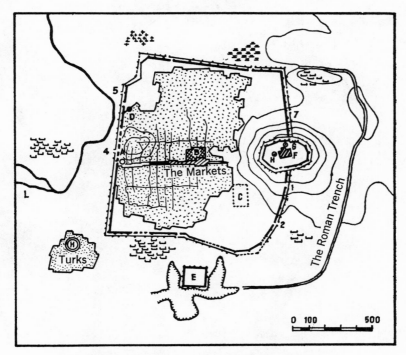

Fig. 4.   Plan of Aleppo (Sauvagel, *Alep*).

as decided by the architect; they are joined by means of wooden crossbeams held in place with ropes or thongs. The space between them is filled in with two other small planks. Into this is poured a mixture of earth and lime, which is crushed in with pounders specially made for this purpose. When the mass is well compressed and the earth sufficiently combined with the lime, a few more shovelfuls of earth are added until the space is completely filled. The particles of earth and lime are by now so thoroughly mixed that they form a unit."

The Eastern masons had several types of mortar: mixtures of clay and lime, sand and lime, and pulverized brick and lime. They continued the Byzantine practice of adding wood ash to the lime.

Methods of construction   The multistoried dwellings of Siraf, on the Persian Gulf, were built of teakwood. In contrast, the buildings of certain important localities in Arabia — Jidda, for example — were reed huts covered with straw.

Rope and felt ladders, as well as the ordinary wooden variety, were used in construction. The codes for the regulation of the markets specified that ladders must be "made of thick, massive wood, with sturdy risers, tightly nailed." Crude scaffoldings can be seen in miniature models of construction scenes; certain materials were carried up on the backs of workers, others were brought up on hand barrows. Plaster was hoisted in a basket by means of a rope, and the plasterers used trapezoid trowels.

Plate 49.
A construction site. From a Persian manuscript, British Museum,
London. *Photo British Museum*

Interior of the building   In some cities we find sanitary facilities such as those excavations uncovered in El Fustat, Egypt. Every house was probably supplied with a "toilet" in the form of a ditch, sometimes connected by suitable piping with a common sewerage system, but in most cases emptied periodically by the homeowner, who made use of the services of concerns that sold this waste material to the suburban market gardeners (a very lucrative business).

Special domestic equipment varied, depending on the climate. Some rich homeowners of Baghdad lived in cellars during the hot season; others utilized punkahs, a kind of ventilator-screen hung from the ceiling. In several regions of Persia, cold water was poured over wall hangings. On the terraces of the houses of Cairo, as in Persian cities, air vents open to the north acted as ventilators during the summer.

Despite their very humble, unimpressive, almost forbidding external appearance, the interiors of the houses that belonged to wealthy middle-class people or to important government officials were of unsurpassed elegance and sumptuousness. Ibn-Khaldun tells us: "Relief figures made of plaster combined with water were applied on the walls. When the plaster had hardened into a solid but still slightly damp mass, it was modeled into the desired form by cutting it with iron awls; lastly, it was given a beautiful polish and pleasing appearance. The walls were also faced on occasion with pieces of marble or with tiles, faïence squares, shells, or porcelains. This gave the wall the appearance of a floor decorated with flowers."

FURNITURE

Lack of furniture   These houses did not contain furniture in the European sense of that term. Most of the money spent on furniture was invested in carpets, tapestries, and fabrics. For the urban middle class, luxury consisted of improving the objects used in everyday life — clothes, beds, dishes, kitchen utensils — and in combating the heat. The furniture in the homes of the urban rich was impressive especially by the quality of the fabrics used. Miniatures show low beds, that is, raised wooden benches; these could be a kind of frame on four legs, with palm-fiber ropes on which a mattress was spread. We are struck by the luxurious appearance of the Moslem *objets d'art* preserved in museums; they are evidence of the splendor of the furnishings used in the mosques of the principal cities and in the homes of important people.

In addition to the low benches, which were covered with long cushions and formed divans against the walls, carpets were spread on the floors and hung on the walls. The chair, well represented in ancient Egypt, did not appear in the Moslem period. It can probably be assumed that the furniture of the lower middle class, as of the working class, resembled that of the peasant — that is, it was practically nonexistent. Armoires and chests were indispensable; they were assembled, not with nails, but with tenons and mortises.

In certain regions locks were made of wood, in the form of a sliding latch, as mentioned in the *Canticle of Canticles* (V. 4): "The keys are pieces of wood

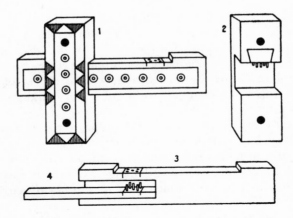

FIG. 5.   Wooden keys (Lane, *Manners . . .*).

with small brass pins. These pins lift similar pins in the lock that have entered certain small holes in the bolt. The pins of the key having pushed the pins of the lock out of these holes, the bolt can be slid back and the door opened."

Abundance of utensils   Upon examining the metal utensils, all the experts agree in according a certain superiority to the Moslem peoples in the industrial arts. We must, however, stress the crudeness of the metalworking equipment. As late as the nineteenth century the brass founders could be seen, like mysterious alchemists, huddled before their ground-level furnace, while others hammered for long hours on cups and pots.

Culinary utensils were notable for their vast proportions; they consisted of basins of various shapes, incense burners, ewers, platters, various kinds of vases, and lamps, and were a major item in wedding trousseaus. The richest articles were damascened. This art consisted ". . . of fixing a metal or silver wire in a groove cut in the metal object to be encrusted. The skilled worker often let this wire stand out in high relief against the background, thus forming a kind of cloisonné that gave an excellent effect." Or, ". . . after the design to be encrusted had been engraved, small slots were raised, with the graver, on the edges and the bottom of the cut, to hold the gold or silver blade or wire. This wire was then fixed in place with the help of the hammer, which forced it into these slots." Ancient writings testify to the quality of the encrusted enamel.

Ceramics, rock crystal, and the precious glasswares of Syria, especially from Tyre, Aleppo, Hebron, and particularly Damascus, also helped to embellish the surroundings in which the wealthy passed their lives.

FOOD

The pre-Islamic Arabs boasted of eating a dish consisting of camel hair and blood, blended with a stone and cooked in the fire. Poetry speaks of Bedouins with empty stomachs, for whom lizard eggs were the only available diet. Mention is made of a companion of Mohammed who did not want to eat a chicken because he had never seen one. Moreover, "the comparison between the coarse food of the Arabs and the rich Iranian dishes is a frequent cliché." In truth, the diet of the nomads was (and still is) camel's milk, cold, warmed, or curdled; they also ate dates, and were especially fond of fat.

Gastronomy Gastronomy became famous during the Sassanid reign (circa 226–circa 640 A.D.); recipes were therefore of great antiquity in the East. In the Abbasside (749–1258 A.D.) *belle époque* recipes were invented by famous people, and the best disciple of the philosopher al-Kindi did not disdain to compose a recipe book with a menu for every day of the year.

Eastern food is highly spiced, requiring the use of condiments, and is served with sauces; oil plays an important role in the cuisine. A great variety of porridges, thick soups of flour or semolina, and beans formed the nutritional base of the poor class, the Sufis (ascetics), the religious orders, and the workers. They supplemented this with chopped meats, jellies made of flour and honey, and fritters. Meat roasters played an important role; sheep was their specialty. Culinary extravagances existed that were worthy of the "Banquet of Trimalchio," for example, the fish tongues of Ibrahim ibn-Mahdi.

One writer testifies to a certain concern for well-being: he speaks of a "well-kept servant who carries in, on a clean plate, fruit covered with a clean napkin." Thus certain milieus sought after comfort in their leisure time, as an expression of *joie de vivre*.

Everywhere, from Spain to the East, instead of cooking its own food the lower middle class and the proletariat purchased it ready cooked in the market: meatballs, roasted meat, fish, fritters fried in oil, pancakes, and nougat.

Tableware was of gold, silver, china, simple faïence, or even wood, depending on its owner's circumstances. Food was usually eaten with the fingers; however, knives were sometimes used, and some diners even used a species of fork.

LIGHTING

The most common source of light was the ancient oil lamp, either terra-cotta, glazed, or even bronze; its shape was not modified by the Arabs. Ancient ceramics and a few miniatures show pictures of glass lamps, splendid specimens of which have been preserved from the time of the Mamluk sultans of Egypt. The mosques were illuminated with bronze rings of lights or large hanging lamps with glass bulbs. There were copper chandeliers, the richest of which were encrusted with gold and silver; these were the pride of the region of Mosul until the Mongol onslaught caused the dispersion of the artists to Damascus and Cairo. These chandeliers, in most cases located near the *mihrab* (a niche in the middle of the rear wall of the mosque, indicating the direction of Mecca) were crowned with long white candles "as large as palm-tree trunks." Tinderboxes of punk, or various woods of known properties, were used to light them.

Plate 50.
Mosque lamp. Persian, circa ninth century. The Metropolitan Museum of Art, Harris Brisbane Dick Fund, 1964. *Photo The Metropolitan Museum of Art*

TRAFFIC PROBLEMS

In the Middle Ages an Eastern city presented the appearance of a compact cluster of fairly tall houses separated at random by small, erratic, airless, dusty streets. The plans of these cities reveal a certain confusion, apart from the major arteries leading to the city gates, and the streets, which sometimes end in dead ends, form a jumble in which a straight road is an exception. True, there was no need of wide streets, since carriages did not exist, but these streets did not even have the width of seven cubits recommended by the *hadith*, the body of religious tradition second in authority only to the Koran, and would not even permit "two camels laden with straw to pass without rubbing up against each other." It was forbidden to tether beasts of burden in, and to transport merchandise which would cause dirt through, the markets, and laws were passed concerning the precautions to be taken by the water carriers.

The princes undoubtedly liked large gardens, and some of these gardens have become famous. We have proof on at least one occasion, in the eleventh century, of the creation of a public park in Córdoba, left to the city by the last will and testament of a rich man.

The cities sometimes spread out on both banks of a river, and Baghdad, Seville, El Fustat, and Cairo had boat bridges; that of Baghdad was still in the existence in the nineteenth century. The boats were linked end to end by iron chains, and were attached on each bank to a large, solidly anchored beam. Almost everywhere these bridges were supplemented by ferry services, and thus boatmen contributed to the transportation of people, animals, and merchandise. In the Egyptian countryside the bridges over the canals were built with movable planks, which were removed at certain hours of the day so that ground traffic would not hinder the passage of boats.

The job of the authorities in charge of postal and caravan routes was to ensure their safety, and to serve travelers by supervising the maintenance of the watering places and constructing hostels at way-stops. In certain Persian caravansaries, underground stocks of snow may have been stored for use during the summer. Permanent watchmen were appointed for the protection of travelers, and on snow-covered roads signs informed travelers the direction to take.

Bridges and military architecture It would be an exaggeration to claim that boat-bridges, which were often swept away by floods, were the only methods used for crossing rivers. The truth is that stone bridges, some of them built in the Moslem period, were maintained in Spain, at Harba in Mesopotamia, and at Lydda in Palestine. At Susiana, a bridge dating from the days of the Sassanids was rebuilt in the eleventh century. Its arch was capable of sustaining heavy weights; baskets and pulleys had to be used to descend to its foundations, and iron and lead were used to bond the stones together.

Many large projects were brought to successful conclusions. Some construction techniques seemed extraordinary, but they do not warrant a belief that the methods used were a radical departure from the traditions of the trade. It is told that when the foundations for the port of Acre were being laid, the engineer requested a supply of large sycamore beams. With them he built two floating platforms in the shape of square towers, each with a large opening on its upper side. Then he built a stone and mortar pillar on each platform, in five sections joined together by large reinforcing tie rods. The platforms descended deeper into the water under the increasing weight of the stone. When the engineer was sure that this apparatus was resting on the sand, he abandoned it for one year to give it time to acquire perfect stability. Then he resumed work, being careful to establish a great coherence between all the sections of his project. He connected it with the old wall inside the port, and threw an arch over the opening.

Military architecture was equally well represented, by the citadels of the major cities (Aleppo, Damascus, Cairo), the castles in Syria, and the fortified caravansaries along the Persian roads. In 1183 the Spanish traveler ibn-Jubayr watched the work in progress on the citadel of Cairo. He tells us: "The men working on these structures, and the people charged with executing the projects and supplying the considerable quantity of material required — for example, sawing the marble, cutting the large stones, and digging the moat around the wall of the fortress (a moat which must be dug right through the rock, with picks), is a work that is truly one of those marvels which leave permanent traces — these workers are Christian prisoners, in incalculable numbers; one can employ no other workers on these projects."

TRANSPORTATION

On the whole, all the Moslem governments understood that the problem of good roads and methods of transportation was basic. There was, for instance, a regular service for the transportation of snow from Syria to Egypt.

The camel   There were two reasons for the disappearance of the carriage. For one thing, it became evident that, given the lack of maintenance of the roads that could support carriages, there was a danger that wagons would become bogged down in fields or break up on the stones. Moreover, the carriage could not compete with the camel, whose introduction into the transportation system was to cause a major revolution in methods of transportation. Between 400 and 1300 the East came to depend on the camel, the fastest and most comfortable animal and the one best adapted for crossing great desert areas. The extreme mobility and unequaled endurance of the camel have been rightly praised.

"The raising of saddle, draft, and slaughter animals is not often mentioned in the descriptions of the geographers." This statement of Lévi-Provençal about the Iberian Peninsula can be extended to include all the Moslem countries. However, horse and camel breeding was regarded by the Bedouins as a noble calling, although the breeder of small livestock was despised. Horses were definitely a subject for experimentation by specialists who hoped to supply high-quality animals to the polo players or to place champions in the races. The pack camel required qualities of strength, whereas the racing camel had to be rapid. There were races for camels, especially for the animals from the region of Mahara in Oman (southern Arabia). Arabia had competent breeders who by skilled crossbreeding were able to develop the most desirable qualities and breed top-notch racers.

Disappearance of the carriage   The use of carriages may have continued in the most easterly portion of the Moslem world, for Avicenna expresses the opinion that traveling by carriage is good for the health. Elsewhere, however, it seems to have been exceptional. The silence of the Arabic writers concerning this method of travel is significant, all the more so in that they were able to say that in India "the only method of travel is by ox-drawn wagons." The fact that the traveler ibn-Batuta devotes a lengthy discussion to the wagon used in the Crimean and the sledges used in the area of the Don proves how rare they were. Carriages are occasionally mentioned in connection with the transportation of large blocks of stone in Alexandria and Tripoli in Africa. In northern Syria carriages drawn by buffalo were occasionally used; in Spain such carriages were used by the army for baggage transportation. Not until the Mongol period was there regular use of convoys of wagons.

Women and feeble individuals traveled by camelback in palanquins, or in covered palanquins balanced on each side of the harness, or on litters with front and rear shafts borne by mules or horses.

Methods of harnessing the horse varied from one region to another, and the quality of the harness depended on the social standing of the rider. The saddle was almost flat, with a raised pommel and cantle. The stirrup appeared in Eurasia in the second half of the sixth century, in Iran in the middle of the seventh. The horseshoe was used in the regions conquered by the Arabs; prior to the Hegira the Arabian nomad had never shod his horse, but had placed leather sandals on their hoofs.

River navigation   River transportation, on the Euphrates as on the Nile, continued to flourish much as it had in antiquity; it supplied food to the cities and provided the provincial trade centers with a method of rapid transportation for their merchandise. This transportation was very costly, however; thus each city had to satisfy its own needs, and even the smallest cities had their pottery and brickmaking factories, their dyeing establishments, their oil and soap industries.

The ancient poets speak of navigation on the Euphrates, on which "large, heavily laden ships and light feluccas rock back and forth." Rafts made of rushes were in use on the Tigris during the twelfth century (Petachia of Ratisbon), while in Egypt the Nile and the canals were used for commercial transportation. This was not a very rapid method; it took six days to travel from Alexandria to the capital, and eight to fifteen days to go down the Nile from Kush to the capital, or vice versa. In Old Cairo the middlemen had booths on the bank of the Nile, and the merchandise was unloaded at their doors.

FIG. 6.   A sailboat (from a miniature in the Bibliothèque Nationale).

FIG. 7.   A ship on the Euphrates (from a miniature in the Bibliothèque Nationale, Paris).

Maritime navigation   Although allusions in the Koran to navigation indicate that the Arabs were familiar with the sea, they had only crude decked boats for coastal traffic. The Red Sea was navigated by Abyssinian dhows, which the Arabs called *aduliya* (from the name of the African port Adulis).

The historian al-Tabari mentions, for the year 865, galley-type vessels operated by gypsies. Their crew consisted of a captain, three naphtha throwers, a carpenter, a cabinetmaker, thirty-nine rowers and fighting men. Seagoing pirogues with shallow draft, equipped with outriggers, must also have existed in the Indian Ocean. In any event, the sailors of these regions had ships capable of facing storms and the monsoon. In the Persian Gulf, near Obolla and Abadan, wooden scaffoldings were erected and fires lighted on them during the night to warn approaching ships of dangers along the coast.

Regular services appear to have plied the principal maritime routes. There were ships "which went to the country of pepper"; others went "to the land of gold"; shipping companies were formed. Siraf was home port for a great many ships; from here they made trips to China, Malacca, and India. Their captains had a professional organization; one captain tells his frightened passengers: "Know ye that travelers and merchants are exposed to terrible dangers. But we members of the captain's organization have taken an oath never to let a ship go down until its time has come. When we pilots go on board a ship, we stake our lives and our destiny upon it."

An old proverb states: "Arriving in China without losing one's life en route is already a miracle; returning safe and sound is unheard of." There was the danger of pirates, notably those from the island of Socotra for vessels coming from the Red Sea, and signal towers were erected on the coasts to warn the inhabitants. Moreover, rough waves and lack of visibility posed difficulties for ships during the monsoon period.

Naval construction   Vessels constructed of pieces of wood sewn (in Syria, nailed) together were a specialty of the outfitters of Siraf and the coast of Oman. Oman shipbuilders cut down the coconut palm, sawed it into planks, made thread with its fibers, and used the thread to sew the planks. Masts and yards were made from the same wood, sails from the leaves of the tree, and cables with its fibers. According to some authors, in the Abyssinian Sea iron nails offered no solidity, because the water corroded and dissolved them and made them fragile, forcing the builders to replace them frequently and to caulk the joints with fibers coated with grease and tar.

Robert Lopez tells us (in *Les influences orientales et l'éveil économique de l'Occident*) that "the lateen sail, despite its name, appears to have been acquired from the East." As for the center or "sternpost" rudder, it is found in a miniature of the School of Baghdad dated 1237. This, however, could be an isolated case; the Spanish traveler ibn-Jubayr, returning to his country in the year 1184 on a European ship, speaks of "its two rudders, which are the two legs by which it is guided." In the writings of Marco Polo we read that "the ships of the Persian Gulf have only a single mast, one sail, and a rudder [therefore a sternpost rudder] assembled in a special manner." We are told that at the end of the thirteenth

Plate 51.
A boat with sewn planking. Thirteenth-century miniature. Bibliothèque
Nationale, Paris. *Photo Bibliothèque Nationale*

century, at the request of the Mongol Sultan Arghun, builders came from Genoa to Baghdad, where they built two galleys for the purpose of breaking Egypt's trade with India by blockading Aden.

In the Red Sea ships the captain sat on the prow, "with numerous useful nautical instruments" (unfortunately Idrisi does not name them). He attentively scanned the waters to discover the reefs, and told the helmsman which course to take. The sailors probably determined position with the help of the quadrant, and we know that the compass was used by Moslem sailors beginning in the thirteenth century. As for the astrolabe, its construction by the Arabs has been the object of numerous studies.

Although there were fishing boats in addition to the commercial fleets, it seems probable that the people were little given to the practice of water sports. On the banks of the large rivers and in the seaports, fishing was a normal trade. It was done with nets and pots. The skill of certain specialists who caught tuna with harpoons was much praised.

Obligatory service continued to be the rule for the personnel of the fleet, and not only sailors but also various kinds of craftsmen were recruited. The first ships of the Moslem Empire were constructed only in Egypt. The raids of the Byzantine corsairs soon caused the Caliphate to increase the number of its shipyards. Egyptian carpenters were called in to construct the shipyard at Acre, and 3,000 Copts were sent to Tunis to install another one. Tyre, whose shipyard was a base for racing ships, was not the only such city on the Syrian coast, and in Spain the arsenal of Tortosa was famous for the quality of the wood in the neighboring forests.

CRAFT AND INDUSTRIAL TECHNIQUES

"The word 'industry' has become synonymous for feverish activity and teamwork in which each member specializes in a single, unchanging task."

Ancient industry   The industry of the ancient Arabs of the peninsula, like that of the great nomadic peoples, was rudimentary; it was limited to the production of basic objects of prime necessity —· pottery, weaving, basketwork — and was complemented by smiths, carpenters and cabinetmakers, tailors, and leatherworkers. Undoubtedly the villages also had professional cartwrights, but every peasant was capable of repairing a tool.

Many families were engaged in baking bread. The oven was a hemispherical vault with a brick lining and superimposed layers of mixed clay and salt, a practice seen particularly in the Maghreb. The salt retained the heat, while the clay held the bricks in place.

The peasant also knew how to use vegetable and animal fibers for many purposes — drawing water, tying down a tent, tethering animals and tying burdens to their backs; everyone, declares William Marçais, was a ropemaker of sorts, and learned in childhood how to braid and twist. Shepherds everywhere spun wool while guarding their sheep.

Industry, therefore, remained at the level of village crafts, the weaving of

wool, and the art of the leather dresser; it did not even include the more complicated technique of leather tanning. Trades more closely concerned with urban
life were confined to the ranks of slaves and freedmen. The Jews of Medina were
smiths, leatherworkers, and jewelers.

An exception must be made, however, in the case of Yemen, famous in pre-
Islamic times for its textiles. Yemen's hides were the principal export of Arabia
and were used in the making of thongs and ropes, sandals, saddles, tents, pails,
and even large pots. Yemen also made perfumes, and probably had shops for the
production of weapons. The preparation of the hides exported to all parts of
the Moslem world was confined particularly to Yemen, Tunisia, Morocco, and
Córdoba, which was the city of punched and embossed leather (and which gave
to the French language its word for shoemaker — *cordonnier*). Markets grouped
together in various cities the light industries that catered to the needs of travelers
and pilgrims for palanquins, tents, and riding and packsaddles.

Impressment of craftsmen   Egyptian papyrus documents of the seventh and
eighth centuries describe the difficulties experienced
by the Caliphate in ensuring the functioning of the economic life of the state.
It is possible, Mr. Rémondon tells us (*Papyrus grecs d'Apollônos Anô*), that the
number of crafts was sufficient for the maintenance of Egyptian society, but the
state could not simultaneously fill this role and support a constant program of
armament, maintenance of the court, and construction — an effort that far surpassed the limits of Egypt. In order to achieve this program, the Ommiad caliphs
sought to control all production: carpenters, cabinetmakers, caulkers, embroider-
ers, and masons were mobilized in teams and sent to the place of work, or were
recruited locally, to work for the government. For this purpose a list of adult
workers was drawn up, containing their and their fathers' names, city of birth,
and their skills. However, these methods were a complete failure.

Equipment   Judging only by pictures from ancient Egypt, we
see that the equipment of the various trades has
hardly been improved upon in recent periods. Late Persian miniatures shed light
on the creation of the crafts by the ancient Persian King Jamshid, but tells us
nothing very remarkable. They depict the equipment of the smith with his powerful bellows, the weaver, the tailor, and the turner. This crude equipment continued to exist almost into modern times. Similarly, certain miniatures with
construction scenes should not be neglected; they show us, for example, a miner's
pick (Figure 8), in a picture that dates from the beginning of the fourteenth century, and a saw with center blade in a frame (Figure 9). The cabinetmakers utilized the pulley, known from ancient Egyptian pictures.

In speaking of the crafts, the turners must be mentioned, especially the skill
with which they made use of their big toe to speed up their work. The stories
of travelers in the East were not tales of bygone customs, for, as we have observed,
methods of work had remained the same for centuries. "The cabinetmaker uses
his feet as assistants to hold and guide the piece of work. He uses sections of
columns as grindstones. His plane is extremely crude; he has a special plane for
hardwood, the bottom of which has the thickness of two fingers and is of tempered

FIG. 8. Miner's pick. (miniature from a manuscript in the University of Edinburgh).

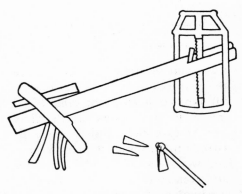

FIG. 9. Saw with center blade (from a miniature in the Bibliothèque Nationale, Paris).

steel; the plane bites easily by its own weight, without any obligation on the part of the worker to lay his hand on it, and because of its hardness it polishes the piece while shaping it." Woodturners had a specialty that was soon to disappear: they were the ones who did the ". . . delicate woodwork, those curiously carved and cut wooden grills, which project over the street and act as windows."

At Ramle in Palestine, marble was cut with a smooth saw and sand from Mecca; the slabs were cut in the same fashion as planks are cut from a tree.

PAPER AND TEXTILES

The appearance of paper　　On the eve of the Islamic age papyrus was being used for writing; it was made in several places in Lower Egypt, and then in Sicily. The Caliphate installed a paper factory in Baghdad. A roll of papyrus was quite expensive, costing approximately the price of renting a *feddan* (about 232 rods) of workable land or the annual rent of a shop. For this reason it was sparingly used, and was washed after use so that it could be reused. Another practice was to write across the roll without erasing the first message. Parchment, which was even less satisfactory, was scraped clean after each use so that it too could be reused.

Arabic tradition claims that a Chinese prisoner captured in 751 at the Battle of Talas taught the Arabs the secret of papermaking. The first paper factory is mentioned in 795 in Baghdad, and from then on paper, the major project of the ninth century, greatly widened the domain of culture; the book industry became organized. For many years Samarkand was the leading center of this trade; paper was manufactured at Damascus, Tiberias, and Tripoli in Syria, in addition to Baghdad. The Spanish city of Játiva, in the province of Valencia, was famous for its factories which produced thick, glazed paper. Between 719 and 815 only papyrus existed; from 816 to 912, paper was still very rare, but between 913 and 1009 papyrus gradually disappeared and ultimately was completely replaced by paper.

Wool　　The textile industry, the glory of ancient Persia and Byzantine Syria, continued its brilliant progress under the Moslem domination. The Arabic authors enthusiastically praise the

Plate 52.
Jamshid giving instructions in the various trades. Firdusi,
History of the Kings, a fifteenth-century manuscript. Chester
Beatty Library, Dublin. *Photo Chester Beatty Library*

magnificent luxury products offered by numerous shops to a clientele eager to ornament itself and to accumulate treasure. "In the Moslem cities, money exchange and arbitration has always centered on stocks of precious garments, gold and silver embroideries, stored in special *sūqs* [markets] called *kaisariya*" (Massignon). Writers mention the grouping together in a single market of the vital trades — weavers, fullers, dyers, darners, tailors, designers, laundrymen, and ironers. Their equipment was very crude; until a very recent period, the manglers were still using the beetle.

Combing of wool was done by hand. The carders of cotton used a gigantic bow, one of those "long instruments that look like the ancient harp." The spinners quickly learned that the spindle had to hang very low to ensure a more even twist. The tailors realized that their work could be done more quickly with short threads, and while they used the thimble (which in most cases was a simple ring without a bottom) we cannot accept the idea that they invented it.

Carpets   The carpet industry was also a very flourishing trade, and we learn from Sidoine Apollinaire that Persian carpets were popular in Gaul. A legend tells of the famous royal carpet, probably embroidered, which was part of the booty of the Battle of Ctesiphon (634). Knotted carpets certainly existed by the time of the Sassanid kings, and Arabic literature speaks of the carpets of Hira. At the beginning of the seventh century the Chinese pilgrim Hsien-Chiang praises the skill of the Persian workers, who knew how to weave fine silk brocade, woolen cloth, and carpets. Geographers praise the beauty and technical quality of the carpets made in the region of Ispahan, used simultaneously with the sumptuous carpets of Armenia. In addition to these written documents, the excavations at El Fustat have uncovered fragments of carpet with Kufic inscriptions. In the twelfth century, carpets from Baghdad, Persia, and Asia Minor are mentioned in France.

The technical information collected some years ago is probably valid to a certain extent for the medieval period. The loom is almost always a vertical loom, and the warp is of undyed wool. The children working on it squat on a plank that rests on the rungs of two vertical ladders, so that it can be raised as the work progresses. A woolen weft thread is passed between each row of knots; the knot is made on the right side, as in the case of all high-warp carpets. Each tuft of wool for the pile is passed with the right hand twice around a warp thread, is knotted around the warp thread next to it, and is then cut with a sharp knife held in the palm of the right hand. The rapidity of execution is extraordinary and the children's deftness surprising. So that the children will lose no time by looking at a drawing, a workman dictates in a singsong voice the stitches required, and the children repeat the instructions in unison while following them.

DRUGS AND PERFUMES

Oil and sugar   Oil presses functioned in the olive-growing countries; the methods of production remained primitive, but nevertheless the results were of excellent quality.

Sugarcane presses existed in Egypt, and played an important role in that country's general economy. Confectioners thrived, and sold all kinds of delicacies made with honey or sugar, almond and fruit pastes, and even (a fact that scandalized austere people) sugar dolls. In the markets the confectioners' quarter was next to that of the producers of syrups and electuaries. Sugar factories also flourished in Syria, where soaps, too, were famous, and southern Persia was a center for sugar refineries.

The druggists formed an important guild in every large city. In Cairo, Leo the African saw "their beautiful shops richly decorated with very pretty ceilings and armoires," though the market police kept a close watch on them in order to prevent frauds.

Perfumes   The dye and perfume industry had always flourished in every region of Islam; the technique was ancient, and was certainly improved upon. We learn from the ancient poetry that the men perfumed their beard and hair, and containers for perfume are described. Elegant middle-class women spent their time applying makeup, using various cosmetics and depilatories, anointing themselves with perfumes, combing and dyeing their hair, and trying on new dresses. Coquettish women used polished bronze mirrors, which are mentioned in pre-Islamic poetry.

In Córdoba the famous Ziryab, the Andalusian Petronius of the ninth century, opened a "genuine beauty institute where the arts of applying cosmetics, removing superfluous hair, using dentifrices, and dressing the hair were taught."

Jewels   Judging by pre-Islamic poetry, wealthy women wore extremely luxurious jewels: earrings decorated with pearls (some of which, it is said, were as large as pigeon's eggs), belts studded with precious stones, bracelets, foot rings that made a pleasant clicking noise, amber, coral, pearl, and topaz necklaces, and black-and-white striped glass beads.

The Islamic religion did not change this taste for luxury, and it is strange to find, in an eighth-century Greek papyrus, an order for ankle rings. The poetry of the Moslem period is filled with recitals of jewels and precious stones. Every large city, from Iran to Spain, was proud of its skilled goldsmiths.

METHODS OF MEASURING

The Sassanids were acquainted with a mechanical device that told the hours of the day. The Moslem world had to know the time of day for the purpose of its five ritual prayers; it had sundials, clepsydras, hourglasses, and mechanical clocks. The Islamic technicians devoted their efforts to the art of automata; a clock was sent to Charlemagne by the Abbassid Caliph Harun al-Rashid, and the Spanish traveler ibn-Jubayr describes at great length the mechanism of the hydraulic clock in the Ommiad mosque at Damascus. The mechanical principles used in these works will be discussed in the second volume of this work. Less costly methods of telling time also existed. We learn, for example, that one part of the revenue of a *wakf* must be used for the upkeep of a rooster "assigned" the task of awakening the muezzins at daybreak.

Moslem metrology was a complicated affair; it was everywhere subject to close supervision. It varied from one area to another, and was influenced both by Islamic tradition and by local custom.

The Roman balance (Figure 10) was known, and balances with arms are seen in thirteenth-century miniatures. One Sanad ibn-Ali, a converted Jew, was a manufacturer of so-called "water" balances used for discovering frauds. The scientist al-Biruni perfected an apparatus for measuring density.

Weights and measures were carefully supervised. We possess official stamps of the early Islamic period that indicate the capacity of glass flasks. Some vases have inscriptions affirming their relationship to a gauge contemporary with Mohammed. A model cubit was installed in a clearly visible spot in the market-place.

FIG. 10. Balance (from a miniature in the Bibliothè-que Nationale, Paris).

THE EXPLOITATION OF NATURAL RESOURCES

Gold and silver

The mineral wealth exploited in the East was considerable; gold mines were located in southern Egypt and Arabia. Various authors describe the methods used by gold prospectors. The prospectors went out into the countryside at night, and looked about for a glitter on the ground. When a man saw gleams in the darkness, he concluded that there was gold at that spot. At daybreak every worker went to work at removing the earth from "his" patch of ground, which he took to a well. The sand was washed in wooden buckets, from which the metal was removed, mixed with mercury, and smelted.

Idrisi, speaking of a silver mine near Herat, claims that its exploitation was stopped because of the depth of the mine and the lack of wood necessary for the fusion of the metal. Elsewhere, near Balkh, small streams with particles of silver in the water indicated the presence of the metal; the miners descended into the bowels of the earth as far as their lamps could remain alight. The silver mines of the Iberian Peninsula and the Upper Atlas, so rich in classical antiquity, were still being worked in the Middle Ages. They were subterranean excavations in which hydraulic wheels had been installed to remove the water; workers then brought the clay up to the surface and washed it. This required large amounts of capital, but it was profitable, since the crude ore furnished a quarter of its weight in pure silver.

Mercury
The cinnabar mines of Almadén (meaning "mine") in Spain provided employment for a thousand workers. Some of them went down into the shaft and worked at stonecutting; others were used to transport the wood needed for the combustion of the ore, others for the making of the vessels in which the mercury was smelted and sublimed. Still others operated the furnaces.

Salts
Other deposits added to the wealth of certain provinces, ranging from precious stones (for example, the emerald of Upper Egypt, the turquoise of Fergana, the balas ruby of Badakhshan, various stones and particularly varieties of cornelian and onyx in Yemen and Spain) to the salt deposits of the Hadhramaut, in Isfahan, in Armenia, and in North Africa. Leo the African tells us that "in most of Africa, the only salt found is that extracted from mines by digging galleries, as if it were marble or gypsum."

Egypt and the Sudan had alum, and natron (carbonate of soda) was sought in certain regions of eastern Egypt; it was used for whitening copper, wire, and cloth, and in the preparation of leathers. It was utilized by the dyers, glassmakers, and goldsmiths, while the bakers mixed it into their dough, and meat roasters used it to make the meat more tender.

Pearls and coral
A major place belongs to the pearl-fishing industry on both banks of the Persian Gulf, in the Abyssinian Sea, at Ceylon near Siraf and the island of Kish, the coast of Bahrein, and the approaches to the island of Dahlak. Ibn-Batuta gives several details on the methods of the divers who did the pearl-fishing:

"The diver attaches a rope to his belt, and plunges. When he arrives at the bottom he finds the shells embedded in the sand, among small stones. He detaches them with his hand or removes them with the help of a knife he has brought with him for that purpose, and places them in a leather bag around his neck. When his breath begins to give out, he jerks on the rope; the man who is holding this rope on the bank feels the movement, and pulls him up on board the boat. His bag is taken away from him, and the shellfish are removed; the bits of flesh inside are removed with a knife. As soon as they come in contact with the air, they grow hard and are changed into pearls. These are all collected together, both small and large sizes."

Coral banks were located near the coasts of North Africa, and also in places near Sicily and Sardinia. Idrisi describes the method of gathering the coral:

"Coral is a plant that grows foliage like a tree and then becomes petrified at the bottom of the sea, between two very high mountains. It is fished with instruments that have numerous flax net bags; these instruments are moved by men in ships stationed above. The net bags become caught in the branches of coral which they encounter. The fishermen then pull the bags up and remove the coral that has been caught in large quantity."

Emery, found in Nubia and Ceylon, was used to polish precious stones.

CERAMICS

Islam, like every other civilization, made extensive use of pottery, for cooking, lighting, and bathing. An eleventh-century Persian traveler tells that in the Cairo bazaar the grocers, druggists, and hardware merchants themselves supplied the glasses, faïence vases, and paper used to hold or wrap their wares. This habit became permanent, an Arabic historian of the fifteenth century tells us:

"Every day they throw a thousand dinars' worth of waste products on the refuse heaps and in the rubbish pits. This is the debris of the red terra-cotta pots in which the milk suppliers sell their milk, the cheese merchants place their cheeses, and the poor consume their meager meals on the premises of the cooks' shops."

Fine ceramics More delicate ware was also produced, however, and almost everywhere in the Moslem world the ceramics industries were kept very busy; the pottery of Persia, Mesopotamia, Egypt, and Syria competed with the faïences of Tunis and Córdoba. The square, glazed faïence tiles of Málaga have become famous under the Spanish name of *azulejos*.

The description of a modern Moslem potter's wheel would probably also be valid for those of the entire medieval period:

"The potter's wheel is composed of an inclined platform topped by a wooden axle, which supports another disk-shaped, round piece. It rests on a crossbar. The worker turns the lower wheel with his foot; no great effort is required for this, for because of its incline the platform passes the point of equilibrium by its weight alone."

The same Persian traveler helps us judge the quality of the Egyptian faïence of his day:

"The Egyptians produce faïence of all kinds. It is so fine and diaphanous that you can see your hand on the outside of the vase through its walls. Bowls, cups, plates, and other utensils are decorated with colors the nuances of which change with the position of the vase."

Perfection of the technique Numerous books have stressed the variety of the Eastern ceramics and analyzed their techniques; here we need only enumerate them. There were ceramics with underglaze embossed decoration, next to objects with glazed decoration. Classifications have been made that include clay pieces varnished with lead, enameled with tin, with metallic reflections, in copper or silver; argillaceous-calcareous faïences with alkaline-earth glazes in various colors; and, lastly, non-enameled argillaceous terra-cottas. Among this latter group belong the *alcarazas*, or water coolers.

We possess thousands of pieces and sherds of pottery. Their artistic qualities are not all on the same level, but their technical perfection argues in favor

Plate 53.
Pottery bowl with lustered decoration. Persian (Kashan),
second half of the thirteenth century. The Metropolitan Museum
of Art, Gift of Horace Havemeyer, 1964. *Photo The Metropolitan
Museum of Art*

of craftsmen who knew all the secrets of their trade. Special mention should be made of the Persian workshops of Kashan, particularly because a Persian-language treatise on the art of ceramics was composed at the end of the thirteenth century by the director of the Kashan factory, the scion of a famous family, certain members of which signed their names to wall tiles. (This work was edited and translated into German in 1935 [*Orientalische Steinbücher und Persische Fayence-technik*].) Each region had its traditions concerning equipment, method of using the wheels, and chemical compositions. Detailed analyses would involve us in needless lengthy discussion, especially since before becoming an industry pottery was probably a spare-time craft for supplying domestic needs.

Lévi-Provençal has called attention to a discovery made by a Córdoban in the ninth century:

"Abbas ibn-Firnas, using tools of his own invention, discovered the secret of making crystal, and put it to work in the glassmakers' kilns in the Andalusian capital. He had a representation of the sky constructed in glass, and could make it light or dark at will, with flashes of lightning and sounds of thunder to add to the effect."

WEAPONRY

The Sassanids had a heavy armored cavalry, a light cavalry of archers, and elephants (the latter disappeared from the Eastern armies until the end of the tenth century, when they initiated a fashion in the easternmost regions of the Caliphate).

Greek papyrus documents from Egypt speak of armored cavalry and infantry, and mention that the invading Arabic army was supplied by a convoy of ships and was followed by teams of smiths to repair damaged equipment.

Armor   The individual soldier's head was protected by a leather helmet or sometimes surmounted by a crest, a metal one. This was often complemented by a hood of mail. Very soon the mail tunic was being made in various models: shirts, habergeons, or simple breastplates, with armguards and leggings. This defensive armor was made either with small iron plaques, horn, or skins; fragments of horn were tied together with gut, or were hollowed out and inserted into each other. Fine tunics of mail were lined with garments stuffed with silk and mastic.

The horse's head was protected by a chamfron; the animal's neck and withers were protected by small plaques linked together, and the center of its chest was also protected.

Various types of shields were used by the fighting man; one type was the roundel or leather buckler covered with metal. For protection during sieges the Arabs soon began to make shelters out of palm-tree branches covered with skins.

Weapons   The most common weapons were the lance, the sword, the bow-ax (sometimes with double-edged blade), for the cavalry, the mace. Joinville saw "a Saracen holding a Danish woodman's ax." At rest the cavalryman carried his lance on his shoulder; his

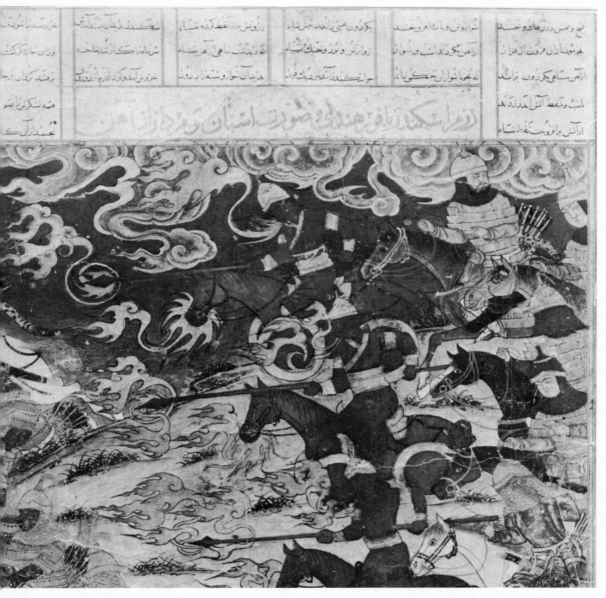

Plate 54.
Persian armor. Fourteenth-century miniature, Fogg Art Museum,
Cambridge. The warriors to the right, part of Alexander's Iron
Cavalry, wear one of the earliest styles of armor. The armor
was made from strips of horn, leather, or metal fastened to leather
or cloth. *Photo Fogg Museum of Art*

sword was probably hung from a shoulder belt. The lance tips were at first made of horn, later of metal. The saber had a straight or scimitar-shaped blade. "A curved blade," Volney tells us, "with a receding cutting edge, moved by the motion of the arm and continued its action in a large area." The javelin, either barbed or double-pointed, was a throwing weapon.

Steelmaking
A treatise on weaponry from the pen of a contemporary of Sultin Saladin (a French translation of it was published in 1948 by Claude Cahen) furnishes valuable information on the preparation of steel. We learn that there is iron in Maghreb, Spain, and Asia Minor and that the best quality steel is found in India and China, while that of the Maghreb and Spain, produced at Bougie and Seville, is an inferior variety. There were numerous variations in the composition of the steel, and its working depended on its composition. According to the treatise, the steelmaker took soft iron (the best type came from old nailheads), threw myrobalan from Kabul and belleric on top of it, then placed it in a tub and cleaned it thoroughly with water and salt. The whole was placed in a crucible and sprinkled with crushed magnesium. The fire of the foundry was then blown on it, so that it was melted; it was collected in the form of an egg. This activity required a period of several days. The metal was then filed down and made into a saber. A good tempering included secret methods, during which often complicated chemical compositions were used. The saber was heated in the fire, and sprinkled with a suitable liquid. After the iron had absorbed the liquid, it was chilled and covered to protect it from dust.

The bow and the crossbow
The Arabs of the Moslem period also had teams of slingers and corps of excellent archers, for they realized the moral effect produced by such devices, and the power of firing in salvos. The rider carried his quiver on his right side. Almost every city had its archery range, and archery conquests became quite complex, beginning early in the Egyptian Mamluk regime. A gold or silver cage containing a pigeon was placed on top of a pole. The contestants had to try to hit the bird while galloping past on their horses.

In time the bow was replaced by the light crossbow, mentioned in the ninth century. This portable weapon, handled by a single man, was soon joined by a heavy crossbow that required the work of several men; it hurled Greek fire or large numbers of darts. (See Figures 11–13.)

Siege and ballistic machines
Siege machines had existed since antiquity; the art of ballistics was developed in the period of the Sassanids. The ballistic weapons have numerous Arabic names, and it is sometimes difficult to determine their exact European equivalent. Translators have used such names as battering ram, ballista, catapult, movable tower, mangonel, stone-thrower, trebuchet. Joinville, in any case, recognizes the superiority of the Saracen machines. These machines, some of which were very heavy and difficult to handle, hurled first enormous stones and then incendiary materials into the enemy camp (Figure 14). The latter were handled

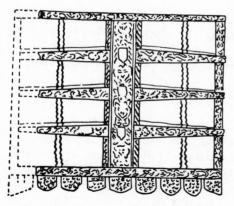

Fig. 11. Triple crossbow (drawing in a manuscript in the Bodleian Library).

Fig. 12. Crossbow for hurling bottles of naptha (drawing in a manuscript in the Bodleian Library).

Fig. 13. Winding mechanism of a crossbow (drawing from a manuscript in the Bodleian Library).

by corps of artificers. Some of these machines were installed on wheels; at the siege of Acre (end of the thirteenth century) the oxen harnessed to one such machine did no have sufficient strength to pull it. In any event, during the Crusades the Moslems seem to have been far advanced in the handling of these engines, and the range of their rocks was more than 960 feet.

Fig. 14. Siege machine (miniature from a manuscript in the library of the University of Edinburgh).

During sieges a corps of sappers dug tunnels, piled wood in them, and set fire to it — a method as old as the world itself. The sappers of Aleppo were famous for their skill.

To close certain access routes to the enemy, a kind of wire entanglement, or road trap, carefully camouflaged, was used.

Incendiary devices The text of Procopius concerning the Median oil of the Persians, already mentioned by Ammianus Marcellinus, is often quoted. The Persians hurled pots full of sulfur, asphalt, and naphtha against the enemy fortifications. It is on the occasion of a raid against India, in 777, that we find mention for the first time in the annals of Islam of the general use of naphtha as an incendiary material. At the siege of Herakleion (803) balls of stone wrapped in linen and coated with flaming naphtha were hurled over the ramparts by mangonels; the fire caused the stones to burst. From then on, the armies of the Caliphate and the Moslem fleet had corps of naphtha throwers. The throwers protected themselves by covering the flammable materials with freshly scraped skins, and the men who handled the naphtha wore garments impregnated with talc.

The glass grenade filled with naphtha appeared in 934 (although it may have been actually an incendiary javelin; in any event the fire grenade was definitely in use a few years later). A dangerous tenth-century improvement gave added terror to the Moslem fleets: copper tubes that hurled a liquid consisting of a mixture of niter and sulfur. Moreover, with the help of chemicals the Arabs had succeeded, at least by the middle of the ninth century, in protecting their ships from Greek fire; we do not know the composition of the mixture used for this purpose. The handbooks give a certain number of instructions for the making of incendiary materials from naphtha, including hurling it with arrows or mangonels, and making a naphtha which continues to burn on the water and ignites ships. Various museums have large collections of terra-cotta grenades, which were shaped to fit the hand and were covered with relief motifs to prevent their sliding unexpectedly while being thrown. The historian of Greek fire, Maurice Mercier, acknowledges that in the course of the twelfth century the Moslems began to use mealed gunpowder grenades ("Syrian pitchers") with explosive power.

We do not know whether these machines were utilized uninterruptedly. In any event, all the sources of the Mongol period (that is, the thirteenth century) speak of Moslem troops hurling fire with machines. Joinville's impressive description should be kept in mind:

"The manner of the Greek fire was such that it advanced as large as a vat of verjuice, and the tail of fire that spread from it was also as large as a large sword. Its approach was so loud that it seemed as if it were the thunder of heaven; it seemed like a dragon flying through the air."

Strategy and tactics appear to have been somewhat crude; in particular, the Arabic conquest does not seem to have benefited by preconceived plans. It was the Mongol army that in the thirteenth century was characterized by "the organization of campaigns, thorough utilization of an information service, and the coordination of actions which in many cases were being carried out thousands of kilometers apart" (Sinor).

GENERAL PICTURE OF THE MOSLEM WORLD

Our impression of the Moslem world is one of stagnation rather than material progress, and it is difficult to discover anything that could be called an enriching of acquired knowledge — unless our judgment has been distorted by the multiplicity of discoveries in more recent periods and is thus more severe. The techniques of antiquity were neither improved nor altered in the centuries immediately following the Arabic conquest. Moreover, the Arabic and then the Moslem civilizations did not have to invent the ruler, the square, the lathe, the lever, the bellows, the potter's wheel, or the winch. In Egypt, for example, the *aratrum* and the sickle of the contemporary *fellah* are the same as those found in bas-reliefs of the pharaonic period.

Moslem society was not interested in the problem of fatigue on the part of the manual workers, and did not trouble to increase the yields of its various industries. First, the continuation of slavery was an insurance against a shortage of labor; second, the use of this subject population was not conducive to research in the area of inventions, an area to which little attention was paid. Young children were present in great numbers in light industry; they operated the forge bellows and hammered copper. Moreover, industry did not progress beyond the craft stage (with the possible exception of the textile industry), and was oriented toward the production of objects used in daily life — another reason why equipment remained unchanged and no technical revolution occurred to change radically the methods used.

Everywhere in the East, yesterday as well as today, "it is the entire village that plays the role of the farm: it is in the village square that the *fellah* deposits his harvests, beats, husks or cleans them, and sometimes remains there to guard them when the community itself does not perform this function. His house is a simple box in the village, a cell in this ancient phalanstery where the spirit and forms of collective work still subsist, although in greatly weakened form. It is essentially a shelter, and no more than that."

In the cities progress was more apparent, in comparison with antiquity. To mention only one example, the increase in the number of public baths everywhere aroused a desire to perfect the methods of supplying water to the cities.

We suspect that there was a corporate form of government, for it is otherwise impossible to imagine the writing of numerous treatises on the supervision of markets that can be compared with our "Books of Trades." These treatises are particularly valuable for the information they give us on religious and foreign matters, and our knowledge of certain bad practices comes from several details on frauds which they supply.

Professional activity was dependent on commands; one wonders if it was possible to find ready-made objects. However, it is difficult to imagine the reason for the increasing role of brokers in business transactions unless we assume that they stored limited stocks in order to fill cutomers' orders quickly.

Agriculture and commerce were the foundation of the economy of the Moslem world; agriculture supplied domestic needs, while commerce was more international in character. To ensure a completely efficient functioning of the system, the program outlined by a great eleventh-century Persian minister had to be put into effect:

"The ruler will attend to the successful accomplishment of everything concerned with the community prosperity. He will install underground pipes to irrigate the lands; he will dig canals, install bridges over the large rivers, assemble the people in the villages, and see that land is cultivated. He will build strongholds, establish new cities, construct noble monuments and splendid residences; he will build caravansaries on the royal roads. A supervisor will be posted in each city to see that weighing instruments are exact and that the prices fixed for each object are not exceeded. He will carefully watch all goods imported for sale in the market, so that there will be neither fraud nor deceit, and he will verify that the weights are exact. If a penniless peasant needs oxen and seed, they must be procured for him on loan, so that he will not take refuge in a foreign place."

In short, the Moslem civilization inherited from antiquity certain procedures and formulas that were respected despite the passage of time, smoothly and without any visible shock from original inventions. In the eastern part of the Moslem world, however, at least until the fifteenth century, the mechanical methods that every society requires must have been continued, possibly even with the addition of new methods.

BIBLIOGRAPHY

ARNOLD, THOMAS W., and GUILLAUME, ALFRED (eds.), *The Legacy of Islam* (London, 1947).

AYALON, DAVID, *Gunpowder and Firearms in the Mamluk Kingdom* (London, 1956).

BURLAND, ARTHUR COTTIE, *The Arts of the Alchemists* (New York, 1968).

CRESWELL, K. A. C., *A Bibliography of the Architecture, Arts and Crafts of Islam to 1st Jan. 1960* (Cairo, 1961).

GILLE, BERTRAND, "Technological Developments in Europe 1100–1400," in Guy S. Metraux and François Crouzet (eds.), *The Evolution of Science* (New York, 1963).

HILL, DEREK, and GRABER, OLEG, *Islamic Architecture and Its Decoration* (Chicago, 1964).

HOURANI, GEORGE F., *Arab Seafaring in the Indian Ocean in Ancient and Early Medieval Times* (Princeton, 1951).

KOHN, M. A. R., "A Survey of Muslim Contributions to Science and Culture," *Islamic Culture*, XVI (1942), 1–20, 136–157.

LEVEY, MARTIN, *Medieval Arabic Bookmaking and Its Relation to Early Chemistry and Pharmacology* (Philadelphia, 1961).

The Medical Formulary of Aqra-badhin of al-kindi, transl. and edited by Martin Levey (Madison, Wis., 1966).

LEWIS, BERNARD, "The Islamic Guilds," *Economic History Review*, 8 (1937), 20–37.

MAYER, L. A., *Islamic Architects and Their Works* (Geneva, 1956).

——, *Islamic Woodcarvers and Their Works* (Geneva, 1956).

——, *Islamic Metalworkers and Their Works* (Geneva, 1959).

——, *Islamic Armourers and Their Works* (Geneva, 1962).

PARTINGTON, J. R., *History of Greek Fire and Gunpowder* (London, 1960).

READ, JOHN, *Prelude to Chemistry: An Outline of Alchemy* (reprinted Cambridge, Mass., 1966).

STILLMAN, JOHN MAXSON, *The Story of Alchemy and Early Chemistry* (New York, 1960).

TAYLOR, F. SHERWOOD, *The Alchemists: Founders of Modern Chemistry* (New York, 1949).

WULFF, HANS E., *The Traditional Crafts of Persia* (Cambridge, Mass., 1967).

CHAPTER 13

THE BYZANTINE EMPIRE
(SIXTH TO FIFTEENTH CENTURIES)

ALTHOUGH the founding of Constantinople took place in 330 A.D., it is the reign of Justinian (527–565) that marks the actual beginning of the Byzantine civilization. For almost ten centuries thereafter, and despite numerous wars and internal dissension, the splendor of a civilization that was highly advanced from the technical point of view radiated from the Byzantine Empire throughout Europe and the Near East. According to Steven Runciman: "Despite its importance and its renown, the Byzantine fondness for theory and culture was sterile; surprisingly enough, it was in the realm of practical achievement that the Byzantine genius manifested itself."

In this chapter we shall give examples chosen from various areas of technology that completely confirm this English scholar's viewpoint. It must be noted, however, that while we possess numerous remains (buildings, *objets d'art*, and so on) of the Byzantine civilization, we have very little information on the techniques employed in their realization.

The Byzantines rarely concerned themselves in their writings with technical questions (with the exception, however, of treatises on military strategy). They preferred abstract concerns.

CONSTRUCTION AND PUBLIC WORKS

The Byzantines were great builders, as is evident from their numerous religious and secular buildings, many of which are still standing. In this regard they were inspired by the Sassanids rather than the Romans; their architecture, as P. Lemerle has noted, is chiefly brick, including vaults and domes.

These buildings were characterized by the groin vault supported by pendentives (a portion of a sphere divided by the vertical planes of the square). This gave them an appearance of lightness and harmony, the best example of which is the basilica of Hagia Sophia in Constantinople.

Houses were built either of stone, marble, or baked bricks ($\beta\tilde{\eta}\sigma\alpha\lambda\alpha$ or $\tau\sigma\nu\delta\lambda\alpha$) that were often stamped with the seal of the factory. A wooden frame ($\dot{\iota}\mu\dot{\alpha}\nu\tau\omega\sigma\iota\varsigma$) was often used to strengthen the walls. The bricks were held together by mortars made with lime, sand, or powdered brick.

Tools used in construction included trowels, saws, hammers, and so on; like the Byzantine agricultural tools (which will be discussed later), they were forged. The square was also commonly used.

373

Plate 55.
"Construction of the Tower of Babel." Twelfth-century mosaic,
St. Mark's, Venice. *Photo Giraudon-Anderson*

The house   Houses generally had one or two stories. Beginning in the fifth century, edicts promulgated by the Emperor Zeno (474–491) regulated the width of streets and balconies, and prohibited outside stairways, which were built of marble (interior stairways were of wood). The façades were whitewashed with lime or covered with polychrome painting. The roofs were covered with tiles separated by sheets of zinc to protect them from the rain. They were often topped by a weathervane; that of the city of Constantinople was of bronze, and had four pyramid-shaped blades.

The doors, which were the swinging type, were made of iron or brass. A Byzantine lock of the sixth century was recently discovered in Serbia; its principal part was a spring that prevented the bolt from moving. The bolt was lowered with a key, which unlocked the mechanism (Figure 15). Padlocks still in use in the monasteries of Mount Athos function on the same principle (Deroko, 1957). Windows had small panes of glass wedged into plaster frames.

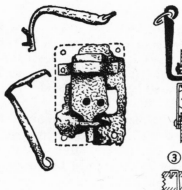
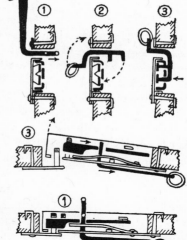

Fig. 15. Sixth-century Byzantine lock (Serbia). *Left,* general view; *right,* diagram of its operation (from Deroko, 1957).

Stoves were heated by wood, and the smoke escaped through quadrangular pipes. The houses had water closets; in the capital strictly supervised sewers carried waste material down to the sea. The public baths consisted of a rotunda covered with a cupola. The water, heated in bronze kettles on an iron stove or on fireproof stones, flowed through a pipe into the baths.

The houses were lighted by means of candles and oil lamps that could be used in candelabras and chandeliers. The furniture (tables, chairs, and beds) differed little from that of antiquity.

Aqueducts and cisterns   In addition to palaces, churches, hospitals, and private housing, in the fourth century the Byzantines began to build numerous aqueducts, on the Roman model, to supply Constantinople with drinking water, which was carried by lead pipes. About forty cisterns, most of them covered, were built for the same purpose. The two most important of these (Jeré Batan Serai and Bin bir Direk) were able to hold 82,400 and 43,200 cubic meters of water, respectively. To make them watertight, these strange structures, which were supported on pillars or columns (Bin bir Direk is Turkish for "Thousand and One Columns"), were lined with a special mortar of lime, crushed brick, and oakum; this mortar in turn was covered with a layer of lime, oil, and cotton wadding.

Byzantium also had wells and fountains or *phiales*. One of the latter has been reconstructed in the Byzantine Museum of Athens.

ARTISTIC TECHNIQUES

The skill of the Byzantine artists and the great perfection of their techniques are by no means negligible factors in the explanation of the complete success enjoyed by Byzantine art.

Mosaic The method used in mosaic work consisted of applying to the surface to be decorated first a preliminary layer of cement, then a second, more solid, layer to receive the tesserae, or small cubes of various materials (stone, marble, mother-of-pearl, glass, and precious stones), which were colored by metallic oxides (copper for the blues, chrome for the greens).

The secret of the superiority of the Byzantine over the Roman mosaics lies in the much greater variety of materials used and the varying sizes and richness of colors (which are still fresh) of the tesserae. Moreover, after the sixth century gold backgrounds were used; they were obtained by applying a very thin gold leaf between two layers of glass, one of which acted as the background.

Fresco and painting In "tempera" fresco work, the colors were blended with eggwhite or gum and applied to a dry layer of plaster; in true fresco, the colors were applied without adhesive to the still-damp plaster. The latter method was the most durable.

In other fresco work the artists used charcoal, brushes made of badger hair, glue (obtained by macerating ox or buffalo hides in limewater), and plaster made from gypsum.

Most of the technical formulas of the Byzantine artists are given in "The Guide to Painting," a manuscript of uncertain but very early (pre-sixteenth-century) date, which was still in use in the nineteenth century in the monasteries of Greece. Here is an example of one of these formulas handed down from generation to generation:

"How to make gilt letters: take paint and mercury, and place them with some tin on a slab of marble. Add very strong vinegar and lead or silver; crush these substances until water forms. Collect this water, and use it with a brush. When the letters are dry you will be able to make them very shiny" (translated by Durand, 1845).

The technique used in painting icons, which were so popular in Byzantium,

Plate 56.
Empress Theodora. Sixth-century mosaic, San Vitale, Ravenna.
Photo Giraudon-Anderson

was simply a variation on the fresco technique. The oldest icons, however, were painted with encaustic, a technique used in Greco-Roman Egypt: the drawing was engraved with a hot needle on a wooden panel coated with colored wax.

Goldsmith's work

The Byzantine goldsmiths had achieved great perfection in the working of the metal used in making secular and religious objects (chalices, pattens, book covers, and so on). Most of the book covers were of silver, often encrusted with precious stones, enamels, or gold filligree work. Under the Islamic influence, damascening, the encrustation of thin gold or silver wires in metal, appeared; this technique spread from Byzantium to the West.

For small objects the technique called "respoussé" (cold hammering of metal leaf supported by a flexible mastic) was the one most commonly used.

Enamels

Enamelwork had been in existence since antiquity (Egypt and Rome). The Byzantines made champlevé enamels (in which the enamel was poured into the hollows in a piece of metal) and cloisonné enamels. The latter technique was more typical of the Eastern Greek Empire: fine strips of gold of from one to two millimeters wide were soldered in sections onto the base (generally a sheet of gold), along the outlines of the desired design, forming compartments into which the enamel was poured.

Glass and ceramics

Glass was used for making various vessels (phials, perfume bottles, and so on), oil lamps, and panes of window glass (several panes dating from the sixth century are preserved in Ravenna). We are more familiar, however, with Byzantine ceramic vessels (bowls, cups, plates, pitchers, and so on). The oldest varieties were made with white clay; beginning in the twelfth century, however, a red clay variety covered with a white slip under the enamel is found; the decoration was made by removing parts of this layer before enameling (the so-called *sgraffito* pottery).

Sculpture

Stone and ivory sculpture was highly perfected, for example in the sculpturing of capitals. Beginning in the fifth century the typical smooth acanthus leaf of the Corinthian capital was replaced by the thorny acanthus leaf. This work was done with the chisel and drill.

In the architecture of the same period, the architrave (a stone beam) was replaced by the springer (a stone cube), which had become indispensable for bearing the thrust of the arches that had replaced the lintel course. However, the springer itself was later eliminated and was replaced in the sixth century by a more massive capital shaped like a cube or a truncated cone.

Manuscript techniques

Manuscripts were written on papyrus, parchment, or paper. The latter, a discovery of the Chinese, may have been introduced into the Mediterranean area in the middle of the eighth century. Cotton rag paper (*bombycinus*), however, was not used in Byzantium until the eleventh century, and did not become common until the thirteenth. Manuscripts were essentially codices, or books, as opposed to the *volumina* (rolls), which disappeared during the sixth and seventh centuries. The binding was known as a *stachoma* (στάχωμα). Valuable books were locked in caskets.

Pens were made of carefully cut reeds. Ink was black, cinnabar, or blue-green, and gold was frequently used.

Fabrics The Byzantine fabrics used for clothing or in the decoration of churches, palaces, and homes were either woven fabrics or tapestry.

The Coptic weavers of Egypt (where the horizontal loom had been in existence since the Twelfth Dynasty, the vertical loom since the Nineteenth), who had been the official purveyors to the Roman Empire, played the same role for Byzantium, where until the seventh century fabrics consisted of a flax weft decorated with woolen designs. André Blum has recently noted the discovery, in a necropolis of Byzantine (that is, sixth-century) Upper Egypt of a wooden stamp for printing fabrics (Figure 16), which shows that printed fabrics were already in existence at this period.

Fɪɢ. 16. Wooden stamp for printing fabrics; sixth century. Found in a necropolis at Achmim (Egypt) (from A. Blum, 1954).

Also in the sixth century, the raising of silkworms was introduced to Constantinople by Emperor Justinian in order to supply work for the Byzantine shops, which for political reasons were no longer receiving raw silk on which to work. Byzantine silk weaving must have corresponded approximately to the following description written in the fifth century A.D. by St. Theodoret, Greek Bishop of Cyrrhus:

"Grasping the very fine threads, the women warp them, then begin to stretch them in regular order, like threads on a loom. They unwind the weft thread, while separating the warp threads with a shuttle, here loosening, there pulling the drawcords that are attached to these threads. Finally, using tools for this purpose, squeezing and as it were fulling the weft, they finish the cloth." This is the very old "simples" loom that, when improved, led to the "draw" loom. Weaving and silk working then became widespread, and a state silk factory was in existence until the tenth century.

Here, as in many other technical fields, the influence of Egypt and Persia was considerable, to the point that it is often very difficult to distinguish a Byzantine from a Sassanid silk fabric. In the words of Algoud:

"An undeniable similarity exists between these fabrics, which belong particularly to the type of production that utilizes the effect of the weft, or of wefts showing the design by touches of color that are in contrast to the (usually satin) background; this involves concealing the weft under the warp threads in the plain-colored portions of the background, thus giving a better result for the final contrast which accentuates the decorative design" (op. cit., p. 34).

The symmetrical motifs frequently seen in the Byzantine silks were obtained by the arrangement of the warping in figure harnesses in which "all that is required is to set up the simples loom so that when the drawcords placed there for this purpose are pulled, they move the threads they control in the desired symmetrical and reverse order, causing the loom to reverse the design (ibid., p. 35).

AGRICULTURAL TECHNIQUES

Plowing tools — Fieldwork is depicted in illustrated Byzantine manuscripts; in one such manuscript dating from the fourteenth century the following tools can be seen: the *aratrum,* the sickle, the scythe, the two-headed mallet, the planter, the pitchfork, and the yoke. The *aratrum* consisted of a curved shaft attached to the yoke of the harness; the plowshare and the handle held by the farmer were attached to the shaft.

Wheat was treaded with the help of a very ancient tool that had already been known to the ancient civilizations of Mesopotamia, the *tribulum (tribolds),* a kind of elongated sledge with flint tips arranged on its inner face in the form of a quincunx.

Harness — The transportation required for farm work was supplied by carts drawn by oxen harnessed abreast under the yoke. The Byzantines still used the ancient system of horse harness in which draft came from the necks of the animals. They did not discover nailed horseshoes until the ninth century, but eighth-century manuscript illuminations show that the saddle of the draft animal was attached to the saddle girth. In the thirteenth century the Serbs became acquainted with the modern type of shoulder-collar harness, which had already been in use in the West for three centuries.

Techniques of cultivation — In the *Geoponica,* a collection of texts relative to agriculture composed in the tenth century by order of the Emperor Constantine Porphyrogenitus, we find original passages on the cultivation of various plants (grapes, olive trees, vegetables, fruit trees), grafting, beekeeping, and economic zoology.

HUNTING AND FISHING

Hunting — Hunting scenes are frequently depicted in Byzantine manuscripts, especially after the twelfth century. Birds and mammals (rabbit, fox, deer, wild boar, and bear) were hunted;

the former were captured with the help of snares and nets, the latter were fought with bows and javelins. Tamed animals (for example, the falcon and the leopard) and dogs were often used to assist the hunters; this was an Eastern influence.

Fishing   As in ancient Greece, fishing was held in high regard in Byzantium. The nets, made of broom fibers, were mended with wool and a wooden needle. The Byzantines were acquainted with the seine (γρίπος) and the landing net (ἀπόχι). In the case of fishing poles, then as now the hook was suspended from a line attached to a slender pole. The fly rod was called an ὁρμιά. Large hooks were maintained in an upright position by inflated waterskins.

Pots were used for crab fishing, for tuna and octopus, the three-pronged harpoon. Octopuses were also captured by the ingenious technique of sprinkling them with fresh water, which they cannot endure. Fish were often attracted by fire-fishing, that is, fires of resinated wood on lighted grills. Diving for fish, which is regarded as a modern invention, may have been practiced as early as the fourth century A.D. (P. Koukoulès).

NAVIGATION

The Byzantine vessels differed little from those of the ancient Greeks, and the techniques of their construction (sawing of the planks intended for the hull, drilling holes, and so on) were practically identical with those of their predecessors. However, the Byzantines borrowed from the Arabs the technique of caulking, or filling up the chinks in the hull, with tar-impregnated fabric or oakum.

Attempts at improvement were concentrated on the military fleet, a perfectly natural procedure given the location of the capital, which made it a prey to frequent attacks. L. Bréhier distinguishes three periods in the Byzantine Navy's technical evolution:

1. Prior to the Arabic invasion (that is, until the seventh century), the Byzantine vessels differed little from those of antiquity: they were powered by oars, and two oars attached to the stern acted as a rudder (the sternpost rudder was still unknown). In the sixth century the dromond, a ship with a single row of rowers protected by a roof, and a light version of the dromond known as a *liburna*, were the most common vessels.

2. The period from the seventh to the twelfth centuries was the golden age of the imperial navy. During this period appeared dromonds with two tiers of sometimes more than a hundred rowers; the flagship was known as the *pamphylos*. Transport ships (*chalandia*) were galleys with forty-two rows of rowers; in addition, there were special ships for transporting horses.

3. At the end of the Empire, biremes and triremes were the most commonly used vessels in Byzantium. Sailing vessels were not greatly developed, but a few pictures of them can be seen in miniatures.

Signaling on the sea was done by flags during the day, by lights and lighthouses at night. Warships were equipped with battering rams and incendiary devices (to be discussed later).

MILITARY TECHNIQUES

The numerous conflicts the Byzantine Empire was forced to sustain explain the considerable development of military techniques. In the words of Steven Runciman, "In the Middle Ages, Byzantium was the only place where tactics, the organization of the army, and strategy were calmly and carefully studied."

Greek fire and pyrotechnics   It was the tremendous superiority of the Byzantines in the field of chemical weapons that permitted them to resist the Slavic and Arabic thrusts for eight centuries. The famous Greek fire (πῦρ ρωμαιχόν) was successfully used against the Arabs in the seventh century by Constantine Pogonatus, and in the eighth century by Leo III the Isaurian, while in the tenth century it was used by Romanus II Lecapenus against the fleet of the Russian Prince Igor.

From the recent studies of Maurice Mercier, we learn that this formidable defensive weapon consisted of crude petroleum from Asia Minor that was transported to Byzantium in waterskins. Under the name "liquid fire" (πῦρ ὑγρὸν), it was elevated to the status of a state secret in the reign of Constantine VII Porphyrogenitus (tenth century).

Certain warships armed with flamethrowers were extremely effective in naval combats (Figure 17). Mercier suggests that the combustible was stored on such ships in oxhide waterskins with a very large capacity. These skins had a cover that needed only to be pressed with the foot in order to draw the combustible liquid into the throwing device. This is an ingenious explanation, but it cannot be verified, for until now no pictures or descriptions of such objects have been found in documents of the period.

FIG. 17. Naval battle with Greek fire (from *Cod. Vat. Graec., 1605*), tenth century.

In addition, in the ninth century the Byzantines began to utilize small grenades (χειροσίφωνα) filled with a paste (πῦρ μαλθάχον) composed of naphtha, saltpeter, sulfur, and a controlling combustion product. There were thrown by hand or with the help of brass tubes. Although these devices, which are lighted before being thrown at the enemy, operate on the principle of the missile, the term "grenade" appears to us more appropriate, at least in the case of those thrown by hand. They also had a pyrotechnical mixture which ignited spontaneously (πῦρ αὐτόματον); it had been known to antiquity (Athenaeus, Xenophon) and was perfected by an engineer named Callinicos. Mention should also be made of the treatise of Marcus the Greek (dates uncertain) on the various Byzantine pyrotechnical mixtures.

The Byzantines, however, did not continue to possess their great secret exclusively. The Arabs, too, had long known of military fire made with naphtha, and ultimately utilized it themselves in their struggles against Byzantium.

Sieges, strategy, and tactics The art of sieges and various other strategic and tactical techniques were the subject of numerous anthologies compiled especially after the tenth and eleventh centuries, when the Macedonian dynasty was at its peak. These treatises are for the most part compilations by Hellenistic authors, but there also exist authentic Byzantine tacticians, the most famous of whom was the Emperor Leo VI (886–912), who authored several works. (Numerous texts of this type have been published in France by Alphonse Dain.)

The weapons used by the Byzantine troops were the sword, the bow and quiver, and the lance for the cavalry, and the ax, the lance, the sword, and the shield for the infantry; the latter also had suits of mail.

The optical telegraph The optical telegraph, already in existence in antiquity, was utilized in Byzantium, especially in the ninth and tenth centuries, when Cilicia formed the most southerly frontier between the empire and the Caliphate. It consisted of a system of fires and lanterns lighted from hill to hill, ending at the lighthouse built on the terrace of the Great Palace in Constantinople. A violation of the border could thus be known in the capital one hour later. Leo the Mathematician perfected the system during the reign of Emperor Theophilus (ninth century.).

MISCELLANEOUS TECHNIQUES

In the following pages we shall discuss various techniques that cannot easily be included under the preceding headings.

Scientific devices and instruments Although the scientific disciplines were not very brilliant in Byzantium, neither were they completely neglected. In particular, mathematics, hydraulics, astronomy, and alchemy continued to be studied. We possess information on certain scientific devices and instruments of this period.

As in classical antiquity, time was measured with clepsydras and sundials. A single Byzantine astrolabe dating from the eleventh century is still in existence, and the Benaki Museum of Athens possesses a compass dating from the tenth or eleventh century (Figure 18).

FIG. 18. Byzantine compass (Benaki Museum, Athens) (original).

In addition, Greek alchemical manuscripts of the medieval period (eleventh to fifteenth centuries) contain numerous drawings of devices (alambics, boilers, double boilers), most of which go back to an older tradition (M. Berthelot).

Automata and hydraulic devices From various tenth-century writings, particularly those of Emperor Constantine Porphyrogenitus and Ambassador Liutprand, we learn that near the throne of the Byzantine emperors stood ingenious devices intended to amaze foreign visitors: a golden tree with metal birds that could "sing," gold lions that could "roar" and stand up on their legs, and a device by which the throne could rise into the air.

If we abide by the exact definition of an automaton — "An organized machine which imitates the movement of a living body by means of springs" — only two of the devices can be considered as automata, and we know absolutely nothing of the mechanism that operated them. We do know, however, that such automata existed in the same period at the court of the Islamic rulers.

In contrast, the tree with the birds can be explained by an application of the ingenious inventions of Hero of Alexandria; air escaping under pressure of water from a sealed compartment through a narrow opening can produce certain sounds. Thus the tree may have been made of hollow pipes, and this device may have been closer to that of the hydraulic organ (which was known in Byzantium, as we shall see) than to a true automaton.

Hydraulic mills are also mentioned in the charters of various hospices and monasteries.

Musical instruments In their religious and secular ceremonies the Byzantines utilized numerous musical instruments, but we have few details about their production. Some of them differed little from those of ancient Greece. Stringed instruments included the psaltry, with thirty or forty strings, and the pandora, an ancestor of the lute. Wind instruments included the salpinx and the aulos, ancestors of the trumpet and the oboe, respectively.

Organs were at first hydraulic, then pneumatic; there are pictures of portable organs on the base of the obelisk of Theodosius I in Constantinople (fourth century). Some large organs were moved about on a platform with small wheels.

Percussion instruments (cymbals) were also known.

Production of metals Brass was prepared according to a method borrowed from the Persians, iron from Indian formulas. Speaking of iron, Stephanides has stressed that many objects manufactured in Byzantium were called "Indian" in order to increase their value.

Jewels and cosmetics In addition to jewelry (pendants, earrings, and so on) decorated with valuable precious stones, there was also an industry that produced imitation stones, using Persian alchemical formulas (M. K. Stephanides).

The women utilized cosmetics made from lead, red lead (minimum), and antimony, as well as hair dyes.

Plate 57.

a. Well. Barberini Psalter (twelfth century). Vatican Library, Rome. *Photo Vatican Library*

b. Construction of a building. *Photo Vatican Library*

SUMMARY

Byzantine technology reflects the threefold Roman, Greek, and Eastern influence the Greek Empire of the East was to experience in many areas.

The Roman influence was felt especially in the architecture of certain utilitarian buildings (for example, aqueducts), as well as artistic techniques (for example, mosaics). The Greek influence predominated in everything concerning techniques of agriculture, hunting, fishing, navigation, and the production of certain luxury items (for example, musical instruments). Brick architecture was a Sassanid contribution, while Islam contributed to various Byzantine techniques, such as damascening, caulking, and weaving of fabrics (Coptic Egypt).

These varied influences, however, were able to blend into a harmonious and successful synthesis, and the Byzantines, who have so long been criticized for the sterility of their culture, were actually very advanced from a technical point of view, very often even surpassing their Occidental and Oriental contemporaries. This is particularly true of construction and artistic techniques, in which they excelled beyond compare.

They were undeniably innovators in the field of chemical weapons, and here as in other fields played the role of intermediaries in the spread of Eastern techniques to the West.

BIBLIOGRAPHY

I. GENERAL WORKS

BAYNES, NORMAN H., *Byzantine Studies and Other Essays* (London, 1955).
——— and MOSS, HENRY ST. L. B. (eds.), *Byzantium, an Introduction to East Roman Civilization* (Oxford, 1948).
HUSSEY, JOAN M., *The Byzantine World* (London, 1957).

II. SPECIALIZED WORKS

ANTHONY, E. W., *A History of Mosaics* (Boston, 1935).
DALTON, ORMONDE M., *Byzantine Art and Archaeology* (Oxford, 1911).
DOWNEY, G., "Byzantine Architects: Their Training and Methods," *Byzantion*, 18 (1948), 99–118.
GRABAR, ANDRÉ, *Byzantine Painting* (Geneva, 1953).
KELLER, A. G., "A Byzantine Admirer of 'Western' Progress: Cardinal Bessarion," *Cambridge Historical Journal*, XI (1955), pp. 343–348.
OSTROGORSKY, GEORG, "Agrarian Conditions in the Byzantine Empire in the Middle Ages," *Cambridge Economic History*, Vol. 1 (Cambridge, 1941), 194–223.
PARTINGTON, J. R., *History of Greek Fire and Gunpowder* (London, 1960).
RUNCIMAN, STEVEN, "Byzantine Trade and Industry," *Cambridge Economic History of Europe*, Vol. 2 (Cambridge, Mass., 1952), 86–118.
TALBOT-RICE, D., *Byzantine Art* (London, 1954).

PART FIVE

PRE-COLUMBIAN AMERICA

PRE-COLUMBIAN TECHNOLOGY

THE TERM "pre-Columbian America" is used in this book to designate the
entire territory of the Americas and the evolutional stage of their inhabitants prior to their contact with the Europeans. This contact took place in the
chronological period between the fifteenth and nineteenth centuries. Our discussion of the technology of each region will end with this confrontation of the
two worlds.

The civilizations of pre-Columbian America were slowly built up on a thin
cultural base formed by three successive waves of Mongoloid immigrants, who
came from Siberia across the Bering Strait between 25,000 and 7,000 B.C., and
reached the Straits of Magellan around the latter date. These nomadic hunter-
fishermen, who practiced food-gathering, invaded both continents. By 3,000 B.C.
agricultural peoples were cultivating land in the Andes, Guatemala, and the
central plateau of Mexico. By the beginning of the sixteenth century, America
(including the islands of the West Indies) was supporting thirteen million human
beings, and was a mosaic of cultures: nomadic peoples, sedentary peoples, fishermen, hunters, peoples practicing agriculture on an occasional basis, farmers,,
and city dwellers. There were even dead civilizations, forgotten or swallowed up
by the jungle: Teotihuacán in central Mexico, the Olnecs on the south side of
the Gulf of Mexico, the Mayan civilization, Chimu in Peru, Tiahuanaco in
Bolivia.

The cultural development of the New World took place over a period of
thousands of years. It was not uniform, especially as regards technical progress,
and this fact makes it impracticable to apply the usual standards of technological classification. For example, judging them on the basis of their tools, we
would place the early American civilizations among the Paleolithic, or at most
Neolithic peoples. Since the Bronze and Iron ages did not exist because these
metals were unknown, must we then conclude that the American never surpassed
the stage of primitive peoples?

The technological evolution of Mesoamerica (Central America) and the
central Andes may be delimited within the following periods (approximately the
same stages, except the fifth, are found in northern Mexico, with a considerable
time lag):

1. Preagricultural stage (25,000–6000 B.C.). Stone tools were made by flaking
   until 12,000 B.C., after which they were produced by pressure and dressing.
2. Preceramic stage (6000–3000 B.C.). Early stages of settlement, domestication of plants.

3. Primitive or archaic stage (3000–1500 B.C.). Horticulture; primitive ceramics, agriculture, basketmaking, weaving.
4. Formative or preclassical stage (1500 B.C.–200 A.D.). Progress in ceramics, agriculture, basketmaking, weaving. Appearance of stone architecture: pyramids, cities.
5. Classical period (200–1100 A.D.). Development of technology, invention of metallurgy; predominance of agriculture as the base of the economy (hunting became a sport); monumental architecture, cities, invention of the Mayan false (corbeled) vault.
6. Imperialist period (1100–1500 A.D.). Military expansion, colonization, decadence.

Since America remained essentially isolated from the rest of the world after the last Mongoloid immigration (except for a few sporadic and limited contacts with Melanesian peoples), Americans were forced to discover for themselves solutions to the ordinary problems of life, the exploitation of nature, and the survival of communities. On the whole they succeeded remarkably well. The creation *ab nihilo* of the three great civilizations — the Aztec, the Inca, the Maya — far from the civilizing currents that fertilized the rest of the world, is an achievement unique in history.

America's technological evolution (agriculture, ceramics, weaving, architecture, metallurgy, basketry) was free of all foreign influence except for the contributions of the first immigrants — fire, the drill, the throwing weapon, and flint- and wood-working. The last wave of immigrants (around 7000 B.C.) contributed two extremely important elements: the dog and the bow for shooting. In the course of this evolution independent discoveries and inventions were made. Once the American continents were populated, symbioses and borrowings naturally occurred among the various groups. The sharing of the latest discoveries, as well as the acquisition of new techniques, was thus facilitated by ethnic movements on the individual as well as the tribal and national level. On the whole, however, America was fermenting in a vacuum. The American was not a born technician.

He did not feel any need to develop the consequences of technical invention to their logical conclusion, and therefore never applied his discoveries to improving the comfort of the masses. Moreover, his natural curiosity was quickly satisfied by metaphysical and magical explanations. The inevitable result of this attitude was stagnation in creative activity together with technical inertia: once arrived at a certain degree of technological efficiency that seemed to him sufficient, he stopped at that point. Of course, this did not exclude ingenuity and increasing manual skill in execution applied to the perfecting of certain techniques.

In several fields the pre-Columbian Americans never took the step that transforms a discovery into a useful invention The *metate* (a small, flat, rectangular stone on which maize is crushed with the help of a stone roller) has been utilized everywhere in America for at least seven thousand years without modification of its shape, despite the urgent need to increase production (maize flour is the nutritional base of the diet): every household grinds the exact amount of flour needed for each meal. The house, the kitchen equipment, and the furnishings have remained the same, except among the dominant classes. In the purely technical field, metal tools and weapons perpetuated the forms of their stone predecessors; only in ornamentation were new technical techniques developed. It must be stressed, however, that metallurgy appeared late (twelfth century) and was practiced sporadically.

Three elements that strongly influenced technological development were lacking. First, there were no domestic animals, that is, no draft, saddle, or pack animals. This lack was revealed in the American diet (absence of meat and dairy products), clothing (lack of skins, leather, and wool), agriculture (impossibility of developing the plow, harrow, and noria, to say nothing of fertilizer), and transportation (which was of necessity limited to human power). However, wherever animal raising became possible it was practiced, if only on an elementary level: the dog, the turkey, and bees in North America, the guinea pig and the alpaca in Peru (stock raising for food purposes); dogs (with travois) among the Eskimos and lamas in Peru (for transportation). Also in Peru, the alpaca and its wild counterpart, the vicuña, were sheared for their wool.

In the second place, there was no application of the mechanical principle of the wheel; that is, there was neither pulley nor potter's wheel, nor lathe nor mill nor vehicles, although the gyratory mechanical principle was utilized in whorls, spindles, and drills. On the south coast of Veracruz, Mexico, prior to the eighth century, children were using strings to pull small terra-cotta dogs mounted on two axles and four wheels. Because of a lack of complete understanding of the mechanical gyratory principle, which transforms rectilinear movement into circular movement and vice versa, the Americans never discovered the screw, the crank, the pedal, the driving belt, the steering wheel, the gear train, demultiplication of power, or lifting and driving machines.

In the third place, there was the lack of metal. Metal was discovered late — for the most part in the period from the eighth to the tenth centuries. Moreover, preference appears to have been given to its artistic or purely decorative aspect. Copper was used at most for axes, adzes, shears, and gravers. Despite the success of several alloys, bronze remained practically unknown. Only the Eskimos made use of meteoric iron.

A description of pre-Columbian technology is influenced by two factors: (1) the absence of contemporary written evidence, since writing as we know it had not been invented in America; (2) the absence of many objects and tools that disappeared because they were made of organic materials. Because of this lack of documents, the description of techniques is often empirical. Archaeological excavations, ethnological research, pictographic evidence, and (in the case of architecture) existing ruins nevertheless furnish precious material for the analysis of the principal pre-Columbian techniques.

AGRICULTURE

In the field of agriculture America invented and utilized all the possibilities offered by both the conformation of the land and wild plants. However, millenniums passed before agriculture was discovered, and it was practiced intensively only in the southwestern United States, in Mesoamerica, and in the central Andes — the same regions in which grain was being crushed seven thousand years ago. The southwestern part of the United States became an agricultural country around 300 or 200 B.C.; the East, and particularly the Southeast, began to cultivate maize between the second and fourth centuries, under the Mexican influence, then developed agriculture; its inhabitants overtook the Mesoamericans in everything pertaining to agricultural techniques between the tenth and fourteenth centuries.

In the Mesoamerican region agriculture appeared between 6000 and 3000 B.C., and flourished between 1500 and 1200 B.C.; maize, beans, pumpkins, tomatoes, and pimiento were cultivated.

Around 3000 B.C. the first farmers in the Andes were growing pumpkins, gourds, beans, and pimiento. Two thousand years later maize, peanuts, avocado, and manioc appeared. Between the seventh and ninth centuries agriculture became the base of the economy: the white potato, the sweet potato, cocoa, and various fruits, including the pineapple and the papaya, complete the list of plants cultivated. Techniques for preparing the ground included fertilizers, the use of terraces, and intensive irrigation.

Agricultural techniques   *Preparation of the ground.* — In the cultivation of the maize-bean-pumpkin complex, the ground was cleared by burning (except among the peoples of the eastern United States), a technique that was complemented by the ax. The destroyed vegetation reappeared after several years, however, because after three or four years the soil became useless for cultivation. This *de facto* fallowing was understood by the Incas, who let a piece of land lie fallow for several years after they had cultivated it for one year. Fertilizers helped in the regeneration of the ground; the Peruvians brought bird manure (guano) from the offshore islands, and the Aztecs used human compost. Elsewhere the peasants were obliged to move as the lands became exhausted.

Plowing. — To turn over the ground, level it, and break up the clumps a single instrument, the hoe, was used. Made first of wood, later equipped with a stone blade, it was a tool of many uses. The digging stick was also used for this purpose, and the peasants of the central Andes broke up the clods with a stone-headed club.

Sowing. — The digging stick was utilized everywhere for digging seed holes; after the seed was buried in them (Figure 1) the holes were filled in by foot. The Peruvian Inca stick (1438–1532) was almost a spade, with a lateral bar to support the foot; the tip was enlarged into the form of a blade (Figure 2). After the appearance of metallurgy the tips of the digging sticks were made of copper.

Irrigation. — Natural irrigation occurred through floods or regular rainfall. In places where these phenomena were lacking, the Americans compensated for the lack by constructing artificial irrigation systems reinforced by reservoirs, especially in the central Andes, where irrigation was systematically practiced. Their complexity and scale often suggest deliberate planning; this is the only possible explanation for the digging of canals, the longest of which (at Chicama) was seventy-four miles long, or the construction of the aqueduct of Ascope, which was almost a mile long and fifty feet high (central Andes, seventh–eleventh centuries).

Care of the crop before harvesting. — As everywhere, the growing crop had

Fig. 1. Planting seed with a digging stick. Aztec (from the Codex of Florence).

Fig. 2. Planting seed with a digging stick with crossbar. Inca (from Poma de Ayala).

to be hilled, weeded (by hand or with the digging stick) (Figure 3), and protected against animals (by means of scarecrows, traps, and fences).

The harvest. — Tubers and root plants were harvested with the digging stick or the hoe; maize was harvested by hand (there was neither scythe nor sickle). Women as well as men participated in all these agricultural activities.

Preserving food. — Whether whole or husked, the cereal grains were kept in enormous pots or baskets and in silos. The Aztecs, Maya, and Incas maintained community silos in preparation for periods of scarcity.

Fig. 3.   Weeding. Inca (from Poma de Ayala).

Special preparation of the ground

Terraces. — Where the ground was hilly or sloping, terraces were built; they were supported by walls of dry stone, and sometimes (among the Incas) were as large as 984 feet long by 22 to 33 feet wide. Terraces were sometimes used as protection against erosion.

Floating gardens. — Due to the scarcity of cultivable ground, the Aztecs conceived the idea of constructing wooden rafts anchored to the bottom of a lake; vegetable matter was piled on them so that they could be used for the cultivation of plants.

Plate 58.
Terrace farming. Urubamba-Pisac, Peru. *Photo Musée de l'Homme—Tracol*

CATTLE BREEDING

Cattle breeding was practically nonexistent, despite the presence in prehistoric times of the camel and the horse, which are American in origin. These animals. victims of collective slaughter and a change of climate, had completely disappeared by 10,000 B.C. Aside from the large and small animals imported after the sixteenth century by the Spaniards and later by other Europeans, the only livestock were the llama (transportation), the alpaca (wool and meat), and the guinea pig (meat) in Peru. North America had the dog (transportation) for the Eskimos, meat for the Mexicans) and the turkey; apiculture was practiced in Mexico. Domestication of the mammals must have occurred between 6000 and 3000 B.C. Everything leads us to believe that breeding was above all extensive, with only the alpaca and the dog being raised intensively.

CERAMICS

Ceramics, like agriculture, architecture, weaving, and metallurgy, should be regarded as a completely American and fully independent invention, developed *ab nihilo* in each region. In Mexico it was invented during the archaic or primitive period (around 2200 B.C.). Its dissemination over the Mesoamerican territory was so rapid that it is very difficult to determine its appearance in each area. The same is true in the central Andes, where it was invented on the north coast of Peru (Chavin) around 1200 B.C. In the United States there were two centers: in the Southwest the Hohokam culture (300 B.C.–500 A.D.) and the Anasazi culture, also called the "Modified Basket Makers" (sixth–eighth centuries); in the Southeast around the third century B.C. and in the Northeast between the sixth and tenth centuries. On the whole the distribution of pottery is more or less the same as that of agriculture; however, its development is difficult to follow, for it is not easy to distinguish the steps in its progression from a crude, badly fired, unpolished and unpainted ceramicware to a more perfected pottery that ultimately achieved an unusually high quality of firing and finishing. It is common to find the same techniques being practiced at different periods in widely separated areas. In addition, it is possible for pottery to appear in fully developed form without known antecedents.

Materials The New World is particularly rich in varieties of clay, most of which are sufficiently pure to eliminate the necessity for cleaning. The clay was extracted in thick lumps from sedimentary deposits near rivers or from veins. The lumps were broken in order to facilitate purification (done by rubbing them on stone mortars), and were then kneaded or trampled. During this operation, and depending on the possibilities offered by the environment, nonplastic or cleaning agents were added: herbs, straw (southeastern United States around 200 B.C., central Andes during the first millennium), sand, or potsherds (southern Louisiana sixth to tenth centuries, Mesoamerica beginning in 1500 B.C., central Andes beginning in the first century A.D.). The slip was generally a simple barbotine to which coloring matter was often added. Completely waterproof objects were rare, for this supposed chemical knowledge and perfected methods of firing which the pre-Columbian civilizations did not have; the result was a complete absence of vitrification. However, the application of several layers of slip was a satisfactory substitute. Beginning in the second century B.C., the use of slip became general in Mesoamerica and the central Andes; it did not appear in the eastern United States (southern part of Louisiana) until the sixth century, in the Southwest, at the earliest, around the end of the twelfth century.

MODELING

Modeling Of the three possible methods of shaping clay, two were used in pre-Columbian America: modeling and molding, the former more widespread than the latter. Turning on the potter's wheel was never practiced, given the absence of the wheel. We feel it is unnecessary to give a detailed description of the techniques practiced by the American potters in the shaping and finishing of their objects, for these techniques differ very little from those practiced elsewhere. Thus each technical term employed here should be understood in its general meaning.

1. Modeling in block. — This technique was rarely found among the less evolved peoples of North and South America.
2. Modeling in individual pieces. — This technique is found in the Anasazi culture (sixth to eighth centuries) in the United States, in preclassical Mesoamerica, and in the central Andes (100 B.C. to 600 A.D.).
3. Modeling by coiling. — This was the typical modeling technique used for American pottery. It appeared around 1500 B.C., but in the United States was practiced only by the Hohokam (300 B.C.–fifteenth century) and Pueblo (tenth to eleventh centuries) peoples.
4. Molding. — This technique was invented at Teotihuacán (central Mexico) in the sixth century, in the Mayan area in the eighth century, and along the northern coast of Peru around the same time. Molding was used especially for the mass production of figurines that had a magical or religious significance. Both closed molds and open or simple molds were known.

Decoration Decoration, like modeling, was done in .various ways. Impression on soft clay. — From prehistoric times motifs had been incised on the pottery by means of punches. Various motifs were stamped with a terra-cotta (or any other hard material) matrix on Mesoamerican and central-Andean pottery beginning in the preclassical period, and later in the eastern United States.

2. Removing or adding clay. — Incised, cloisonné, and champlevé decoration (Teotihuacán, sixth century). This technique was used all over the Americas from the beginning of pottery making, but was utilized particularly for emphasizing details of physique or dress in preclassical figurines.
3. Application. — Beginning in the preclassical period, the typical method of decoration was by painting on the slip before firing. Negative painting very soon came into general practice. For both negative and positive painting the lost-wax and batik-coloring methods were frequently utilized.

The fresco method was also used: the motif was painted, after firing, on a very fine layer of stucco. This technique was strictly localized, both temporarily and spatially. Around the sixth century it appeared among the Zapotecs and Teotihuacáns of central Mexico, and the Guatemalteco Maya of Kaminaljuyu. There was also a "fresco-type" painting that was executed before firing on an unslipped object (Jalapazco, Mexico, and eleventh-century Maya). Objects were also lacquered. The potters of Cholula (Puebla region, Mexico, eleventh–thirteenth centuries) gave their ceramicware a gleaming surface reminiscent of that of lacquer. In reality it was a second firing that fixed several layers of slip applied to an already fired vase.

Polishing	Polishing was done with stone (in most cases quartz), pieces of leather or cloth, or fragments of calabash.

Firing   Firing has in all times and places been the touchstone of the potter's art. Their firing practices thus reveal, on the one hand the technical inadequacies of the pre-Columbian Americans, on the other hand their ingenuity combined with tireless patience. Direct evidence that would permit us to attribute a given technique to a given people at a precise time is lacking. We shall therefore limit description to the "kilns," which were vertical rather than horizontal; the "corridor-kiln" characteristic of the Eurasian peoples is never found in pre-Columbian America. However, the quality of the ceramics leaves nothing to be desired in either firing or finishing. The potsherds are sometimes so fine that the term "eggshell pottery" had to be invented to describe them. We need only look at a Zapotec urn to realize the dexterity required to successfully complete such complex objects.

In the American territory firing was done for the most part in one of two ways: with an open fire or in unlined kilns that in most cases were also open.

The first technique consisted simply of piling wood on top of the objects and maintaining as steady a heat as possible for several hours. This method resulted in temperatures of from 400 degrees to 600 degrees, which permitted the firing of unslipped objects to a dark red. This technique was very widely used in all periods wherever the kiln was unknown. A shaft was sometimes used, a method halfway between the pile of wood and the kiln; it was shallow, but nevertheless offered the same possibilities for regulating the draft as the semisubterranean kiln. A strong draft gave a red color, while black or gray resulted from a weak draft. The hearth was on top of the kiln, on a layer of dung sprinkled with sherds, which covered the objects to be fired. Both types obeyed the principles of a reducing atmosphere.

The unlined kilns, constructed on the ground, had openings for the injection of the air required to create the oxydizing atmosphere that permitted the obtaining of light colors.

ARCHITECTURE

Architecture has obviously left more traces than all the other techniques combined. It is also undeniable that construction techniques are sufficiently well known to dispense us from the obligation of explaining architectural terms and their use.

Depending on the available material, the pre-Columbian Americans built various types of houses in or on top of the ground.

Shelters: made of hides (the Onas of Patagonia), fibers (the Botocudos of Brazil), bark, pisé, or thatch (tribes of the North American prairies).

Tents: domed (in hide, bark, or thatch among the Alkalufs of Patagonia) and conical (hide or bark among the hunters of the northern United States and the prairie tribes (Figure 4).

FIG. 4.   Tepee (from Underhill).   FIG. 5.   Wigwam (from Underhill).

Huts: circular, domed: bark or hide (the wigwams of the tribes of the northeastern United States) (Figure 5), or blocks of ice (the Eskimo igloo). With wooden walls and thatched roofs (in the Amazon region and the Guianas). With thatched roofs and walls of stone, pisé, dried bricks, or rush (preclassical Mesoamerican period). Four posts, pisé walls, and a thatched roof (southeastern United States). (Among tropical farming peoples the posts were so short that the two slopes of the roof almost touched the ground.)

Independent adobe houses with flat roof: Hohokam people, southwestern United States; the Caddo people of the Southeast and Mesoamerica. Adobe and stone, thatched peaked roof: central Andes and Mesoamerica (this is the "typical" pre-Columbian American house). Stone, thatched roof: northwestern Argentina.

Independent wooden houses: the Iroquois "longhouse" (Figure 6); houses of the northwestern United States (Figures 7*a* and 7*b*); the Auracans, southern Chili.

Grouped multistoried houses. In dried or adobe bricks: Pueblo and Zuñi peoples of southwestern United States. The upper stories were reached by means of ladders (Figure 8).

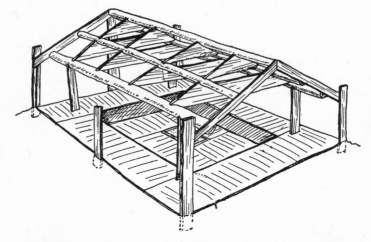

FIG. 6.   Framework of a Chilkat house (from Underhill).

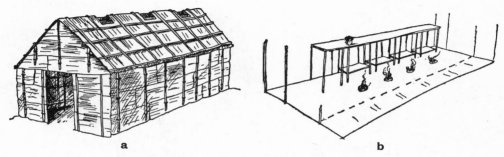

a   b

FIG. 7.   Iroquois longhouse (from Underhill).

Semiunderground houses: built of large stores — central Andes; with wooden and pisé walls — southeastern United States (Figure 9).

All these types had several traits in common: a single main room, doors on one or two sides, absence of windows, floor at ground level, holes for ventilation drilled in the roof. The result was great economy of materials. Foundations were nonexistent; the framework supported the entire building, and the roof was made of whatever material was available in the area.

In contrast, the architecture of the religious and urban centers furnishes technical details that can more easily be verified. In general these cities, which were built on a predetermined plan in accordance with magico-astronomical requirements, reveal a unity of concept that was not seriously weakened by the diversity of styles. Groups of buildings with unequal sides were built around squares (*patios*), while the pyramids rose in isolation from immense parvises. Between these various groups of buildings, and asymmetrically arranged, were gardens, cultivated fields, swamps, cemeteries, and lesser buildings. The streets were in general well laid, at right angles, and were paved. Water was supplied by aqueducts (Figure 10); gutters, troughs, drains, and pipes carried the waste to sewers.

Fig. 8. Old Pueblo-Zuñi village (from Underhill).

Fig. 10. A Maya sewer at Palenque, showing drainage gutter.

Fig. 9. Mogollon house (from Underhill).

The following descriptions relate especially to the preclassical and classical peoples of Mesoamerica and to the Incas, for their techniques were adopted by all the surrounding peoples who possessed a stone architecture.

Tools Their implements were not a great improvement over those of the Neolithic peoples. There were the basalt chisel, hard stone and wooden mallets, obsidian and flint knives, the adze (whose stone blade was replaced by a copper blade as soon as metal appeared), a wooden scoop that replaced the trowel, stone and wood polishers, and the plumb line. Although the pulley and the crane were nonexistent, the lever was definitely known. Lacking mechanical devices for lifting, large crews of workers succeeded in lifting enormous blocks of stone to the top of the pyramids by making use of rollers, ropes, and ramps.

Materials Wood was used for framework, lintels, and roofing. Small stones formed the central core of the pyramids, and were used for terraces, fortifications, walls, columns, and pillars. They were often replaced by flat stones, particularly stones of volcanic origin; these were piled up without mortar, notably at La Quemada and at Tzintzuntzan in northwestern and western Mexico. Buildings were faced with squared, polished blocks of basalt, granite, or sandstone.

Dried or adobe bricks had been widely used since ancient times. Their production was extremely simple: the earth (often argillaceous) was kneaded or trampled underfoot, and water, straw, or grass was added to it. This mixture was poured into a rectangular or square wooden frame, and dried in the sun.

Genuine brick was extremely rare; it is found only in central Mexico (Tula, ninth to twelfth centuries), the Mayan "Old Empire" (fourth to tenth centuries), and in Aztec buildings at Tlaxcala.

Lime appeared in the third century; it was obtained by burning shells. Mortar was a mixture of lime and sand to which fine gravel was sometimes added; it was already in use two or three centuries before Christ. Frequently it was so hard, particularly in central Mexico, that it can be compared to cement. When sufficiently blended with water, mortar was used for the same purposes as stucco.

Foundations and basements

In mountainous areas terracing was an absolute necessity, while in the plains the terraces became platforms for supporting the weight of the buildings. The first platforms were held in place by vertical walls, but as the mass of the building increased the materials slipped toward the bottom and spread out. To remedy this, the walls followed the natural slope of a hillock of land or gravel, and thus efficiently held the small stones of the interior. This made possible the construction of increasingly tall pyramids. The American pyramid was built in stepped tiers.

Masonry

For the most part polygonal stone masonry with either sharp or blunt angles predominated in foundations. In contrast, the upper walls were built in either a trapezoidal or rectangular coursed ashlar (Figure 11) (Teotihuacán, Maya, Aztec, Tiahuanaco, Machu Picchu, Cuzco). When small, evenly cut stones or bricks (adobe) were used, the masonry became almost Hellenic. In almost all cases the walls were not bare but were faced with stucco (usually polychrome) or geometrical mosaics made of stones.

Windows and doors

Temples and palaces, like domestic architecture, had very simple openings. A few buildings had windows which opened to the outside (the Maya buildings at Palenque, Chichén

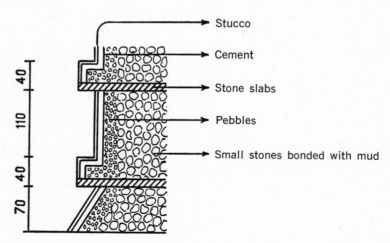

Fig. 11. Cross section of wall, early period of Teotihuacán (Mexico), third century (from Marquina).

Itzá, and Tikal) and, more rarely, T-shaped or trefoil windows in the inner partitions (Palenque). Ventilation, however, was ensured by very small openings at the top of the walls and partitions. The doors had stone sills (often raised), doorjambs, and lintels; the method of latching them is unknown to us. Doors of a rigid material were nonexistent, and in their place mats or other materials attached on each side of the opening were used.

Roofing   Apart from the typical Mesoamerican thatched roof, in which the peaked roof was covered with thatch or with sisal leaves arranged like tiles, there were two other techniques that were used within clearly defined geographic limits.

Outside the Maya region, where the false or corbeled vault came into use beginning in the fourth century, the flat terrace roof was used on all buildings that were not roofed with an imitation of the thatched roof. The terrace roof, which was sloped to facilitate the draining of the rainwater, consisted of wooden planks and mortar. The beams were supported directly on the walls; smaller beams were placed crosswise on the large beams to hold a thick layer of stones and lime or mortar. When the area to be covered was very large, one or several rows of columns or pillars were used.

From the fourth to the seventeenth centuries, the Maya constructed both false vaults and arches — one of the best examples of the stagnation of the creative impulse of the pre-Columbian Americans. Although they possessed all the elements needed to create the arch and the vault, they were satisfied with the elementary structure despite such immediately obvious disadvantages as the impossibility of roofing areas more than $16\frac{1}{2}$ feet wide (the average varied between $9\frac{3}{4}$ and $11\frac{1}{2}$ feet), the thickening of the walls, and the limitations it imposed on the height and width of the rooms. The method of construction was extremely simple. Stones were piled up vertically in the form of piers. Then, at the desired height, flat stones were superimposed on the pier, each slightly projecting over the one beneath it. In this way the space separating the piers was gradually closed over. At the very top, a single stone covered the remaining space and thus closed the arch. In addition to roofing, the false vault made it possible to construct aqueducts, sewers, and arches of triumph.

Pilasters, pillars,   The first pilasters were used at Teotihuacán before
and columns   the tenth century. Serpentiform pillars are found at Tula in central Mexico around the same period and slightly later at Chichén Itzá (Maya). The stone column was a Toltec invention (tenth to twelfth centuries) which made possible the construction of hypostyle halls (Figure 12). At Mitla (near Oaxaca, Mexico, fifteenth century) the columns were monolithic, with neither base nor capital; the shaft gradually narrowed toward the top. At Tula the Toltecs cut their columns in five parts: one for the base, three for the shaft, and one for the capital. The parts were assembled with tenon-and-mortise joints. These columns, $19\frac{1}{2}$ feet high, are genuine caryatids. Assembly by tenon-and-mortise joint was widely used. In most cases, however, only the tenon was used, particularly when it was a question of embedding mosaics or other decorative elements. At Tiahuanaco stones

Plate 59.
Arch of Labna, Mexico. *Photo Musée de l'Homme—Renaudet*

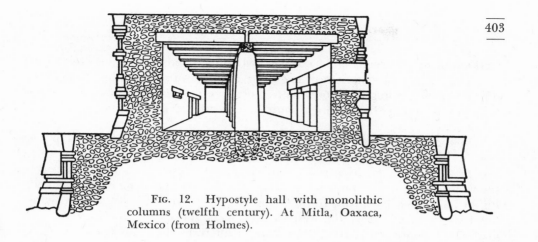

FIG. 12. Hypostyle hall with monolithic columns (twelfth century). At Mitla, Oaxaca, Mexico (from Holmes).

FIG. 13. Southeastern temple (from Underhill).

and slabs were assembled with notches and joints; copper clamps were inserted in T- or I-shaped slots.

Multistoried buildings and staircases Most palaces and public buildings had only one story. However, at Chavin central Andes, seventh to eleventh centuries) and among the Aztecs there were two-story houses. In the sixth century the Mayas began to build two-story buildings with receding upper story; by the eighth century the two stories had become four (Tikal). Palenque had a four-story tower. Interior stairways sometimes led to the upper floors, but this was not the general rule (Figure 13).

In contrast, the exterior stairways, which were always straight, were numerous and monumental. Built of stone, faced with stucco and paint, and often decorated with bas-reliefs, they formed an indispensable part of the esthetic ensemble of the monument. All the pryamids had axial stairways leading to the top. The monumental staircase was everywhere flanked by two large ramps; the Aztecs added a third in the center. The steps were so narrow and the risers so steep that the stairs had to be mounted and descended sideways.

METALLURGY

In order to understand the development of pre-Columbia metallurgy, we must distinguish between two distinct phenomena: first, the accidental discovery and working of native metals (gold nuggets and copper were hammered everywhere, as well as meteoric iron wherever it existed), and second, the deliberate exploitation and conditioning of metals, that is, genuine metallurgy. Consequently, the origin of metallurgy has no single source; there were several independent centers of invention.

In the Great Lakes area of North America a date for Carbon 14 proves that by 1500 B.C. tools were being made by hammering native copper, which was also annealed. Smelting was unknown; nevertheless, this is the only case in the world where metal was worked before the appearance of ceramics. Hammering of native metals spread through the eastern portion of the United States, but never reached either the western portion of the country or Mexico. This method of production continued unchanged for three thousand years, a striking example of the stagnation characteristic of the pre-Columbia Americans. The other center for working native metals was in northern Peru, in the period of the Chavin culture (1200 to 400 B.C.). Here gold was the first metal to be hammered in order to obtain thin strips that could be cut and embossed. Objects which were a combination of silver and gold appear in the Necropolis of Paracas (400 B.C.–400 A.D.).

Metallurgy was invented around the eighth century in two places quite distant from each other: Colombia and southern Peru. Before the year 1000 the entire central coast of northern Peru had adopted metallurgy, which under the impulsion of the Incas developed in Colombia, Argentina, and Chile. Gilding and silver-plating was invented between the fifth and eleventh centuries in the Mochica culture of northern Peru; bronze between the eleventh and fourteenth centuries at Tiahuanaco, from where it reached the northern coast of Peru (Chimu, fourteenth and fifteenth centuries). The Incas (1438–1532), who inherited all these techniques, became very skilled goldsmiths, but their only contribution to metallurgy was encrustation of one metal in another, different metal. From Colombia metallurgy reached Central America, where it was concentrated in Panama (at Coclé and Veraguas). It was also practiced in Costa Rica, and finally reached Mexico. It should be pointed out that metallurgy appeared in Mexico fully developed, around the tenth century, and thus did not evolve from the working of native metal, which was discovered early in the ninth century. The Maya were an exception, for until the arrival of the Spaniards they acquired through trade not only metal tools and objects but also strips of metal which they shaped as desired. Only a few Maya groups in the high plateaus of Guatemala practiced smelting.

The metals utilized   The metals used in pre-Columbian America were gold, copper, silver, lead (Peru and Mexico), platinum (Ecuador and Colombia), tin, and mercury.

Alloys   These metals, even gold and copper, were rarely found in pure state, and it was probably during the process of extracting them that alloys were discovered. Moreover, since the same furnace was used for smelting various metals, accidental alloys must have occurred. In the central Andes, around the sixth century, we find the gold-copper alloy. But it was in tenth-century Colombia that alloys were systematically obtained. The chief alloy was *tumbaga,* which quickly became the typical alloy both by its frequent use and by its diffusion in all the metallurgical centers. *Tumbaga,* a mixture of 82 parts of gold and 18 of copper, had two special features the Indians did not fail to exploit. When an object was subjected to the action of heat and was hardened in an acid bath, the copper on its surface disappeared, leaving the gold to show through; in other words, the *tumbaga* made coloring possible. When subjected to the action of annealing and hammering, it became as hard as bronze or soft steel.

The other alloys were more localized: gold-silver (35 percent to 50 percent silver), gold-silver-copper, and silver-copper (20 percent copper) on the south coast of Peru between the eleventh and thirteenth centuries; copper-arsenic (5 percent arsenic) on the north coast and around the same time; copper-lead among the Aztecs, and lastly copper-tin. Bronze appeared between the eleventh and fourteenth centuries at Tiahuanaco, and then reached the northern coast of Peru. There were two types: the first contained very little tin, was suitable for hammering, and was used in the making of tools; the second variety, which had a larger proportion of tin, permitted the casting of complex ornaments with a minimum of imperfections. Only the first variety was utilized in Mexico.

Extraction   Gold was obtained either by digging shallow galleries along a visible vein or by using placers containing nuggets or silver-bearing river deposits. Copper deposits involved mining, with open shafts and galleries, the deepest of which reached twenty-three feet. Like copper, silver was rarely found in pure state. Mercury was extracted from cinnabar or collected from underground pockets; it was not used in Mexico. Platinum, which was worked only in Ecuador and Colombia, was often found in the same rivers that contained the silver deposits.

Smelting   We know almost nothing about the preliminary processing of the metals before they were smelted. The ore was crushed in a stone mortar or on a metate (the last Peruvian smiths, in the seventeenth century, mixed the powder thus obtained with charcoal and lama dung). Native gold and copper, like the other metals, were refined before being smelted in open furnaces. The Andean or Mesoamerican furnace, which was made of terra-cotta, was extremely rudimentary. Small and portable, it looked more like an enormus triped pot with a globular bulge. The combustibles used were wood and charcoal. The air needed was blown in through copper blowpipes if the furnace had suitable openings; otherwise the furnace was placed on the side of a hill in the direction of the wind. However crude, these "blast furnaces" performed their task efficiently. Platinum alone had too high a fusing point to be treated in these furnaces, but this difficulty was overcome by combining the platinum with powdered gold in order to obtain a malleable alloy.

Using equally simple methods, the goldsmiths of the Central Andes, Panama, Costa Rica, Oaxaca (the Mixtecs), and central Mexico did work whose technical difficulty of execution and esthetic qualities command our admiration (Figures 14 and 15). However, pre-Colombian metallurgy was decorative rather than utilitarian. The description of the objects themselves is outside

FIG. 14. Mexican goldsmiths
(from the Codex of Florence).

FIG. 15. Aztec goldsmith
(from the Codex Mendoza).

the limits of this study, and we shall therefore limit ourselves to a consideration of the techniques used.

Cold-working. — This included hammering and rolling. The sheets were then cut, embossed, and incised.

Hot-working. — The heat treatment consisted of shingling, annealing, and welding. Casting could be done in open or closed molds, but the lost-wax method, which was well developed in Colombia, Peru, Panama, Costa Rica, and among the Mixtecs and Aztecs, was preferred. The piece to be cast was modeled from a clay and charcoal paste, then covered with beeswax and enclosed in a thick envelope of the same clay mixture. The rest of the operation was similar to the lost-wax techniques practiced elsewhere. Appearances to the contrary, genuine filigree work did not exist; wires coated with wax and clay were placed on the object to be cast, and the casting was done in one operation, the imitation filigree forming a single unit with the object it was to decorate. In Mexico and Peru objects of combined gold and silver were cast; the mold was cut in half, and each half was filled with metal. After casting, the two pieces were joined together. They were covered, as usual, with a paste made of clay and charcoal, and the whole was recast so that the metals were melted at the point of contact. This method could be used for objects made of more than two different metals.

However, when adding to an object elements that had been cast separately, either riveting or soldering was used. Soldering was invented in the central Andes between the seventh and eleventh centuries, but it also existed in Mexico. The technique of soldering with mercury probably existed there as well, although this cannot be proved. As for gold, soldering was done by means of the combined action of hammering and heating. As soon as the part to be soldered was malleable, it was coated with a paste of gum and crushed copper salts. The heat consumed the gum, and the resulting copper combined with the gold to form a solid, almost invisible joint. The Peruvian and Mexican goldsmiths assembled complex objects by means of soldered rings, a technique that permitted them to create movable objects.

There were three techniques used for gilding. One was coloring; the second consisted of lining the mold with thin hammered gold leaf before pouring in the copper. In the third technique the copper or silver object, which had first been engraved with acid, was coated with gold crushed with mercury; the action of the heat eliminated the mercury and the gold remained on the metal. The third technique was also used for plating stone, wood, and leather. Encrustation and enchasing were common.

Finishing When necessary the metals were hardened before the finishing of the object: tempering and cold-hammering for copper, annealing for *tumbaga* and bronze. Objects were generally polished with plants (*Curatilla americana, Tetracera volubilis*) or siliceous sand to give them their finished appearance.

FEATHERWORKING

Featherworking, essentially an indigenous technique, became an accomplished art among the Aztecs, but it was also practiced by the Peruvians, Maya, Olnecs, and Tarasco. It originated in the eighth century with the Olnecs and Maya living on the Atlantic coast of southern Mexico, which abounded in birds with colored plumage. In the tenth or eleventh century they began to execute genuine feather mosaics by means of two techniques.

1. *Sewing.* — In the beginning the feathers were probably interwoven and tied with pieces of string. The next improvement was sewing the feathers directly onto a piece of material, and this was the method used by the featherworkers in composing their mosaics (central Andes, thirteenth to fifteenth centuries; Aztecs, fourteenth to sixteenth centuries). In addition, the Aztecs sewed feathers on a framework of hide or woven straw, and made use of the Gobelin weaving technique to insert feathers in the body of the weft.

2. *Gluing.* — Gluing, which was utilized especially by the Aztecs, required greater skill and special equipment. First the pattern was traced on a piece of wood or on a sisal leaf that was colored and cut out. The feathers, first sorted according to the colors in the pattern, were cut to the desired length, soaked in warm water, and glued to the surface to be decorated, whether fabric, paper, or hide. This was done with the help of a T-shaped instrument made of bone or copper. The glue used was made from the orchid flower. A first layer of ordinary feathers formed the background for the second layer of fine feathers arranged so that the barbs always hid the shafts. Gold thread and gold leaf highlighted the designs.

MOSAIC

Like featherworking, the mosaic technique was an indigenous discovery that appeared in the ninth century (Teotihuacán, Zapotec, and Maya). The art of the mosaic flowered after the Toltec dispersion in the twelfth century, reaching its peak under the Aztecs. Elsewhere we find turquoise mosaics in the Pueblo culture of southwestern United States, and also in the central Andes.

The encrustation of pieces of jadeite seems to have been the origin for this purely decorative technique. Production included three steps. First the motif was drawn either on paper or directly onto the wood or stone to be decorated. The stones were cut, and were assembled and glued with the sap of the *tzauhtli,* a species of tree-perching orchid (*Epidendrum pastoris Ll.*).

In contrast to Western mosaic technique, which utilized small cubes, the Americans cut the stones into flakes. The surface of each flake was carefully leveled with a flint scraper. Then fragments were carefully cut from the flakes in accordance with the drawing of the model, and their facets were cut so that they could be assembled. This was done by gluing them onto a wooden or other support. The mosaic was then polished with very fine sand. When necessary the Aztecs perforated and grooved the flakes with a tubular copper drill.

The range of colors included turquoise for blue, jade for pale green, shells and coral for red, obsidian for black, mother-of-pearl for white, and gold leaf for yellow.

PAPER

In Mexico the first traces of the existence of paper date from the fifth century (Teotihuacán), a fact confirmed by the presence of stone beaters in central and western Mexico and the Gulf Coast beginning in the fifth century. Paper increased in importance under the Aztecs, to such a point that three types of paper are known from this period: palm (*izotl*), sisal leaf (*metl — Agave sp.*), and fig bark (*amatl — Ficus sp.*), the latter being the most common. In all three cases, the methods of production were practically the same. The fibers or bark were first retted in a river. Cooking or scraping followed, then a preliminary beating to loosen the fibers or bark. Next came a preliminary polishing with a sizing of *Epidendrum pastoris Ll.,* a second beating on a plank in order to give shape and consistency to the sheet of paper, and a final polishing with a stone polisher.

The beaters were made of hard or porous stone, and were rectangular or oval in shape; they were corrugated on one or both faces, and had a groove around three sides so that a reed or osier handle (Figure 16) could be added (some beaters had a handle carved in one piece with the stone itself). There were also polishers made of ears of corn, and mallets with grooves for beating.

Fig. 16. Stone beater for making paper.

Paper was made in several sizes: 18 inches by 14 inches for ordinary sheets, 27½ feet by 4 inches for the codex, which was accordion-folded in small leaves eight inches long; 33 inches by 55 feet for the paper used in festivals, and 110 feet by 5½ feet (thickness 15 millimeters) in the case of paper required in religious ceremonies.

COLORS AND PAINTING

Considered from any point of view — ceramics or painting, dyeing of textiles or painting of the body — the native palette was one of the richest and most varied in existence. Moreover, since certain pigments had to undergo chemical transformations that were always constant and uniform, we are justified in stating that the American techniques do not belong solely to the area of mechanical acquisitions and their development.

Pigments Some of the colors used were of animal origin, for example the cochineal of the nopal, which gave carmine and purple. Vegetable colors included the flowers of *Guaiacum coulteri Gray* or *sanctum L.* for blue; the cosmos (*Cosmos sulphureus Gav.*) (orange); the leaves of *Indigofera suffruticosa Mill* (indigo blue); the leaves of *Tinta capichaba* for violet. The fruit of the genipap (*Genipa*) gave blue-green, the seeds of the annatto (*Bixa orellana L.*) red, the wood of the paper mulberry (*Broussonethia tinctoria K.*) yellow, *Genipa americana L.* and also the soot obtained by burning fir trees, black. Mineral colors included ocher, calcinated lime, chalk, and manganese. In order to give them consistency, the pigments were combined either with the sap of *Epidendrum pastoris Ll.* and *L.,* the oil of *Salvia hispanica L.* (sage), or the gum of the orchid *Bletia automnalis.*

Painting Mural painting was done in two techniques that were related to the fresco: either on a wet surface, a technique practiced in the early period of the three civilizations, or on a dry surface, a technique particularly characteristic of the second period (seventh to tenth centuries). In the first case the wall was prepared with a very thin layer (two to four millimeters) of lime or stucco, on which the drawing was traced while the material was still damp. The drawing was darkened or corrected with red lines, and the color was applied. Often the wall was polished with a very fine stone polisher while the paint was still damp. In the second case the wall was polished and coated with a very fine layer of stucco; once the drawing was done and the artist was certain that the wall was dry, the color was applied with brushes.

In painting the human body, the colors were applied to the skin with the fingers or brushes, or with stamps and ink pads (*pintaderas*).

TEXTILES

The pre-Columbian Americans excelled in weaving, particularly in Peru, where almost all the techniques in existence today were developed together with others that it is impossible or impracticable to reproduce on mechanical looms. The distribution of weaving (or textile techniques) coincides with that of advanced agriculture: in North America the tribes of the Southwest and Mesoamerica, in South America the tropical agricultural peoples and the Andean farmers. The peoples who practiced agriculture on an occasional basis possessed the techniques of plaiting (hair and fibers) and also looms on which narrow strips could be woven. In contrast, the nomadic hunters of the South (Patagonia, Argentina, Chili, southern Brazil) and all the hunters of the North (Algonquins, Iroquois, Great Lakes, Plains, and Canadian tribes, and the Eskimos) were ignorant of weaving.

Several of the oldest textiles in the world (2500 B.C.) — corded fabrics of the preceramic period — have been found along the northern coast of Peru. Weaving made great progress after 1000 B.C., and three centuries later had become firmly established in the region of the central Andes. Prior to the seventh century B.C., the techniques of twisting, knotting, braiding, and knitting with needles were used; they produced only tulles and gauzes, the heddle being unknown. In addition, the region of Nazca and the Necropolis of Paracas (400 B.C.–400 A.D.) possessed all the techniques of weaving that would be used in the surrounding areas several centuries later. With the invention of the heddle, brocade, embroidery, and painted fabric made their appearance.

From the seventh to the eleventh centuries simple cotton fabrics predominated, but wool was also woven. Colored threads were rare, and the decorative motifs were quite simple. Tapestry, gingham, embroidery, brocade, and twill existed. Tapestry became the characteristic technique of the eleventh and twelfth centuries, along with complicated fabrics — brocade, double fabrics, braids, painted fabrics, knitted fabrics, velvets, and garments made of individual pieces. Batik dyeing (in which the fabric or the threads of the warp were knotted and plunged into a vat of dye; the parts protected by the knots remained uncolored) was in existence. All these fabrics were woven in cotton or wool of various colors. From the thirteenth to the fifteenth centuries, in contrast, certain overelaborate techniques were abandoned. Tapestry was declining; knitting, gauzes, and brocades were abandoned. Fabrics were made of cotton, sometimes double, and were painted and batik-dyed.

Although the Mesoamerican civilizations and cultures remained unsophisticated in weaving techniques by comparison with Peru, weaving began in approximately the same way in both areas. Beginning in the eleventh century we find the techniques of tapestry, brocade, twill, embroidery, gauze, and — a technique peculiar to Mexico — the incorporation into the fabric of rabbit fur and down in order to obtain velvet fabrics. The technique of knotting was practiced everywhere; the thread was handled with the fingers, as in the manufacture of Gobelin tapestries.

Fibers utilized　　　Cotton (*Gossypum barbadense*) was the most important fiber. It gave a color scale ranging from white to dark brown, and was utilized by all the pre-Columbian American weavers, in many cases even before they had knowledge of pottery. Next came sisal fibers (Mexico, pre-tenth century), rabbit fur (Mexico), human hair (Peru, but sporadically used for rope), sedge and reed fibers in all regions, and often prior to genuine fabrics), and lastly wool. The Peruvians had the possibility of weaving woolen cloth because of the existence of the llama and alpaca (especially the latter), which supplied wool. However, they preferred the wool of these animals' wild cousins, the *huanaco* (wild llama — *Lama huanacus*) and the vicuña, which is finer and more flexible (the wool of the vicuña is the finest in the world:

almost 2,000 hairs to the inch) and is easier to dye than cotton. Only silk was missing.

Spinning

In the preceramic period thread was often coarse and uneven; the spindle was a simple twig, and the whorl probably did not exist at all. Later we find a genuine spindle of wood or thorn, very fine and perfectly balanced, and small ceramic whorls that were used all over the continent. Strangely enough, both possible methods of turning the spindle are found in America. The Inca Peruvians, like the spinners of the other continents, let the spindle dangle at the end of the thread, and the rotating movement was slow but even (see Figure 21). The typically American technique, however, was to spin the spindle at top speed in the bottom of a saucer or on the ground. After turning on its tip, the spindle slowed down and stopped for winding. The advantage of this system was that it reduced the tension and vibration of the thread. Strictly speaking there was no distaff, notwithstanding the use of forked sticks by the sixteenth-century Incas for winding up the thread. As an example of the degree of skill reached with mechanical methods that appear insufficient in our eyes, we may mention that in Incan Peru it was customary to make wool and cotton threads so fine that modern spinning machines are unable to duplicate them.

Dyeing

Although we are almost completely ignorant of the methods used for dyeing, it can be said that beginning in the second century vegetable colors, cochineal, and shells supplied the colors of the dyer's palette; blue was the principal color, with red and brown next in importance. The other colors were the same as those mentioned under the heading "Colors and Painting." Alum was a universally used mordant. Cotton was often colored while still on the plant, which is a very modern technique.

Looms

Since the American looms present no special features, there is no need to give a detailed description of them. We need only mention that the harness, the treadle, and the shuttle with spool were lacking in the American loom. All the other essential parts of the loom had been invented: the beam, the breastbeam, the heddle, the leash rods, the beater with comb. The back-strap loom (Figure 17) was the most widespread; low-warp looms were found among the civilized peoples, but the

FIG. 17. Inca weaving loom (from Poma de Ayala).

Plate 60.
Mexican woman spinning. *Photo Musée de l'Homme—Treznem*

high-warp loom was quite rare, even in Peru. Fabrics more than six feet wide were nevertheless woven, which is very large for a hand loom. In Peru fabrics intended to be used as ponchos were woven with a slit for the passage of the wearer's head.

BASKETRY

It is possible that basketry was imported by some of the early immigrants. In any event it spread rapidly over the entire continent, in all regions which possessed adequate quantities of the appropriate raw material. Thus basketmaking exists neither among the Eskimos nor in the Northwest. From information presently available to us, it seems to have appeared in the central Andes between 3000 and 1000 B.C., in Mesoamerica around the third century B.C., and in the southwestern United States between the first and the eighth centuries; it characterized the "Basketmakers" culture, whence it spread among the tribes of the Southeast.

Techniques Techniques included coiled basketry — the true coil of the Eskimos, and the woven coil of North America, the Amazon area, and the Chaco. Coiled basketwork was found in Alaska, British Columbia, among the Plains Indians, in tropical America and Argentina. Woven basketry was universally practiced, as was diagonal weaving for straw mats. The finished products included walls, screens, roofing, seats, shelters, fans, personal ornaments, mats, pots, and traps.

STONEWORKING

Since exact information regarding the extraction of blocks of stone is lacking, we can only note the existence of quarries and mention the tools and methods used for stonecutting. Flint-cutting techniques were identical with those of the Paleolithic and Neolithic peoples. The production of flakes by pressure was widespread among the hunting peoples but was unknown among the preceramic agricultural peoples.

Implements were made chiefly of stone, but also of wood, reed, bone, leather, and (at a later period) copper. The stone tools — mallets, axes, wedges, chisels, scrapers, drills, and gravers — permitted the craftsman to crush, break, bend, crumble, sharpen, scrape, perforate, bore, incise, and polish. Wood and bone tools were used for scraping, polishing, and retouching by pressure; reed, bone, and metal instruments were used to perforate stone, leather to cut it. Ultrahard abrasives being unknown, the workers used stone of the same variety as the object waiting to be polished, on the theory of using diamond to cut diamond. Sand and water were also commonly used as abrasives. Around the end of the fifteenth century and until the Conquest, the Aztecs cut rock crystal with emeries and a hardened copper tool, the only instruments that could carve this quartz, which was harder than steel. Elsewhere, amethyst, chalcedony, agate, sard, cornelian, bloodstone, jade, and turquoise were worked.

In the third century B.C. the Olnecs began to cut jade with a virtuosity that has never been surpassed. This was not the true Asiatic jade but a local jadeite (hardness index: 6.5–6.8). The inhabitants of Colombia, Panama, Costa Rica, and Guatemala, as well as the Maya and almost all the peoples of central Mexico, also made ornaments, pendants, and statues of jade; jade axes have been found in Alaska. The block of jade was sawed into slabs by means of leather strips, with only water and sand as abrasives. The slab was shaped by detaching flakes from it. When it had been refined to the maximum, the unwanted areas were removed and the rough places were refined away by rubbing and polishing with a stone first moistened and then rubbed against a bamboo. For finer cuts the leather strip was abandoned for a simple rope or even a string. Sawing was begun with a perforation near the edge that was cut very gradually. Starting out with this technique, the Olnecs and the early Maya succeeded, several centuries after Christ, in cutting and detaching pieces from inside the object. The area to be cut out was marked off by a dotted line of tiny perforations, and the saw (in this case a piece of string) cut through the jade from one hole to the other with the help of an abrasive (sand, emery, or pumice stone). This was an extremely laborious method, for the string had to be replaced frequently; however, it was very efficient, since it permitted straight and curved cuts inside the object, far from the edges. The Olnecs and the Maya made little use of this technique, but it was the customary method of the Guapiles of northeastern Costa Rica. The drilling tools were equally simple: originally made of bone and corncob, by the thirteenth century tubular copper drills (Figure 18) were in use; we do not know how they were operated. The holes were drilled on every face of the object, meeting inside at the center.

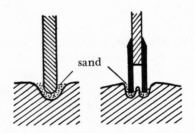

FIG .18.   Drills. *Left,* bone; *right,* tubular copper drill.

A similar technique was used for making onyx vases. The size of the block was reduced by pressure, and flakes in the shape of small blades were detached. The inside was hollowed out by means of holes drilled with a tubular drill. This left small columns that could easily be broken. Next came the scraping and polishing of the irregularities, then the decoration and final polishing.

WOODWORKING

Wood was worked at a very early period, but owing to organic decomposition there are few remains of the first wooden tools and objects made in the Meso-american and South American regions. In contrast, the peoples of the regions that are now part of the United States and Canada furnish considerable material regarding the techniques of the carpenters and cabinetmakers. Each ethnic group was able to turn to advantage the types of wood available in its territory: hardwood for major construction, softwood for domestic objects. The bark, roots, leaves, fruit, trunk, and branches were utilized, but it would be both tiresome and useless to list and describe the use of the wood and the objects made from it.

Tools from Mexico to South America were fairly simple: wooden mallets, chisels, knives for carving, awls, scrapers, polishers, drills, and especially the adze (Figure 19), which was universal. Its blade, which was of polished stone, iron (Eskimos), or copper (Mexico and Peru), was bound to a curved or T-shaped handle. It is often difficult to distinguish the shape of the tool in relation to its function — a problem we invariably encounter in the course of every investigation into pre-Columbian technology. The chisel, the knife, and the scraper resemble each other, and often a single tool filled all three functions. We do not know how objects were put together, but we do know that nails were nonexistent and that fish glue was used.

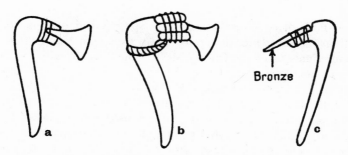

FIG. 19. Adzes. *a* and *b,* polished blades in handles; *c,* bronze blade.

In America, north of Mexico, two regions excelled in woodworking: northeastern United States and the northwest coast of the continent. Assembly-line production and constructon also existed. Two examples will suffice to give an idea of the skill of these craftsmen, who used approximately the equipment described above well into the eighteenth century.

The Iroquois built their long, rectangular houses by forcing posts into the ground at intervals of between five and eight feet. Horizontal beams were attached to this framework. The framework for the roof was nothing more than poles bent into arches; this frame was covered (except for airholes along the ridge) with sheets of bark, which were held in place by a second, outer framework.

On the northwest coast the Chilkat people assembled their houses without using nails or even pegs. The Haida made chests by cutting a fairly thin, long plank to form the four sides of the box. At the places destined to form the corners of the box V-shaped grooves were cut into the wood. The plank was softened by steaming so that the three sections marked off by the grooves could be bent at right angles. The two ends of the plank were brought together and sewn with fibers made from the root of the fir tree. The bottom of the chest was cut from another piece of wood, and was grooved to receive the sides; the bottom and sides were then sewn or joined with pegs. The box was then firmly tied and left to dry.

RUBBER

Rubber was being exploited in tropical America before the modern period, but little is known about the methods of extracting the substance, the preparation of the latex, or the making of objects from it. Several latex plants were exploited: *Sapium jenmani S. Cladegyne Sapium eglan dulosum, Castilla elastica Cerv., Hancornia sp.,* and especially *Hevea brasiliensis.* In Mesoamerica balls and balloons for the religious game of pelote were manufactured. Chicle (chewing gum) is also of American origin, but only the Aztec prostitutes chewed it.

COMMUNICATION AND TRANSPORTATION

Artificial methods of communicating ideas, speech, and information were nothing more than mouth-to-mouth reporting by men who transmitted the message verbally, sometimes using mnemonic devices such as drawing or knotted cords (the *quipu* of the Peruvians), or delivering objects that had symbolic significance. Here as everywhere, acoustic and visual signals were also transmitted.

Two methods of communication were available to pre-Columbian man: aquatic and terrestrial. In the first case he used simple boats without superstructures, propelling them usually by pole or paddle, on rare occasions by means of a sail. The Eskimo canoes, the *kayak* and *oumiak,* were made of hide. The hunters of the northern United States made bark canoes by joining several frames and covering them first with thin planks and then with a sheet made of several pieces of birch bark sewn together; the joints were caulked with resin and wadding (Figure 20). In the islands of the Magellan Archipelago the canoes were cigar-shaped, and were made of three pieces of beech bark sewn together.

Fig. 20.   Algonquin birchbark (from Underhill).

The agricultural peoples of the tropics cut pieces of bark into the desired shape and size, sewed them together, and turned back the edges before inserting the benches and the framework; some of these canoes were from forty to fifty feet long and could carry thirty men. The tribes of the southeastern United States and the Mesoamerican peoples used rowboats and one-man pirogues hollowed out with fire and axes; this type was also used by the Arawak people of the northeastern coast of South America, the Maya, and the Caribs of the Antilles, who added two or three masts and sails made of mats. Pirogues made from wooden planks were used by the Auracans of southern Chili, rafts by the prairie tribes of the United States, the coastal peoples of Peru and Ecuador, and the tropical agricultural peoples of South America. The balsa, found on Lake Titicaca in Bolivia and along the coasts of Peru, was a cross between a raft and a boat; reeds tied together in bunches were utilized like planks, and a sail was often added despite the narrowness of the balsa, thus forcing the solitary passenger to remain on his knees. Lastly, there was the circular, raw leather coracle of the prairie tribes. However, maritime navigation was limited, and artificially created ports were unknown.

Bridges and footbridges were used for crossing rivers; in the Andes they were suspended from cables, while in Peru they were raised on piles and were sometimes paved. At Tenochtitlán, the Aztec capital, movable wooden bridges crossed the canals.

FIG. 21. Man carrying a burden on his shoulder blades. He is carrying a forked stick on which to wind thread; the spindle is hanging from the end of the thread (from Poma de Ayala).

Only two possibilities for land transportation existed: portage and hauling, either by animal or human power. Human portage had no special features: objects were carried on the head, back, or shoulders (Figure 21), by means of a headband across the forehead (this was the method most frequently used in Mesoamerica), on the shoulder blades, with a crossbelt, with two straps over the shoulders (to carry baskets), and by balancing the object on the shoulder. Animal portage was extremely limited, for the burdens that can be carried by dogs and llamas are much lighter than those that can be borne by a man. Thus the packsaddle was reduced to a simple padded protective device attached by one or two straps, while the harness was simply a halter. Hauling was done only among the northern peoples; among the Indians a dog with travois was used, among the Eskimos a dog hitched to a sledge (Figure 22). This lack of animal transportation forced man to supplement it with his own strength: either he pushed toboggans or sledges, or else he simply pulled the objects he wanted to move.

In general, roads were footpaths indicated by posts, or broad tracks indicated by a double line of stones. By the fifteenth century the Incas had covered their territory with a genuine network of roads provided with inns. The state of the roads, certain sections of which (on the outskirts of cities and monuments) were paved, must have been satisfactory, for the official messengers, working in relays of 1¾ miles, were able to cover 149 miles in one day. The remains of the 62-mile-long road that linked Cobá with Yaxuna (11 miles southwest of Chichén-Itzá)

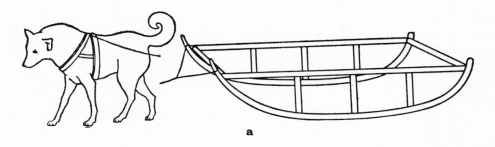

a

FIG. 22.   Eskimo devices for hauling. (*a*) in Greenland and Canada;
(*b*) in Alaska (from Lerio-Gourhan).

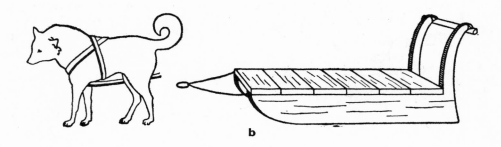

b

after the seventh century, as well as those of the network of sixteen roads that radiated from Coba, can still be seen in Yucatán. These roads, which were 14¾ feet wide, were between 2 and 8 feet above the ground level, and were as straight as arrows. The foundation was a heap of stones, while the visible sides of the road were lined with crudely cut stones. The surface of the road was not paved, but was covered with a layer of pebbles flattened by a compressor roller made of calcareous stone; the roller was 13 feet long with a diameter of 2 feet and a weight of 5 tons, and was pushed by fifteen men.

BIBLIOGRAPHY

In addition to the works listed below, consult the *Anthropological Papers* and *Handbook Series* of the American Museum of Natural History (New York), the *Middle American Handbook* of the Bureau of American Ethnology, and the reports of the Robert S. Peabody Foundation for Archaeology (Andover, Mass.).

ADAMS, ROBERT McC., *The Evolution of Urban Societies; Early Mesopotamia and Prehistoric Mexico* (Chicago, 1966).

AMSDEN, CHARLES A., *Prehistoric Southwesterners from Basketmaker to Pueblo* (Los Angeles, 1949).

BENNET, WENDELL C., *Ancient Arts of the Andes* (New York, 1954).

BENSON, ELIZABETH P. (ed.), *Dumbarton Oaks Conference on the Olma* (Washington, D.C., 1968).

BERGSE, PAUL, *The Metallurgy and Technology of Gold and Platinum Among the Pre-Columbian Indians* (Copenhagen, 1937).

BUSHNELL, G. H. S., *Ancient Arts of the Americas* (New York, 1965).

——— and DIGBY, A., *Ancient American Pottery* (London, 1955).

COE, MICHAEL D., *America's First Civilization* (New York, 1968).

FLANNERY, KENT V.; KIRKBY, ANNE V. T.; KIRKBY, MICHAEL J., and WILLIAMS, AUBREY W., JR., "Farming Systems and Political Growth in Ancient Oaxaca," *Science,* 158 (Oct. 27, 1967), 445–454.

GIBSON, CHARLES, *The Aztecs Under Spanish Rule* (Stanford, Calif., 1964).

HAGEN, VICTOR VON, *Ancient Sun Kingdoms of the Americas* (Cleveland, 1961).

JENNINGS, J. D., and NORBECK, E. (eds.), *Prehistoric Man in the New World* (Chicago, 1963).

KELEMEN, PAL, *Medieval American Art* (2 vols., New York, 1943).

LANNING, EDWARD P., *Peru Before the Incas* (Englewood Cliffs, N.J., 1967).

LEROI-GOURHAN, ANDRÉ, *Évolution et techniques: l'homme et la matière* (Paris, 1943).

LOTHROP, SAMUEL K.; FOSHAG, W. F., and MAHLER, JOY, *Pre-Columbian Art* (New York, 1957).

MacGOWAN, KENNETH, and HESTER, JOSEPH A., JR., *Early Man in the New World* (New York, 1962).

MARTIN, PAUL S.; QUIMBY, GEORGE I., and COLLIER, DONALD, *Indians Before Columbus* (Chicago, 1946).

MORLEY, S. G., and BRAINERD, G. W., *The Ancient Maya* (Stanford, 1956).

MULLER-BECK, HANSJURGEN, "Paleohunters in America: Origins and Diffusion," *Science,* 152 (May 27, 1966), 1191–1210.

OUTWATER, J. OGDEN, "Possible Evidences of a Textile Manufacturing Center at Palenque," *Technology and Culture,* 3 (1962), 161–166.

SELLARDS, E. H., *Early Man in America* (Austin, Texas, 1952).

UNDERHULL, RUTH MARY, *Red Man's America: A History of Indians in the United States* (Chicago, 1953).

WILLEY, GORDON R. (ed.), *Archaeology of Southern Mesoamerica* (Austin, Texas, 1966).

WILLEY, G. R., and PHILLIPS, P., *Method and Theory in American Archaeology* (Chicago, 1958).

PART SIX

THE MEDIEVAL AGE
OF THE WEST
(Fifth Century to 1350)

INTRODUCTION

THE substitution of the "great light" for the "night" of the Middle Ages in the minds of historians required only a few decades. But while medieval esthetics and literature rapidly regained their rightful place, historians generally refused to consider the medieval period as an age of progress, particularly in the areas of science and technology. Greek science and the Roman achievements, in their opinion, remained without successors until the Renaissance. The history of science has now reestablished a balance, but the history of technology, being less known and less thoroughly studied, is still wavering between two extremes: at one end stand those who claim that the Middle Ages were centuries of barbarism, while their opponents counter with lists of "inventions" the extent of which often confounds the imagination.

It seems opportune, therefore, to establish a general outline of the technology of the medieval period in the West, and to determine what we know and do not know of this period. Describing the efforts (if not the inventions) of the Middle Ages appears to be a hazardous undertaking. We have hardly begun to lay the groundwork for such a study, and there is a great risk that the hypotheses will dominate what is definitely known. Despite the difficulties, however, such an effort can nevertheless serve a useful purpose if only to emphasize the gaps that still exist in our knowledge, and to indicate areas for future research. Our methods of research and exposition also warrant several explanations.

Written sources are the first ones that come to mind. Ordinary documents are numerous, at least after a certain period in time. The texts that establish laws usually mention only the objects of these laws: they serve chiefly to date the appearance of a tool, a method of production, a product. Contracts between individuals can be more descriptive, particularly notarial documents, which contain lists of equipment, reports on the condition of dwelling places, or specifications for a piece of work. Public ordinances also supply important information while guild rules and mining regulations supply numerous details, not so much on the tools or products used as on methods of work.

Texts are of two types. Memoirs, memorandums, and chronicles, like literary works, are difficult to use, often because of their lack of precision. Descriptions are generally absent, even in the works of authors like Marco Polo, whom we would expect to be interested in material details and in any case interested by certain technical problems. Technical treatises were not yet very numerous, and a large number of them, following an ancient tradition, are concerned with agriculture; they are often of great interest. The works of Theophilus Presbyter, which were undoubtedly written in Germany, deal only with very special techniques of limited range. Toward the end of the period, Guido da Vigevano's treatise on the military art appears as the first example of a genre that was to continue to be perfected in the fifteenth and sixteenth centuries and that would later supply the historian of technology with a wealth of information envied by the medievalist. These two works are separated by a multitude of manuscripts containing practical instructions with regard to a large number of techniques. Some of them had been handed down for generations. A distant ancestor of the engineers' notebooks of the fifteenth century, the notebook of Villard de Honnecourt, is the only specimen of its kind, and even that is damaged.

The written sources available to us are relatively rare, and are often difficult to interpret. Numerous factors are at the origin of the lack of precision of the texts. Vagueness in form frequently occurs in the literary works: they speak of objects or machines not only without describing them but, on occasions, even with certain distortions. From Ordericus Vitalis we learn of a machine for cutting water pipes that was utilized at the siege of Alençon in 1118. How was it constructed? — Like the ship-launching device mentioned at Marseille at the end of the thirteen century! Marc Bloch recalls the monk who ordered the construction, in 1295, of a new type of mill that a single horse was supposed to be able to operate. Experience proved that four vigorous animals could not operate it. We know nothing of this device, but the detail is important because it indicates a desire to improve techniques and an inclination to experiment, although probably the chronicler's intention was merely to mock one of his contemporaries.

Lack of knowledge of technical vocabulary is not the least source of error. A recent work on the vocabulary of the textile industry has shown how the historian of technology can benefit from such works. The history of technical vocabulary, although important in itself, should not encourage illusions; it is doubtful that it can answer all our questions, even those concerning the dissemination of certain techniques. A semantic borrowing does not necessarily conceal a technological borrowing, and distortions are numerous and frequent.

Pictures are certainly an important source material (at least after the twelfth century), and one to which the historian who is trying to translate into reality a word or an expression found in a text will gladly turn. But it can also be a dangerous source if we fail to apply to it the elementary laws of historical criticism. Concerning the *aratrum* and the plow, for example, A.-G. M. Haudricourt and M. Brunhes-Delamarre have given an excellent demonstration of the limits of the information gained from miniatures, drawings, and sculpture. It is nevertheless to be hoped that a vast corpus of these pictorial sources will be formed. There are many cases where we could benefit from comparisons and detailed studies.

Archaeological remains are extremely few. With the exception, of course, of everything concerning architecture, we can be sure that our museums and collections possess only an infinitesimal number of the implements — even the weapons — in use prior to the middle of the fourteenth century. The predominance of wood in their construction has resulted in the loss of almost all this equipment. The discovery of Norse ships is an exceptional privilege. A disaster like the one that destroyed Pompeii has preserved more remains of antiquity than we have for the entire period of the Middle Ages. Here, again, inventories of these rare archaeological remains would be extremely useful.

Basing our work on the accumulation of all these sources, however, we could attempt to lay the groundwork for a history of medieval technology. It is the proliferation of incomplete and sometimes uncertain evidence that makes possible the working hypotheses to be discussed here. To be avoided at all costs are questionable statements whose only supporting references are those of an *a priori* logic. The simplicity of a mechanism, the ease of production of an object, the empiricism of a process, are not rigorous proofs of their existence in the period under consideration. In the history of the human intellect, the history of technology is perhaps the one most lacking in logic.

These very illogicalities and the great diversity of techniques render a general study of medieval technology particularly arduous. For the sake of convenience it has been necessary to introduce a system of division that does not always correspond to the actual problems. It is true, however, that all techniques are influenced by more general problems, including in particular those related to transportation and the utilization of power. These techniques, which are interrelated, are necessarily subordinate to many others, woodworking, for example, having been, as we have stressed, of primary importance in the Middle Ages. The general problem of mechanization are more particularly related to the utilization of power.

The Middle Ages, lacking what we call by the ambiguous name of "industrial civilization," continued to attach great importance to the elementary exploitation of nature, to the techniques of acquisition, agriculture, cattle raising, fishing, and mining. The techniques of working raw materials by mechanical, chemical, and thermal processes were still relatively crude. In contrast, the techniques of construction and joining, which come the closest to being intellectual activities, were quite well developed. These improvements, however, do not appear to have originated with the Middle Ages. On the contrary, they were part of the immense body of knowledge acquired by the human race in the course of the preceding millenniums. It was the techniques of organization that were, perhaps not necessarily perfected, but transformed following the end of the Roman period. It is in this domain that we can best determine the original contribution of medieval civilization.

We realize, then, that this discussion will of necessity be brief and, if not vague, at least conjectural. Moreover, there is a hiatus of approximately five centuries between the end of the Roman or Gallo-Roman civilization and the first definite glimmerings of the medieval civilization, a period of primordial importance in certain fields, and one about which we know almost nothing, though a small number of archaeological discoveries, skillfully interpreted, have been able to lead to interesting conclusions.

CHAPTER 15

TOWARD A TECHNOLOGICAL EVOLUTION

T HE GREAT invasions had brought about profound changes in the Roman world, and new political, economic, and social structures were being built on its ruins. Moreover, the transformation was not only European. It was Mediterranean as well. By the seventh century the Arabs had ended their conquest, and had just clashed with the Franks on Gallic territory. The Norsemen in turn were about to inundate the northern coasts of Europe. Invasion almost inevitably brings about a revolution in, and new contributions to, material civilization. The technological aspect of civilization, therefore, was about to be considerably modified.

Undoubtedly the Roman expansion to the north had already brought new ideas into the ancient world. Even Caesar was astonished by Gallic metallurgy and the ships of the Venetes, while Tacitus describes the Viking ships, which must have amazed him. Certain words current in the late Latin period already reveal such influences, for example the word for plow, which may have designated a chariot rather than the *aratrum*. In any case it was probably military conquest, in the Iberian Peninsula as well as in central Europe, which made it possible for the Roman civilization to make more general use of iron.

Influence of geographical conditions The transition from a relatively warm, dry climate, shallow arable soil, and stunted forests that could be regenerated only with difficulty, to heavy, oily lands, abundant rivers, and thick forests rich in beautiful trees could not have been accomplished without numerous modifications in technology. The Romans probably adopted part of the primitive equipment of the local populace, especially since their own equipment was not yet well developed, and modified and adapted those techniques that could be modified. The conquerors could not have failed to notice that some of their techniques could not be successfully adapted, not only among peoples who already possessed well-developed tools but also in countries whose natural conditions differed considerably from those of the Mediterranean basin. To mention only one example, they soon discovered that in the northern countries the paved Roman roads were broken up by infiltrations of water that froze in winter.

In agriculture the native tools were undoubtedly retained. The Romans introduced other techniques they had perfected, such as the systems of roads, bridges, law, and a strong government, and others that, precisely for climatic reasons, did not become widely diffused — for example, the water mill.

The geographer Jules Sion has forcefully argued the relationship between technology and physical environment. Noting how seriously the ancient world suffered from the lack of livestock, or had only an emaciated livestock that even with the use of the modern system of harness was incapable of major efforts, he has shown that the water mill could not possibly be developed in countries where the flow of the rivers was often scanty and generally irregular and that fireplaces, which were more hygienic and efficient than the Mediterranean braziers, would have caused the complete disappearance of the forests. In addition, no iron was available for agricultural tools.

The material existence of the northern regions, which had running water, abundant forests, and fast-growing vegetation, was inevitably different from that of classical antiquity. While the possession of areas favored by relative flatness, vast forests, peaceful rivers, abundant ores, and power was undoubtedly not the only reason for the quickening of the inventive spirit of the Middle Ages, several of its discoveries corresponded in these youthful countries to increased needs and possibilities.

The transformations taking place in the Gallo-Roman world were, moreover, being felt even before the barbarian invasions, for these very same climatic and natural reasons. To consider only one example: The transportation of wine was facilitated by the invention of the wooden vat, which effectively replaced the fragile amphora. Transportation of merchandise became more important. It also owed much to inland waterways, which were practically unknown in the ancient world and which were perhaps being utilized in the northern countries even before the Roman conquest. We know of such riverboats from the specimen preserved at Utrecht, which dates from the third century. It is possible that triennial crop rotation began to be practiced in the northern provinces during the Gallo-Roman period. Biennial rotation, with its shallow and repeated plowings was characteristic of the dry, thin soil of the Mediterranean basin. Many medieval innovations probably originated in these climatic differences, which even in modern times still constitute a predominant factor in technical development. Failure to recognize this state of facts led during the Renaissance (which sought only to copy antiquity) to eccentricities that were sometimes extremely ineffectual.

Perpetuation of ancient techniques

Differences in natural resources thus led to technological modifications. It can be said that classical antiquity was a civilization of dry-farming, stone, bronze, and textiles of vegetable origin, while the medieval West had drainage farming and utilized chiefly wood, iron, leather, and textiles of animal origin. A great number of the technical procedures practiced by the ancients were nevertheless perfectly capable of being adapted to other civilizations.

This was undoubtedly perfectly natural on the shores of the Mediterranean, which to a great extent retained the ancient techniques. Medieval Mediterranean agriculture utilized the same implements, the same *aratrum* without wheels, treading, and the biennial crop rotation so perfectly adapted to the soil and climate of this region. In certain areas (Spain, for example) the Arabs even took over the irrigation system perfected by the Romans, making a few improvements of their own. In fact, certain provinces of North Africa were more prosperous under Roman domination than at any succeeding period.

Despite the radical transformation of the ships plying the northern and western coasts of Europe, the Mediterranean long remained faithful to the ancient types of vessels. Not until the fourteenth century did the Nordic ship penetrate these latitudes.

Apart from this geographical continuity, it is impossible to enumerate all the ancient techniques and machines still in use in the Middle Ages. We shall see that most of the products cultivated, despite climatic differences, had already been highly valued by the Romans and that in this regard the Middle Ages made few innovations. The same was true of almost all the techniques of acquisition. Mining, for example, in which the Romans were past masters, appears to have been less exploited at the beginning of the medieval period than in antiquity. In the working of raw materials, thermal techniques underwent few modifications, in contrast to the major improvements made in mechanical and chemical processes. The simple construction techniques (textiles and basketry, for example) in some cases dated from the Stone Age and were not modified until modern times. Others, in contrast, underwent those transformations that were indispensable in different natural conditions.

As for machinery, the Greeks and Romans had achieved theoretical and practical results on which it was difficult to improve. Most of the organs of transmission and transformation of motion (toothed wheels, cams, endless belts, the screw, and so on) were known and used by the ancients. This is why the machines that are depicted on the bas-reliefs of the Late Empire (the squirrel cage, capstans and winches, screw presses, and others) continued to exist for a long time. In its use of hydraulic power (which was in short supply on the shores of the Mediterranean) and its adaptation of this power to a large number of machines, medieval man in many cases was simply turning to advantage earlier discoveries.

Modifications in small tools do not appear to have been of major importance. The equipment pictured on the funerary stele of the smith of Ostia undoubtedly differs little from that of the medieval smithy. Tools used in leatherworking and the textile industry (including the weaving loom) were not greatly changed. The same is true of wooden tools: it appears certain that the plane, with which the fifteenth and sixteenth centuries have so often been credited, was known in the first century of the Christian era. Drills had been perfected in the prehistoric period, and the lathes that we shall later describe were already in use at the peak of the Greek classical period.

Thus the techniques of antiquity by and large continued to exist throughout the medieval period — an inevitable development, given that the methods and equipment corresponded perfectly both to the materials utilized and to certain social structures.

The contribution of the great barbarian invasions The material civilization of the barbarians who overran the Roman Empire is not sufficiently known to permit us to determine its exact influence on the medieval West. Perhaps technology, or at least certain phases of it, tended to return to its pre-Roman level. This phenomenon was partly justified by the fact that for the reasons we have just mentioned, certain techniques carried north by the Roman conquest had not been completely assimilated by the area.

On the other hand, the "barbarians" could not fail to profit by a technical civilization that often was clearly superior to their own. Their assimilation of it was all the more rapid insofar as a certain penetration must have occurred even before the great invasions, which contributed to the expansion of Roman techniques in regions the Roman legions had never conquered.

Analyses of samples of iron from the Merovingian period reveal few differences in metallurgical production from what is known of the metal industries of the protohistoric period and antiquity. The metallurgists progressed from a semisoft to semihard steels. Hard steels were nonexistent, owing to the difficulty of soldering them. The Merovingian furnace of Marishaufen, near Schaffhausen, is very similar to the protohistoric furnaces that have been preserved in greater numbers. Fundamental differences are particularly noticeable in the technique of forging. The Merovingian smiths, according to Edouard Salin, certainly experienced difficulties in drawing relatively large masses of metal. They therefore juxtaposed small quantities of metals of different grades by soldering them to produce a laminated structure, and inserted hard tempered steel at the ends or on the edges. This technique, which was widely utilized in the manufacturing of weapons and probably also of tools, resembled the methods used in early ages. Intense rolling produced a metal that was both strong and flexible. The masterpiece of this metallurgy was the long damascened sword, obtained from a bar formed by the superposition of three strips of soft iron between four strips of carburized iron. The whole was then soldered together and the cutting edges were added. Both in clamp-joint soldering and in the use of nitrogenous organic materials (for the hardening of steels, which causes nitriding of the metal), feats were performed that indicate a perfect mastery of this craft. The homeland of these techniques appears to Salin to have been Noricum, located between the Danube and the Carnic Alps. Production later spread over a wide area, particularly in the region of Siegen, Germany, about fifty miles north of Cologne, whose ores were perfectly suited to this treatment.

This barbarian metallurgy, which was very superior to Roman metallurgy, produced stronger weapons that quickly overcame the occupying Legions, the Frankish battle-ax (the typical weapon of the Frankish invasion) and the long sword being the instruments of this conquest.

The barbarians also brought with them their agricultural methods. It is certainly difficult to say whether they practiced triennial crop rotation, since their agriculture was undoubtedly more extensive than intensive. But in the ninth century, in the great estates of the Austrasian regions, we find both the practice of triennial crop rotation and the large wheeled plow with moldboard.

In contrast, the Germanic peoples seem to have contributed no new species of plant to Roman agriculture. At most we find, undoubtedly coming in the wake of the invasion, the appearance of breeds of horses stronger than the breeds already in existence; they were heavy animals with arched noses, of which the Frisian horse is probably the most direct offshoot.

In the course of this study we shall mention certain techniques that may reveal a Germanic contribution: for example, the plans of various cities. However, it would require more, and more intensive, investigations to determine exactly what medieval technology owes to the barbarian civilizations. It is certain, on the other hand, that between the fifth and tenth centuries the Germanized countries utilized the ancient techniques that had not yet been fully developed. We need only consider the expansion of the water mill, which was already known in Germania in the eighth century. The Salic Law mentions it in the sixth century, while the Alemannic and Bavarian laws of the first half of the eighth century show that this instrument had been in use for some time.

The contribution of Eastern technology	A study of the Eastern contribution appears even more difficult. It is a dual contribution, since it could have come either from the Arabic invaders or

from techniques of Byzantine or even Asian provenance transmitted by the Arabs.

Being a nomad and a warrior, the Arab brought little with him, since nomadic techniques found scant use among sedentary peoples. He disrupted the North African agricultures the Romans had developed to a high degree of perfection. In these same regions he discovered the camel, which had been in use since the Late Empire. In Spain, however, the Arabs skillfully adopted the irrigation network, which was better supplied here than in Africa, and undoubtedly adapted to it the idea of the hydraulic noria. We probably also owe to the Arabs the importation into the West of the windmill, which they had discovered on the plateaus of the Middle East.

Certain ancient mosques, as we shall see, have ribbed vaults; those of Córdoba date from 965. However, the Arabs never adopted the rib to the groin vault. These ribbed vaults or cupolas gradually traveled north. A magnificent example of them north of the Pyrenees is in the Hôpital Saint-Blaise on the route of the great pilgrimage to Santiago de Compostella. Some authorities see in its vaults a distant ancestor of the Gothic vaults.

The only constructive element of the specifically Arabic contribution was its breed of horses, which began to be greatly prized by Western horse-breeders in the early medieval period. In the case of the other Eastern techniques, it is a question rather of the transmission of Byzantine or Asian methods.

Byzantium had direct communication with western Europe, by way of Venice and other Italian cities. However, the technology of the successors of the Roman Empire does not appear to have been very innovative. Even in the case of weaponry, the idea that the new artillery that dominated the entire medieval period came from Byzantium appears to be incorrect. In any case, Greek fire was a military secret of such importance that it seems not to have reached the West, where, moreover, the sources of raw materials it required were lacking.

The Asian contributions that came by way of the great plains of northern Europe are also very difficult to determine. We shall see this in the case of the modern horse harness. Time lags, in any event, are not necessarily a proof of importation: parallel evolution may very easily have existed.

It is quite difficult, as we have just seen, to distinguish the various technological influences. It is equally difficult to pinpoint exactly the inventive role of the medieval West or even post-tenth-century medieval research. We shall see, for example in the case of the compass and the pulley wheel, how difficult it is to determine the birthplace of an instrument or a technical process.

By the tenth century the West nevertheless was in possession of techniques that were more highly perfected than those of the Romans: triennial rotation of crops, a well-developed metallurgy, more widespread use of the water mill, and the modern harness represent gains that are far from negligible. Undoubtedly, however, these improvements were still the monopoly of favored regions and individuals, areas open to the invasion of foreign influence, and of large estates.

The factors in a new evolution   Between the tenth and twelfth centuries there occurred a very visible evolution in all areas of activity. The most important phenomenon to be noted in this period was an intense demographic movement of twofold significance.

In the tenth century the population began to increase throughout Europe. This large "population explosion" continued through the eleventh century, and appears to have culminated in the first half of the twelfth century. There was also a qualitative aspect to this increase. The population shifted and concentrated, and cities began to reappear and to increase in size. A study of the elements of this urban renaissance has recently been made; it appears that the first increase in growth of the city occurred in the tenth century and culminated in the eleventh and twelfth centuries, a period in which the second series of fortified walls appeared in city plans. Moreover, these cities were not simply centers for fairs. Industry, more than commerce, caused them to become permanent establishments and permitted them to develop. The technological expansion, reinforced by the demographic expansion, certainly contributed to this exceptional urban activity.

There was also a social mobility, which was expressed in the gradual emancipation of the rural populace and which resulted in the clearing of vast tracts of land, rebirth of cities, and tentative outlines of new political organizations.

These transformations inevitably influenced technological developments. There can be no doubt that technological progress is linked to demographic expansion; we shall see that both phenomena came to a halt at approximately the same time. That this progress benefited from the breakup of a social and political organization based on an economy of large estates is no less undeniable. The estate, with very few exceptions, could not easily become a center of industrial technological innovations. It was the growth of the cities and the repurchase of industrial rights (which became characteristic in the ninth century, in the register of Prüm, Germany), which favored the growth of industrial markets, which in turn served as the basis for a certain concentration of manufactories.

With the arrival of political stabilization, commercial relations could now be resumed. This commercial renewal was to cause the renaissance of a monetary situation favorable to the investments required by every technological evolution. At the end of the eighth century, at the moment when the commercial contraction reached its peak, gold money had disappeared, only to reappear with the resumption of minting in the twelfth century. The reestablishment of industry was probably the basis for the reestablishment of the commercial equilibrium made possible by Eastern gold. The social transformations we have mentioned permitted the formation of a wealthy class among whom a more abundant supply of capital was accumulated.

These diverse circumstances greatly favored the progress of technology between the tenth and the thirteenth centuries. The simultaneity of all these phenomena strongly indicates their interaction.

CHAPTER 16

THE PROBLEM OF TRANSPORTATION

ALL ECONOMIC development and expansion is extremely sensitive to problems of transportation. It is even possible that these problems condition industrial and agricultural development in large measure, and thereby contribute to technological progress, for example, in the birth of the railroads.

During the early Middle Ages transportation was almost exclusively overland. The only major ocean voyages consisted of crossing narrow arms of the sea. Overland transportation was not necessarily limited to roads; navigable waterways, especially in northern Europe, were widely used. The twelfth century witnessed a genuine development of maritime transportation that in the case of some merchandise supplemented land and river transportation.

LAND TRANSPORTATION

Portage

In order to understand the conditions under which the technology of transportation was transformed, we must remember that several very special factors were involved in this problem. Long-distance transportation was rare. To us the figures for international traffic appear extremely low, a normal situation in a compartmented economy in which each region was almost completely self-sufficient. This long-distance transportation handled chiefly luxury products and a few major consumer products (for example, salt). Large-scale methods of transportation would, moreover, have required better lines of communication than the medieval roads, which were certainly not so well maintained as the Roman roads. This explains to a certain degree the increasingly widespread use of river transportation, which was more convenient, safer, and faster.

Those regions in which the river system did not permit easy transportation had only carting services, which were few, and especially animal transport. We have visible proof of this in the records of tolls, especially those in mountain regions (the Jura, for example, or the Alps); crossings of laden animals are far more frequent than those of wagons. Nothing need be said on the subject of this technique, for it had remained completely unchanged since classical antiquity; we need only note that the number of horses was greater than it had been during the Roman Empire.

Concerning short hauls, we have little information, for it is very rare to find either references to or pictures illustrating this method. It was probably very difficult, for while the peasant could easily bring home his harvest in the classic small two-wheeled cart, such was not the case with, for example, the large construction sites of the cathedrals, where the loads to be transported were enormous. Certain indications permit us to believe that these loads posed difficult problems. Thus we find that in order to avoid carting heavy loads the stones were cut and even sculpted in the quarries. Transporting supplies in wagons was impossible for technical reasons that will be explained a little later; in the case of products that could easily be divided into small loads, it was useless to employ wagons, which were awkward to handle. The texts show that a quite well-developed human portage system was used on cathedral construction sites, to which more studies have been devoted than to any other construction projects. It consisted of people who had volunteered for this work for religious or psychological reasons. At Monte Cassino, at Saint-Trond near Liège, at Lindisfarne Abbey in England, at the collegiate church of Châlons-sur-Marne, it was regarded as a pious action to contribute to the construction by carrying building materials, with the help of wheelbarrows, to the construction site. At Royaumont and Vézelay even prominent people performed this work. At Santiago de Compostella pilgrims carried the stones to the lime kilns.

Numerous miniatures testify that in most cases ramps and wooden scaffoldings bristling over the façades under construction were used to raise the materials to the building itself. (Pivoting cranes and large lever devices did not appear until later.) These materials were carried either in hods with a single handle, as depicted in the ninth-century manuscript of Hrabanus Maurus — a rather rare implement, for the classic hod had two handles — or in wooden vats, as seen in Herrad of Landsberg's *Hortus deliciarum* in the twelfth century and the mosaic in the atrium of Saint Mark's at Venice, or in hand baskets (Psalter of Canterbury) or baskets carried on the back and used to remove excavated materials (thirteenth-century miniature of the cathedral of Modena) (Figures 1 to 4).

Transportation on construction sites must have been greatly simplified with the invention of the wheelbarrow, an instrument that considerably facilitated leveling. Undoubtedly there will be much further discussion of this instrument, which is so practical and so simple in its conception; the most glaring errors have been repeated *ad nauseam*. The texts are made more difficult to interpret by the fact that the French word (*brouette,* from *birouette,* "two-wheeled") does not correspond to the actual object. A thirteenth-century miniature in a *History of the Holy Grail* discovered by Viollet-le-Duc seems in fact to depict a wheelbarrow with a single wheel. Other extant pictures of this implement date from the first half of the fourteenth century. It must, however, be noted that such pictures are still relatively few in number before the fifteenth century.

Wagons

Here again there is a lack of studies; in addition, the texts relating to wagons are not very explicit. In the granting of rights of usufruct, for example in the case of removing wood, it is not so much the type of vehicle as the type of harness that is specified. Our most exact information must therefore be garnered from the miniatures in manuscripts.

There are two types of wagons, the four-wheeled and the two-wheeled.

Fig. 1. Hod with one handle. Beginning of the eleventh century (Du Colombier, *Les chantiers des cathédrales*, pl. 1).

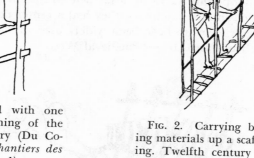

Fig. 2. Carrying building materials up a scaffolding. Twelfth century (Du Colombier, pl. III).

Fig. 3. Carrying burdens on the shoulder. Thirteenth century (Du Colombier, pl. VI).

Fig. 4. Shoulder basket. Thirteenth century (Du Colombier, pl. VI).

Fig. 5. Wheelbarrow. End of the thirteenth century (Singer, p. 641, Fig. 580).

Fig. 6. Wheelbarrow. Middle of the thirteenth century (Massa, *"La Brouette,"* in *Techn. et Civ.,* II [1952], p. 93).

P. Deffontaines notes that "the forms of wagons are to be related not so much to ethnic divisions, as to types of rural civilization." The four-wheeled wagon is supposed to have been characteristic of disciplined, community-oriented countries in which homes were grouped together and fields were left unenclosed. At least as regards wagons for transportation, the ancients were acquainted almost exclusively with the four-wheeled wagon. The modern cart required a different harness, since the harness of antiquity did not permit harnessing of a single animal, that is, did not permit the use of a carriage with shafts. Two-wheeled wagons nevertheless existed in large numbers; they had a center pole and an ox harness. This type of wagon seems to have been widely used in the Roman countryside, as well as in the rural areas of the medieval West.

FIG. 7.   Wagon pulled by men. End of the eleventh century
(Bayeux Tapestry).

The four-wheeled wagon, which was convenient for the transportation of large objects (however, wood — even large pieces — could always be transported in two-wheeled carts), had one major disadvantage: the movable forecarriage that makes maneuvering these wagons practicable does not appear to have been known in western Europe until the second half of the fourteenth century. Before 1350, in any case, we have no picture of it. This disadvantage greatly restricted the use of this type of transportation.

We also have very few details on the construction and shape of the wagons. Disk wheels, which had been common in antiquity, seem to have gradually disappeared (although numerous examples remain), and the wheel with rim and spokes was much more widespread. The medieval users do not appear to have bound the wheel with iron to eliminate wear. It is more probable that this role was filled either by iron plates or even simply by nails with large heads. The axle must have had not the slightest resemblance to our modern axles. The entire

FIG. 8.   Wagon with tandem harness. Around 1338 (Singer, p. 548, Fig. 499).

FIG.   9. Saxon   two-wheeled
cart. Eleventh century (Singer,
p. 547, Fig. 495).

FIG.   10. Public   carriage.
Around 1317 (Singer, p. 547, Fig.
496).

axle supporting the wheels probably pivoted and passed through journals placed in the pieces forming the frame. These details seem to be clearly apparent in the small wagons depicted in the Luttrell Psalter (first half of the fourteenth century).

The problem of the harness

The problem of the harness has been studied much more thoroughly. From the early works of Lefebvre des Nouettes to those of Haudricourt, we possess extremely precise information based in some cases on an iconography that seems to be complete, in others on linguistic factors that permit the establishment of certain dates.

The modern horse harness is essentially composed of a shoulder collar with a rigid frame, nailed horseshoes, and the system of harnessing in tandem. Ancient harness was based on neck traction and the system of harnessing the animals abreast. The modern harness is therefore more powerful, by virtue of the fact that the animal's respiration is no longer hampered. A twofold problem now arises: What is the origin and exact area of expansion of this invention?

The origin of the modern harness is difficult to pinpoint. The oldest type is the breast harness, which is seen in Han bas-reliefs of the second century B.C. For linguistic reasons Haudricourt places its appearance in Europe in the sixth century. The shoulder collar may not have reached Europe until the eighth or ninth century.

Analyzing the numerous pictures he had collected, Lefebvre des Nouettes thought in any case that the great expansion of this technique dated from the eleventh and twelfth centuries; the first miniature depicting the shoulder collar dates from the tenth century, while the breast harness appears for the first time in a twelfth-century miniature. We must also take into account the number of extant pictures: they do not become really abundant until the eleventh century. In any case, it is certain that even if this invention were known prior to this time, it could not be fully developed until the trade routes began to revive. Oxen rather than horses were undoubtedly used for transport in the rural areas, and the ox harness had no need of further perfection.

The exact origins of the nailed horseshoe will undoubtedly be a topic of discussion for a long time to come. It was not linked with the problem of the harness, and was aimed only at accommodating the foot of an animal that was quickly fatigued by the road. It would surely be surprising that no text or picture of the nailed horseshoe is extant, if this invention had been known to the ancients. Nevertheless, some historians continue to believe that it was already in use in the late Roman period.

Not all historians agree on the expansion of this invention of the modern harness. Some have adopted Sion's opinion that possibly the breaking-in of the horse in antiquity may have rendered its equipment less burdensome. Harnessing in tandem is also mentioned by Pliny. But the principal point appears to be the importance of road transportation, which was extremely poor and was generally dependent on animals. The difficult problem of transporting certain merchandise (for example, large vats of wine) was ultimately solved only by an increase in the areas of production. Particularly in the northern countries, waterways constituted a major supplementary method for the circulation of merchandise.

There is one last matter on which agreement has not yet been reached: the whiffletree. It seems, however, that this piece of equipment, so essential for proper harnessing, appeared in the twelfth or early thirteenth century.

Waterways — Today it is difficult to imagine the importance of waterways in the Middle Ages. We would better understand the problem if we were to draw a map of the waterways in use in this period. Even the smallest river was pressed into service for navigation; for example, one could travel up the Seine as far as Troyes, or even to Bar-sur-Seine. This navigation was certainly impeded by natural conditions; in addition it was hindered by numerous obstacles, mills, and fisheries, as numerous documents testify. Nevertheless, it rendered considerable service to medieval trade.

We have little information on this method of transportation, to which scant study has been devoted. It certainly required that a certain amount of work be done on the waterways — control of the water level, and sometimes even dredging, on which we have practically no information. We shall later examine the work done in the plain of the Po River, both to regularize torrential rivers and to create other navigable waterways. Nevertheless, the idea of an extended network of artificial navigable waterways did not appear until the end of the fourteenth and beginning of the fifteenth centuries.

Our knowledge of the ships utilized for this purpose is no greater. Probably even before the Roman conquest the rivers of Europe were being utilized by small vessels, a practice that continued for centuries. Did the Romans modify these small boats? Two discoveries help us to imagine these boats, which were very flat and bulged in the center, so that they must have looked almost like rafts. One such boat, dating from the first century, was found in the Thames; the other, discovered at Utrecht, dates from the third century.

There are few miniatures to show us what these medieval skiffs and barges looked like. The manuscript of the *Vie de Saint Denis* (around 1318) shows us very flat boats maneuvered with poles.

Descending the river is understandable, but it is difficult to imagine going upriver. In most cases the boat was undoubtedly demolished upon arrival and its wood sold. Towing was hardly possible. The tolls indicate that this was in fact the case.

River ports can have been neither very numerous nor very well equipped. The boats were probably beached on small beaches like the Strand at Paris.

NAVAL TECHNIQUES

During the Middle Ages the naval techniques of western Europe developed along two lines. It appears that despite definite contacts between north and south, the Mediterranean did not adopt the North Atlantic types of ships until quite late. It is difficult to discover an explanation for this situation, other than the fact of the existence of deep-rooted traditions, and it is equally difficult to explain the abrupt Mediterranean reversal of this policy.

The Norse and Atlantic navies, which had been little influenced by the traditions of antiquity, were oriented in a clearly defined direction that was totally different from that of the navies of antiquity. Their modifications, which were of major importance, were concerned essentially with the hulls and the rudder.

The Scandinavian ships   The origin of the Norse and Scandinavian ships that conquered the coasts of northern and western Europe still presents many problems, although recent discoveries have made great progress possible. The Als ship (southern province of Jutland), which dates from the fourth century B.C., may be the offspring of the one-passenger pirogue, like the two ships of Tuna discovered in the Swedish province of Upland: planks may have been added to the edges, and the primitive trunk may have gradually become a keel. The intermediate stages between this ship and those of the Vikings are better known. The work of the builders, who must have profited by greater experience on the sea, increased in skill. While certain defects were corrected, serious disadvantages continued to exist, particularly as regards the stability of the boats. In the Nydam (third century) and Galtaback (Swedish province of Halland) ships, each plank had projections that permitted the planking to be joined to the frame; the overlapping planks were fastened together with clinched iron rivets. The Nydam ship was tapered both front and rear, a practice which Tacitus had noted; it had no deck, and its length was 75½ feet overall, its beam 11 feet, and its depth 4 feet. This ship had 14 oars on each side, and a single quarter rudder. The Sutton Hoo ship, discovered, in the small English village of that name, dates from the middle of the seventh century. It presents an improvement over the Nydam ship: in the latter the strakes of the planking were single timbers stretching from stem to stern, while in the Sutton Hoo ship the strakes are jointed rather than single timbers, its more skilled builder having been not at all hesitant to use joints. It is about 80 feet long and 14 feet wide, and is still a flat ship (its depth is 4½ feet). The hull is stiffened by 36 timbers. The ship had 19 oars, and still had the quarter rudder on the starboard side.

The Gokstad and Oseberg ships, which are the best preserved, date from the ninth or tenth century. These ships already reveal great mastery and a better knowledge of stresses. The keel is built in a single piece, slightly curved in the middle. The length varies from 69 feet to 78¾ feet over all. These ships had no true decks, but simply movable rabbeted panels resting on a ledge of the planking. The Gokstad ship is built of oak, and has a projecting keel. The planking consists of 16 strakes, which is more than earlier ships had; 17 ribs support the planking. The eight lower strakes are still bound to the timbers by pine-root thongs passing through gudgeon pins in the planking, while the other strakes are nailed to the ribs with iron pegs; permanence was ensured by clinching (hammering the point of the nail down over a rove or washer), a method that is still in use. In the center was a stump for a movable mast. The ship had 16 oars on each side and a single quarter rudder. The Oseberg ship, which was slightly smaller (it had 15 oars), reveals exactly the same technique.

Several basic characteristics can be distinguished. All of these ships are clinker-built; that is, the planking consists of narrow strakes laid one on top of the other with simple joints, a method unknown in the Mediterranean. The caulking was done with braids of animal hair inserted when the strakes of the planking were being laid on top of each other. This system permitted the use of very thin pieces of wood. The strakes were attached to ribs of curved wood, which were also very thin. Lightness of weight was thus the principal goal; binding by tying is another proof that this was in fact the quality the builders were seeking. This type of binding may perhaps have given the hull an elasticity that facilitated the boat's progress through the water.

There were two types of ships, differentiated by size: the dragon and the serpent. Numerous intermediate models must also have existed. The "large serpent" of Olaf Trygvason (end of the tenth century) was 137¾ feet long overall, therefore similar in size to the galleys of the seventeenth and eighteenth centuries. However, the increase in the size of the boat, which became necessary when the Norsemen's raids required transportation of cavalry, posed new problems that were only gradually solved.

These ships had one movable mast. Tacitus states that oars were at first the only method of propulsion. No indication that the square sail was used is found until the seventh century; definite proof of its use appears in the eighth century. The mast was removed when the oars were being used. This implies a rather limited number of ropes: shrouds and halyards to hold the mast in place and raise the sail, sheets to orient the sail. The deck mentioned above permitted the combatants to look down on the enemy, and protected them from heavy seas. Texts indicate that "castles" — simple platforms supported by openwork scaffoldings — may have existed. Shields were placed along the edge, like battlements. Some boats may have been protected above the waterline by iron or brass plates nailed to the planking.

From the Scandinavian boat to the medieval ship	These are the ships that profoundly influenced all the naval techniques of western Europe before penetrating the Mediterranean in the fourteenth century. Successive modifications caused a slight

loss of trim, but the principles of their construction remained the same. The oar was abandoned, to be used only in exceptional cases; the hull became round, precisely because sailing had been definitely adopted.

The Bayeux Tapestry depicts the construction of the fleet of William the Conqueror (Figure 11). The ships are of exactly the type we have just described. The installation of the sail has been changed, for the fleet (which consisted of ships for transport rather than for battle) had only auxiliary oarsmen who also served as soldiers. There is a single sail, a single quarter rudder, and lapped joints — in short, they are built completely in the Viking technique. They are already rounder in shape, for they had to transport a numerous cavalry. These ships were deeper, less tapered, but had greater tonnage. Their shape and their flat floor timbers accentuated their cargo-ship character, and also facilitated beaching and stability, since they were pulled up on the beach for disembarkation. These ships, which could have had decks, were shorter and wider than they were high. They were less easy to handle, but were more useful.

The transition from the Scandinavian boat to the medieval ship is difficult to pinpoint; documents and even pictures are not very precise. According

Fig. 11.   Norse ship. End of the eleventh century (Bayeux Tapestry).

Fig. 12.   A late model of the Norse ship. Fourteenth century (fresco in the Skamstrup church, 1375).

to some authors, large "ships" may have been in existence by the eighth century. In the ninth century the Frisians of Sluys invented a large decked boat, the *hogge* (cog).

A cast bronze bas-relief on the baptismal font of Winchester Cathedral (probably a Flemish work of the end of the twelfth century) depicts a sauceboat-shaped ship whose prow and stern are decorated with figures of animals; it is still very close to the Norse ships. In the thirteenth century this type of boat may have conquered the entire northern and Atlantic regions of the West. It also appears on seals (Calais, 1228; Gravelines, 1328; Santander, 1228; Pamplona, 1238; La Rochelle, fourteenth century), as well as in numerous miniatures, all of which give a quite clear idea of the final product of the evolution of the Scandinavian ships. The heavy, round hull is still upswept at both ends. The planking thus has a pronounced curve; the prow is very round; the stem is convex.

Fore and aft castles were becoming very frequent, and tended to become permanent. Extant representations of them, which date from the end of the twelfth and beginning of the thirteenth centuries, are all English: the seals of Dunwich (1199), Hythe (end of the twelfth century), Pevensey (beginning of the thirteenth century), and Sandwich (1238). By the second half of the thirteenth century castles were the general rule. Nevertheless, for a long time they con-

Fig. 13.   Norse ship with reef. Thirteenth century (seal of La Rochelle).

Fig. 14.   Norse ship with castle. Thirteenth century (seal of Pevensey).

tinued to be nothing more than light platforms resting on scaffoldings composed of pillars, with ogive or round arches. The seal of Dover (1284) is indicative of an evolution toward the permanent castle as an intrinsic part of the vessel.

However, tapered boats were still in existence. In the boat shown on the seal of Sandwich (1238), the stern and bow are symmetrical, and the rudder is still located on the side of the ship. The portholes for the oars appear in the third plank under the ribband. Thus this ship is still very close to the Scandinavian boats. On the seal of Damme (1328), we see a sternpost rudder and a cutwater front (an overhanging prow with a concave line), an unusual shape that did not become frequent until later.

The rigging generally consisted of a single mast with square sail. Two-masted ships may have existed in the twelfth century and even earlier, but

FIG. 15a. Norse ship with castles added, 1199 (seal of Dunwich).

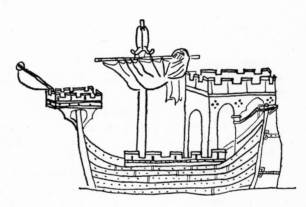

FIG. 15b. Norse ship with sternpost rudder and permanent castles. Fourteenth century (Oxford, Ms. 666, folio 23).

of this we have no definite proof. The bowsprit had appeared very early; it is seen chiefly in Baltic seals (Elbing, 1242; Wismar, 1250; Kiel, thirteenth century), where it projects more or less horizontally from the forecastle. It was not connected with a more complete suit of sail, but was used only as a place for the attachment (or rather, return), in a pulley, of the bowline of the single sail. The sail was hoisted and held in place by ropes; we know very little about the methods of maneuvering it. Reef points appear to be a Scandinavian invention; they are mentioned for the first time by Wace (1120–1183), a native of Jersey raised in Caen. They were located either near the bottom (as seen in the thirteenth-century seal of La Rochelle), which supposed that the yard was lowered, or at the top, as appears in a Parisian manuscript of the thirteenth century; the former position appears to have been the most frequent (seals of Hastings; Bergen, 1278; Dublin, 1297). In some seals (Dunwich, 1238; Dover, 1284; Hythe; Winchelsea) the sail appears furled on the top yard, in the Roman manner, so that the sailors were obliged to mount it astride. The crenellated tops are located at the end of the mast, either symmetrically or hanging on one side.

The sternpost rudder   The second invention is perhaps less important than has been claimed, but is no less controversial: the sternpost rudder, which in these Norse navies replaced the single quarter rudder. Where was it born? In England, where the first permanent castles are found? In the Baltic? The texts are silent, and many pictorial representations, such as the seal of Ipswich (around 1200) and the baptismal font of Winchester Cathedral (end of the twelfth century), are questionable. Apart from these two examples, we have the seal of the Baltic city of Elbing (1242) and a miniature from an *Apolcalypse* in the library of Breslau, which dates from the same time. From the thirteenth century we have the seals of Stralsund, Wismar, Kiel, and Stubbekobing, all cities on the banks of the Baltic. Not until the fourteenth century do we find examples in the North Sea (seal of Damme, 1309).

So we see that it is particularly difficult to pinpoint this improvement. Except for the first two examples, we note, however, that the sternpost rudder appears to have been born at the far end of the Baltic, in the first half of the thirteenth century, and to have slowly traveled to the North Sea by the beginning of the fourteenth century. In England the first definite example may be that of Poole (1325), while Spain may already have had the sternpost rudder between 1282 and 1297. This rather early date may perhaps explain the term "Bayonnaise tiller" by which this type of rudder became known to a large part of Europe. If, as is generally admitted, the ships were then decked both fore and aft, the quarter rudder did not have far to go in order to reach the stern. Maneuvering it, however, required greater strength, and in the early pictures we find a very large lever being used for this purpose. To avoid the spinning caused by the lines of the hull, the ship's stern had to be transformed by raking and flattening it.

A Persian miniature of the first half of the thirteenth century (1237, to be exact) attributed to the miniaturist Wasiti, of the School of Baghdad, shows a Persian ship crossing the Persian Gulf — a ship that has a sternpost rudder resting very visibly on its pintles. It is difficult to speak of "influences" between the Baltic and the Persian Gulf, and thus, as in the case of the water mill, more probable to suppose in parallel evolution.

The Mediterranean ships   The traditions of the ancient navies disappeared from western Europe at the time of the barbarian invasions, but were preserved in Byzantium, where the Italians came at the end of the early Middle Ages to search for them. As early as the sixth century, however, Venetian ships came to the assistance of Belisarius, who was besieging Ravenna.

The Byzantines had preserved, with a few slight modifications, the types of ships in service at the end of the Roman era. The *dromond* was the principal warship. This was a long (some 131 feet), oar-propelled (each side had 25 oars, in two rows on each side) vessel with two decks; it was lighter than the ancient galley, but used the same oars. It still had a ram, but more frequently utilized propulsion weapons. It had raised castles and, like the Norse ships, two platforms on pillars — one in the bow, the other at the level of the mast.

One of the basic modifications of the Mediterranean navy occurred in the

FIG. 16. Elbing ship with sternpost rudder, 1242 (seal of Elbing).

FIG. 17. Ship with sternpost rudder, 1250 (seal of Wismar).

FIG. 18. Ship with sternpost rudder, 1180 (?) (Klemm, *Technik*, pl. 4a).

FIG. 19. Mediterranean ship. Fourteenth century (Sottas, *Les messageries maritimes de Venise aux XIVe et XVe siècles*, p. 60). →

FIG. 20.   Mediterranean ship. Fourteenth century
(Sottas, p. 56).

sails. The ancient navy had had only square sails. Lateen sails appeared at a very early period, undoubtedly before the sixth century; definite proof of their existence exists in the form of miniatures in a Greek ninth-century manuscript. This transformation may have come about gradually, the yard simultaneously turning around the mast and tilting. The strengthening and, finally, the definitive establishment of this change inevitably led to a parallel transformation of the sail, in which a triangular type came into existence. The interpretation of a text of Prokopios would lead us to believe that lateen sails were already in existence at the time of the expedition of Belisarius against the vandals in Africa in 533.

In 877 Pope John VIII ordered the construction of dromonds at Civita Vecchia, thus marking the beginning of the renaissance of navies in the western Mediterranean. In the eleventh century (1081) Robert Guiscard was conquered by the Greek navy at Durazzo. In 1084 he was already in a position to avenge his defeat. A quarter of a century later, Bohemund repeated the same achievement.

Although Richard the Lion-Hearted sent a fleet of ships into the Mediterranean, and although Mediterranean galleys ventured north in such numbers that a *clos des galées* (a special shipyard and shelter intended chiefly for galleys) was built at Rouen in 1373, there was no interpenetration of the two shipbuilding techniques until the beginning of the fourteenth century. It was in the Mediterranean, however, that the galley maintained its leading role. The thirteenth-century galley was a ship some 131 feet long and approximately 16½ feet wide, dimensions similar to those of the galleys of the end of the eighteenth century. The ram had disappeared, but the two quarter rudders were still present. These ships had one, two, or three masts, the foremast being the largest.

Venice had two other types of ships that were derived from the galley. The first was born of the Adriatic ships, which continued the tradition of the low galley with a single tier of oars. In contrast, the *gat* or *chat*, mentioned for the first time in 1122, had 25 oars on each side, two oars to each bank and two men on each oar — as it were, a large, much more compact galley.

The Mediterranean ships were direct descendants of the round cargo ships of antiquity. These were round, wide, short ships with high sides. They too retained the two quarter rudders. Our knowledge of these vessels comes from the negotiations of St. Louis in Genoa and Venice, at the end of the thirteenth century, to obtain the ships he needed for his crusade. The Venetian ship was 118 feet long (excluding the castles), the keel being only 75½ feet; it was 43 feet wide and 42 feet deep. It was thus a very round ship which had two full decks and a third, incomplete, deck under the castles and along the sides. The castles were two-storied and were surrounded by a corbeled bulwark. The two ends were practically symmetrical, the timbers having approximately the same width at both ends. The Genoese ship was very similar. Both types had one major fault: They drifted a great deal.

There also existed a multitude of various types of ships that in most cases derived from traditions of individual shipyards or cities. Thus it is sometimes difficult to determine exactly what is meant by a certain name, for it may have been simply a local term. Fairly large ships existed, the memory of which is preserved by the annals: a ship with three sails that evacuated the Venetian colony of Constantinople in the twelfth century, a ship whose castles were as high as the fortifications of Ancona (1172), a ship that by its sheer mass broke the chain across the entrance to the Golden Horn.

Ships intermediate between the northern type of vessel and the galley also existed. The *tarida* had two masts with lateen sails, like the northern vessels, but its sides were lower, it was narrower, and it carried, or could carry, 10 to 20 oars on each side. The *buysse* had three masts and 80 oars, reminiscent of the Byzantine *chelandion*. A similar vessesl was the *selandra,* a cargo sailing ship approximately 65½ feet long, and ranging in width from 19¾ to 23 feet; its mast was not so tall.

It is certainly difficult to establish an exact date for the invasion of the Mediterranean by the Atlantic ships. The Atlantic type of hull appeared in the Mediterranean at the end of the thirteenth or beginning of the fourteenth century, and this type was then adapted to the local traditions: at first this ship had as many as three rudders, the Atlantic sternpost rudder and the two quarter rudders of the ships of antiquity. A large central mast supported a square sail, while a small mizzenmast held the lateen sail. The Mediterranean hull gradually evolved from this somewhat bastardized version whose essential features nevertheless were now part of the northern, western techniques rather than of those of antiquity. Only the galley, with its sternpost rudder, preserved to a certain extent the memory of the ships of antiquity.

Problems of navigation   Our knowledge on this subject is somewhat scanty. We should remember, however, that at this period navigation was almost exclusively coastal, and thus did not present major difficulties. The crossing of the Mediterranean was not very long, and since land was reached quite quickly ships did not get lost. Rocky coasts and pirates were more dangerous than accidents of navigation. The question of the ocean navigation of the Vikings, who may have reached Greenland and even America, remains an enigma.

Scientific navigation involves two basic elements: the ship's bearings, based on latitude and longitude, and their notation on a map that shows the route to be followed by the ship. Maps from the medieval period are practically non-existent. The map of Peuntinger was much more a nomenclature of routes than a genuine map; a thirteenth-century copy of it is still extant. The first map is mentioned by Guillaume de Nangis; it was on the Genoese ship on which St. Louis traveled in 1270. The oldest known map is a Pisan map (probably of Genoese origin) that dates from the very early years of the fourteenth century, and which was a departure from the *portolano,* or simple description of the coastline. Not until 1354 did Peter IV of Aragon order his ships to have sea charts. These were in any case flat projections, and thus were neither uniform nor to scale; they noted the features of the coastline, landmarks, distances, and rhumb. Tables, which undoubtedly appeared at the end of the thirteenth century, made it possible to lay out routes.

The astrolabe was described in the sixth century by John Philoponos. The oldest definitely dated instrument extant is a Persian tenth-century object. Treatises on the making of astrolabes began to be written in the tenth and eleventh centuries (Hubert of Barcelona, Hermann the Lame). The instruments that have been preserved have many errors. There was a confusion between the nautical astrolabe and the astronomical astrolabe; the former did not appear until the early years of the sixteenth century (it is mentioned for the first time in 1529, on a map now preserved in the Vatican). No treatises on astrolabes makes the slightest mention of their use for nautical purposes. The quadrant was not used by sailors before the end of the fifteenth century.

The compass with magnetic needle was known in China at the end of the eleventh century (1089–1093). The advantages that this object, which was intended for religious use, could offer to sailors were quickly understood. It may

Fig. 21.   Raising the mast on a Norse ship. Thirteenth century.

already have been in use for navigational purposes by the beginning of the twelfth century. It was definitely being used in 1122, and its use became common in the waters of eastern and southern Asia in the twelfth and thirteenth centuries. The Arabs learned of it in 1242. It was already known in Europe around 1190, although we do not know how it reached Europe. It is possible that compasses were placed on board ships in order to determine direction. In any case, there was no way of noting the information thus obtained on a map.

CHAPTER 17

THE PROBLEMS OF POWER AND MECHANIZATION

THE PROBLEMS of mechanization and power are closely linked. While the use of a new source of energy often tends to modify the structure of machines, the continuing existence of certain mechanisms has on the other hand contributed to an equally great extent to the maintenance of the traditional sources of energy.

We have already noted how the shift in economic life had revolutionized the factors in the utilization of power. The abundance of waterways throughout western Europe guaranteed the ultimate triumph of the water mill, which the Mediterranean basin had not been able to develop fully. Despite a considerable increase in cattle breeding, which can also be attributed to climatic conditions, the water mill ultimately became capable of powering all the machines that, in the heyday of the Roman Empire, had been operated by animal-driven tread-mills; the latter continued to exist only in dry regions where they had long constituted the principal source of power. In many cases, however, human energy remained the only form of power that could be used, and in such cases the machines involved underwent very few modifications.

New machines and old mechanisms

Machines are closely dependent on devices for the transmission of motion. The medieval period does not appear to have made great innovations in this field. The most important progress was in the use of materials of better quality; thus the solutions found for the problems of the rubbing and resistance of driving shafts were improved.

The ancients were acquainted with most of the devices utilized in the medieval machines; the mechanicians of the School of Alexandria had described almost all these devices, and some writers had even attempted a scientific theory of them. Undoubtedly it was the lack of adequate materials and abundant natural energy that prevented them from making the transition from the "toys" which appealed to these authors to genuine machines. Even if we omit the "simple machines" (the wedge, the ramp, the screw, the lever) that had long been in existence, we know that the transmission of circular movements in several planes had been widely utilized in classical antiquity: toothed wheels and lantern gears are mentioned at the beginning of the Christian era.

The problems of demultiplication of power had also been solved. The theory of the lever was centuries old; pulleys and pulley blocks were in common use, as were winches and windlasses; Pappus had given a perfect definition of the role of the gear train. Thus the Middle Ages had only to choose from this theoretical and practical arsenal it had inherited. Let us immediately stress that the crank-

and-connecting-rod system, which made it possible to transform a continuous circular movement into a straight up-and-down alternating movement did not appear until the end of the fourteenth century, and thus was still unknown at this period. This lack of knowledge considerably limited the applications of mechanization. To remedy this disadvantage, extremely simple systems were devised that operated basically on the spring or the rope-and-reverse-pulley principles, both of which permitted only circular alternating movements of limited range. However,

FIG. 22. Potter's wheel. Thirteenth century (Singer, p. 288, Fig. 270).

FIG. 23. Hand mill. Fourteenth century (Bennet and Elton, *History of Corn Milling*, p. 163).

FIG. 24a. Grinding wheel. Around 1338 (Trevelyan, *Illustrated English Social History*, Fig. 64).

FIG. 24b. Grinding wheel. Eleventh century (Singer, p. 651, Fig. 593).

by placing cams on a driving shaft, other tools (almost exclusively percussion tools such as hammers and mallets) could be operated.

The originality of the Middle Ages was thus quite limited. The use of hardwoods and lead plates on moving parts eliminated the rubbing action to some degree. All these machines were nevertheless short-lived, and repairs were frequent, a condition to which the number of rights of usufruct accorded for this purpose in the forests testifies.

Lathes The study of lathes is complicated by the lack of exact descriptions and pictures. In many cases ancient models continued to be used without modification; such was the case, for example, with the potter's wheel, which certainly had no need of further perfection. The crank-turned wheel, which generally had an endless belt and was turned by an apprentice (many examples of it can still be seen in Diderot's *Encyclopédie*), is pictured only in connection with pulley wheels. This machine, with its two wheels acting as demultipliers, does not appear to have been utilized in antiquity; at least no mention of it is made in any form.

The pole lathe, also unrepresented in classical antiquity, appears to have been the most common. Numerous examples still existed in French country districts in the nineteenth century. A strap (in most cases made of leather) connected a treadle with the pole suspended from the ceiling. The strap wound around the axle of the lathe. The successive action of the treadle and the pole thus gave the lathe an alternating movement sufficient for working small objects. We find a detailed picture of this tool in a Parisian manuscript dating from the middle of the thirteenth century (Figure 25).

FIG. 25. Pole lathe. Thirteenth century (Singer, p. 645, Fig. 586).

FIG. 26. Pole lathe. Fourteenth century (Klemm, *Technik*, pl. 7b).

The bow lathe is sometimes seen even today. It operates on exactly the same principle, but in this case the spring is formed by a bow suspended to a beam; the cord of the bow holds a small idler pulley. The transmission cord is attached at one end to the treadle, at the other to a wall, at a level lower than that of the pulley when the bow is stretched. It, too, winds around the axle of the lathe and slips into the groove in the pulley.

The double-treadle lathe is seen in a thirteenth-century stained-glass window in the cathedral of Chartres. The ends of the strap were attached to the two treadles (Figure 27). In an early system it wound around only the axle of the wheel, but later it seems to have passed through a pulley hooked to a ceiling beam. It is this latter system that is depicted in the Chartres window; the reverse pulley can clearly be distinguished in the picture.

All this machinery appears quite primitive; there was certainly no question of turning hard metals with it. However, it was not so much the machinery as sufficiently strong tools that were lacking. These lathes could be used only for working relatively soft materials, and chiefly wood. The number of medieval turned objects still extant is, moreover, very limited.

Power devices   Here, again, the Roman equipment was by and large still in use. The lifting devices that are pictured in numerous miniatures showing the construction of the Tower of Babel and religious edifices are no different from those pictured on bas-reliefs of the Roman period. Simple winches with a reverse pulley hooked to a scaffolding at the top of the building are the most numerous, and squirrel-cage cranes also appear frequently. Pivoting systems that could carry even large weights were quickly perfected. Such was the squirrel-cage crane of the port of Bruges, which probably dated from the mid-fourteenth century and which was still an object of admiration in the fifteenth. Viollet-le-Duc gives a drawing of a complex device consisting of a large winch turned by an enormous lever lowered by a man's weight; moveable disappearing valves prevented it from turning in the wrong direction. Unfortunately, Viollet-le-Duc does not mention his sources.

However, the importance of these machines should not be exaggerated. Numerous late miniatures still show materials being carried up to the top of a building under construction by means of scaffoldings and ramps, and this method appears to have been the one most frequently used.

It is to Villard de Honnecourt, in the second half of the thirteenth century, that we owe the first drawing of a screw jack (Figure 29). A large vertical wooden screw, terminating at the bottom in a winch, pierces the bottom plate and turns by means of pivots in the groundplate and the cap; two inclined posts hold three horizontal pieces in place. Two sliding pieces hold a large screw nut made of hardwood, with iron clamps, which supports a ring with hoisting scissors. When the windlass is turned, the nut rises between the two grooves. This device made it possible to lift enormous burdens, provided that the machine was sufficiently large. It remained long in use at construction sites, and appears to predate Villard's period. In his chronicle the monk Gervais mentions screws for lifting burdens, at the end of the eleventh century or the beginning of the twelfth.

The medieval presses also owed a great deal to classical antiquity. Screw and winch systems were commonly and simultaneously jointly used. Those most frequently pictured are the winepresses (for example, the theme of the mystical winepress). An illustration in an eleventh-century commentary on the Apocalypse, preserved at the National Library in Madrid, and the press depicted in Herrad of Landsberg's manuscript (beginning of the twelfth century) (Figure 30) are perfect examples of these large screw presses, a magnificent thirteenth-century specimen of which still exists at Chenove, near Dijon, in the wine cellar of the Dukes of Burgundy. Particularly worthy of note is the fact that the beam that acts as a lever can be placed at different heights.

Other power devices are known to us from the texts of chronicles, but no details are given. The *Vita sancti Arnulfi* (end of the eleventh century) mentions the memory of a lifting machine for straightening and realigning a wooden

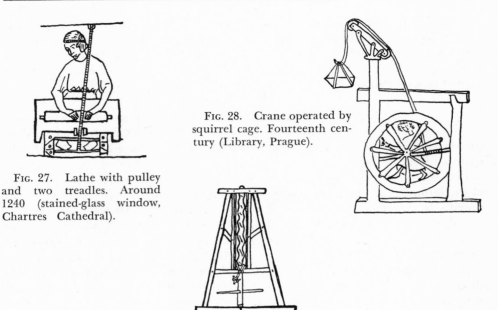

FIG. 27. Lathe with pulley and two treadles. Around 1240 (stained-glass window, Chartres Cathedral).

FIG. 28. Crane operated by squirrel cage. Fourteenth century (Library, Prague).

FIG. 29. Screw jack. Around 1270 (Singer, p. 647, Fig. 589).

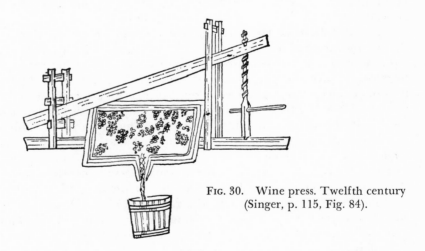

FIG. 30. Wine press. Twelfth century (Singer, p. 115, Fig. 84).

building that had become warped or had collapsed. At Marseille, in the thirteenth century, ships were launched with the help of a special engine that in the city statutes is called a *vasa;* judging by several extant texts, this device consisted of a number of pieces of wood joined together with pegs, and maneuvered by ropes and winches that alternately supported the galley and pulled it so as to bring it out of its shelter for launching. These engines belonged to the commune, which fact permits us to suppose that they were very large and must have been very costly.

Machines for hurling stones, also built of wood, will be studied later.

The ingenuity of the medieval builders of machines must have been very great. It is not rare to find a chronicler expressing astonishment as he contemplates them. Some of these machines — we have already mentioned the case of the crane of the port of Bruges — retained their reputation for centuries. Moreover, there was a constant search for greater power and efficiency. It is for this reason that certain machines that to us seem rudimentary did not appear until very late. Such is the case of the jack, an instrument that was so practical for lifting very heavy burdens, but which could be constructed only of iron. We see it for the first time in the crossbow, and this not until the early years of the fifteenth century. The problem of hoisting devices greatly occupied the attention of engineers and architects, and this attention remained unflagging during the succeeding years; the Renaissance engineers resumed the study of these engines.

The water mill:
its general forms

Of all the technical achievements realized between the tenth and the thirteenth centuries, perhaps the most remarkable and spectacular one was the expansion of the water mill and its adaptation to a multitude of industries. It is possible, although difficult, to name ten (or even fewer) hydraulic wheels in Gaul in the sixth century (the lack of documents is undoubtedly to a certain extent responsible for our inability to find others); in the twelfth century they were counted in the hundreds, in the thirteenth by the thousands. At the end of the eleventh century the Domesday Book listed 5,624 in England. This indicates the importance of this new source of power.

Many of the technical details of water-mill construction are unknown; most of the mills, being constructed of wood, have disappeared. While several stone buildings — fortified mills, mills of abbeys — are still standing, their works have either disappeared or have been completely rebuilt in the course of the centuries. One of the basic problems, that of the position of the hydraulic wheel, therefore inevitably remains obscure. Some historians believe that horizontal wheels were the first to be used. They are still utilized in several regions which, it is true, are small and widely separated in space: the Dalmatian coast, the Massif Central and the Pyrenees regions in France, and northern Europe. One fact is certain: in medieval pictures of water mills, only the vertical wheel appears. In the Toulouse area the horizontal wheels of the Bazacle became famous in the eighteenth century. We have no proof of the existence of this type of wheel before the seventeenth century.

In contrast, all possible methods of utilizing the vertical wheel existed; undershot wheels, overshot wheels, and side wheels appear in miniatures and drawings. Their use depended on environmental conditions. In streams that had a steady flow and a slight slope, they were installed downstream, often under the piles of a bridge. Some examples are found in the tenth century. These dikes consisted for the most part of wooden cofferdams with stone fill, as in the Bazacle mills at Toulouse; the wooden piles were driven in with hand-powered pounders. In regions where the current of the streams was scanty and irregular, larger dams were constructed and ponds were formed. In this way landscapes were quite visibly changed; examples of these changes can be seen on a large-scale map of the Périgord and Nivernais regions of France. In the mountains,

FIG. 31.   Mill with undershot wheel. Around 1338 (Singer, p. 596, Fig. 543).

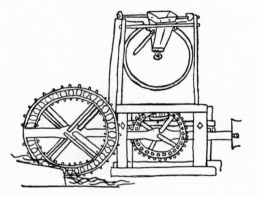

FIG. 32.   Millworks. End of the twelfth century (Singer, p. 648, Fig. 590).

FIG. 33.   Millworks. Second half of the thirteenth century (Singer, p. 648, Fig. 591).

where the rapidity of the streams did not lend itself well to the installation of mill wheels, channels along the side of the slope, or small wooden ditches (called béalières in the Alps), were built. Sometimes the mills were even fed by long supply pipes, as in the Roman era; for example, at Obazine, in France, the Cistercians dug a canal approximately 4,920 feet long through rock.

In order to provide the power required for the mechanized trades, some cities built numerous canals that actually turned them into miniature versions of Venice.

The mill wheels were generally built of oak wood. The iron-banded portion of the shaft rested on a wooden bearing covered with lead. The internal gear

FIG. 34.   Mill with overshot wheel. Around 1270 (Verriest, *Le vieil rentier d'Audenarde*, p. 87).

FIG. 35.   Floating mills. Around 1317 (Singer, p. 608, Fig. 552).

Plate 61.
The Mystical Mill. Twelfth-century capital, Vézelay. *Photo Mélie*

trains were also made of wood, frequently elm wood. Transmission was achieved by wheels and lantern gears with pegs, and the relations of demultiplication were learned by experience. Mills built at the end of the eighteenth century still had much the same appearance as their medieval counterparts, and thus an eighteenth-century engraving gives us a good idea of the appearance of a medieval mill. The mills were extremely fragile structures that had to be repaired and replaced quite frequently, and rights of usufruct in the forests were often granted for this purpose.

Two particular types of mills deserve special mention. Floating mills had existed since antiquity, and they were undoubtedly very numerous in the Middle Ages. Until the twelfth century, at Toulouse for example, this was the only type in existence. There were approximately sixty of them, and they may have been the first floating mills on the major rivers of France. In some cases they constituted a major hindrance to river traffic.

Tide mills obviously appeared later, and could not have been known to the Roman world. The oldest may be those of the port of Dover, which were in existence in the time of William the Conqueror. The mills on the Adour River in France are mentioned around 1120–1125; the principle on which they were established was extremely simple, since all that was needed was to dam up an indentation in the coastline with a dike. However, this type remained relatively rare.

Circular systems   Early uses of the mill were naturally applied to engines with continuous circular movement. All the treadmill-type machines invented by the Romans could therefore be applied to them, with the necessary modifications. The first such adaptation appears to have been for the grinding of grain. Only the millstones were modified. They became (or, better yet, returned to their original form of) two flat, circular stones, replacing the nested cones of the treadmill-type millstones of Pompei. Making exception for improvements in the cutting of the millstones and certain mechanical improvements (to be discussed later), the drawing of a twelfth-century mill — that of Herrad of Landsberg, for example — differs very little from those of the seventeenth or eighteenth century.

The old oil press, with its large round millstone turning on its edge, was also able to profit from hydraulic power, but it utilized somewhat more complicated mechanisms. The Romans do not appear to have been acquainted with hydraulic oil presses; we first find them in the eleventh century in the Gréisivaudan area. The use of this device was extended to include all the oleaginous plants, including mustard (Forez region, thirteenth century) and poppy (in the north, same period).

The problem of the hydraulic noria is considerably more difficult to solve. Though treadmill-type norias had existed in antiquity, a text of doubtful interpretation, and the very mutilated second-century mosaic of Apamea (Dinar, Turkey), do not appear to be sufficient proof that hydraulic norias did. The norias of the mines of Tharsis in Huelva Province, in the Sierra Morena of Spain, could not have been hydraulic, since they were used for removing water. These lifting wheels were able to operate either with treadmills or by squirrel cage. It is very possible that the system of hydraulic wheels was introduced into Spain by the Arabs; it has retained the Arabic name. The wheels of Hama, so well known from photographs, probably date from the eighth or ninth century; they are mentioned by ibn-Jubayr in 1184. The gardens of the estate known as the Noria (perhaps because the instrument was still an exception), built near Córdoba by the Emir Abd Allah (883–912), were watered by one of these hydraulic machines.

The Albulafia of Córdoba, which was depicted in the fourteenth century on the city seals, was constructed in 1136–1137 by the Emir Tashfin, the Almoravide governor of the city. This does not appear to have been an isolated case. At the time of the King of Badajoz, an Arabic geographer who died in 1094 noted the great variety of hydraulic wheels in Spain. In a text that tells of an agreement between the Archbishop of Toledo and an Archdeacon of Segovia, mention is made of a noria; it is again mentioned in 1143. Other wheels of this type appear to have been installed in the twelfth and thirteenth centuries.

FIG. 36.   The noria of Córdoba. Fourteenth century (seal of Cordoba).

Similar machines are mentioned in Sicily, at Palermo, in the twelfth century. In the north the need for such machines was less evident, for water was much more abundant. There was one exception, however: the lifting wheel installed between 1095 and 1123 by Lambert, Abbot of the monastery of Saint-Bertin, near Saint-Omer.

Lathes could be directly adapted to hydraulic wheels, but places with this equipment do not appear to have been very numerous, at least in the period under consideration, undoubtedly for economic reasons. We have but few examples of such a *tornaglium* or *tornallium:* one in 1347 near Vizille (Dauphiné), and another one somewhat later in the same region.

Cam systems   In order to adapt waterpower to other machines, an intermediary mechanism was required to transform the circular movement of the driving wheel. This mechanism was basically the camshaft, a simple invention of many uses that even today continues to regulate numerous machines, including the most highly perfected. This device, or at least its conception, is relatively ancient, since we find it in use in certain automata of Hero of Alexandria. A crude type of gear with projecting pegs or cams at regular intervals on its outer surface was attached to a driving rod, either to a square section of the rod or with the help of wedges. These pegs pressed on the handle of a tool that pivoted around an axle or raised a sliding shank in a groove. The weight of the tool attached to the handle or the shank caused the latter to return to the position of equilibrium after the pegs had passed. By this means it was possible to move hammers, mallets, and pounders by hydraulic power.

The fulling mill seems to have been the first such system to appear. At first the cloth in the vats was fulled by beetles lifted by the cam, replacing fulling by foot, which had been known and practiced for a long time. The oldest fulling mill is mentioned in Normandy at the end of the eleventh century (1086); at the end of the same century it is found in the Forez and in Champagne, and in the twelfth century it came into general use in France. In England the first such mill dates from the end of the twelfth century, and by 1327 there were between 120 and 130 in the entire country. Evidence of it in Italy exists for the end of the twelfth century, for the very early years of the thirteenth century (1212) in Poland.

Hemp mills were of the same type; in many areas, in fact, their terminology is the same as that of the fulling mill. They could be found in the Dauphiné in 1040, in the twelfth century in the Lyons region (Vienne, 1184), and almost everywhere in the thirteenth century. Tanning mills, which were similar, date from the same period; they were found at the end of the tenth century (990) at Romans, in 1142 at Pontoise and in Burgundy, in 1154 in Italy, in 1162 in Normandy; by the thirteenth century, they were in existence everywhere in western Europe. Color mills were known in the fourteenth century. Beer mills are much older (1088 at Evreux, 1100 at Lillebonne). All these mills operated on the same principle, and thus it is natural that they all came into use at the same time.

The iron mill, however, operated on a slightly different principle: the cam pressed on the handle of a hammer, which fell on an anvil. The dates that have been established for the iron mill are approximately contemporary with those of the other instruments that utilized hydraulic power. The principle of the medieval iron mill seems to have been unique: the cam acted on the handle of the hammer, beyond the pivot axle (the terminal type). The two other systems — lateral and frontal — do not seem to have appeared before the fourteenth century.

The first known iron mill may be that of Cardedeu in Barcelona Province, Catalonia, mentioned in 1104; it was followed by the French iron mills of the thirteenth century. Thus a considerable distance separates these practically contemporary mills; the instrument must already have been known for some time. An iron mill is supposed to have been installed in 1197 at the abbey of Soroë, in Sweden. From this we can probably conclude that by the end of the twelfth century every region of Europe had the iron mill. This spatial expansion was in addition accompanied by a fairly rapid conquest in depth, for in 1151 fourteen forges are already mentioned as existing in the Catalonian Pyrenees, and in the second half of the century references to them become very frequent.

Dates given by documents for other regions are later, but we learn that in the thirteenth century the hydraulic hammer spread throughout France: Champagne (Ervy, 1203; Nogent, 1249), the Dauphiné (1226), the Massif Central (Besseges, 1237), and the Montagne Noire (1283). It must have become known in England around the same period (1200). It is found in the Rhine Valley in 1226, in Calabria in 1274, in Silesia at the end of the thirteenth century, and in Poland in the fourteenth century. From this period down to modern times, although the extant pictures date only from the fifteenth century and later, we can say that few changes have been made in this mill. It was probably the West that adapted waterpower to the papermaking technique it had learned from the Far East. The four hundred paper mills that existed at Fez in 1184 obviously could only have been manually operated. In the twelfth century Idrisi mentions the paper mills of Játiva near Valencia in Spain; he does not mention whether they were hydraulic machines. Rights granted to these factories in 1238 and 1273 attest, in any case, to the fact that at that date they were using waterpower.

The technique may have been perfected at Fabriano, in Italy, where paper mills are mentioned in 1268, 1276, and 1283. Here the waste products of flax and hemp, which at Játiva may have been simply pressed under a simple millstone, were crushed in vats by pounders raised by cams. With little modification, pounders and vats continued to form the basic elements of the paper mill until the end of the eighteenth century, when the "Hollander" appeared.

By the end of the thirteenth century many Italian cities already possessed their own paper mills. From Genoa, where the paper mill is mentioned in 1292, it spread at the beginning of the fourteenth century to Bologna, Padua, Treviso, and Venice, and from Italy to France; in 1338 Troyes built the first French mill, followed by Grenoble (1344). After 1350 the paper mill spread rapidly across France. Not until the fifteenth century did it reach Switzerland and Germany (Nuremberg, 1390) and the Nordic countries.

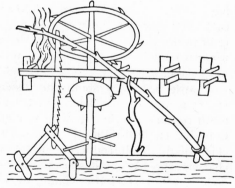

FIG. 37.   Hydraulic saw. Around 1270
(Singer, p. 644, Fig. 584).

Given the absence of the crank-and-connecting-rod system, the saw posed a difficult problem. The cam pulled the saw in one direction; a spring pulled it back in the other direction. A splendid picture of a saw appears in the notebook of Villard de Honnecourt (around 1270), except that the artist drew the paddles of the hydraulic wheel backward (Figure 37). Examples are mentioned in the Swiss Jura in 1268. A machine of this type was purchased in 1303 by the Abbey of Saint-Sernin at Toulouse; it is mentioned slightly later in the regions of Aude and Isère.

Thus we see how the use of the mill gradually spread to an entire series of industries. The water mill and the machines it operated played a very important role in the medieval period of the Western world. We shall later discuss its legal and social repercussions. The water mill, which represents the typical medieval machine, constituted an important technological revolution that culminated in the twelfth and thirteenth centuries.

The windmill — The origin and spread of the windmill still poses numerous problems and equally numerous enigmas to historians. Was it in fact known in antiquity? The texts attributed to Hero may be later glosses, and no other texts and no archaeological remains exist to confirm this hypothesis that the windmill was known to the ancients.

In his *Golden Prairies* the Arabic historian Ali al-Mas'udi states that the assassin of the Caliph Omar, who was a Mazdean (Zoroastrian) from the Iranian plateaus, had boasted of his ability to construct a windmill. This would seem to prove that there were windmills on the plateaus of Iran (where they may have begun to spread in the seventh century) at a time when they were still unknown to the Arabs. Around the middle of the tenth century, two Persian geographers again mention the windmills of the province of Seistan (Shahri-i-Zabul). As in the case of the hydraulic norias, it was the Arabs who disseminated the knowledge of the new machine.

Here, again, the specifically technical problems are difficult to solve. The first Persian mills, like almost all the mills of Asia, probably had vertical axles, as is still true today. The Arabs borrowed the same system, as can be seen from a thirteenth-century drawing. The first Western mills that we see in pictures have wings and a horizontal axle. Consequently a system had to be discovered that would permit the wings to be always facing the wind. All the drawings indicate that in this period only the mill that pivoted as one on an enormous wooden tripod — the type of mill known to the Anglo-Saxons as a "post mill" — was in use.

Little is known about the dissemination of the windmill in Europe. In Spain, windmills may have been turning in the region of Tarragona since the tenth century; the first general reference to them is found in a decree of Pope Celestine II (1191–1198). It is very evident that the legend that credits the Crusaders with the introduction of the windmill in the West, as well as several charters whose falsity has been amply demonstrated, must be rejected. Ambrose's *History of the Holy War* claims, in contrast, that it was the Crusaders who may have brought this invention to Syria.

Texts mention windmills in Normandy around 1180, in Brittany shortly thereafter. By the thirteenth century this device was in use almost everywhere in France in regions that had steady winds — Normandy, Brittany, Champagne, Blésois, Artois, Picardy, and Flanders. In England windmills were to be found at the extreme end of the twelfth century in Tanrigge, Canterbury, and Bury Saint Edmunds. In Belgium we find the first windmills around the same period, which seems to prove that this new instrument actually appeared at that time. By the thirteenth century it had reached Holland, eastern Pomerania, Denmark, and Bohemia, in the fourteenth century Poland (1330–1340) and Sweden (1334). However, our map of the progress of wind power still contains numerous white areas, and a systematic exploration would undoubtedly make it possible to draw more exact conclusions.

Medieval pictures show us only one type of windmill, and documents men-

FIG. 38. Windmill. Around 1270 (Verriest, p. 67).

FIG. 39. Windmill. Fourteenth century (Singer, p. 624, Fig. 563).

FIG. 40. Windmill. Fourteenth century Oxford, Ms. Bodley 264).

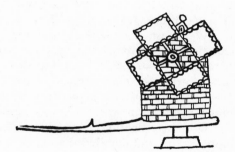

FIG. 41. Windmill. Fourteenth century (Singer, p. 623, Fig. 562).

FIG. 42. Windmill. Fourteenth century (Skilton, *British Windmills*, pl. I).

tion only one use for them — the grinding of grain. During this period the windmill was never put to the variety of uses to which it is still applied in modern times, for example in Holland. The only extant references relate to discussion on the right of grinding by windmill and the windmill's integration into the feudal system of obligatory use that applied to the water mills. Moreover, in the mid-fourteenth century the spread of the windmill was still completely relative, and it does not appear to have really begun to develop until the fifteenth century.

CHAPTER 18

TECHNIQUES OF ACQUISITION

THE TECHNIQUES of acquisition were certainly the most important and the most widely practiced, and it is undoubtedly for this reason that the heritage of preceding centuries was most in evidence in this area of technological development. While urban activities were admittedly the most conducive to progress, at least in these early days, it is easy to understand why the inhabitants of the rural areas felt neither the need nor the desire for a radical transformation of their techniques, some of which dated from the Stone Age.

We must not believe, however, that tradionalism or obscurantism systematically dominated the techniques of exploitation of the land and its subterranean resources: it is simply that innovations in these areas appeared far more slowly and more gradually than in other areas of activity. This is perhaps due to the fact that the products of these activities had relatively limited consumer markets, in which changes must therefore be neither harsh nor sudden.

Our investigations into the techniques of acquisition must therefore be more careful and detailed than is the case in other areas. It is facilitated, at least in certain aspects, by a very abundant documentation. However, while historians have devoted much attention to agricultural problems, the same cannot be said for mining techniques or even for such activities as fishing, hunting, and cattle raising. Here, again, there are major gaps in our knowledge.

AGRICULTURE

Agriculture continued to be one of the principal activities of the medieval economy, and the one that occupied the largest segment of the population. Despite its apparent stagnation, there was visible development: the enormous demographic growth of the twelfth and thirteenth centuries and the noticeable improvement in the diet of the average man are more than sufficient evidence of progress in agricultural techniques. While it is not possible to speak of an agricultural revolution similar to that of the eighteenth century, it cannot be denied that medieval agriculture clearly differed from that of Roman antiquity, if only because of those climatic and geographic conditions to which we alluded at the beginning of this study. Thus, one of the original contributions of medieval agriculture was that it succeeded in reconciling an ancient (and certainly not negligible) heritage with barbarian traditions that were undoubtedly better adapted to local conditions, building onto this synthesis by an undeniable attempt at originality.

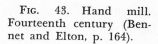

Fig. 43. Hand mill. Fourteenth century (Bennet and Elton, p. 164).

Treatises on agriculture The ancient knowledge of agronomy had never completely disappeared from the Western memory. Manuscript copies of the Roman authors continued to be made during the Middle Ages, and despite the general illiteracy it would be a distortion of the truth to believe that the influence of ancient theories played no role whatsoever in medieval thinking; quotations from works of antiquity in the medieval treatises prove that a tradition was being continued. Certain regions were privileged in this regard, for example the tenth-century Hispano-Moorish city of Córdoba, which replaced Baghdad as a center of arts and sciences, and where botanical gardens were installed and experimental fields for seedlings were created. The development of agronomy in Córdoba was based on the one hand on the traditions of the classical agronomists, on the other on the *Agricultura nabatea,* a work that was Arabic and Persian in inspiration. Treatises proliferated after the eleventh century, and were circulated throughout the Christian world by means of Latin and Castilian translations. These treatises, however, were written with the agriculture of the Mediterranean countries in mind, a fact that greatly limited their area of application.

Other themes were developed, and other treatises were written, in the northern provinces of Europe, chiefly in England and France. In the thirteenth century in the Anglo-Norman lands there appeared not only scientific treatises but also practical handbooks, the reading of which is perhaps even more instructive; there were, for instance, Walter de Henley's *Husbandry,* the *Seneschaucie,* the *Fleta,* and Robert Grosseteste's *Rules,* all of which date from the first half of the thirteenth century, and the number of manuscript copies of which is unquestionable proof of their success.

The entire medieval agronomy and all the recollections of the Latin authors are united in the *Opus Ruralium Commodorum,* which an Italian author, Pietro Crescenzi, composed between 1304 and 1306 — a work that is both erudite, with original theoretical essays such as the *De Vegetalibus* of Albert the Great, and practical, like the Anglo-Norman treatises. Pietro Crescenzi seems, however, to have preferred experimentation to the authority of the experts — a sign of a new type of thinking. This work appears to have enjoyed a certain amount of success, which increased when Charles V had a translation of it made in 1373 (unfortunately by a translator who was both unfaithful to the original, and incompetent).

*Improvement of
agricultural techniques*   Substantial transformation occurred in agriculture in the Middle Ages, but it is difficult to determine origins, period, and dissemination in depth. For the period prior to the thirteenth century we have only scanty and fragmentary information. A primitive agrarian system utilizing chiefly manpower and possessing for the most part wooden and stone equipment had undergone very few modifications since prehistoric times. This system was replaced by extensive spatial and temporal cultivation of a variety of crops (fallowing and biennial crop rotation), and thence by a well-developed system characterized by the harnessing of natural and animal aids, the increased use of iron in tools, and the reduction of unproductive areas and periods of time. It is this transformation that made possible greater yields, which in turn meant a more abundant diet better adapted to the needs of the populace.

Thus medieval agriculture contributed, not new crops or new techniques, but a series of major improvements whose dates are difficult to pinpoint because they were gradual and even imperceptible. This agricultural evolution seems to have begun at the end of the eighth century and to have been particularly intense between approximately 950 and 1050; the centers of its dissemination seem to have been concentrated between the Loire and Rhine rivers. Its origins are believed to be twofold: on the one hand Germanic, represented by more extensive use of iron and an increase in the number of spring sowings, on the other hand Roman, including, for example, regular fallowing.

Large estates, which were more receptive to innovation than small farms, were naturally the first to be influenced. The great aristocratic, royal, and ecclesiastical holdings of the Carolingian age undoubtedly were experimental areas. Progress slowly spread from them piecemeal to the small farms and to less-favored regions. However, small "islands" and sometimes entire regions remained untouched by progress.

The existence of an agriculture carried on chiefly by manual labor, as is seen in so many medieval pictures, is probably indicative of both an overpopulation of the cultivated areas and a rich rather than a primitive type of agriculture. The problem of overpopulation on small manors in the ninth and tenth centuries was solved by working the fields with spades, which indicates an extremely careful cultivation whose yield was probably much higher than that of systems that at first glance would be considered more highly developed. In

Fig. 44. Peasant digging. Early eleventh century (Singer, p. 98, Fig. 66).

Fig. 45. Peasant digging. Early eleventh century (Singer, p. 98, Fig. 66).

any case it is certain that this gradual transformation of the technological conditions of agricultural production was to play an important role in the demographic, social, and political development of the entire medieval West.

The nature and mystery of the soil

It was undoubtedly in the increase of the lands under cultivation that the twelfth and thirteenth centuries made the greatest progress. Most of these new lands resulted from clearing, since the draining of swamps and reclamation from the sea were the fruit of work that was very limited by comparison with the enormous increase of arable land won from the forests. The uprooting of large trees, the clearing of underbrush, and removal of tree stumps were widely practiced with very old and simple techniques, until sturdier, more highly developed, and more widely used equipment was invented to sustain the efforts of the clearers of the land. We have already remarked this interdependence of techniques, and we shall see that the rebirth of the metallurgical industry exactly coincides with this period of conquest of new lands. The cinders left by the burning of the land acted as an excellent fertilizer for the soil.

The medieval peasants were completely capable of recognizing the good features and deficiencies of their land, as is amply demonstrated by the Anglo-Norman handbooks we have just mentioned. It was very soon realized that intensive cultivation exhausted the soil. Methods had to be found to return to the soil a fertility that disappeared quite rapidly, and these methods were of two types: crop rotation and fertilizing.

Crop rotation

The Roman farmers and the barbarian peasants had already realized that it was impossible to grow the same soil-exhausting crop on the same soil two years in succession. Some groups practiced fallowing when the density of population permitted; others invented biennial rotation, which often proved to be insufficient. Triennial rotation must have represented a great step forward: it made better use of the area cultivated, and permitted a better rest for the soil. Undoubtedly there will be much continued discussion of its origin, the date of its appearance, and the stages of its expansion, for documents are too few in number and too incomplete to resolve these questions with any degree of certainty. The Romans certainly practiced regular rotation, but biennially. A barbarian form of triennial rotation may have been adapted to this rhythm. Triennial rotation was based on the alternation of a winter grain, a spring grain, and a year of fallowing. Traces of it are found in Germany by the eighth century, and the *Polyptyque de Saint-Germain-des-Prés* (a kind of property register) seems to hint at its practice in the Paris basin. Gradually it completely penetrated all western Europe in the succeeding centuries — probably not without resistance, for in England, for example, it was not commonly practiced until the twelfth century. It may have been brought to Poland in the thirteenth century by German colonists, while in Russia it was seldom practiced before the end of the fifteenth or beginning of the sixteenth century. Thus it probably originated on the great estates of Austrasia — parts of eastern France, western Germany, and the Netherlands, with its capital at Metz.

Triennial rotation brought about great changes in the rural world. Cattle breeding profited greatly by this new system, on the one hand because of the possibility of pasturing animals on the fallow lands, on the other from the fact that oats was the best spring grain. The spread of the horse was thus linked in large part with the spread of the triennial system. The southern countries nevertheless continued for the most part to practice biennial rotation as it had been practiced in the late Roman period; in medieval Roussillon and Catalonia, for example, it was the only system of cultivation used.

In the northern regions, in contrast, triennial rotation was practically universal in the fourteenth century. It thus contributed to the definitive adoption of wheat and consequently the disappearance of millet, which had previously been more common because it did not exhaust the soil as much as wheat. The most widely grown spring grains were, first oats, then spring wheat, and lastly leguminous plants. Grapes, grasslands, flax, and hemp were not included in this system of rotation. At Gournay-sur-Aronde (Île-de-France), for example, by 1162 crop rotation was based on wheat, oats, and a fallow period.

Fertilizers

The problem of fertilizers was not solved at all in the Middle Ages, or was solved only in very incomplete fashion. This was unavoidable, in view of the absence of chemical knowledge. Many practices that today seem completely scientific undoubtedly were born of fortunate mistakes or a widespread empiricism. For example, when a flock of animals was unwell, there was a tendency to blame "poisoned" pasturelands. The fields were therefore limed to burn off the impurities, and since the animals were usually suffering precisely from a lack of lime, the problem was in this way solved.

There was no possibility, at least on a large scale, of utilizing manure (which was much in demand and very expensive) to fertilize land. Because of transportation difficulties the manure available in large cities could not be hauled into the country. Livestock was not yet very numerous, and in particular, no system of stalls existed; the animals lived out in the open on collective pasturelands that often consisted of uneven terrain or woods, where it was impossible to collect the manure. In some areas, nevertheless, the animals were tethered in order to enrich the fallow lands section by section. The maure was carefully collected from stables and cowsheds, and was gathered in heaps that were mixed with soil, grass clippings, and barnyard droppings.

The practice of liming had been known to the Romans, but seems to have disappeared at the time of the barbarian invasions. It may have been rediscovered and resumed under Charlemagne, at least if we can judge by the decree of Pîtres (864). It was generally done every fifteen years. In the coastal areas river mud or conchitic sand was also used. According to certain English texts, manuring appears to have sometimes been done by plowing under certain fodder crops before they had matured in order to bury them in the soil. Manure from pigeon coops was in most cases used on the meadows.

Thus the medieval peasants were able to distinguish from among the various fertilizers those that could give good results; they also knew in what quantities to use them and which lands needed them. It was known that marine algae were dangerous for certain types of soil, that sandy soil should not be limed or burned, that the benefits of manure lasted no more than two or three years, that manuring should only be done immediately before sowing. Experience had taught the medieval peasant, at the price of many disasters, some of the information that appears in our most modern handbooks.

Techniques of cultivation   It was in techniques of cultivation that changes were undoubtedly most apparent. Here again, the Mediterranean regions clung to the techniques of antiquity. In most cases there were three plowings, each of which had a specific purpose. The first one was the most important, for the stubble had to be turned over. The English agricultural treatises recommend that it not be done too deeply. The second was merely a surface plowing to eliminate the thistles and weeds. The third, which prepared the ground for sowing, should be only slightly deeper than the second plowing. In many cases the clods had to be broken up by hand after the first plowing (Luttrell Psalter); a large cylinder attached to a pole was also used. After the sowing, which was done broadcast, came the harrowing.

The activities that could be termed "maintenance" were few in number. The Roman authorities agreed neither on the advisability of weeding nor on the period when it should be done. The English agronomists, in contrast, recommended weeding. Some counseled that it be done in the spring so that the weeds would not have time to grow back; others advised the month of June. It was done, as can be seen in a miniature in the Luttrell Psalter, with a fork, which grasped the thistle, and a sickle attached to a pole, which cut it without damaging the ears of grain.

The harvesting of wheat and other grains was done in two stages. The top of the ear was first cut off with a sickle, after which the straw was mown — two ancient operations that had not changed. The ears were laid down in loose sheaves or heaped up in the fields. The wheat, when it was brought in, should be neither too dry (which would cause the grain to fall off) nor too damp (which would cause it to rot). A picture in the Luttrell Psalter depicts the two-wheeled cart used to haul the heads of grain and the sheaves into the barns.

Threshing methods were radically transformed. In the southern lands, as in those that had formed part of the civilizations of antiquity, the practice of

FIG. 46. Peasant winnowing. Around 1250 (Singer, p. 98, Fig. 65).

FIG. 47. Cutting off ears of wheat. Fourteenth century (Bibliothèque Nationale, Paris, fr. 9219.)

FIG. 48. Pruning canes of grape vines. Fourteenth century (Bibliothèque Nationale, Paris, fr. 9219.)

treading the grain had long been the only method used. The northern countries utilized the flail, an instrument whose origin is still obscure. In the South threshing was done on the ground in an open space, while in the North, where the climate was more inclement, it was done in the barn, where treading was difficult. It would certainly be possible to make a map showing the extension of these two methods, and it would be interesting to compare it with a modern map, but our sources are undoubtedly too scanty to permit the drawing of definite conclusions.

The techniques used in the cultivation of other crops remained almost unchanged. The grape was essentially a Mediterranean crop, but the difficulties of transportation caused it to be taken up farther north, following the same techniques that were used in the South. In the Bordeaux region the vines were planted in rows, or were combined with other crops which in the Champagne area were many and varied. In the Cerdagne arbors were used; in the Chartres area the vines were planted in rows but were later flattened out, which caused them to lose their alignment. Manure came into use in the vineyards of the Paris basin in the ninth century, in Lorraine in the tenth century. The number of plowings varied: three in Burgundy at the beginning of the eleventh century, two in the Paris basin and in Normandy. The peasant of the Bordeaux region sometimes plowed four times; the first one being done in May. Plowing was done with the hoe rather than the plow. The light plow used in vineyards did not appear in the Bordeaux region until the fifteenth century. Cutting was done at the end of the winter (February or March) with a *sarpa*, a tool with a large flat blade, used to cut moderately thick branches, or a *vidubium*, a crescent-shaped tool used for pruning trees; these tools had remained unchanged since Roman times. This cutting operation can be seen in numerous pictures. Removal of superfluous tendrils and leaves was done later in the year. Staking the vines was widely practiced. The stakes were pulled up in November, and new ones put in the following April, and the vines were attached to them with pieces of straw.

The other crops required less work. The meadows were mown; leguminous crops were generally cultivated by hand.

Tools   The question of the equipment used in medieval agriculture is practically insoluble, given the lack of exhaustive documentation. We must therefore limit our remarks to a brief outline, which nevertheless indicates that major transformation took place during the Middle Ages in this domain. It has often been stated that a distinction must be made between the southern countries, which utilized the ancient *aratrum*, a symmetrical instrument without wheels, and the North with its dissymmetrical, wheeled plow (*charrue*). The problems raised by the *aratrum*, which lightly scratches a shallow vegetal soil, found easy solutions, for the instrument changed little from Roman antiquity down to the nineteenth century. The problem of the dissymmetrical *charrue* with moldboard and colter is more complex. The *charrue* is probably of Celtic origin, and was known to the Romans, although they made little use of it. Between this version, of which very little is known, and the plow shown in numerous miniatures, there must have been various stages of evolution whose dates cannot easily be fixed.

ENGLISH AGRICULTURE IN THE MID-FOURTEENTH CENTURY

FIG. 49.   Hay wain with tandem harness. Around 1338 (Singer, p. 549, Fig. 500).

FIG. 50.   Plow without forecarriage. Around 1338 (Singer, p. 89, Fig. 54).

FIG. 51. The Sower. Around 1338 (Singer, p. 100, Fig. 68).

FIG. 52. Weeding. Around 1338 (Singer, p. 177, Fig. 144).

Some authorities believe that the *charrue* with its dissymmetrical plow-share was an outgrowth of the shallow scratching of the *aratrum*. The diffusion throughout Europe, in the first century of the Christian era, of the plowing instrument with wheels (a two-wheeled forecarriage can be adapted to an *aratrum*), which gave greater stability, favored the transition from the *aratrum* to the *charrue*. It was the combined presence of the colter, the dissymmetrical plow-share, and the moldboard that gave birth to the *charrue*. The action of this

FIG. 53. *Aratrum* with handle and sole. Before 830 (Library, Vienna, MS. 387).

FIG. 54. *Aratrum* with forecarriage. Tenth century (Haudricourt and Delamarre, *L'homme et la charrue*, p. 361).

FIG. 55. *Aratrum* with forecarriage. Tenth century (Haudricourt and Delamarre, p. 361).

FIG. 56. *Aratrum*, tooth type. Eleventh century (Pentateuch of Tours).

FIG. 57. Plow (*charrue*) with forecarriage. Eleventh century (Singer, p. 88, Fig. 52).

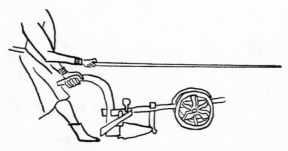

FIG. 58. *Charrue* with forecarriage. Twelfth century (Bibliothèque Nationale, Paris, lat. 15675).

new tool opened the ground to the air and water, and permitted the soluble elements that benefit plants to reach deep levels, while the roots of harmful plants were pulled out and died in the open air.

It is much more difficult to date the stages in the expansion of the new tool. Basing their judgment on linguistic factors, some historians place the period of popularization of the *charrue* and the stabilization of its vocabulary

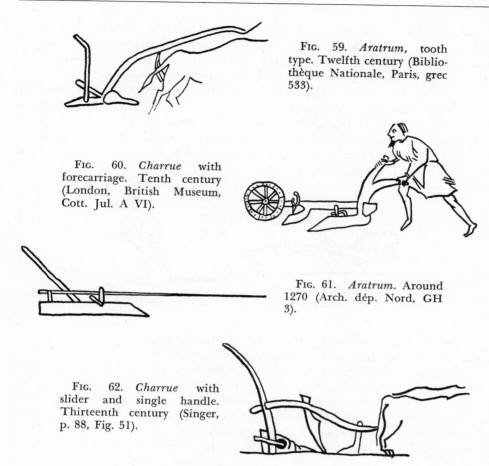

FIG. 59. *Aratrum,* tooth type. Twelfth century (Bibliothèque Nationale, Paris, grec 533).

FIG. 60. *Charrue* with forecarriage. Tenth century (London, British Museum, Cott. Jul. A VI).

FIG. 61. *Aratrum.* Around 1270 (Arch. dép. Nord, GH 3).

FIG. 62. *Charrue* with slider and single handle. Thirteenth century (Singer, p. 88, Fig. 51).

approximately between the sixth and tenth centuries (the Salic Law mentions only the *aratrum*, while the *De Villis* capitulary already speaks only of the *charrue*). The invention appears to have originated to the north of the Alps, on the eve or at the beginning of the Christian era. By the sixth century the *charrue* was already in general use in central Europe, and between the sixth and tenth centuries it spread through northwestern Europe.

The new instrument also had an effect on the conditions of cultivation. Only a heavy tool of this type could have made possible the exploitation of the cleared lands without the use of a large supply of manual labor. In addition, the fields could now be laid out in elongated strips, a form in which it was difficult to maneuver with the *aratrum*. In the tenth century, English, northern French, and Flemish miniatures began to depict instruments with a forecarriage, colter, plowshare, and two handles. The first plows with wheels appear in Germany in the fourteenth century; in the fifteenth century we find them as far as northern Italy. Instruments of the same type but with a single handle existed in eastern France, southern England, and, at the end of the thirteenth century, in Denmark (see the seal of Fros, South Jutland). Northwestern Europe had plows with runners beginning in the thirteenth century. The two-way plow with movable moldboard and symmetrical plowshares appears in a Belgian manuscript of the last quarter of the thirteenth century. Low forecarriages, which originated in eastern Europe, are the type most frequently seen.

In the earliest pictures the animals, principally bovids, were harnessed in pairs to the forecarriage. Modern harnessing with a single equid animal appears in the Bayeux tapestry (eleventh century); the harness is still the breast-strap type. The use of horses for plowing seems to have continued in the northern regions, insofar as we can be certain of the provenance of those manuscripts that have decipherable miniatures.

FIG. 63.   Harrow. End of the eleventh century (Bayeux Tapestry).

The cutting instruments utilized by medieval agriculture — the sickle, the *sarpa,* and so on — had undoubtedly remained unchanged since antiquity; the most we can presume, given the absence of precise texts, is that the steel used was of better quality. The flail, which originally consisted of a single flexible pole, became in the fourth century the instrument we know today: two pieces of wood connected by leather thongs. Threshing was done by the feet of animals or with rollers or wooden sledges, which had also been used in Roman antiquity; in the Grasse region in the thirteenth century, it was done by mares. The abbey of Saint-Jean-de-Sorde, in the Béarn, had regular labor crews to do the threshing with the help of mares. After threshing, the grain had to be washed.

Plants cultivated in the Middle Ages   For the most part the botanical history of medieval agriculture has yet to be written. While we know that most of the plants cultivated in the Middle Ages were known in Roman antiquity, we know almost nothing about the few new plants or, in particular, about the tremendous efforts made to improve the existing species. We have one piece of early testimony concerning these efforts. At the time of the Caliphate of Córdoba (tenth century), there existed botanical gardens and experimental seedling stations; a number of the seedlings came from the East. Although most of these plants were medicinal, they did nevertheless include other, more important, plants. Thierry d'Hirçon, an Artois farmer who lived in the first quarter of the fourteenth century, constantly recommends what seeds to choose in order to obtain the best harvests. A half-century earlier, the author of the *Fleta* had recommended variety in the sowing of grains. It was also known which species were best suited to the various types of soil. As for wheat, a distinction was made between the bearded type (especially red, and a white variety that is now extinct), which were best for light soils, the beardless varieties (red and white) for the heavy soils, and gray wheats for argillaceous soils. Most varieties of wheat were winter grains. As for the spring grains, they were chiefly oats (the so-called "black oats" variety), rye, a grain appreciated for its qualities of resistance, and barley. Oats, which had been little cultivated for grain in antiquity, had since then become a grain valued for the feeding of livestock.

Stability of the species under cultivation was thus characteristic of medieval architecture. However, some major developments did occur, although we have very little information on their extension. The first was the gradual disappearance of millet, the cultivation of which seems to be typical of primitive economies. The substitution of bread consumption for that of porridge brought about its disappearance, with the development of triennial crop rotation.

The appearance of new cultivated plants was rare; that of buckwheat is still surrounded with much mystery. Delisle's work fixed its date as no earlier than 1460, but more recently evidence of its existence in Germany around 1436 has been found. The implantation of buckwheat, which probably came from eastern Europe, must have occurred, if we follow the conclusion of certain historians, between the thirteenth and fifteenth centuries. In the middle of the fourteenth century it was still of minimal importance. The first reference to hops appears in a document of Pepin the Short, in 768; it may have followed in the wake of the great invasions, between the third and seventh centuries. However, for a long time it appears to have remained a minor crop used principally for flavoring beer; only later did it come into use for breadmaking. In northern Italy and southern France, in the twelfth and thirteenth centuries, appeared a plant called *surgum* or *suricum,* which appears to be sorghum; it is mentioned in Padua in 1191, at Fréjus in 1276. A few historians, who refuse to regard maize as an import from America, believe (without sufficient evidence) that this is the plant in question.

In the Middle Ages several crops characteristic of the warm countries became established on the Mediterranean shores. Rice was cultivated after the thirteenth century in Italy and in the southern portion of the Hungarian plain. Sugarcane appeared in 1420 in Spain, Madeira, and the Canary Islands. These crops, however, were very slow to establish themselves.

The historian is astonished by the limited number (ten at most) of vegetables cultivated in the medieval period. The vegetables cultivated for their roots — rampion, parsnip, caraway — disappeared with the arrival of the potato in succeeding centuries. Starches were supplied especially by such products as acorns and beechnuts, a swell as chestnuts, which continued for a long time to form part of human nourishment. Several new vegetables appeared in the Middle Ages: scallions are mentioned for the first time in the *De Villis,* and there were also less important newcomers — caraway, watercress, horseradish. The Arabs were cultivating spinach in Seville in the eleventh century, a plant that entered Europe at the time of the Crusades. Lettuce also appeared in the Middle Ages. As for the bean, historians do not agree. Some believe it was brought from America and spread during the sixteenth century, while others think that the white pea, which is mentioned in the Middle Ages and is also an Arabic importation, should be equated with the bean.

There appear to have been no innovations in fruit cultivation, although much attention was given to improving the grape. Philip the Bold's very detailed decree of 1395 is an example of attempts at improvement, and was undoubtedly the fruit of much patient research. The black grape of Burgundy seems to have been very widely cultivated in the West; exact information on this matter is unfortunately very meager.

The distribution of cultivated plants is still more difficult to study. It is evident that the difficulties of transportation inevitably required a mixed agriculture. Single-crop cultivation is of very recent date, and the few cases that did exist must be regarded as very exceptional. For example, at Collioure in the eastern Pyrenees area of France, by 1290 the cultivation of grapes had completely replaced that of grain, and the same was true at the end of the thirteenth century in the region of Auxerre, the reason being that the first-named place is next to a port of exportation, and the second is located on the banks of a river that brought the local produce to a large consumer market. The importance of the sowing of mixed grains has often been overlooked. Specimens of such mixed-grain sowing in Poland, dating from the period between the sixth to twelfth centuries, include rye, barley, wheat, leguminous plants, and many weeds. If one species did not succeed, the other might develop. This combining of crops was also important for the soil; it produced a straw stronger than the product that resulted when only a single species of grain was grown, and the plant was less susceptible to blight and to being beaten down.

Without going thoroughly into the problem, we should mention the delicate question of yield. In many cases we do not know to which lands the rare extant figures apply, and we should therefore be prudent in drawing conclusions. An English handbook of the thirteenth century lists the following yields: rye, seven for one; barley, eight for one; peas, six for one; wheat, five for one; oats, four for one. Some private accounts, which, however, apply to relatively rich lands, definitely surpass these estimates. In the cae of wheat, on his Roquetoire estate in the Artois in the period 1319–1327 Thierry d'Hirçon harvested 8.6 for one, at Gosnay in 1333–1336 12.9 for one.

The problem of yield was not the only problem faced by the medieval peasant. Good harvests also had to be stored as a hedge against possible failures in succeeding years, and this technique probably did not come into existence until the beginning of the nineteenth century. Grain stored in granges becomes hot, ferments, or rots. In certain cases famines can result not so much from a catastrophic decrease in the harvest as from lack of distribution of a very variable quantity of produce. Unfortunately, we lack exact information on the manner of preserving grain, which was for centuries a key problem of rural economies.

The forest   Nowadays it is difficult to imagine the role the forest, despite clearing and immense depredations played in the economic and technological life of the Middle Ages. When we consider that wood was the principal construction material, when we recall that wood was one of the few available fuels, when we learn that the forest was perhaps the largest pastureland, it becomes easier to realize its exceptional importance. This importance, moreover, involved more than just its immediate use; its determining role in hydrology was not a negligible one. Studies on the forest of Orléans have demonstrated that the decrease of forested area was accompanied, beginning in the Middle Ages, by a considerable diminuation in the water supply.

What was this forest, and of what trees did it consist? It is very probable that its composition had not changed since prehistoric times. During the period under discussion, practically no new species appeared, and the only arrival after the ninth century, the plane tree, is not a forest tree. The chestnut retreated considerably, disappearing from the Orléans region in the twelfth century. Other trees, which later came to be highly valued, appear not to have been sought after, such as the poplar, which was little known except in its forest version, the aspen.

Strictly speaking, the forest was not "maintained"; it was cleared and pushed back, and was not replanted. While a definite regrowth of forest land occurred in post-Carolingian France, this was due more to the abandonment of cultivated land than to a definite policy. The same phenomenon occurred in Normandy and England after the Conquest, when the feudal lords sought to increase their hunting lands. We know of only one example of artificial seeding — in 1285 in the Orléanais, if we are correctly interpreting the texts.

A technical study of the forest must therefore be concerned essentially with its exploitation. It is indeed regrettable that not a single recent work has been devoted to this important subject; existing studies are interesting but very much outdated. The forest had two uses: on the one hand its products were exploited much as we use them today, and on the other it was the object of rights of usufruct concerning the trees and the pasturing of animals. These rights authorized the removal of certain wood for a clearly defined purpose in fixed quantities. A distinction was made between "dead wood," which bore no fruit (yoke elm, aspen, alder, maple, elm, hazel) and "living wood." The quantities were determined either by the number and nature of the tools of exploitation or by the composition of the teams of animals that hauled the wood away.

The pasturing of animals was very widely practiced. Goats were very often prohibited, and the number of sheep sometimes limited. Pasturing of flocks during certain seasons of the year (chiefly spring) was soon forbidden, as well as their presence in copses.

In addition, the forest was a means of livelihood for a multitude of industries that utilized either the wood itself (cartwrights, carpenters), the bark of trees, or charcoal (metallurgists, tilemakers, brickmakers).

To the modern observer the methods of exploitation appear very rudimentary and irrational. The gathering of fallen and dead branches and the cleaning out of the underbrush were frequently done. The former permitted the development of the tree trunks, which were so badly needed for the construction of buildings, while the latter was indispensable to the pasturing of animals. The small pieces of wood supplied the faggots necessary for heating and the fascines so often used in improving roads and major highways and even the foundations of buildings. The only other type of exploitation known concerned full-grown trees. Reserving of young trees was practically unknown. Until the fourteenth century the felling of isolated trees was the principal method used. This meant that only the stunted trees were left standing. The sole purpose of sectioning, the existence of which is attested in the thirteenth century, was to reduce and gradually abolish the rights of usufruct. The same purpose was served by repurchases of forest rights, which were equally numerous. The royal ordinance of 1346 indicates, at least in the case of the forest areas owned by the French crown, a concern with regulating tree felling, with a view to a constant supply. The fellings, which were too close together in the copses, were on the contrary very far apart in the stands. Cutting back was the only way of restoring vigor to a section that had been badly exploited, stunted, or accidentally burned. Identifying marks very soon came into use for felling, and a special constabulary was formed to guard, if not to maintain, the royal forests.

The exploitation of the forest also posed a major problem: that of transporting the wood to the large consumer markets, and particularly fuel to the large cities. Carting was difficult and costly, but in some regions it was the only method available. Contrary to a deeply rooted tradition, floating was in fact practiced in the Middle Ages, in the forms of single logs and rafts. It is mentioned in the eleventh century in the region of Namur. In the twelfth century, Limoges was supplied with wood by way of the Vienne River.

By the thirteenth century all the forests were under exploitation, and there was not a single remaining example of virgin land. The forest was a technological reservoir of prime importance. However, it was gradually realized that it must not be decimated, and an attempt was made not so much to replenish it or improve its exploitation as to suppress or diminish abuses. It was difficult to suppress them all, and the rights of usufruct were tenacious, for the pasturing of the flocks and the gathering of nuts, firewood, and bark for ropemaking were too vital to existence.

These in brief constituted the various technical occupations of the medieval rural world. There were undoubtedly others that were related to the difficulty of obtaining certain industrial products. The birth of a rural craftsman class (which has yet to be studied, for our knowledge in this area is meager) showed that the peasant of the twelfth and thirteenth centuries turned his hand to very numerous and varied occupations. In any event he cannot have been an innovator, since very often those who exploit the earth are little inclined to be interested in technological progress. But the Middle Ages must have had its share of great landowners who were primarily interested in acquiring large revenues and were concerned with improving agricultural techniques. There were secular landowners, like Thierry d'Hirçon, who carefully supervised the exploitation of his lands, and especially great ecclesiastical landlords. The latter had the advantage of being able to know the results of their various experiments and to make large-scale application, if not of new then at least of improved techniques that had been perfected in their numerous monasteries. It was undoubtedly through these examples, which were eventually imitated in the rural areas where they had originated or had been applied, that agriculture was slowly transformed. Particularly careful study of the methods of exploitation used by such religious orders as the Cistercians, and an understanding of why these religious orders were in demand throughout Europe and how they were able to influence the surrounding population, should be devoted to this problem.

THE EXPLOITATION OF THE ANIMAL KINGDOM

The exploitation of the animal world had several objectives. The first was to supply power, chiefly for transportation and agricultural work. This was the principal use to which livestock was put. Equally important was the problem of food supply, whether for meat (the consumption of which was still limited) or for dairy products and their by-products. In addition, livestock furnished two essential products: wool and leather.

Two techniques helped to supply the medieval population with the animals needed: cattle raising, which was by far the most widely practiced, and the techniques of hunting and fishing, the latter being particularly important.

Cattle raising Little is known of cattle raising in the Middle Ages. While we possess fairly numerous treatises on hippology which present a certain interest, the extant sources concerning farm animals are much more meager. In any event no major innovations appeared in this period, for all the animals raised in western Europe between the tenth and fourteenth centuries had been known in antiquity.

There is no doubt that medieval man attempted to improve the breeds; we have numerous evidence of struggles against their extinction. The Englishman Walter de Henley paid particular attention to this problem. As regards horses, historians have stressed the importance of certain stud farms, which existed on a sufficiently large scale to permit the Abbey of Saint-Amand in Normandy to receive, in 1070, a tithe of mares. The abbeys themselves had similar establishments. The superiority of the Arabic horse was already widely known, and crossbreeding had been practiced. The Percheron and Boulogne breeds were being developed. We have less information about the other animals, but the widespread existence of the boar and the high price of certain rams indicates that the improvement of stock was carefully supervised. As for sheep, the Spanish and English breeds were already famous. But we are not able to pinpoint the birth and development of such breeds as the merino. Profiting by his rotations with the Moslem princes of North Africa, Charles the First of Anjou attempted to improve the sheep of southern Italy by bringing in Barbary sheep and rams (1278). From pictures we learn that the pigs of this period generally had pointed, straight ears, in contrast to the drooping ears of the modern pig.

Experience had taught the medieval breeders to supervise the development of their animals and, in particular, to eliminate the frequent diseases. An English treatise of the thirteenth century recommends that the breeder always try to discover why an animal has died. Remedies, however, were completely empirical, as, for example, the above-mentioned liming of the meadows, which compensated for certain deficiencies. Liver ailments in the bovine animals and in sheep were in most cases attributed to a plant that grows in damp soil; actually they resulted from a worm whose larvae multiply in a certain freshwater mollusk. The English *Fleta* (thirteenth century) recognizes that this illness can attack animals that go into unhealthy pasturelands, where they swallow either a mist or "white turtles" — the latter being apparently a reference to these mollusks.

The chief problem was the feeding of a sometimes large number of animals.

There is no doubt that most of the medieval livestock was of very inferior quality, for the reason that for a great part of the year it was difficult to feed the animals. The meadows were not very large. Certain pasturelands — those of Normandy, for example — were already valued for the production of the most handsome animals, and the "salt meadows" were famous by the eleventh century. The two large pasturelands most frequently used were fallow lands and the forests.

The fallow lands were used principally for pasturing sheep and bovine animals. At the end of the summer the aftergrowths of the mown meadows, the stubble fields, and the stubble of the plowed lands were utilized, unhealthy areas being carefully avoided. In the fall the sheep were led out only after the sun had risen and the hoar frost had disappeared, in order to avoid scabies; lowlands and damp places were avoided to combat flukeworm. The forests were the principal grazing areas for the pigs; goats and sheep were usually excluded in order to protect the young shoots. The pig was undoubtedly the most common animal, and was in any event the animal whose meat formed the basis of popular meat consumption. These pigs were long and had flat flanks, with very hard bones; they were omnivorous, but preferred acorns and the various forest products.

A supplementary food supply was absolutely necessary, especially during the winter, which most of the animals undoubtedly spent in a state of semi-hibernation. For this reason the old cows, useless oxen, and toothless sheep were killed in the fall, especially since the general scarcity of grasslands throughout western Europe did not permit mowing and storage of hay. The horses were fed oats and vetches; pigs destined for fattening were fed coarse barley, peas, beans, and skim milk. Sheep were kept in the stable from the feast of St. Martin (November 11th) until Easter, and were fed hay and stalks of bean plants.

Mention should be made of the practice of transhumance, the seasonal migration of livestock and the people who tend them between the lowlands and the adjacent mountains. This practice raises questions that are still far from solution. It is difficult to know whether transhumance was already in practice during the Late Roman Empire. Many historians believe that it was begun by monks in the Alps, some attempting to find traces of it in the Testament of Abbon in 739, and then in documents dating from the first quarter of the eleventh century. Migrations are specifically mentioned at the beginning of the twelfth century for the Abbey of Bonnevaux (1122); the clarity and precision of this charter indicate that the technique was already very ancient. It may have also been in practice in Auvergne by 1292. In the Pyranees it seems to have appeared at a later period and to have combined transhumance toward the plains and toward the mountains. It appeared in Italy around the same time, where the shepherds of Apulia moved up to the Abruzzi, joining those coming from the Papal States. Several decrees of King Roger II (1101–1154) deal with this practice.

The custom of stabling the animals was very poorly organized, and the equipment was primitive and dirty. Although a few rare specimens of thirteenth-century granges, especially in the Cistercian abbeys, are still in existence, we have no medieval stables, which must have been built of wood.

Barnyard animals were very numerous, at least if we can judge by the number of feudal taxes concerning them. We have very little information about the techniques of their breeding, but we can assume that fowl was raised in quantity both for eggs and meat. Geese are also mentioned frequently, while domesticated ducks, in contrast, were less numerous.

Cattle raising in the Middle Ages thus appears to have been quite difficult, chiefly because of the feeding problem, for which a solution was found only much later. Moreover, these technical difficulties indicate an organization that was still primitive, the chief consequence of which was a clearly insufficient consumption of meat on the part of the populace, as well as limited transportation.

Beekeeping was widely practiced. On one of the capitals in the cathedral of Vézelay (1120–1138) we see small beehives, shaped like very pointed cones and made of closely twisted bunches of straw placed one on top of the other and bound with brambles or honeysuckle vines. The same type of beehive is seen on a capital from Cluny (around 1150). The English *Housebondrie* (fourteenth century) speaks of hives furnishing up to four swarms a year. Such frequent swarming indicates that they must have been very small hives. The hives were emptied only once every two years when the swarm was killed by smoking so that the honey could be collected. Swarms in the forests were also exploited, but it was forbidden to fell a tree in order to get the honey.

Hunting

Hunting was chiefly a sport for a certain class of medieval society; it also furnished a supplement to the medieval diet. When he was preparing to go on a Crusade, Alphonse of Poitier had two thousand wild boars hunted down so that he could have them salted and included in his supplies.

Though some of the techniques of hunting were inherited from classical antiquity, others were unquestionably barbarian in origin. The former included the use of dogs (the future riding to hounds), in which packs of dogs forced the animal to run. Almost nothing is known about the origin of the dogs utilized for this purpose. A bas-relief from Herculanum and a marble in the Vatican show a dog with drooping ears, the model of the hunting dog and perhaps the ancestor of the Italian hound. The Gauls probably had such breeds also, since they too hunted in this fashion. The Gallo-Roman period was a brilliant age of hunting to hounds: Sidoine Apollinaire makes mention of it in the fifth century. The Norsemen contributed the Norse breed, later brought to England by William the Conqueror. The most common dogs were the hound with long, drooping ears and fine snout that flushed out the animal, and the "boarhound," which pursued it. The hunted animal was generally dispatched with the sword, if we can judge by the miniatures.

Treatises on venery proliferated in the thirteenth century; there were, for example, *La chace dou Serf,* by Guy de Châtillon and Henri de Vergy, and *L'art de vénerie,* by Guillaume Twici. The very successful *Les chasses du roi Modus* appeared at the beginning of the fourteenth century.

Shooting was, naturally, infrequently practiced. It was done with the bow, and later with the crossbow, which continued to be used even after the appearance of firearms. There is not much to be said about this technique, which differed very little from modern hunting except that it must have been much less profitable. In *Les chasses du roi Modus,* we see the hunter, installed on a small platform, watching the wild boar, which has just sprawled in the mud.

Falconry, which had already been practiced by the Gallo-Romans, was restricted to the upper classes. There appear to have been several schools; for example, the treatise of Frederick II belongs to a Sicilian and Eastern tradition modeled after that of the twelfth-century Moslems Mohamin and Chatrif, while Daudé of Prades, Canon of Maguelonne, represents an Anglo-Norman tradition. Falconry was used only for birds, and was not done only with the falcon: there were birds of swift, powerful flight (falcon, lanner, saker, hobby), and birds chiefly noted for endurance (goshawk, gerfalcon, sparrow hawk, merlin), each of which was used for a specific type of hunting. All were caught in traps and tamed. A good falcon had to have large shoulders, wings which were not too long, and long, slender talons. Their training was relatively long, and required exceptional qualities on the part of the trainer.

Hunting was also practiced with numerous devices restricted to the peasants, and consequently scorned by most of the treatises on hunting. However, *Les chasses du roi Modus* contains much information about them, usually accompanied by miniatures. There were innumerable kinds of traps, most of which probably dated from prehistoric times. They were generally based on the principle of the spring (such as an enormous trap for catching deer), or consisted of skillfully placed pits. The largest group of these devices, however, was represented by nets of all sizes and shapes, corresponding to the instincts and sizes of the animals to be caught. There were small nets, purse nets to be placed over the entrances to a rabbit burrow (shown in the Luttrell Psalter), and, for catching birds in trees, pocket nets, suspended from poles, which could be pulled shut by means of a rope. Large nets hung on posts hammered into the ground were used to catch large animals or flocks of birds. Most of these nets, which were woven of hemp or flax, were coarse so that they could not easily be seen. Possibly the net itself was not as important as the art of placing it in a good location.

A decree made by Philippe le Long in 1318 forbade the use of snares for rabbits and hares. The statutes of Vercelli in Italy forbade hunting partridge and pheasants with nets and dogs.

As for the animals hunted, the deer was not valued very highly, and the hunting of pheasant did not become common until the thirteenth century. On the other hand, some animals that are no longer hunted today, for instance the heron, which was specially bred for that purpose, were hunted in the Middle Ages.

Fishing

The consumption of fish served a dual purpose. First, there was the obligation of fulfilling religious requirements, which were strictly observed. In addition, in the absence of the consumption of meat, fish was one of the few foods that could combat certain deficiencies. In any case, the importance of fish in medieval life is a fact that has frequently been emphasized.

Early in the Middle Ages the practice of ocean fishing spread over a wide

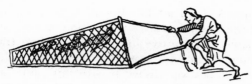

FIG. 64. Net for capturing birds on the ground. Early fourteenth century (Trevelyan, . . . *English Social History,* Fig. 31).

FIG. 66. Net for capturing rabbits. Early fourteenth century (Trevelyan, Fig. 32).

FIG. 67. Fishing with nets. Early fourteenth century (Trevelyan, Fig. 30).

FIG. 65. Net for capturing birds in the trees. Early fourteenth century (Trevelyan, Fig. 33).

FIG. 68. Fishing with nets. Fourteenth century.

area. Almost every variety of fish was caught, and some became very popular, chiefly because of their abundance and the ease with which they could be preserved. Included in this group was the herring, which had been unknown in classical antiquity. It was caught with large, vertical, rectangular nets ranging in length from 65½ feet to 164 feet and 33 feet high, to which floats were attached. The nets were placed in the water at night and were gathered in in the morning; the fish were caught in these nets by their gills. In season the herring were caught near the surface. As soon as they began to disappear, the fishermen were permitted (for instance, by act of a count of Boulogne in 952) to use a large pouch known as a dragnet, a primitive form of the modern trawl, which was dragged along the bottom. This net was also used .for catching quantities of other, chiefly flat, fish. Sardines were fished by stretching fine mesh nets along the bottom, as is shown by a law of Charles II of Provence dated 1298. The seine, a triangular net, was used for catching porpoise. Cod and tuna were fished with primitive, rigid lines. Red tuna was caught in madragues (walls of nets attached to posts), as had been done in antiquity. This technique had been abandoned by the Arabs, who did not like tuna, but was soon (in the thirteenth century) reestablished by the Normans in Sicily and by the Spaniards after their victory over the Moors. Porpoises were caught in Brittany by the same method. Sharks and seals were hunted with harpoons.

FIG. 69. A whale hunt. Four-teenth century (seal of Fontarabie).

FIG. 70. Fish pots in a stream. Around 1338 (London, British Museum, Add. MS. 42130).

The whale was hunted in the Atlantic, and especially by the Basques, beginning in the seventh century. These animals are particularly fond of sun and light, and therefore traveled south, drawn to the Gulf of Gascony; the females made use of the natural indentations in the coastline to bear their young. Whales played on the sand and often ran aground on the beach at Anglet, not far from Bayonne. At first the fishermen were probably satisfied simply to make use of the animals caught on the sand; in 1315 Edward II reserved the products of such catch for his own use. Gradually attempts were made to catch them in the sea around the Honce, at what was then the mouth of the Adour River. Towers were built on the coast, and watchmen reported the presence of the whales. The slender lines of the whaling longboat can be seen on the seals of Biarritz (1351) and Fontarabie (1335); all the oarsmen rowed on the side opposite the animal. The harpoon, thrust in by a vigorous arm, caused the whale to lose blood; the animal struggled, dragging the boat, whose weight fatigued it. Its flippers and tail were one by one immobilized with lances. Some authors believe (without satisfactory proof, however) that the harpoon was hurled by catapults; its rope sometimes had a float at the end, so that the animal would not be lost from view. The whale was towed to shore and cut up, and the pieces were heated to draw out the oil. In the twelfth century whalebone was used to make luxurious plumes for helmets. Genuine whale hunting probably did not begin before the tenth century; the first indications of feudal taxes on this activity date from 1059.

The movements and timing of the fish were soon learned. From May to July the herring lay off the Orkneys and Shetlands and the coasts of Norway. It then traveled south, arriving in January at the mouth of the Seine. Variations in temperature and salinity could cause it to disappear. After the ninth century we find an improbable abundance of this fish in the Baltic and the North Sea. It was preserved by smoking with beechwood and oakwood; it could also be salted. Sour herring was practically incorruptible.

Cod had not been known to the Romans. In the Middle Ages it was still abundant in the Atlantic, the English Channel, the North Sea, and along the coasts of England, the Netherlands, and Germany. In the ninth century it began to be fished off the coasts of Norway, in the twelfth century off Scotland. It was split, salted, and stored in casks. In Flanders in 1143, dried codfish and haddock were also preserved.

Beginning in the tenth century, the sardine was fished in the Atlantic waters; few texts mention it in the Mediterranean, where it had been very abundant in antiquity. Mackerel, hake, coalfish, and turbot were also dried, as well as the conger eel and tuna (Spain, 1133). An attempt was also made to extract oil from all these fish, even from herring; mention is made of this in Brandenburg in the twelfth century. Sharks and seals were caught for their oil in the sixth and seventh centuries in England; rope was made from their hide, while their fur decorated the borders of tenth-century garments.

Shellfish do not appear to have been highly valued in the Middle Ages. Oysters, which were much appreciated by the Romans, are mentioned only at the very end of the eleventh century (1098). Mussels may have been accidentally discovered, either in 1045 or 1235, on posts holding the nets used to capture birds.

Sea fish could be transported only when dried or salted and stored in vats. For this reason river fishing was equally essential and equally practiced. Dams of various shapes were built, and pots were used, as well as various types of stationary nets, including the *guideau* or *dideau,* a bag-shaped, conical net with a large opening, placed near mills or arches of a bridge; the stake net, stretched between two islands; the hoop net, installed along the banks of the river in the spaces between clamps of vegetation; and *tramails* (or *entremailles*), consisting of a fine-mesh net placed between two coarse nets. There were nets with names that no longer mean anything to us; there were seines (still in use), cast nets, dipping nets, gill nets, and small shrimping nets.

A fairly detailed body of laws concerning fishing was gradually developed, which indicates that the importance of this activity was perfectly understood. Some of these laws regulated the seasons for fishing; for example, in Béarn in 1279 it was forbidden to fish for salmon at spawning time; in Vence in 1303 the same law was made for trout. An ordinance of 1289 forbade fishing for roach in April and May. Other articles dealt with fishing equipment. An ordinance of Philippe le Bel fixed two sizes of mesh for nets: as large as a large silver Tours franc from Easter to the feast of St. Remi (January 19), as large as a Parisian *denier* the rest of the year. The statutes of Vence of 1303 forbade the use of the *entremaille;* the ordinance of 1289 forbade the use of the *guideau* in April and May.

Until the eleventh century the salmon and the eel seem to have been the principal fish sought, although beginning in the tenth century the sturgeon, trout, pike, and lamprey are also mentioned at Marmoutier. In the tenth century salmon was fished with a line and live flies in Scandinavia, while in Scotland around 1214 the rivers were dammed with osier fences to trap this fish. As in antiquity, the eel was highly valued; it was dried or salted. Lampreys and Alpine lake whitefish were taken with a net. Pike were stored in special fishponds because of their voracity. Carp may have been introduced into Germany from the East around the eleventh century, and may not have reached England until the sixteenth century; this fish was especially valued because it multiplied rapidly in the fishponds.

How was all this fish preserved? This question is easier to answer. Large establishments and manors had their own fishponds, which were regularly stocked. While it appears difficult to speak of actual pisciculture, restocking was definitely done. The fry were on occasion reserved when the ponds were emptied, according to the accounts of Thibaut of Champagne for the years 1258 and 1259. The *Fleta* advises stocking with carp and perch, and recommends the exclusion of such carnivorous fish as pike, tenches, and eels — which indicates a certain distrust of these species.

This information should certainly be clarified, for the important role played by fish in the Middle Ages merits a study that would include its social, economical, and technical aspects. The tremendous importance of salt in medieval life is a direct consequence of medieval dietary habits.

THE FOOD SUPPLY

Little study has been made of the medieval food supply, although sources on this subject appear to be relatively abundant. Undoubtedly the medieval diet was to a great extent vegetarian, and this may be one of the reasons for the general weakness of the population. We shall later see what means were used to compensate for the obvious lack of animal proteins, the price paid by less evolved civilizations.

Being a continent of countries with well-developed techniques of cultivation, at least in its most westerly region, medieval Europe theoretically was no longer one of the food-gathering civilizations. While the rule of St. Chordegand, Bishop of Metz at the end of the seventh century, still mentions acorns and beechnuts as part of human nourishment, after this they are no longer mentioned except as a food generally restricted to animals; only nuts and chestnuts retained their dietary role in human nourishment. Hempseed soups existed, and some wild foods were gathered — raspberries, strawberries, leeks, asparagus, and even the hop; only later were these plants cultivated. As we have seen, only a few vegetables in limited quantity were available as a substitute for the potato.

The method of preparing foods is indicative of the level reached in the development of human nourishment. The primitive method was soup — either sweet, which led to the porridges and then to bread, or acid. The consumption of acid soups is very ancient, especially in the plains of eastern Europe. However, we are not sure whether certain foods, such as sauerkraut, are medieval inventions or the heritage of the cuisine of the Germanic peoples. The preparation of alcoholic beverages will be considered later.

Porridge and bread   Porridge was definitely the principal culinary preparation of the medieval population: broiled food, which still existed, was beginning to disappear. Porridge was itself being transformed by the disappearance of certain plants like millet, various species of maize, and certain leguminous plants, and their replacement by the grains — barley, rye, oats, and, to a lesser extent, wheat. Wheat porridges were limited to the wealthiest classes, as is seen at Saint-Deuis in 862. From porridge the transition was easily made to the flat cakes that were so common in the Middle Ages; they were cooked on the iron grills that had been in existence since classical antiquity. The next step after the flat cake was bread.

According to recent writers, medieval breadmaking was the heir, not of Roman techniques, but of barbarian practices. The cereals that can be made into bread (we could even use the word "plants," flours being composite substances) are numerous. Wheat appears to have been utilized more rapidly in some regions (for example, England) than in others. Elsewhere, oats and barley were used; they produced a black bread that was the basis of the common people's diet. The leavens used were exclusively beer leavens. The ovens for bread baking were domed, like the Roman ovens, but products that were later to become pastry were still being baked under a layer of ashes, as is mentioned in a text of Raimbold, Abbot of Saint-Thierry near Rheims (died 1084). In Old German the word for bread does not become differentiated from other culinary terms until the tenth century.

Alcoholic beverages   Alcoholic beverages had long been in existence. Wine continued to be the most important of these; its techniques of production had changed very little. Storage in wooden vats was the only innovation, but this practice dated from the end of the Roman Empire. Beer had been known since Egyptian antiquity, and it was disseminated by the Romans after their conquest of western Europe. It was made with all kinds of grains: barley (Artois, sixth century), wheat (Auvergne), and elsewhere spelt (ninth century). According to Gregory of Tours, the grain was first roasted, then thrown into boiling water to obtain a liquid similar to modern coffee. Later, oats were utilized, and finally hops, which had the advantage of giving a pleasant flavor to the brew. Cider and perry were known at least after the fifth century, but did not spread very quickly. References to cider do not become numerous until the twelfth century, at which time it reached England.

Fats and oils   Fats, so necessary in human nutrition, came from two sources: animal (from slaughtered animals and dairy products) and vegetable products. In southern France olive oil was preferred, but other oils — colza, poppyseed, flaxseed — are mentioned in texts of the twelfth and thirteenth centuries. The same oil was also used for lighting.

Among the fats of animal origin, butter and lard were the most frequently used. We do not know whether the butter, which was generally salted, was made with cream or whole milk, nor do we know how the cream was skimmed off.

Meat and fish The deficiencies of a diet based on purely vegetable elements supplemented with fats were aggravated by the fact that medieval agriculture had nowhere near the variety of plants that are part of the modern diet. Combating these deficiencies was difficult. Meat production was limited: only the pig appears to have supplied meat on a fairly large scale. The salvation of the medieval world was probably fish, which was kept in fishponds and thus could easily be caught for consumption. These fishponds were installed in the rivers themselves or along the riverbanks, and were fed by canals. Charles Du Cange also mentions special boats that were used for the purpose. Portable reservoirs for transporting fish for great distances are mentioned in 1239, but this must have been an exceptional step taken at the behest of some powerful individual.

Meat and fish could not be preserved indefinitely. They were therefore salted, a method that had been known in antiquity and which was widely practiced in the Middle Ages. Salted herring was sold throughout western Europe in considerable quantities. Animals, especially the pig, were prepared in the same way.

Thus the medieval populace was, if not undernourished (which may have been the case), at least badly nourished. However, more thorough research is needed to permit definite conclusons.

THE EXPLOITATION OF THE SUBSOIL

Because of the lack of documentation, our knowledge of the conditions of mining and the exploitation of the subsoil is meager. A large number of texts on the subject would be needed to compensate for the total absence of pictures. Certain sources mention inventions of which it is impossible to give a technical description as for instance the artesian well drilled for the monastery of Lillers in 1126.

Mining had reached a relatively high level in the Roman period, and undoubtedly declined slightly at the time of the great invasions. We know of numerous mines that were exploited in antiquity but that, although not exhausted, were never again mined. In the Merovingian and Carolingian periods probably only a few iron or tin mines, which posed no overwhelming technical problems since they were open-pit, continued to be exploited. In many cases, moreover, as late as the eighteenth century iron was extracted from open-pit alluvial placers by turning over a shallow layer of loam. Underground iron mines probably continued for a long period of time to be the exception rather than the rule.

Iron mines One of the oldest extant documents on underground iron mining is a text concerning the surface mines of the upper valley of the Breda, taken from the second book of miracles of Peter the Venerable, Prior of Domène in the Dauphiné, around 1120. From it we learn that cave-ins were already frequent occurrences. This industry required an audacity and tenacity that appear to have amazed our prior. Unfortunately, there exists no later document that would enable us to learn exactly which iron-mining practices were used in this region. We know, however, that at the end of the thirteenth century the *seigneur* was obliged to supply the wood required for shoring up the shafts, a practice that because of the lack of wood was rarely followed by the Romans. Timbering was therefore probably a common practice during, but not an invention of, this period. Subterranean mining also appears to have been practiced in the forests of Othe in Champagne. In both of these cases, as well as at Bonivente in the Dauphiné in 1284, lighting was done with candles; the rock was dug out with iron picks, and the ore was brought up to the surface in sacks or baskets hoisted by means of ropes and winches.

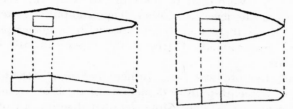

Fig. 71.   Miners' picks, from Massa. Thirteenth century (Simonin, "De l'exploitation des mines et de la métallurgie en Toscane," in *Annales des Mines,* 1858, p. 615).

Lead and copper mines   The lead and copper mines in the region of Massa in Italy are better known to us, thanks to the survival of a very detailed mining code that dates from around the beginning of the thirteenth century. In addition, the mines, which were undoubtedly abandoned around the middle of the fourteenth century because of the great plague, remained undisturbed until their exploration and excavation in the middle of the nineteenth century, and therefore furnish us with extremely important archaeological evidence.

At Massa the shafts were numerous and closely spaced, an arrangement that facilitated ventilation. The veins, which were generally found in layers in quartz rock, were mined by means of vertical shafts; at various levels the miners penetrated the deposits, which were mined in large slices or by means of pillars and levels. There was no water, and therefore no need of drainage channels or levels leading outside the mine. The diameter of the shafts, which were round and perfectly vertical, varied from 3½ to 4 feet; and some of them were walled to resist the thrust of the earth. Crosscuts, in many cases very long, were driven across the veins, and their layout indicates that the miners had very definite ideas on the extent of the pay streaks both as to direction and depth. The crosscuts served as exits; their width varied between 5¼ and 5½ feet, and their height sometimes reached 6½ feet. The slopes are at the correct angle to ensure good drainage if necessary, and are very straight; the mining levels, in contrast, are much less carefully laid out. The veins were attacked by underhand or overhand stopes, and waste rock material was used as filling after the ore had been removed.

The statutes of Massa tell us that the classical instruments, such as levels and surveyors' squares, were used to lay out the level; compasses also appear to have been used. The rock was broken with picks. Specimens of these picks have been found: elegantly shaped, tipped with good-quality steel, and of various sizes, probably depending on the type of rock for which they were intended. Levers, sledgehammers, and wedges were also used. Thre are indications that fire was often utilized; before departing in the evening, the miners heaped faggots at the bottom of the levels and set fire to them. In this way the rock was broken up and was easier to work. Fragments of timbering have been found; they are reminiscent of those utilized in the nineteenth century.

Inside the mine, material was generally carried on the backs of the miners, in sacks made of buffalo hide (fragments of these sacks have been found). Windlasses and hemp ropes were used to hoist the sacks up the vertical shafts and to hoist the ore from one level to another. Rope-and-pulley systems also existed. In the horizontal drifts there were large ropes with reversing pulleys that were probably used to drag baskets. Traces of the ropes can still be seen on the walls of these drifts.

Aboveground, the ore was first crushed and then carefully hand-picked. The fifteenth- and sixteenth-century depictions of these operations are probably valid for earlier centuries as well. More detailed descriptions of these methods will be given later in this work.

Coal

A number of historians have fixed the first discovery of coal in the region of Liège at the end of the twelfth century (in 1195, to be exact). It seems to have been found in England in the ninth century, in the Zwickau area of Germany in the tenth century. A charter of the Priory of Saint-Sauveur-en-Rue, in the Forez district, mentions coal in 1095. It is possible that surface deposits were mined without digging shafts, and the extraction of the new combustible definitely began to be developed in the twelfth and thirteenth centuries, at the time of the great flowering of medieval commerce. At the beginning of the thirteenth century references to levels appear in a written document concerning the mines of Boussagues in the province of Languedoc (1206): it authorizes two parties to excavate one level by extending it into another. Between the beginning of the thirteenth and the middle of the fourteenth centuries, numerous levels were opened throughout this region, followed by those of the Loire River basin.

Certain mining problems were difficult to solve. We have some (not very precise) knowledge of them from the mining codes mentioned above in connection with the mines of Massa, the codes of Iglau approved by Wenceslas in 1248 or 1253, and the codes of Allevard (1395) and Vicdessos (1414), which simply confirmed previously granted rights. The deepening of the mines involved increasingly complicated operations. In shoring up the levels, timbering methods were used that undoubtedly remained almost unchanged until the eighteenth century. The water-drainage problem was much more complex, in addition to being one of the most dreaded. In Bohemia at the end of the thirteenth century, and slightly later in Saxony, the Harz region, and southern Bavaria, hydraulic or treadmill machines for draining the water were installed. These probably consisted of bucket belts similar to the norias of the Mediterranean basin and to certain sixteenth-century devices that will be discussed later.

Salt mines   The search for sources of salt water was also part of mining. Rather narrow square or circular shafts were dug down to bedrock to collect the salt water and protect it from the ever-present danger of infiltrations of fresh water. Those of Salins were highly developed; they were considered quite old as early as 1409. Archaeologists disagree in dating them: some claim that they date from Roman times, others that they were probably dug between the tenth and thirteenth centuries. The levels were now spacious vaulted caverns; those at Amont had barrel vaults 33¾ feet high from floor to keystone. The rock shaft with its groin vault was 36 feet high, 174 feet long, and 52½ feet wide; the shaft for the salt water had several floors with ogive vaults. The water from the various springs trickled through channels separated by low clay walls, and was collected in fir-wood vats. The water was then brought to the surface in a *griau,* a large wooden bucket suspended from an enormous beam supported by a wooden column that was forked at the top, and was poured into a trough.

Quarries were numerous, and were exploited with ancient techniques. The stones were sawed, as appears in a miniature of the manuscript of Hrabanus Maurus of Monte Cassino, dated 1023. Slate quarries are also mentioned in the twelfth century, without details of their exploitation. In any case, the monuments of this period show that the stones were correctly cut, which indicates that the art of quarrying beds of stone was well known.

FIG. 72. Sawing marble. Around 1023 (Singer, p. 385, Fig. 348).

CHAPTER 19

THE TRANSFORMATION OF RAW MATERIALS

Only a portion of what was directly utilizable was obtained through the techniques of acquisition; other techniques transformed the raw material into objects suitable for consumption. Unfortunately, our information in this area is still very meager. We are no longer dealing with the exploitaiton of nature, which gave rise to numerous contracts and to an extremely well-developed and detailed body of law, but with human labor, for which written documents are few in number, and more substantial traces have for the most part completely disappeared.

The modification made in these techniques of transformation since the end of the Roman Empire were probably much more numerous than developments in the techniques of acquisition. Here it was no longer nature that imposed conditions and determined to a great extent man's activity and methods. It was man who confronted matter and attempted to shape it to his needs. Thus he himself completely determined his work, methods, and tools. Progress lay in his mind and at the tips of his fingers.

The techniques of transformation comprised three basic groups: thermal techniques, chemical processes, and mechanical operations. We shall consider each group separately.

THERMAL TECHNIQUES

A careful study of thermal techniques would have to include elements of which we have but fragmentary knowledge. It is evident that because of their lack of the indispensable scientific instruments the medieval artisans knew nothing of the temperatures they had to obtain for certain processes. The same was often true of the structure of the furnaces and especially of the materials used in their construction. Archaeological investigations and a more systematic examination of texts would perhaps make it possible to obtain more complete information on this subject.

In the great majority of cases wood was the only combustible used in the Middle Ages. Peat may also have come into use at a fairly early period. Definite proof exists for its use in Normandy, where it was obtained from the Troarn swamp, by the end of the eleventh century. The use of coal as a combustible began concurrently with its extraction from the earth. Many impossible dates have been argued for the first use of this source of heat. A charter of 852 shows that the monks of Medhamstead, in England, received rents in the form of coal and peat, both of which could only have been used as combustibles. In any case the use of coal definitely appears to have spread at the end of the eleventh century and especially in the twelfth. The tremendous land-clearing activities of this period had diminished the forest areas available for easy exploitation, as witnessed around 1260 by Albert the Great.

The battle against the use of coal continued until the middle of the eighteenth century. Some of its opponents were afraid to use it for domestic heating; others claimed that its use in industry resulted in faulty products. If the Provost of Arles forbade (in 1306) smiths to use coal in their forges, it was undoubtedly because the forges utilized lignites rich in sulfur or phosphorous, both of which caused iron to become brittle. The "impurities" of certain types of coal hindered the development of this combustible for a long period of time.

Glass   Glassmaking, like almost all the industries dependent on heat, long remained a secret, if not mysterious or even magical, process. This industry had reached Gaul around the first century A.D. thanks undoubtedly to several immigrants from the East, who had obtained a quasi-monopoly of its production. A funerary stele of Lyon contains the first extant reference to glassmaking. By the third and fourth centuries it was already in progress along the banks of the Rhine, while northern France and Belgium were by now major producers. The installation of the industry in France coincided with the invention of the blowing iron. Glass objects from the fifth or early sixth centuries discovered in Lorraine were made with Mediterranean natron, and still show traces of complex but definite Asian influences. In the Merovingian and Carolingian periods there was a serious decline in glassmaking, and from the eighth to the tenth centuries references to it are very rare. However, its existence appears definitly proved in the Roman era and slightly later in the Merovingian period. One text indicates that an English abbot had manufacturers of plate glass brought over from Gaul at the end of the seventh century.

At the beginning of the twelfth century Theophilus Presbyter regarded the French as masters in the art of glassmaking. According to a rather uncertain tradition, in the eleventh century a group of Norman glassmakers emigrated to Genoa and founded an industrial center which became prosperous in a very few years.

White glass was obtained by melting two parts of beechwood ash with one part of washed sand. In the case of colored glass, Theophilus is less precise; it seems that its production was purely accidental. For green, blue, and violet glass, materials were used about which we have little information. For hollow glasses, the techniques and successive operations of blowing and reheating are reminiscent of modern methods of artistic glasswork. The glassmaker mixed his glass, heated it, and blew it. He then detached this glass "bag" from the blowing iron and corrected it by attaching the blower to the bottom of the vase. The object was then finished and decorated when required. For long-necked objects, the worker waved the blower around his head. A miniature

FIG. 73. Making glass. Around 1023 (Singer, p. 329, Fig. 309).

in the manuscript of Hrabanus Maurus (beginning of the eleventh century) shows a glassmaker's furnace, the details of which are not very exact, and a worker blowing a vase. The technique of plate glass had changed very little since the Carolingian period.

The art of painting on glass must have eventually become distinct from the craft of the glassmaker and more similar to that of the enameler. The glass used was made with potassium, obtained by burning vegetable matter, particularly ferns. In the thirteenth century soda began partially to replace potassium. In addition, the glass contained a large proportion of alumina and iron oxide; thus it was more resistant, its colors more brilliant. It was small — 4 millimeters thick and ¾ to 1¼ inches square. The crucibles, according to Theophilus, contained approximately 143 pounds of glass, and produced small cylindrical objects. Bubbles and scratches indicate incomplete refining; crude stretching resulted in bumps. Cobalt oxide supplied a blue color; copper gave red or green, manganese purple or violet.

Thirteenth-century glass, which was made from soda, was not as strong as twelfth-century varieties. The technique was approximately the same in both centuries, and still produced only small objects. A type of decoration consisting of pieces of colored glass glued to the stone with mastic, in imitation of mosaic, was even invented. In the second half of the thirteenth century Vincent of Beauvais makes the first known reference to a glass mirror: "Of all mirrors, the best is the one made of glass and lead." Not until the fifteenth century was the lead in the recess of the sphere replaced by a tin coating in the form of an amalgam.

At the beginning of the fourteenth century a new technique was perfected (if not invented) in France for the production of window glass: crown glass. This process consisted of transforming an open glass ball into a disk by rotating the pontil. The glass was thicker at the point where it had been attached to the pontil — this was the "crown." These disks could not have been much more than 2½ feet wide. They were cut into flat pieces to eliminate the crown. One tradition attributes this invention to a Philippe Caquerel or de Caqueray, who settled in Normandy in 1330. However, the older technique continued to be commonly practiced in Lorraine, Bohemia, and Venice.

Ceramics

We have very little information on medieval techniques of potterymaking, and very few examples of medieval ceramics, except for a few wall and floor tiles. The Middle Ages certainly had unglazed pottery, made of a mixture of clay and sand kneaded together. The manufacture of lusterware, so common in antiquity, appears to have been abandoned after the great barbarian invasions.

To remedy the very obvious disadvantage of the porosity of terra-cottas, the idea of covering objects with a coating made of lead, incorrectly called varnish or glaze, was conceived in the West, probably in the Carolingian era. Since this coating was transparent, every imaginable type of decorative treatment could be given to the terra-cotta. Varnishing was the usual technique practiced in the Middle Ages, both for tiles and for vases and pots. The nature of the glaze covering the terra-cotta was the distinguishing feature that differentiated faïence from the pottery objects just mentioned. The clay still consisted of varying proportion of clay, sand, and calcereous marl. Firing transformed it into terra-cotta, but the glaze coating was turned to an opaque white by the presence of tin salts (stanniferous enamel). The decorative motifs were first placed on the fresh enamel which, when dry, took on a powdery form. The metallic oxides that produced the colors combined with the enamel during the firing; the oxides that could resist high temperatures were very few in number. This technique, which was known to the Arabs in the eleventh century, achieved its highest perfection in the Moslem areas and in Spain; it was not adopted until rather late by Christian Europe (its great success at Faenza, in Italy, dates from the fifteenth century).

Medieval Europe never developed porcelain, which had to be fired at very high temperatures; such temperatures required not only suitable kilns but also technical precautions that were only slowly perfected in the course of centuries. Stanniferous enamel, for example, was not known in France until the fourteenth century.

G. Fontaine notes that great uncertainty still surrounds the subject of medieval ceramics. By the end of the twelfth century the use of lead varnish was becoming general. A very small number of pieces may date from the thirteenth century. During the thirteenth and fourteenth centuries a pottery technique was being developed that continued in use without major changes until the end of the eighteenth century and even later. The ceramic wares of the Beauvaisis and the Saintonge were famous. Crude, poorly fired pottery was probably produced almost everywhere; no specimens have survived. In any case, peasant inventories of the first half of the fourteenth century indicate a noticeable predominance of wooden or even tin objects over ceramic wares. Here, again, there is great need for a thorough inventory of all the ceramic objects in our museums which are known to date from the medieval period.

Brick   Brick presents a difficult problem. It was widely used in Roman antiquity for several reasons, the first and most basic one being the lack of wood available for construction. If one did not wish to build in stone, a costly material, brick was the only alternative. It was used particularly for leveling courses and facings. Since both were eliminated in medieval architecture, brick was thus automatically excluded from most buildings. In addition, as we have already noted, the abundance of wood in the northern countries made possible extensive development of wooden construction with earth or plaster filling.

Thus architectural modifications and environmental conditions made brick unnecessary in a very large area of the medieval Western world. Certain regions still felt a need for it, however (at least in important buildings for which wood could not be used), notably the colonized lands wrested from the Slavs, between the Elbe River and the Baltic Sea. This country was denuded of wood, and as soon as the builder decided to eliminate wooden construction he was forced to fall back on brick. This appears to have occurred around the middle of the twelfth century (cathedrals of Lübeck and Ratzeburg, around 1173). In other regions where stone was equally scarce, ease of transportation (chiefly by water) made it possible to overcome this disadvantage. Was not the Tower of London constructed with stone brought from Caen?

Brick buildings nevertheless existed at a fairly early period. At Foigny (Aisne), for example, the Cistercians began in the twelfth century to produce large bricks, called "Saint Bernard's bricks." They were made with forms, and were usually perforated to facilitate firing and better utilization of the cement. Frequently they were made on the construction site, which proves that their firing must have been extremely crude (this was also true in Frisia). The same method of preparation was used at all the Cistercian construction sites in France, Italy, and Germany.

There is no surviving specimen of the medieval brick kiln; the only kilns that have been preserved are lime kilns. The kiln of Commelles (Oise), in the middle of the forest of Chantilly, dates from the thirteenth century, and we are not certain whether it was used for lime or for roofing or wall tiles; the first hypothesis seems to be the most plausible. It is a large pyramidal fireplace about thirty-nine feet high, and is strangely reminiscent of the kitchen fireplaces of the abbey of Fontevrault and the cathedral of Pamplona. References to tileworks everywhere in Europe are very numerous in this period. The Romans had utilized two types of lime kilns, one built aboveground, the other buried. In the first case the limestone was piled along the walls, and the combustible was placed in the center; a hearth gave access to the bottom of the kiln and supplied the air required for the fire. Until the twelfth century makeshift cylindrical kilns of the second type, similar to furnaces for reducing ores, were in use everywhere, either in the quarries or on the construction sites, to fire the debris from the stone cutting. The lime, which was placed along the walls or in layers alternating with charcoal, often became mixed with the charcoal.

Brick had thus almost completely disappeared from construction by the end of the early Middle Ages, and production remained sporadic, continuing, like lime and tile production, techniques that had been utilized by the Romans.

Salt   Salt was obtained by evaporation of water, either through the heat of the sun, in countries with a sunny climate, or by boiling the water, in the northern regions and in the case of salt springs.

The technique utilized for seawater has remained the same since its origin. Concentration was done by running the water through a succession of increasingly shallow basins, concentration ponds, or compartments divided by dikes or ridges of earth. It next passed over flat tablelands on which deposits of salt formed. Pumping is mentioned in Provence in the tenth century. At Istres, in 1067, mention is made of a special machine for leading the water into the basins. Salt marshes existed in certain countries of scant sunshine, as on the banks of the Baltic, where saltworks continued to exist until the salt production of the Atlantic saltworks became sufficient to supply the demand. Here caldrons and procedures similar to those practiced on the Seille were utilized. Terra-cotta rods supported flat tiles with spouts that were arranged in rows over a fire. The water flowed from row to row, depositing its salt on the tiles. It is possible that the sun may have been stronger then in these northern lands, permitting the installation of saltworks.

Such techniques could not be utilized for salt springs, which were generally located in regions of insufficient sunshine. In this case the water was brought to the surface and led through a very complicated network of wooden containers into vats, a process that caused a certain amount of concentration. From here the water passed into round or oval caldrons, made of pieces of iron riveted together (eleventh century) and limed to prevent the fusion of the salt; each caldron held approximately forty-eight large barrels of salt water.

Evaporation was done in several stages, some of which consisted of boiling the water. In the first stage, the caldron was filled and placed over a fire (the furnaces were buried in the ground), which was progressively increased and maintained at peak for several hours. Concentration was accomplished in several hours over first a moderate and lastly a very low fire. The complete operation required from twelve to eighteen hours.

Metals   The reduction of ores was undoubtedly one of the most important of the thermal techniques. Here, again, both texts and archaeological evidence are lacking, and recent excavations have indicated that archaeological research in Europe has perhaps been carried out in insufficiently systematic fashion.

The prehistoric methods of reduction employed in antiquity were undoubtedly still in practice at the beginning of the Middle Ages. The few furnaces of this period that have been studied reveal no major modifications. The Romans had made some progress in iron metallurgy, which was of benefit to the peoples they subjugated: more regular use of bellows, exclusive use of charcoal, the use of fluxes, and production of case-hardened steel. By the fifth century these techniques were in practice almost everywhere. Steel and its properties were known, as well as thermal treatments, tempering, the hardening of metals by cold-hammering, and the development of the structure by forging.

A foundry consisting of twenty-four furnaces arranged in two rows in the loess, and dating from the eighth and ninth centuries, has been found in Czechoslovakia. The system faced toward the West, which proves intentional use of the prevailing winds; there was also a system of artificial blasting. A blast pipe, or nozzle, with a diameter of from 1 to $1\frac{1}{2}$ inches opened into the center of the lower section of the furnace. In this way steels with low or moderate amounts of carbon were obtained. Very refractory clays were utilized, and temperatures must have been very close to fusion point.

During the Carolingian period an attempt was made to increase the pro-

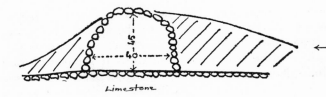

Limestone

Fig. 74. Iron furnace, Marishaufen. End of the seventh century (ed. Salin and France-Lanord, *Le fer à l'époque merovingienne,* Fig. 7).

←

Fig. 75. Iron furnace, Landerthal. Eleventh century (Gilles, "Les fouilles aux emplacements des anciennes forges . . .," in *Le fer à travers les âges,* p. 59).

→

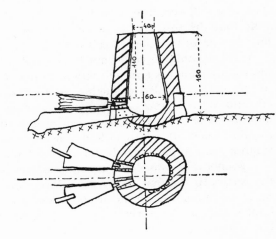

Fig. 76. Forge. Around 1250 (Verriest, p. 90).

Fig. 77. Forge. Fourteenth century (Singer, p. 396, Fig. 362).

duction and the volume of the blooms by modifying the shaft of the lower hearth. It was difficult to increase its horizontal dimension, since the air would no longer react sufficiently in the center. In order to increase its vertical dimension, it was raised above the ground, and the opening was surrounded with a small embankment or a wall of stones covered with clay. The structure gradually became completely external: the small buried furnace had now become a tall structure, a transformation that appears to have been already well advanced in the eighth century. A large piece of wood was used as a core to form the interior space. Clay walls were raised around it, temporarily supported with pieces of wood, and were then built up with earth and stones. The core was finally removed, and an opening was made at ground level. The chamber of these furnaces, which was approximately 6½ feet high, was in the shape of a prism whose bases are parallelograms, or cylindrical, and slightly conical. (Various shapes have been discovered, and we may suppose that the form of the wooden core was the determining factor.) The semicircular crucible was placed in the ground and covered with a layer of clay. Various openings were left for the penetration of air, the removal of the slag, and the supervision of the functioning of the apparatus. Between the eighth and thirteenth centuries the furnaces continued to increase in size, and soon became masonry structures with an interior envelope of refractory bricks. The furnace then acquired the shape of a truncated pyramid installed on a large base.

Factories of the twelfth and thirteenth centuries have also been uncovered. Or, to be more exact, traces of exploitation have been discovered, the buildings themselves having been long since destroyed. The remains of a Polish establishment of this period indicate that the Slavs were familiar with the action of lime on the fusion of acid and semiacid ores. Circular blooms are cut almost in half, so that the quality of the metal can be determined. Russian metallurgical techniques were approximately the same, until the Mongol invasion of the thirteenth century interrupted the development of this industry. The remains of the twelfth-century reducing furnace with several radically placed draw holes and conduits for slag, discovered near Kiev, can be regarded as a transitional stage pointing toward the blast furnace. The Landerthal (Saar) furnace, which dates from the eleventh century, was still built of siliceous alluvium plastered over a lath basket; the subsequent firing left on the alluvium the imprint of the woody fibers. Its conical shape resulted from the fact that the laths were not parallel. The hearth was lined with stones taken from ruins of Roman buildings. Two scorified nozzles on the opposite side prove that there were holes for blasting either to the right or left of the dross hole or that both bellows were worked at the same time. A more thorough investigation of this archaeological evidence, and the comparison of chemical analyses, would undoubtedly permit us to acquire greater knowledge of the metallurgies of these periods.

The raising of the furnaces aboveground and the increase in their size necessitated an increasingly powerful draft. The only satisfactory solution possible, which continued to exist down to the middle of the nineteenth century, was soon discovered, namely, the hydraulic bellows. Much discussion has centered upon their origin and date; here, again, thorough investigation has not yet been carried out in a systematic manner. In the district around Siegen, Germany, bellows were being worked by hydraulic wheels as early as 1311. This invention appears to have caused the abandonment of all the small furnaces still widely dispersed in the forests where both ore and the combustible were found, without the costs of transportation. As soon as a more productive instrument came into existence, it was possible to seek out the raw materials at a greater distance. We possess a contract for the construction of a furnace in Lorraine, the bellows of which appear to have been moved by hydraulic power (1323, region around Briey).

We have less information about the other metals. In the thirteenth and first half of the fourteenth centuries, copper was being worked in the Massa region of Italy, in cupola furnaces of which remains have been discovered. These furnaces were square inside, and were almost 10 feet high, and 2¼ feet wide. The interior of the chamber was lined with bricks and refractory stones, especially granite and porphyry. The locations of the blast pipes were clearly marked. Two smeltings were made, one for crude copper, the other for refined copper. The proportion of impurities in the latter was not permitted to surpass 2½ percent (in 1310 this tolerance was raised to 3½ percent). The copper was made into sheets 4 inches by ½ to ¾ inches thick. The specimens, at least, are of excellent quality. Traces of water supply pipes lead us to believe that here, too, hydraulic bellows may have been used.

Almost nothing is known concerning the extraction of the other metals — lead, tin, zinc — the reducing furnaces of which must have borne a strange resemblance to the primitive furnaces. An agreement of 1180 between the Count of Provence and the Viscount of Marseille, setting forth the conditions of exploitation of a mine of argentiferous lead, tells us nothing about the methods of mining used. Only in the following period do documents appear that permit us to determine the exact differences in the treatments of these various ores.

There is the same absence of precise information on the foundry methods of the various alloys, particularly bronze. The founding of bronze was already known in Roman antiquity, whence it was transmitted to Byzantium. This latter city produced the bronze doors that between the tenth and twelfth centuries came to grace a certain number of Italian churches (Verona, Rome, Amalfi). Little is known about the casting of bells. Recasting was frequent. The artisans who specialized in this work had one or two formulas that were mysteriously transmitted from generation to generation. The work was generally executed on the site, using a mold made from argillaceous earth. Until the thirteenth century the letters of the inscription were prepared with hand-rolled strings of wax; later, letters carved in wood were used. The fusion of the metal was aided by hand-operated bellows. Specimens from this period are very rare. A thirteenth-century bell, an almost cylindrical section of a cone topped with a semi-circular crown one foot high and ten inches wide, has been discovered at Marais (Deux-Sèvres). The bell of Fontenaille (1202), which is preserved in the Bayeux Museum, has a better shape, but the crown is still semicircular. Their molds cannot have been very complex in shape.

Distillation

Distillation is a very old technique apparently invented by the School of Alexandria. The Arabs practiced it widely, particularly in making perfumes. The Arabic treatises were quickly translated and put to use by the Christian West, which by the twelfth century was already in possession of the basic works. In the second half of the twelfth century original writings appeared, which indicates that distillation was already highly regarded in the West.

The devices utilized differ little from those still in use in the nineteenth century. The few extant drawings depict alembics of the most classic type. The heart-shaped Alexandrain condenser, poorly cooled by damp cloths, had been abandoned in favor of the modern type of alembic with a tubular, serpentine, or coiled spout, placed in a vat full of circulating cold water.

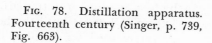

Fig. 78. Distillation apparatus. Fourteenth century (Singer, p. 739, Fig. 663).

This device made possible the appearance, at Salerno around 1100, of alcohol, the basic product of the chemical industries; its production was later improved, thanks to the use of such dehydrating substances as carbonate of potassium. *Aqua ardens* was obtained at about 60 degrees, aqua vitae at 90 degrees. By 1160 nitric acid, a useful substance for separating gold and silver, was being produced by distilling a mixture of saltpeter, alum, and vitriol. The distillation of essence of roses gave birth to a flourishing perfume industry.

The natural consequence of distillation was the development of the chemical rather than the thermal industries, although it is most convenient to include a reference to it among the latter.

CHEMICAL PROCESSES

Medieval chemistry was undoubtedly the most rudimentary of all the exact sciences. It is not surprising, therefore, that the industries which utilized chemical processes passed through long periods of groping experimentation and almost total empiricism. We are often astonished by the utilization of certain processes and products. Since chemical analysis was practically impossible, the chemist was obliged to follow "formulas," some of which were ancient, without understanding their exact purpose.

As long as it was a question of a simple mixture of products, matters were still relatively easy. But just as soon as proportions were involved, the difficulties were vastly increased, and solutions became almost impossible. While they had succeeded in one way or another in learning the properties of a relatively large number of substances, combinations were completely beyond the grasp of the medieval chemists.

Chemical processes were thus limited to the exploitation of the qualities of the products; it was known which ones were noxious or useful in certain clearly defined cases, and the skill of the chemists consisted of understanding approximately what were their external effects.

Textile dyes   Drugs and coloring matter constituted perhaps the most important part of what we now call the chemical industry. However, no decisive discoveries appear to have been made in this area during the Middle Ages; the products utilized in classical antiquity were almost all in use during the medieval period.

Blue was sometimes supplied by lapis lazuli, a hard, imported stone. The natural product consisted of a mixture of aluminosilicate of soda and sodium sulfide. The preparation of the dye was done in the East. After calcination the stone was chilled in vinegar, and pulverized. It was mixed with a mastic made of resin, wax, and linseed oil, and was kneaded under a stream of warm water. This decanted emulsion constituted the highest quality of ultramarine blue. The formula undoubtedly did not become known in Europe until the beginning of the fifteenth century. This material was rarely used for textiles, however; here it was replaced by indigo, or indigotin, made from woad, which was cultivated for this purpose over a wide area of Europe, chiefly in France and Germany; its leaves contain glucosides of indoxyl, from which indigo is derived. The dye is obtained by two successive actions: a hydrolysis separates the indoxyl from the glucose, and an oxidation transforms the indoxyl into indigo. At least this is the process implied in extant documents. The leaves were first crushed by millstones, and the paste was then placed in tumblers and left to dry. The chemical operation had already begun; fermentation permitted the complete separation of the raw coloring matter. The color was sold in both raw and fermented state. After drying, the last step — pulverization and sifting of the color — was completed. In the early medieval period the dye bath was prepared in a caldron heated directly over the fire, and the dyeing was done in the same caldron. The water, softened with bran, was heated separately. The tinctural products (the dye and ashes) were now added, and the bath was covered. The fermentation was completed, and after several hours the bath was ready. Dyeing could be done only while the bath was quite hot. The entire chemical process had been perfected only after very long trial-and-error experiments.

Red and yellow dyes "took" only after a preliminary operation of "mordanting," which removed impurities and grease from the fabric; it was done in large boiling baths. The fibers had to be thoroughly coated with a binding agent, which reacted with the dyes to form a pigment called "lac." Alum was the mordant most widely used. Tartar, which was also used, was prohibited until the fourteenth century, although it was used at Douai until 1250.

Red dye was produced especially from madder; however, in 1204 the dyers of Montpellier may have used only kermes (for economic reasons?), a unique dye of animal origin about whose preparation nothing is known. Brazilwood gave rose or pink; archil gave red or violet. Red-dyed fabrics were put into an acid or alkaline bath, which brightened the color. Yellow was supplied by the yellowweed; after mordanting, the cloth was soaked in a decoction of yellowweed brought to the boiling point. For fawn color, which did not require mordanting, a decoction of the bark and roots of walnut or alder (similar to walnut stain) was used. Black was produced from iron oxide, gallic acid, and tannin. Nutgall must have been utilized before the thirteenth century, but it was later prohibited.

The secondary colors were naturally obtained by the superposition of two colors. Black was obtained by dyeing successively with red and woad blue; the shade of blue determined the final color. Very often green was the result of a correction made on a yellow that had failed, rather than a deliberate creation.

Thus, while the most efficient dyes and often the manner of utilizing them had been learned, there were still gaps in the medieval dyer's knowledge, and many texts speak of major failures. The reputation of certain centers of the dyeing industry reveal, in any case, that these techniques were not matters of common knowledge; there was much competition for the services of good dyers.

Stained glass   Stained glass posed different technical problems, for the product had to be able to resist heat. Theophilus Presbyter is somewhat vague when he speaks about these techniques. The colors appear to have been obtained by an accident in production, but in the case of green, blue, and violet, sticks of color seem to have been melted with white glass.

In the twelfth century blue was obtained with copper carbonate or cobalt protoxide made from ores of Bohemia or Saxony, which gave a superb, very delicate color. In the thirteenth century the cobalt ore was less pure, and violet-blue, greenish blue, and gray-blue included elements of manganese oxide, copper dioxide, and nickel.

Red, obtained from copper protoxide, was so intense that the glass is supposed to have become completely opaque. It was the only color obtained by covering a piece of greenish glass with a thinner layer of red glass. Another technique consisted of alternating layers of color with layers of colorless liquid glass, which produced "veins" in the body of the glass.

Green was produced from dioxide of manganese or copper, purple from manganese oxide brought to maximum oxidation, yellow sesquioxide of iron added to dioxide of manganese, or antimony. At Chartres, in 1328, we find the first use of the famous yellow obtained from the use of oxide that was to give a new orientation to the art of the stained-glass window.

In all these cases the glass was melted at high temperature in large pots made of refractory earth, and was colored by means of the metallic oxides just mentioned. As in the case of textiles, there must have been a large proportion of failures that are unknown to us. The manual operations and combining of materials were perhaps more difficult here than in the textile industry. In any case there was an increasing decline in the twelfth and thirteenth centuries, the principal causes of which may have been the dispersal of the workshops and the uneven quality of the coloring materials.

The chemistry of fats   The importance of fats in medieval life was very great, not only from a dietetic point of view but also for the production of a number of raw materials used for lighting, among other purposes.

The fact is that the chemistry of fats was quite rudimentary, and had made little progress since the time of Pliny, usually because of lack of the necessary scientific knowledge.

Oil was extracted from a number of plants that varied from one region and climate to another. Olive oil was common in all the Mediterranean countries. Farther north, flax, poppy, walnut, colza, rape, and mustard were utilized. Poppy seed oil was used especially in cooking, hempseed oil for lighting, and mustard-seed oil for tawing and ropemaking. Theophilus Presbyter supplies several details on the manufacture of linseed oil. The seeds were separated out with the flail, and were then dried in the stove, ground in a mortar, mixed with hot water, placed in a cloth, and pressed. Olive oil was obtained with methods that had been in use since antiquity. The basic step in all these operations was thorough pressing and decanting of the products.

Animal fats were generally melted; butter was produced by churning milk, following a technique still utilized today. Lard was widely utilized. In contrast, whale oil, which was obtained by heating pieces of the animal's flesh, was quite rare.

The ancient oil lamp was still widely used in the Middle Ages, although resinated torches were also used in the countryside. Wax and tallow candles existed; the technique of coating a woolen wick with these materials had been learned.

Soap had also been known in antiquity, when it had been made from soap-wort roots or mineral matter, tallow, and wood ash. Manuscripts of the tenth and eleventh centuries supply details that indicate that improvements on these techniques were very gradual. It was learned how to liberate the oxide needed for the saponification of fats by adding lime to alkaline lye, which was obtained from wood ash and was thus full of alkaline carbonate. The long boiling of the lye with tallow or oil resulted in an emulsion that was inconvenient to use and had little cleaning power. Soaps made from potassium had a pasty consistency and were thus equally inconvenient. The first cakes of hard soap were made in the twelfth and thirteenth centuries at Genoa or in Spain, with olive oil, natural soda, and a little lime; these hard cakes were infinitely more practical. However, the efficiency of medieval soap was very relative.

The chemistry of saltmaking

The preparation of salt was such a vital activity that its production probably continued without interruption during the great invasions. The salt obtained from seawater or from salt springs must have contained a certain amount of impurities.

Salt was a preservative as well as a food. Salting was extensively practiced in the Middle Ages, both for fish and meat, and raw hides were also salted to prevent them from decaying.

The chemistry of hides

Hides were subjected to major chemical preparations. Few improvements were made during the Middle Ages in the methods and materials used for this purpose. To remove the salt, the raw hides were steeped in sour milk or in a brew made from sumac branches; they were then dried in the open. Fleshing and the removal of the hair were mechanical operations, facilitated by the use of lime. After prolonged beating the hides were impregnated with grease, oil, or tallow.

The ancients had become proficient in tanning, for which they had used beech, linden, fir, and picea bark. The medieval tanners were partial to the barks of the chestnut and oak; they also used bark from the pine, alder, and pomegranate trees. Aside from this difference, the techniques had remained unchanged. Tanning was a very slow process.

In order to obtain supple hides, the medieval tanners used alum, and it is this product which served as the base of the so-called "Córdoban" preparation of leather; it was also used by the tawers. Some authors have claimed that tawing was born in the Poitou region around the tenth century, but it appears certain that it was already known to the Romans. It consisted in removing the hairs with lime and tanning with alum; the hides were then softened in warm water before being fulled in oil. The hides were all coated and impregnated with grease, generally oil and tallow.

The birth of explosives The origin of explosives is a question still widely debated. Extant texts and pictures are not sufficiently precise to permit very clear-cut conclusions.

Was gunpowder imported from the Far East or discovered independently by the West? We shall probably never know the answer to this question. Despite theories to the contrary, military firepower, especially "Greek fire" (although there were numerous types), was known to Western technicians; if it was not more commonly used, this was because most varieties were made from naphtha, a product unknown in the West. Saltpeter was frequently used for various purposes prior to the fourteenth century. Albert the Great appears to have envisioned its use only for the making of nitric acid; we find no mention in his writings of a process for the preparation or purification of saltpeter. It has often been claimed that Bacon may have known of cannon powder, but this appears to be a misinterpretation of his text.

The Arabs were perhaps the disseminators of an invention accomplished in China between the eighth and tenth centuries. They devoted much effort to transforming into potassium nitrate the mixed nitrates they found. The work done in the West was principally an attempt to eliminate from the saltpeter the sea salts it contains. It is thus possible that we have here again a parallel development of two lines of research, rather than a borrowing of an Eastern technique via the Arabs. The treatise of Marcus Graecus undoubtedly had greater influence in western Europe than a hypothetical contribution of the East. Only in the first years of the fifteenth century do exact instructions appear on the methods of producing the powder.

MECHANICAL TECHNIQUES

The study of mechanical techniques raises the problem of small tools, of which little is known. We lack an inventory of the tools used in the Middle Ages, and it would be difficult to make such an inventory.

Little is known even about the production of tools; in many cases we do not know which workers were charged with this task. It is probable that a large part of the equipment was made by the same workers who used it, which considerably limited its development and improvement. Agricultural tools were unquestionably made by the peasants, but there must have been centers for the production of certain tools, a fact proved by the existence of a guild of manufacturers of cutting tools.

There is no doubt that the use of metals of better quality contributed considerably to the improvement of tools. The use of steel, in particular, certainly led to the development of certain activities. Land clearing and the flowering of construction in stone went hand in hand with the renaissance of metallurgy beginning in the twelfth century. Metal, which was still rare and undoubtedly expensive, continued for a long time to be used sparingly. For centuries agricultural tools continued to be made of wood; the functional parts were generally covered with a thin sheet of metal (this was the case with spades and plow-shares). Wooden joinery, as we shall later see, was done with pegs rather than nails.

Simple tools had changed very little since the end of the Roman Empire, and perhaps even since prehistoric times; it is relatively easy to incorporate in a tool the characteristics best suited to its use, and there is little reason to change them. The implements pictured on the tomb of the smith of Ostia are those of the medieval smith.

Percussion tools perhaps stood least in need of modification, and therefore little need be said about them. The sledgehammers used by the miners of Massa in the twelfth century are exactly like those that have been found in ancient mines and that appear on the bas-relief of Linares, in Spain. The raw metals were in the form of small ingots. In most cases the metal was hammered by hand, at least until the invention of hydraulic forging equipment, which made it possible to obtain iron of better structure and quality. The hammers employed by the iron mills remained unchanged until the middle of the nineteenth century.

The same observations are valid for cutting tools. The saw had been in existence since antiquity, with all the characteristics of the modern saw, as is shown in a medieval picture of a man sawing stone (beginning of the eleventh century). The steel blades probably wore out very quickly. As far as we know, there were no circular saws; all were of the band-saw type, even in Villard de Honnecourt's picture of a hydraulic saw. A stained-glass window in Chartres that dates from the middle of the thirteenth century shows the saw in use by the carpenters; the vertical blade is lowered by means of a treadle, and is lifted by means of a pole or bow acting as a spring.

Wood — at least pieces of a certain size — was shaped with that very ancient tool the adze. This can be seen in miniatures that depict carpenters, as well as in those that show the construction of ships (the scenes in the Bayeux Tapestry are unmistakable on this point). Extant roof trusses of this period — quite a few still exist — prove that the wood was first cut with the saw, then chiseled with the adze. Planes had existed since antiquity; the first pictures of them appear in the twelfth century, particularly on a capital found near Barcelona, which depicts the construction of Noah's Ark.

The production of these iron tools must have been quite difficult. They were cut by the makers of cutting tools, and then sharpened on a stone. Methods perfected for sharpening weapons may have been used for tools as well: the cutting edge was attached to the body of the tool with clamp-joint solderings. Nothing leads us to suppose that the excellent barbarian techniques had been abandoned. Concerning files and rasps, we have very little information; the first extant pictures of them date from the second half of the fourteenth century.

Drills were apparently not perfected, if we are to judge by the very fragmentary remains. The auger was used for drilling large holes; for small holes, tools known since prehistoric times continued to be used — the bow drill and the pump drill. It is difficult to ascertain whether the bitbrace had been known in antiquity; while certain writings contain drawings that may have been made from archaeological evidence at Pompei, we are obliged to note that no pictorial representation of a bitbrace appears before the fifteenth century. Only red-hot tips could have been used for drilling holes in metal.

In stonecutting, which was done with the saw, as is indicated in a miniature dating from the beginning of the eleventh century, a major modification appeared at the end of the twelfth century. It resulted from the appearance of the notched bushhammer, a tool still in use, whose striking surface had a series of small teeth that greatly facilitated work.

Small wooden and metal objects were often turned with the machines we have described. For this purpose modern-style chisels were utilized. Stamping was used, beginning in the thirteenth century, for the foliated hinges of cathedral doors. A mold of case-hardened metal was cut out with the graver, and the object to be stamped was molded after it had been heated red-hot and struck with heavy blows of a mallet. Leather was also stamped.

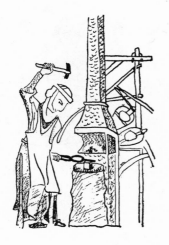

FIG. 79. Forge. Around 1325–1330 (J. R. T. Schubert, *History of the British Iron and Steel Industry,* pl. XIV).

FIG. 80. Metalworking. Thirteenth century (Schubert, pl. VII).

When we recollect that wood was by and large the most commonly used material, it is easy to understand that the problem of shaping it posed no insurmountable difficulties. Very simple equipment was more than sufficient. The same was true for the precious metals, which could also be easily worked. Ironworking was reduced to its simplest level, and was used only for certain tools; its principal use continued to be for weapons. Not enough specimens of weapons of this period survive to permit a complete understanding of the technical conditions of this work. The armor preserved in museums clearly postdates this period, being scarcely older than the beginning of the fifteenth century. A twelfth-century sword found in Italy and scientifically analyzed proves in any case that the techniques had not changed since the barbarian period; the same methods of welding, and the alternation of soft and hard iron that had been perfected the fourth and eighth centuries, were still being used. We have little information about breastplates and helmets, but they appear to have been made of pieces riveted together, for stamping large pieces with complicated shapes could not be done. This equipment was probably expensive, and therefore reserved for an élite group. The ordinary enlisted men were undoubtedly satisfied with leather garments, helmets, and shields.

It is nonetheless true that our knowledge in this area is subject to revision. It is urgent that investigators set to work to collect both written and pictorial documentation on this subject. Certain iconographical themes — for example, St. Joseph's workshop, the construction of Noah's Ark — would undoubtedly supply important elements for such a study.

Fig. 81.   Fettling of a bell. Fourteenth century (Louis F. Salzman, *English Industries of the Middle Ages,* p. 152).

Fig. 82.   Striking coins. Twelfth century (Salzman, p. 11).

Fig. 83.   Polishing of precious stones. Fifteenth century (*Ars memorativa,* 1480).

CHAPTER 20

THE ASSEMBLING OF RAW MATERIALS

WHILE A LARGE quantity of the raw materials processed are able to satisfy human needs without further preparation, often it is still necessary to combine them. The techniques of assembly answer this purpose, which is to say that these techniques are as important as any of the others. Perhaps it was these techniques that experienced the greatest modification during the period between the fall of the Roman Empire and the beginning of the Renaissance.

Some of these techniques are unknown to us; they were practiced by so many private individuals on such a day-to-day basis that they have left no trace of their existence. Such is the case, to take only one example, with basketry, for which we would seek in vain for precise concrete evidence. Objects of woven straw appear in miniatures, but we have no text or picture to describe their production.

TEXTILE TECHNIQUES

Perhaps it might have appeared much more logical to include in preceding chapters those details of textile production that involve in the preparation of their raw materials some of the methods we have had occasion to study briefly. If we have included in this chapter everything related to the techniques of the textile industry (except for the somewhat secondary operation of dyeing) it is because in the Middle Ages this industry formed an extremely close-knit economic and technical complex.

Various textiles were utilized in the Middle Ages: wool, linen, canvas, silk, cotton. Wool, however, was by far the most important. Perhaps it was this preeminence that caused the gradual predominance (with a few exceptions) of its techniques over those of the other textiles. We shall thus concentrate our attention first on the techniques of the woolen industry, and then briefly outline the practices peculiar to the other textile fibers.

In the words of de Poerck, one of the latest authors to study this important subject, "The technique of the cloth-manufacturing industry, as it was known in the Middle Ages, is certainly ancient; it dates at least from classical antiquity." He adds, "However tenuous the relationship, it permits us to unite the information supplied by the medieval sources with the archaeological and literary remains of the Roman period." Thus there was a renewal of a tradition the barbarian invasions had weakened without completely destroying. The clothmaking industry was reborn in completely natural fashion in the second half of the eleventh century, in the same cities where it had already prospered during the Late Empire. It was in Flanders that the techniques of the woolen industry achieved their highest degree of perfection, first, because of the great density of population in Flanders, and, second, after the end of the eleventh century, because of the proximity of England, where certain centers (particularly the Cistercian and Premonstratensian monasteries) carried on intensive raising of breeds of sheep that yielded wool of superior quality. The raw material available to the Italian clothmaking industry (which was being developed during the same period in Florence) was too inferior in quality to lend itself to technical refinements. In the thirteenth century the situation changed considerably. England acquired industrial equipment between the end of the twelfth and the middle of the thirteenth centuries; the fulling mill rapidly turned this industry, which had begun as an urban, luxury industry, into an ordinary rural activity. The rupture in Anglo-Flemish relations in turn diverted the English wool toward Italy and Florence, whose workers, once they had perfected the dressing and dyeing process, set to work to improve the production of the fabric. In addition, Florence succeeded in attracting expert Flemish workers driven from their country by the social struggles.

Preliminary operations   The quality of the wool was naturally of great importance, especially for the centers of luxury production, and this was the reason for the success of the English wools. The wool from adult animals was preferred to that taken from lambs. When a transition began, at the beginning of the fourteenth century, to more ordinary and cheaper products, lamb's wool appeared in fine clothmaking. (We shall find other examples of this technical decline, which was probably the result of an economic recession.) The animal was washed standing up, and was sheared in the month of May. Carcass wool was generally avoided in quality cloth; to obtain it, the hides were soaked, and the wool was pulled off. The same prohibitions were generally applied to recovered wool, which was used only for the weft of the coarsest fabrics.

Inferior or damaged material was eliminated by a preliminary sorting. The wool was then carded to loosen it. A final sorting separated the wool into various categories: fine, average, coarse, and lastly the least desirable types. (English wool arrived already sorted.) Beating with sticks on a hurdle is mentioned only once, at Valenciennes; its purpose was to expand the wool, which thus lent itself better to combing. The wool was next scoured in a series of carefully controlled baths. To soften the wool and make it silky and easy to comb, it was greased or coated with oil or, in the case of high-quality cloth, butter and lard. (A "dry" method also existed.)

The processed wool was now combed, which freed the long fibers and arranged them in parallel rows by the combined action of two slightly heated combs. These combs, which were of various sizes, were used for the various categories of wool, and were always kept in good repair. The combed wool had the appearance of stripes. Carding did not appear until after the first half of the fourteenth century.

Fig. 84. Combing wool. Fourteenth century (Singer, p. 194, Fig. 156).

Fig. 85. Spinning with spindle. Fourteenth century (Singer, p. 203, Fig. 166).

Fig. 86. Spinning with spinning wheel. Around 1338 (Singer, p. 302, Fig. 167).

Spinning   The next step was spinning. Until the end of the thirteenth century there was only one technique, which had been practiced unchanged for millenniums: that of the spindle and the distaff. The bunch of combed wool was attached to the distaff, which was rested on the shoulder and in the fold of the arm. The wool was drawn out in a thread and twisted between the thumb and index finger. When a certain length of thread had been made, it was wound around the spindle, which had a weight to stretch the thread.

The origin and early models of the spinning wheel are difficult to determine with certainty, owing to a lack of precise documentation. The spindle had to be conceived of as an axle, and a wheel acting as flywheel and source of power had to be substituted for the weight that stretched the thread. The spindle was then mounted on a frame, the balance being shifted onto a pulley connected by an endless rope to a large power wheel. This first hand spinning wheel undoubtedly owed much to the winder, which had long been in existence and with which it is often confused.

The appearance of the spinning wheel in western Europe is a complete mystery. According to some authors it may have been born in India between the fifth and ninth centuries, for spinning cotton. In Europe the first texts to mention it are few in number and difficult to interpret. In 1288 its use was forbidden at Abbeville, while at Speyer it seems to have been permitted around 1290, but only for spinning weft thread. These are the only known references to the spinning wheel in the thirteenth century. The oft-cited stained-glass window of Chartres seems to represent a simple winder rather than a spinning wheel. Documentation becomes more abundant in the fourteenth century, and moreover reveals that transition to easier, supposedly inferior techniques we mentioned earlier. Around 1349 the Book of Trades of Bruges indicates that the use of the spinning wheel was still forbidden for warp threads; the thread obtained from the spindle was judged to be finer and free of knots. The spinning wheel was also forbidden at Provins. At Douai it appeared only in 1305, the next reference to it coming more than a half century later (1362).

The medieval spinning wheel had a crank and was turned by hand, as is clearly indicated in all the extant pictures. In most cases the operator worked standing up, in contrast to the Oriental custom. The treadle did not appear until much later, with the invention of the crank-connecting-rod system. Nor did the spinning wheel of this period have a flyer, the twisting of the thread being done solely with the fingers. Thus this device was not so much a spinning wheel, in the modern sense of the term, as a simple mounted spindle.

The finished yarn was now spooled to be made into hanks. Some yarns were even reinforced by twisting. The worker juxtaposed and twisted the slightly dampened threads several times, in the direction opposite to the direction of the spinning. If they were to be used for weaving, the hanks of yarn were wound onto a bobbin for the warp, onto a spool for the weft. It is these winding wheels that are often depicted next to weaving looms, as in the window of Chartres or the Book of Trades of Ypres (thirteenth century).

Warping was a very simple operation. Two posts with pegs were driven vertically into the ground or supported against a wall, the distance between them being fixed by law. (Round warping frames also existed.) Originally the pegs may have been simply fixed to a wall. A two-tiered rack with vertical pegs held the bobbins of yarn. The worker twisted the threads from the rack around his fingers, and arranged the crossover thus obtained on the three top pegs of the warping frame. He then guided his ribbon of thread from one post and from one peg to another, until he reached the last peg, to which he attached it by means of a small crossover. The threads thus arranged formed a *demi-portée* (a *portée* being a group of warp threads); by repeating the operation in the opposite direction a complete *portée* was formed. The warping frame could hold a certain number of *portées*. The whole was fixed with a tie; a warp was made with a certain number of ties. The number of warp threads was strictly regulated for each type of cloth, being much higher for quality cloth. Sizing appears to have been unknown or forbidden (for example, by the Statutes of Comines in 1366), and only the comb was oiled. Here, again, however, during the fourteenth century the rules became increasingly lax. Improvements in the raw material undoubtedly compensated to a certain extent for what were considered inferiorities in the new techniques.

Weaving

The problem of weaving is a complex one. The various looms are known to us in fairly precise fashion through pictures. Apparently no major modifications had been made since classical antiquity. The heddle system is easy to grasp; it served to separate

FIG. 87. Warping. Around 1210 (Singer, p. 209, Fig. 176).

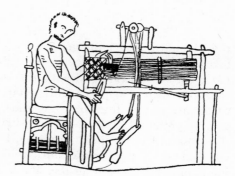

FIG. 88. Weaving loom. Thirteenth century (Singer, p. 212, Fig. 181).

FIG. 89. Loom being worked by two men. Around 1310 (Singer, p. 213, Fig. 183).

the various threads of the warp. However, we lack detailed information on complex patterns of weaving. The two-shaft loom is obviously well represented and well known; the three-shaft system did not appear until the beginning of the fifteenth century, but may have been practiced earlier, perhaps at Rheims in the thirteenth century. The medieval comb and beater resembled the classical models. The structure of the loom differed very little from the looms pictured in eighteenth-century treatises.

The operator depressed one shed, and then threw the shuttle across. To ensure the strength of the fabric, the maximum number of throws permitted before changing the shed was restricted to seven. (In the case of the fine woolen linings of Saint-Omer, four throws were the rule.) On very large looms two weavers were needed to throw the shuttles; this can be verified in numerous pictures. The tension of the yarn had to be carefully watched, and breaks that might occur had to be repaired. Many defects in workmanship were known by names that are now completely meaningless to us.

The fabrics were carefully examined after the weaving; they had to meet the requirements of the statutes as to weight, size, and weaving. Holes caused by breaks in the yarn were discovered by stretching the fabric on two poles. The pieces that were satisfactory were then weighted.

Preparation   Between this examination and its final sale, the fabric was subjected to a certain number of operations included under the general term "finishing." Impurities and imperfections had to be eliminated by washing and mending, and the fabric had to be prepared for sale. Washing cleaned the fabric, and in particular removed the oil. It was done with a soapy product obtained by adding various ingredients (fuller's earth, urine, the dregs of wine, lime, sand) to fat; each center of production had its preferred and rejected methods, but fuller's earth seems to have been the most widely accepted. Scouring, by which the cleaning agent was removed, was done by trampling the fabric underfoot and giving it several rinsings. Burling eliminated knots from the fabric. The cloth was sometimes passed under the fuller's thistle before the fulling to felt it.

Fulling increased the body and homogeneity of the fabric. This required the compression of the cloth through the action of heat, humidity, and movement. Fulling with the feet was the ancient — and most inexpensive — method. The more recent method of mechanical fulling dates, like the mills, only from the twelfth century, and required large amounts of capital, to say nothing of favorable environmental conditions to supply the necessary power. Mechanical fulling also appears to have been considered in the beginning as too harsh a method for high-quality fabrics. Certain areas returned to fulling by foot after a period of using mechanical fulling.

Fulling was done in three steps. For the first operation oatmeal vat gruel was used; for the second, butter (soap was generally prohibited) or lard. (The first operation was made unnecessary by washing with fuller's earth.) The fabric was dampened, folded, and sometimes coated with various ingredients. Only then was the series of fulling begun, the water being changed each time. At the end of the operation the fabric was given a final rinsing. All fabrics were fulled, whereas in the modern industry this step is eliminated for combed fabrics. There was a prescribed period of fulling for each type of fabric.

Teaseling was done on wet (or at least very damp), never on dry, fabric. The cloth was stretched over a rod and was napped (teaseled) on both sides with cards (wooden frames with spikes); at the same time any remaining knots were removed. The piece of fabric was by now smaller (sometimes much smaller), thanks to the fulling; the threads were felted and stuck together, and the surface was already fluffy. Carding increased this fluffiness and straightened the tangled fibers. Teaseling was always done from bottom to top, by passing the cards (old, worn cards were used for the first few motions) over the fabric. Often there were one or two teaselings across the grain to remove the rough fibers. After teaseling, the fabric was given a first cropping (shearing), often done by the same worker who had just teaseled it.

Tenting was the stretching of the fabric to remove wrinkles and give it uniform, exact dimensions. The tenter frames were posts planted vertically in the ground; the length and height of the device was slightly greater than the length and width of the piece of fabric. After tenting, the fabric was dyed. Another series of operations was used for better-quality fabrics: a second cropping, additional burling to eliminate knots, final teaseling, repairing of the small holes made by the cropping shears, smoothing, and glossing.

The cropping technique was carefully regulated. The piece of fabric was stretched on a slightly tilted table that was padded with waste wool to give it a convex curve. The fabric was attached to it by the selvages (which were specifically designed to endure the strain of cropping and tenting), and the nap was raised by carding across the grain with a small card. The lower, heavier blade of the shears was placed down on the fabric, and the upper blade was brought down to it; a small spring then spread the two blades apart. The fabric was sheared from selvage to selvage. The first shearing was done with rather blunt shears.

Numerous changes must undoubtedly be made in the description we have just given (most of it taken from de Poerck), for there were sometimes considerable variations in techniques from one city to the other. Each center of production attempted to utilize tried and tested techniques that would yield products of superior quality — whence the regulations concerning the prohibition of a product or a tool. Technical uniformity was nevertheless achieved in the fourteenth century in processes that were judged to be less perfect. This high degree of technological development — for the clothmaking industry undoubtedly was the most highly developed technique — led, as we shall have occasion to see, to an extreme division of labor. It must be admitted, however, that these techniques applied only to quality production, that is, to high-quality fabrics produced in small quantities, whose value certainly was a contribution to progress. In most cases the rural populace made its own cloth and garments with techniques that had undoubtedly changed very little since prehistoric times. The spinning wheel, to take only one example, remained unknown in the countryside for a long period, and much difficulty was experienced in introducing it in the middle of the eighteenth century in the rural districts. Weaving looms (the oldest specimens in our museums are no older than the eighteenth century) were for the most part extremely primitive. It is possible that the falling off of urban standards and the production of very cheap merchandise to a certain extent effectively contributed, if not to the disappearance then at least to the decline, of this rural industry.

Silk

The history of silk still poses many problems that have not been solved. Silk, an arrival from the East and the Far East, was first introduced to Byzantium. The culture of the mulberry tree and the raising of the silkworm were established several centuries later in western Europe. Historians have been able to establish only tentative dates for the importation (probably by way of the Arabs) of these techniques — for example, in the tenth century in Andalusia. Elsewhere Greek workers may have come to Palermo in 1146, after the fall of Corinth, bringing with them this hitherto unknown industry. Shortly thereafter, around the middle of the twelfth century, a group of workers may have brought it to Lucca. Venice may have established its first looms only after the conquest of Constantinople, in 1204. The silk techniques are assumed to have spread from Italy throughout Europe. They were in use in France by the middle of the thirteenth century, by the beginning of the fourteenth century in southern Germany (1300 at Augsburg and Ulm, 1313 at Nuremberg). Mulberry trees were being raised at

Modena by 1300. Bologna, Florence, and probably Provence learned of the mulberry culture in the succeeding years. In 1360 a Bolognese named Bonafido Paganino wrote the first treatise on its cultivation. The mulberry tree in question was basically the black mulberry (*Morus nigra*).

Silk spinning was a very special technique. The cocoons were collected and smothered (so that the butterfly would not break through the envelope). They were soaked in fairly hot water, and the threads were unwound by hand. This resulted in a preliminary, untwisted thread composed of a certain number of basic strands. The thread was next reeled to make it uniform, and then had to be twisted to give it the necesary strength, first singly, then by juxtaposing and twisting two threads simultaneously, and lastly by twisting several threads separately and then together. In the beginning these twisting operations must have been done by hand, but their mechanization obviously did not pose a very difficult problem. The first hydraulically operated silk mills may have been installed at Bologna in 1272 by a man named Borghesano. The raw silk was placed on a certain number of spindles from which the threads, after the necessary twisting, were wound onto frames. One mill, mentioned in 1331, had two rows of 120 spindles each, and thus was already a large instrument. This invention, which gave a certain superiority to its owner, may have been kept a secret for a long time. The twisting mill, however, may have been introduced in Florence and Venice in the middle of the fourteenth century.

Silk weaving followed the same principles as that of wool, and used similar devices. Very soon, however, complex fabrics gained precedence over ordinary weaving. The Chinese silk looms probably reached western Europe in the thirteenth century; it would have been astonishing if travelers like Marco Polo had not brought with them from the Far East quite simple improvements that could be used on the looms utilized in Europe. The Western weaving looms also appear to have known an evolution of their own. The so-called "Jean le Calabrais" weaving loom apparently dates from the beginning of the fourteenth century, and is one of the oldest known Western looms for weaving figured material. The leashes (loops of string tied around the warp threads and used to lift the threads) were grouped according to the designs to be made, and were attached to buttons that could easily be pulled and hung in a creel. This loom did not lend itself to the use of a drawboy; its operator worked alone, pulling the leashes by means of the buttons, which he hung on the creel; then, moving the treadles, he passed the number of weft threads required over each leash. He then unhooked the button just pulled and hung up the next button, and so on. The relatively limited space available for the cords and buttons did not permit the weaving of large designs on this loom.

Figured fabrics of this period are few in number, the principal reason probably being the difficulties of weaving them. Moreover, silk fabrics were expensive and thus of limited sales appeal. This explains in part the reasons for the medieval tardiness in perfecting the techniques transmitted by the Far East.

Other textiles   Three other textile fibers were also used in the Middle Ages, all of which had been known in classical antiquity: hemp, flax (which was widely cultivated in western Europe), and cotton. Some historians believe that cotton growing was introduced into Spain by the Arabs in the eighth century. Its cultivation remained quite limited, however, and most of the cotton used in medieval Europe was imported. The spread of cotton fabrics was, moreover, still quite limited.

Flax requires a rich soil, which it quickly exhausts; it was cultivated in enclosed gardens near farms (in order to have the necessary manure), or on newly cleared lands. The male hemp flowers mature faster than the female flowers, which supply the seed capsules. To separate the woody tissues from the fiber itself, both flax and hemp were retted. This operation unfortunately polluted the water, and ponds or retting pits filled with diverted water were therefore established. In southern Italy it was forbidden to establish these retting pits less than a mile from a city — a prohibition that was renewed in 1307. Spinning and weaving of these fibers, as well as the dyeing and finishing operations, were the same as for wool.

We have no detailed information on the manufacture of rope, which was generally made with hemp. Ships' sails were often made from a combination of fibers, generally cotton and hemp, at least in the Mediterranean basin. In Marseille, at the end of the thirteenth century, they were made of bombazine; the pieces of fabric had to consist of ten pounds of imported cotton to six pounds of hemp yarn. The cotton industry flourished in Marseille between 1250 and 1350; linens for domestic purposes were made there, in addition to ships' sails.

In 1318 we find the first sales of alfa in the Marseille market. In the beginning the use of this fiber was exceptional, and it did not really become common until 1350. Alfa could not be used in its natural state, it was crushed under a heavy millstone to soften and separate the fibers, an operation that created much dust and noise. This was followed by a kind of retting and a drying period in the sun. Alfa was used as an edging for fishing nets, for making ropes, and for the baskets hung on the backs of beasts of burden. A factory for processing camel's hair was installed in Naples in 1313.

Thus textile techniques were not revolutionized in the Middle Ages. The only fact worthy of note is the important role played by industries of this type in medieval commerce. This importance caused the manufacturers to pay extremely close attention to the methods used; whence the multitude of regulations governing the clothmaking industry. The development of ordinary production, like that of flax and hemp, being more dispersed in the country districts, was undoubtedly more limited.

There is little to be said concerning clothing manufacture; here, again, the techniques were not greatly modified. Cutting and sewing techniques were still much as they had been at the end of the Roman Empire. What did change was on the one hand styles, and on the other the manner of dressing, which was adapted to the climate. This subject, however, is outside the limits of our study. There was an increasing predominance of cotton for undergarments, which led to an increase in the quantity of rags available, and consequently greatly facilitated the expansion of paper, and with it the appearance of printing. But this development occurred chiefly after the first half of the fourteenth century — although cotton began to become common in the reign of Louis VII, generally in the form of a flax-and-cotton mixture called fustian. Felt and furry fabrics utilized for hats appeared at the end of the twelfth century.

Skins, whether in their ordinary use in the form of leather or, especially among the wealthy classes, in a profusion of all types of fur, were used concurrently with fabrics in the making of clothing.

WEAPONRY

Major changes in weaponry, and consequently in all the military techniques, were made during the Middle Ages. The most important of these modifications were realized in offensive weapons by means of a complete renewal of the machines inherited from antiquity. The armies of antiquity had possessed a well-developed mechanical artillery based essentially on the slackening that could be produced by twisting torsion cables (torsion artillery) by means of a winch. The best example of this type were the ballista and the catapult. These appear to have been the only instruments used by the Byzantines; at least the Byzantine authors do not mention any other war machines.

By the early Middle Ages (ninth century) the Carolingian armies possessed a perfected arsenal, especially as regards siege machines. By the twelfth century the portable armament also included an instrument of exceptional power, the arbalest, or crossbow. The origin of these new weapons has been the subject of much discussion. Viollet-le-Duc believed they were Byzantine in origin, which is unlikely. It is certain, in any case, that all references to torsion artillery cease after the tenth century. The new weapons continued to be utilized even after the appearance of the cannon, which required more than a century to supplant its predecessor.

The crossbow   The crossbow was an improved version of the bow, in the sense that the spring was no longer made of wood but rather of metal; it consisted of one, and later several, superimposed strips of steel. The operator could no longer stretch the bowstring by hand; he needed a tool to draw it into the retractable lock. On the other hand, the weapon shot much father and its fire was more precise. The dates of appearance of this new weapon are highly disputed. The third-century bas-reliefs of Le Puy could be a depiction of an intermediate stage. Some authors claim a Chinese origin for the weapon. A manuscript that dates from the end of the tenth century shows bowmen firing on the ramparts of the city of Tyre with completely identifiable weapons. In any case the use of the crossbow was sufficiently general by the beginning of the twelfth century to justify the Lateran Council's prohibition, in 1139, of its use, on the ground that it was too deadly.

The early pictures supply little information on the method of tightening the bow. It can be agreed that until the end of the thirteenth century the system remained unchanged. At the top of the crossbow was a stirrup. The user placed his right foot in this stirrup and, bending over, caught the bowstring in a hook suspended from his belt; then he quickly straightened up and brought the cord into the locking device in the groove. The trigger was merely a small lever. This type of crossbow is still frequently depicted in miniatures as late as the thirteenth century. The beginning of that century saw the appearance of a simple forked lever that was fitted to two points on the stock of the weapon. The lever was engaged, the crossbow being held on the ground by means of the stirrup, and the operator had only to draw the lever toward himself to stretch the bowstring. While this weapon had a greater range (the spring could be strengthened), its major disadvantage was is slow fire, a defect that became steadily worse.

Although John Garland mentions the windlass crossbow in the middle of the thirteenth century, this does not appear to have been a portable crossbow. In any case, it does not appear in pictures before the beginning of the fifteenth century.

Trebuchet artillery Trebuchet artillery seems to have appeared in the second half of the ninth century. The first definite (or almost definite) extant reference to it dates from the siege of Paris by the Normans in 886; the machine was then used by the besieged. In 873 Charles the Bald used "new, improved machines" to capture the *place d'Angers* where the Normans of the Loire were barricaded; it can be supposed that this was an allusion to the trebuchets, which were still little known.

Trebuchet artillery utilized the ballistic properties of the sling but on a much larger scale, thanks to an appropriate mechanism. There were two types, trebuchets as such, and mangonels, each type being distinguished by the manner in which the ball was shot, which directly influenced the form of its trajectory.

Viollet-le-Duc's description of the trebuchet is excellent, and is accompanied by very clear drawings. It consisted of two posts between which pivoted a large beam with a counterweight at one end and a sling at the other. When the beam was free, the counterweight caused it to assume a vertical position, and considerable strength was required to lower it; a much greater pulling effort was necessary because of the narrowness of the angle formed by the pulling ropes and the beam. Recourse was had to two large wooden springs modeled after the drawing of Villard de Honnecourt (middle of the thirteenth century). The springs were stretched by winding a rope attached to them on an auxiliary winch.

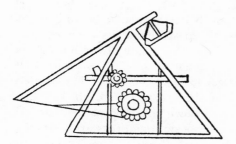

FIG. 90. Arabic trebuchet. Thirteenth century (Mercier, *Le feu grégeois*, p. 120).

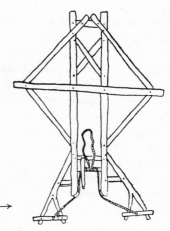

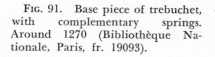

FIG. 91. Base piece of trebuchet, → with complementary springs. Around 1270 (Bibliothèque Nationale, Paris, fr. 19093).

These springs were then released, which pulled the main winch, thus assisting the men who were maneuvering the other winches. The difficulty decreased as the counterweight rose. Eight men were used, two for each lever. When the operation was finished, the pocket of the sling was laid along the trough and the ball was positioned. A pin maintained the system in position.

The launching of the ball was somewhat different from the use of an ordinary sling. A second strap, clearly visible in certain miniatures, linked a point on the beam with the tail of the sling; a picture in a German manuscript shows the operator attaching this second stretcher. The pin was freed by a blow from a mallet, and the beam then abruptly returned to its balanced position. Experience had taught the military technicians how to calculate the length of the sling and the weight of the projectile required to permit the ball to follow the desired path. The projectile was freed when the sling came into line with the beam, which by now had almost completely returned to its vertical position. A shock was now needed to regulate the departure of the projectile, which otherwise would have left its pocket in an almost vertical direction. Here the second strap came into play: it had now reached its maximum tension, and caused precisely this abrupt jolt. The greater the tightening of the strap, and the closer its position to the point of attachment of the sling, the more horizontal was the path taken by the projectile. Thus it was possible to regulate its aim. In addition, to prevent the apparatus from getting out of line because of its abrupt return to the vertical position, the counterweight was doubly mobile in relation to the beam. It was attached in movable fashion to a connecting rod that in turn was attached, also in mobile fashion, to the end of the beam. This connecting rod was supplied with a weight of its own, whose descending movement partially eliminated the shock by pushing the counterweight beyond its position of equilibrium.

Although similar in its general principle, the mangonel presented important differences of detail. One of the cords of the sling was attached to the end of the beam, the other was simply passed through a stylet arranged in such a way that when the beam reached vertical position the arm of the sling left its stylet and the projectile was hurled like the stone of a hand sling. The trebuchet, because of its abrupt, jerky movement, was useful for hurling projectiles over high walls onto roofs, but it could not give the projectile a very long, almost horizontal parabola. The aim of the mangonel could be better regulated because it described a larger arc and because it was possible to accelerate its movement by pulling on the counterweight. The beam was beyond the axle of the journal on which it pivoted, and was thus weighted on the enemy side in its position of equilibrium. A winch operated by squirrel cages lowered the beam. The acceleration of the movement had a special effect: the sling turned more quickly than the beam, which, being placed off center, slowed down as it approached vertical position. When the sling came into line with the beam, one of its arms left the stylet and freed the projectile, whose path followed a much flatter curve than that of the trebuchet projectile because of its lower angle of departure. The operator changed the movement and regulated the straps of the sling in order to regulate the firing. The fire of the mangonel was relatively rapid; it could hurl twelve projectiles per hour. The trebuchet was less docile, but also required less practice, while the mangonel had to be guided by trained engineers and utilized by specially trained men; numerous references are made to accidents involving mangonel operators.

It is extremely difficult to say whether other varieties of these machines existed. A treatise on armament manufacturing written for Saladin mentions Arabic, Turkish, and Frankish mangonels, but we do not know to what, exactly, this division corresponds. In any case, Reynaud mentions (in 1123) the Arabs' use of counterweight engines at the siege of Edessa (Urfa), Turkey. Mangonels may even have been utilized by the Moslems in 806; the stone balls used with them were covered with naphtha-impregnated oakum. At the siege of Tyre in 1124, the Crusaders sent for an Armenian engineer named Havedic, who had a great reputation for his skill in the art of supervising these stone-throwing machines. All these devices were certainly improved upon between the date of their appearance and their abandonment. A manuscript preserved in London depicts a rotating mangonel; other types were even mounted on ships.

FIG. 92.   Trebuchet, thirteenth century.

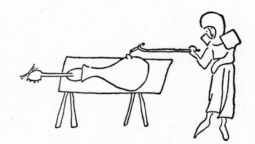

FIG. 93.   The first cannon, 1327.

Firearms — Firearms inevitably caused the disappearance of all these instruments, which were heavy to move and whose only advantage was that they could be constructed right at the scene of the battle. The origin of gunpowder has already been briefly covered.

The appearance of the first firearms on the battlefields of Europe has been much discussed. The two most recent authors, Messrs. Lot and Sarton, who summarize the abundant literature on the subject, are not completely in agreement. A single fact seems to be definitely established: the first picture of a cannon is found in a manuscript of the *De officiis regum,* dedicated by its author, Walter de Millinate, to Edward III in 1327. The reference in the *Registro dello Provisioni* of Florence, under the date of February 11, 1326, is perhaps the first definite mention of *canones de metallo* in a text. Lot considers the oldest example of the use of cannon powder to be that mentioned in Flanders in 1314; the invention immediately thereafter became known in England (1321), the Netherlands, Germany, France, and Italy (1326). Sarton raises many questions; he lists a series of sieges of strongholds and dates when the use of cannons can be inferred from the texts. Powder was known, he believes, at the end of the thirteenth century, and its application to cannons may have followed immediately upon its discovery. Was it employed at the siege of Berwick in 1319? At the siege of Metz in 1324? By the Moors at Baza in Andalusia in 1325? At Martos in 1326? At Alicante in 1331? By the King of England against the Scots in 1327? By the German knights at Udine in 1331? By the king of Castille, Alphonso XI, in 1342? There exists no absolutely precise information that would permit us to answer. Froissart mentions two English *bombardiaux* at Tournal in 1346.

Given these conditions, we can easily understand why it is very difficult to say how these early pieces of artillery were made. If we judge them by the cannons of a slightly later period, it can be said that they were not cast but were built up by soldering rods together and shrinking iron bands around the tube thus formed. The powder chamber was screwed on the tube.

The cannon was a permanent and costly weapon that had to be transported. Transportation equipment, however, was still crude. For this reason the use of the cannon became possible only in combination with a gun carriage, or at least wagons with a movable forecarriage. This topic will be discussed later.

Imagination and war   The human imagination has always been attracted by the problems of warfare, and particularly by offensive weapons. In three centuries there were two complete revolutions in artillery equipment. Moreover, the first half of the fourteenth century saw the birth of those technical works of imagination, unmistakably directed toward military problems, that were to flourish in the following century and lead to the great Renaissance engineers and the genius of Leonardo da Vinci.

Around 1335, Guido da Vigevano composed for Philippe VI, who was planning to go on a Crusade, a treatise in two parts. The first section gave valuable medical information about the manner of caring for oneself in these distant countries. The second section described the engines that would facilitate the conquering of the "infidels." Thirteen chapters give exact instructions, accompanied by drawings, for instruments, machines, or processes that may be useful. A portion of these explanations concerns means of crossing rivers and ditches by classical methods which were to reappear in most later treatises; these include prefabricated, jointed bridges thrown rapidly over rivers, and every known system of inflated waterskins which can be adapted to both infantrymen and knights.

Other machines are more interesting, as for example a ship that has floats made of vats, as well as Archimedean screws or paddle wheels (it is difficult to determine which) operated manually by means of cranks. There are attack towers that are to be brought up to the defensive wall and that can be hoisted to the desired height by means of a mechanism with ropes, winches, and sliders. There is also a complicated chariot moved by the wind, with a "great fury," against an enemy that would undoubtedly be disconcerted by its appearance rather than actually hindered by the machine itself. It is possible — the drawing is difficult to interpret — that the engine had a type of steering device with a gear.

Bizarre as these "inventions" may seem, they nevertheless indicate a change of mentality. The slow, gradual process of successive improvements was about to be replaced by the spirit of imagination that was to give birth to a great part of modern machinery. From this point of view Villard de Honnecourt and Guido da Vigevano, who are separated by an interval of fifty years, mark two stages that appear to have been decisive. While Villard de Honnecourt was still only a simple master builder, Guido da Vigevano, like his fifteenth-century successors, is basically a military engineer. The art of warfare, it is worth pointing out, had now come to the forefront of technology, and was to supply the impetus needed for its future development.

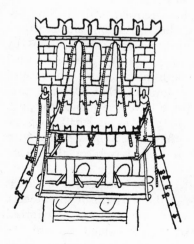

FIG. 94. Submarine. Circa 1327 (Singer, p. 651, Fig. 594).

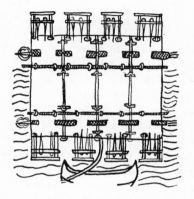

FIG. 95. Wooden bridge. Circa 1327 (Singer, p. 725, Fig. 658).

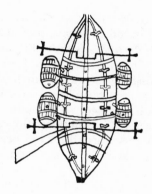

FIG. 96. Assault tower. Its height could be adjusted. Around 1327 (Singer, p. 725, Fig. 658).

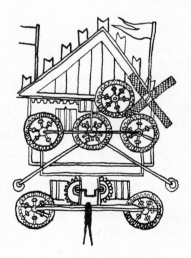

FIG. 97. Wind-propelled chariot. Around 1328 (Singer, p. 726, Fig. 659).

JOINERY

The possession of appropriate materials accounted for much in technological progress, but it had to be supplemented with a knowledge of how to assemble them. The methods utilized for this purpose naturally varied, depending on the nature of the materials and the use to which they were to be put. Thus it will be easier to review those aspects of the problem whose solutions, for readily understandable reasons, appear quite far removed from those discovered by classical antiquity.

Wood: cabinetwork   We have already mentioned the instruments used in the preparation of wood. Various precautions had to be taken if the objects built were to have the desired strength. The wood was used only when dry, that is, only after it had been aged for at least six years. The planks were left in a damp place (possibly in water), then were loosely piled up in dry shelters, where they were frequently turned; sometimes they were even smoked. Excessively aged wood, being subject to cracking and attacks by worms, was rarely used. Oak was the wood most widely employed in construction work because of its fine grain, even hardness, durability, and beauty. The trunks of oak trees were sawed into four pieces, at right angles, and each piece was then sawed according to various methods, always taking into account the texture of the wood and the cracks caused by the drying. We shall not discuss these techniques at length; they are still in use.

Until the thirteenth century straight-grained wood was the type most frequently used; it was worked with the chisel and the gouge. Cabinetwork bore a strong resemblance to small-scale carpentry by virtue of its combining of joints. Surviving medieval objects demonstrate a perfect knowledge of the various types of wood, the principals of good layout, and judicious use of the material.

Pegged and doweled joints were the types most frequently used. The uprights were generally held in place by pegs running through tenons and mortises. The panels that filled the intervening spaces were fitted into the uprights by means of tongue-and-groove joints; in this way, although its edges were narrowed to fit into the groove, the panel lost none of its strength in the center. For greater flexibility it was left free in the groove. It is true, however, that wherever possible the carpenters avoided large panels, which always have a tendency to warp. The panels themselves were not joined together by grooves or small tongues, which did not appear until the fifteenth century; they were joined either with dovetails, embedded rods, or studs of hardwood or iron. Layers of glue made from hides or cheese were spread along the edges.

Cabinetworkers, like carpenters, had learned to end bevelings and moldings at right angles to the joints, in order not to diminish the strength of the objects. They were careful to avoid large curves, which wasted much wood and forced them to cut across the grain. Donkey hide or canvas was often glued to furniture.

These joints had their weaknesses, however, and it is not unusual to see furniture literally bound with bands of iron, just as the jambs of certain doors were held between iron bars riveted together. In this way these objects were given the rigidity they lacked.

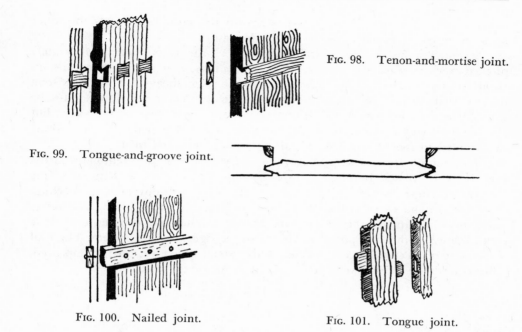

Fig. 98. Tenon-and-mortise joint.

Fig. 99. Tongue-and-groove joint.

Fig. 100. Nailed joint.

Fig. 101. Tongue joint.

When we study the construction of certain objects, we find in addition that these techniques developed in the direction of a greater economy of material as methods of joinery were improved. Until the eleventh century doors were formed of contiguous planks backed by other planks laid crosswise, and nailed. In several doors found at Gannat and Le Puy, the backing consists of a wooden lattice.

On the doors of the Sainte-Chapelle, the sections are firmly jointed with two uprights, three crossbars (battens), and braces that bring the weight of the door to bear on its hinges. The battens are jointed to the uprights with dovetails. In addition to tenons the braces also have tongue-and-groove joints to strengthen the whole; furthermore, the battens are held between two iron rods. The battens and braces are beveled, and lighten the framework. This is the type of door most frequently found during the closing years of the medieval period.

Roof trusses   The ancients had already mastered the art of the framework, and no major innovations in this area appear to have been made until the twelfth century. It should be noted, however, that the Norsemen contributed innovations that were closely related to shipbuilding techniques. Framework techniques were, however, closely linked with building construction. The truss systems of the Roman basilicas were utilized in the unvaulted churches of the Romanesque era. As in Syria, the alternation of stone arches and trusses was frequently used. The Romanesque truss, like that of Roman antiquity, included tie beams. These trusses were sometimes alternated with small trusses without tie beams; in this system large square beams split along the diagonal were used.

Light vaults also had to be protected. At first the roof rafters were rested directly on the vault, thus increasing its thrust. The builders raised the front and rear walls, and placed collar beams against the vault. Ultimately the collar beams were passed clear across the top of the vault.

Several reasons favored a transformation of the truss system. Moderately pitched roofs were perfectly suited to dry countries, and the enormous beams amply supported the considerable weight of the very heavy tiles. It was soon realized that steeper roofs were better suited to the temperate climates. In the old systems the roof truss rested directly on the outer surface of the vault, but this required very thick walls. In the Gothic period it was necessary to combine steeply pitched roofs and a narrow surface on which to rest them.

The common rafters were then supported with the help of templates in the case of small trusses. For large trusses the tie beams acted as templates. Tie beams and templates were jointed to the wall plates with dovetails; a stanchion ensured the bond between the rafter and the template or tie beam. The rafter was generally jointed to this template by two tenons and two mortises. The collar beam and the king post contributed to the rigidity of the system. The wall plates were often cross-braced together. This system, which was further improved, appears to date from the very end of the Romanesque period.

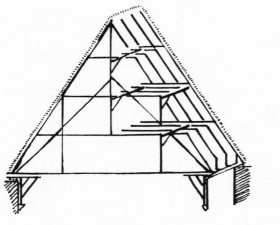

Fig. 102.   Roof truss with collar beams.

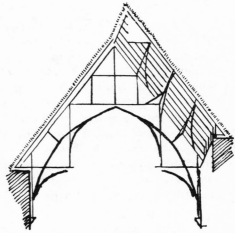

Fig. 103.   Roof truss with curved struts.

Some roofs were jointed in a system of two principal rafters supported by cross braces. Such braces were sometimes used even in tie-beam trusses. Roofs with small tie beams became the typical system of the Gothic period. Trusses with tie beams existed in Paris by 1180: the two cross braces and false tie beams, supported in the center or at their ends, lessened sagging. The stress developed by the small trusses without tie beams was later transmitted to tie beams by sills under the truss; now only the ridgepole needed to be added. The braces, which were no longer needed, then disappeared.

The same systems were employed in open timber roofs with large tie-beam trusses. A wainscoting of grooved planks was nailed under the rafters, and the curved shape given to the cross braces made it possible to build the inverted ceilings that are so frequently seen. The space between the wainscoting and the roofing formed an insulating space against the cold.

In order to eliminate the tie beams, it was necessary to discover a system that would prevent the principal rafters from spreading apart. This role was filled by collar beams and curved struts. At Westminster, where the truss has a span of more than 65½ feet, the builders left a large spece between the arch and the principal rafter, and filled it with a kind of latticework that turned it into a panel almost impervious to warping. In England, too, we find the hammer-beam roof, a type also frequently seen in the Scandinavian countries. Very often, however, the English failed to understand the role of the tie beam, which should always be absolutely horizontal and should not support anything. In the thirteenth century the tie beam often supported the king post.

Thus it is evident that the carpenters not only knew the art of joinery but were also able to understand the roles and stresses of the various components of their constructions. All these roof trusses were jointed either with dovetail joints (when stresses were involved) or tenon-and-mortise joints; the structure was put together with pegs. In theory, not a single piece of iron was used in the early trusses.

Metal joinery

Metal objects have not been as thoroughly studied as the creations of the carpenters and cabinet-makers. Very few metal objects of the medieval period have survived.

Grilles generally consisted of a frame, sometimes divided by crossbars into compartments; frame and crossbars were soldered together, and the spaces between them were filled with scrollwork made of small iron bars. At first these scrolls were held together by simple heat-shaped rings, as can be seen in the magnificent grille of Conques. Later the various bars were soldered together to give greater strength (the rings were retained, however), as can be seen on the grille of Ourscamp preserved at Rouen.

Sheet metal was generally shaped with appropriate hammers and anvils. The making of "seams" had to be learned; they were generally made by riveting rather than soldering. Examples can be seen in helmets of the period; the screw nut did not appear until the fifteenth century.

It is extremely difficult to determine the method used in making coats of mail. A few specimens dating from this period are still in existence, especially in England. The iron wires were probably soldered link by link.

Masonry

The practitioners of stonecutting during antiquity, who had reached a wonderfully high standard, had used no mortar, and their successors in the southern countries continued this practice for part of the medieval period.

Most medieval masonry, however, was constructed with the help of mortars made of lime and sand. Between the seventh and tenth centuries, very thick layers of mortar were used; after this they became much thinner. The quality of the mortar varied considerably during the entire period, being dependent especially on the good or bad firing of the lime; that of the thirteenth century was definitely the best.

Just as the cabinetmakers knew how to utilize wood, so too the medieval masons knew the proper techniques for laying stones. The cabinetmakers avoided cutting across the grain; the masons knew exactly which stones could be bedded against the grain.

The craftsmen of medieval western Europe were not always perfectly skilled in preparing their materials, especially when it was a question of thermal or chemical processes. On the other hand, they knew exactly how these materials should be used. While much of their knowledge had been inherited from their ancient predecessors, the medieval craftsmen had nevertheless learned to adapt it to different climates and to improve upon it.

CHAPTER 21

THE ORGANIZATION OF SPACE

I T IS OUR intention to discuss in this chapter all the techniques related to human habitation and modes of travel — in short, construction and public works. These techniques are all closely related, and largely surpass the stage of individual activity. We propose to proceed from the simplest to the most complex, beginning with urban and rural dwellings, continuing with buildings for special purposes (generally constructed of stone), and ending with major constructions — first the city with all its problems, and then the great hydraulics, port, and highway projects.

While detailed investigations in these areas are relatively numerous, good general studies are often lacking (with the exception of city planning). In this chapter we shall summarize these investigations. Documentation is scanty; much archaeological evidence exists, but technical interpretation of it has often ignored its detailed aspects in favor of artistic or economic considerations.

THE PRIVATE DWELLING

Medieval archaeological remains are very few in number, and most of the old houses still in existence are no older than the fifteenth century. Studies such as Commander Quenedey's work on the typical dwelling of Rouen would have to be broadened, in order to furnish us with a general picture of private urban construction in western Europe. However, we can hope that his conclusions are more or less valid at least for a large area of medieval northern Europe (the southern part of the continent having undoubtedly retained the traditions of antiquity).

Wooden construction in the cities
Wood predominated in construction for a long period of time. Stone did not begin to regain its former place until the sixth century, and the use of wooden trusses was not abandoned until the twelfth century for religious architecture, the thirteenth for military architecture, while in private construction it continued to be the general rule for centuries.

In wooden construction there are two organisms that are related but have different functions. There is the frame, which acts as a support, and the studs, which serve to ensure stability but whose principal object is to contribute to the support of the plasterwork, that is, the filling of the walls.

In medieval Europe, only one system of construction appears to have been used. Its basic principle was the use of upright posts extending in one piece from the base to the top of the building. The horizontal pieces are jointed in the vertical pieces, and the uprights support the roof truss. If the gabled end wall faces on the street, these posts form two lines enclosing the house; if the front wall borders on the street, only four posts are needed. The length of these posts is the most important point: in some cases they must be some thirty feet long. All these pieces of wood are very square, to ensure sufficient rigidity.

The horizontal beams of the façade strengthen the structure and serve as a support for the uprights of the studwork. The internal beams (joists or girders) support the flooring of each story. All these horizontal pieces end in tenons that enter the mortises in the posts, and the structure is firmly bound together with pegs. The sill rests on the stone filling between the joists; it generally consists of two pieces of wood, with several posts dovetailed between these pieces. The girders at the top are jointed in various ways, ranging from a simple bracing to a jointing of the horizontal plank in the vertical plank, and finally (the rational solution) to the resting of the joist on the post.

The strength of the girders, which supported the joists, raised a serious problem. Not only did they have to support heavy burdens (this problem was met by using various sizes, depending on their span), but their ends, which were tenons, were dangerously weak. This was remedied by the use of supports.

The studwork, which forms the wall, has the triple roll of closing, supporting, and strengthening. With thick walls of clay-and-straw mortar the medieval builders could be satisfied with a simple framework of posts and horizontal beams (girts). Thin walls made matters difficult. A more closely woven framework was required, which in addition helped to support the girts and prevent them from sagging; it was also needed to avoid that distortion of planks called "bowing," which occurs in frameworks built solely of right-angle joints. There was a disadvantage in this method, however: the builder ran the risk of lessening the strength of the vertical pieces by cutting them to receive the elements of the studwork, especially since the water ran down these vertical pieces and collected in the cuts, causing the wood to rot. In most cases plaster was used for the filling; grooves cut in the boards ensured its adherence to them.

According to Commander Quenedey, this method of construction — the primitive method — was related to methods of construction in use in the countryside during the same period. Few traces of clay-and-straw mortar remain in Rouen, a disappearance that is explained by the fact that when Normandy was attached to the crown at the beginning of the thirteenth century plaster of Paris came into common use. The jointing of the upper girders with cross braces is proof of the purely rural origin of this type of wooden construction. The urban modifications of this style were necessitated by the decrease in space caused by the increased height of the house. This forced the builders to lengthen the posts, which in turn favored the use of horizontal joints.

The type of construction we have just briefly described on the basis of studies made in Rouen does not appear to have been merely a local type. The system of posts extending the entire height of the house appears to have been widespread in France; it may have been even more extensively used in other cities outside Normandy. English specimens dating from the thirteenth and fourteenth centuries prove that its use extended to England and western Germany, and as far as Sweden; thus the system was widespread throughout all northern Europe. While it appears to have been modified at the end of the fourteenth or beginning of the fifteenth century, it nevertheless remained common in all the rural areas of those countries that had abundant rainfall and supplies of wood. We shall later see that there were certain elements that corresponded to this primitive type of wooden construction and that changes in this construction brought about modifications in these elements; one of the most important was the roof truss.

The interior It is very difficult to group together diverse techniques that today belong to very different trades, and it is no less impossible to dream of systematically surveying in this chapter all the problems posed by the construction of the interior of the house. We shall therefore discuss here only a few of these techniques.

In stone churches, both the cubic air space and the openings for ventilation were sufficiently large to eliminate the need for window sashes which could be opened and closed. This was not true of private dwellings. The arch did not lend itself well to these movable windows, and the lintel was in danger of warping. The medieval builders therefore resorted to relieving arches and frames. Until the thirteenth century the windows had no sashes; the glass pane was attached directly to the stone, which was a major inconvenience in the cold northern climates. The use of sash frames spread slowly, and did not become common until the fifteenth century. The windows were similar to church windows; they consisted of small panes of glass set in lead. In general the houses were better lighted than the Romanesque churches. In the warm countries the ancient system of glassless windows closed with wooden shutters was undoubtedly preserved, while in the north the windows were often closed with reed grilles or with transparent oilcloth stretched on sashes. The "window cloths," as they were sometimes called, were coated with a compound that included white wax and resin or terebenth; oiled paper was also used. These fragile shutters often had to be protected by grilles.

In Roman antiquity the only method of heating the average dwelling had been the brazier, which was very satisfactory in countries where the winters were not excessively cold. The hypocaust was used only in the homes of the rich or in certain public buildings. The tradition of the central fireplace had been preserved in scattered areas. The open fireplace is characteristic of the Middle Ages. Some historians have sought to see in it a post-Crusade import from the East, but it seems certain that this technique was utilized sporadically by the Romans. It did not come into general use until the eleventh century, the northern countries being endowed with much more abundant forest than those of the Mediterranean coasts. The back wall and bottom of the hearth were built of flat tiles to withstand the fire, and hoods were generally utilized. Large rooms had double fireplaces. All these fireplaces were conceived to burn large pieces of wood laid on small grilles, as appears in a sculpture in the cathedral of Rheims. Outside, the chimney was surmounted with a hollow column capped with an open, lantern-shaped stone or terra-cotta chimney pot, of a type that disappeared in the fourteenth century.

Water piping in private houses was extremely primitive. Most large houses had a well in the courtyard. In the southern countries rainwater was collected in cisterns which were sometimes located in the roof; Gothic specimens still survive in Apulia in Italy. Most private houses obtained their water from public fountains. Contrasted with these private arrangements, we possess a plan of the piping that supplied water to the main buildings of the cathedral of Canterbury; the various outlets are indicated on it. Such pipes were constructed either of terracotta tubes or hollowed-out tree trunks.

It would undoubtedly be useful to compare certain monastic installations which had greater space and facilities with private urban dwellings. One example is the piping of water inside the building. In the eleventh century the Abbey of Cluny had a series of twelve vaulted cells, each of which contained a wooden bathtub. Equipment of this type probably existed only in very large buildings.

The same is true for the methods of removing waste water. The abbeys had highly developed sewer networks. There were also public sewers, but frequently the cities did not have the resources necessary for such projects. In most cases they resorted, at least in private houses, to cesspools, provided the ground was sufficiently permeable to absorb the water. Failing this, it was necessary to build stone ditches, which were emptied from time to time. In cities where the contours of the land and the type of soil permitted, drains similar to those used in antiquity were installed, and strict regulations forbade the throwing of waste matter into the street. The medieval cities cannot have been paragons of cleanliness, and perhaps this was one of the causes of the epidemics that were so frequent in this period. However, large houses often had many latrines. The drainpipes were made of stone, even in wooden houses.

The kitchens of private houses utilized ordinary fireplaces, but the ovens of the monastery kitchens were genuinely independent buildings with pyramidal vaults surmounted by chimney pots. Remarkable specimens of the twelfth and thirteenth centuries can be seen at Fontevrault Abbey and the cathedral of Pamplona. Many others of the same type are known from old drawings. These ovens are reminiscent of the primitive central hearth; they were sometimes copied in industrial buildings like the lime kiln of Commelle (Oise), which dates from the early years of the thirteenth century.

No detailed study has been made of the medieval dwelling, and the techniques of certain regions are still completely unknown. It is certain that progress in housing was undoubtedly slower than in any other area. Man modifies his habits of private life more slowly than he changes his working methods. The urban house was transformed under the pressure of external circumstances, namely, the diminishing supply of wood, a situation which began at the end of the fourteenth century and grew steadily worse, reaching its peak in the sixteenth century.

LARGE STONE BUILDINGS

A completely different set of technical problems was involved in large stone buildings. Most of these were churches, which, because of the expansion of the faith, reached such proportions that completely new solutions had to be found to construction problems. In addition, geographical conditions imposed upon the builders new materials and special arrangements, for which the ancient techniques were no longer valid. We shall briefly mention these problems, many of which have not yet been completely solved.

Romanesque techniques François Choisy defined the basic problem of the Romanesque technique as the vaulting of the Roman basilica. The roof truss, which was common in all the early Romanesque buildings, was on the point of disappearing, and the Romanesque architects were about to demonstrate a twofold originality. The method of building up the vaults with horizontal courses, which had been the method of the Roman vaults, was replaced by a system of radiating blocks. The pillar that received the thrust began to be broken up, while still playing the roles of pier and abutment. As a counterbalance to the thrust of the vaults, however, was the idea of supporting them with buttresses, an imperfect solution that nevertheless revealed the first glimmerings of a spirit of analysis foreign to Roman antiquity.

The building materials used were more or less correctly dressed stones bonded with mortar. The leveling course became superfluous, and facings were eliminated when the stones had been carefully cut. The mortar now served to transmit pressure; it was no longer merely a binding, but a plastic material that determined the balanced distribution of burdens from one course to another. The stone was laid in place already cut — redressing *in situ* is disastrous in masonry construction. The arches were round, like their Roman predecessors, and their backs were generally curved. Until the eleventh century the ribs were constructed of cut stones, while the rest of the arch consisted of rubble; by the twelfth century the arches were being built completely of small, regularly shaped voussoirs.

The arcades rested on square abaci. The arches were back to back, and the tympanum tended to be crushed at the point while exerting a thrust on the haunches. The builders therefore tapered the arches vertically, which reduced the sharpness of the tympanum, or gradually narrowed them so as to leave a space between their backs. A third method was to slightly corbel the intrados; the arch was then supported on an engaged colonnette, while the archivolt rested on the main body of the pillar.

The Romanesque vault was a light, keyed, generally semicircular vault constructed of small stones; it was formed on a centering, and was combined with a protective roof. The ribs placed at regular intervals along the vault certainly acted as reinforcements to strengthen the thin vaults (although the arch also worked for itself), but were especially intended as an aid to construction, specifically to prevent the distortion of the tie beams during the centering.

For a long time the medieval builders avoided the groined vault. Instead of two intersecting barrels, the builders stepped them so that one began above the top of the other. The groined vault was gradually approached, but it was used only over the side aisles, where the spans were not as great. The diagonal rib was semicircular, while the curved transverse ribs were of almost the same height as the diagonal rib; either raised or ogive arches were used for this purpose. In certain regions (Auvergne, Normandy, Poitou), however, the Roman system of intersecting barrel vaults was retained. The barrel vault extended the entire length of the nave; it was broken by lunettes. A transverse rib between two bays was required only in dome-shaped groined vaults (for example, in the Cluniac and Rhenish vaults). Before deciding on these difficult and costly (because of the centering) solutions, the medieval builders tried domes, which had already been utilized by classical antiquity. The Romanesque cupolas were spherical or ogival. The cupola of Périgueux was constructed up to a certain point of horizontal relieving courses; this reduced the opening of the keyed portion and thus lessened the thrust. The early domes directly supported the roofing, but later builders were careful to pitch the slopes in order to ensure good drainage of rainwater. The cupola of Périgueux was topped with a cone, to cause water and snow to slide off. At Loches the sides were pitched so steeply that they ended in conical sections. The pendentives were spherical triangles or squinches.

The principal problem was the thrusts of the vaults. The Romanesque architect's only method of coping with this difficulty was to support the vault directly with buttresses or to subtend it with tie beams. In the case of the side aisles, thick walls with small windows were sufficient support. These side aisles in turn buttressed the nave, at least up to a certain height, above which the thrust was received on the inside by the pillar, on the outside of the building by a buttress supported on the haunches of the vaults of the side aisle. For barrel vaults, longitudinal beams sunk at the height of the haunches received the thrust exerted along its length, and in this way the thrust, distributed by isolated buttresses or tie beams, was better met. In the groined vaults of Vézelay can still be seen the fastenings of the tie beams, which were later made unnecessary by the construction of flying buttresses. The simplest idea was perhaps to construct (as at Tournus) a series of transverse barrel vault buttresssed on one hand by a massive façade, on the other by the semicircular dome of the apse.

Thus the Romanesque architect was not rebuffed by difficult solutions. But he was also wise enough to limit his ambition to relatively small buildings whose interior lighting was dim. His principal "intervention," if this word can be used, was the counterfort acting as a buttress — which the Romans had seldom used except in supporting walls — and the idea of placing these counterforts on the outside of the building in order to increase the area inside. The small size and high position of the windows were an advantage at a period when there were few means of closing them; they were protection against cold, drafts, and thieves.

The first vaults the Romanesque builders dared to throw over large naves were barrel vaults, with the springings placed as low as possible — so low that they could not admit the light. The solution adopted at Tournus — the cupola, but a cupola made of unsuitable materials — was limited in its possibilities. The early naves were practically blind naves. After building two-storied side aisles in order to open upper windows, the builders succeeded (at Saint-Étienne de Nevers) in opening genuine high windows; they had now reached the limit of their possibilities. This system became general in the Burgundian school when continuous tying and external abutments were invented. Finally, at Vézelay, groined vaults were used for the first time over the main nave; the tie beams were to prove insufficient, and flying buttresses had to be used. Along the banks of the Rhine the builders adopted very high groined vaults that were almost calottes, and massive pillars — this despite the cumbersomeness of this arrangement.

The birth of the pointed arch The question of the pointed arch has not yet been completely resolved. Numerous theories have been suggested, none of which are completely satisfactory. The dispute centers not simply around the origin but around the very principles of this new method.

Viollet-le-Duc affirmed the preponderant role of the ogive and the flying buttress, which were the innovations and basic elements of Gothic architecture. Others have demonstrated that the ribs were purely decorative and that the ogive vault was actually only a groin vault; the construcitve innovation, according to these authorities, was the widespread use of intersecting vaults. The flying buttress is supposed to have been developed later, for the purpose of supporting excessively daring vaults. Between these two positions adopted by numerous authors, a number of intermediate doctrines have been elaborated.

The difficulty involved in building the Romanesque groin vault was the cutting of the voussoirs used for the ribs that held the cell (the body of the vault) together: the slightest defect weakened the rib. Part of the problem could be solved by using diagonal arches to support the ribs; it would then matter little whether the lapping of the courses was correct or even existed at all. Supported diagonal ribs and independent cells were to be the inevitable conclusions of later research. The ribs were constructed of large stones; they were less compressible, strengthened the entire structure, and collected the major part of the weight by converting it into thrusts that were distributed along the vertical planes of the ribs. The cells had a slight tendency to slip to the side of the wall ribs. The strains were thus much better localized; as Choisy says, the play of equilibrium was completely in the hands of the architect. Formerly, the rib voussoirs had to be heavy, which in turn meant that the entire vault also had to be heavy. This was no longer necessary; the cells were now very thin, and the vault, being infinitely more flexible, could accommodate itself to distortions.

The plan followed for the ogive vault was that of the groined vault. The diagonal arch continued to be round in almost every case. Pointing the transverse arches and wall ribs (inevitably more narrow) sufficed to make their tops, like each point, equal in height to the homologous points on the diagonal arch (Figure 104). In a few cases the diagonal arch was stilted (early period of the Gothic arch) or in contrast raised, as at Chartres and Rheims.

FIG. 104.   Diagram of the ogive vault (F. A. Choisy, *Histoire de l'architecture*, II, 271).

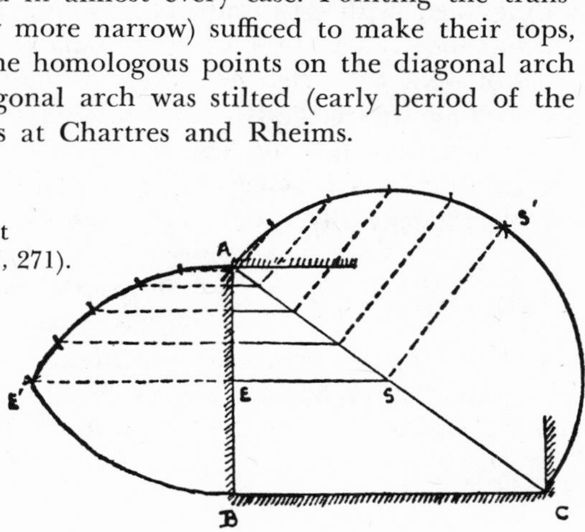

We shall undoubtedy never know the exact origin of this method of construction. The use of ribs under vaults has been applied to vaults of very diverse systems. The Arabs in particular used vaults over ribs, generally cupolas, which appeared around the tenth century, but they never applied this system to the groined vault, which was so well suited to it. In western Europe the system was applied in the Christian regions, using the same principle (for example, the vault of the Hôpital Saint-Blaise in southwestern France). The architects of the Île-de-France province and the area belonging to the French crown may have received the rib as applied to pointed arches around the first quarter or first third of the thirteenth century, from the Anglo-Norman archtiects who had begun to make systematic use of it around 1100, and the French may have systematized its use in an ensemble of pointed arches.

It is evident that the rib possesses an undeniable artistic effect, and it is no less obvious that a well constructed vault has no need of supporting ribs, as has been demonstrated by cases in which ribs were blasted away by aerial bombardments, leaving the cell intact. The rib certainly originated, however, as the support for a vault that was difficult to construct. In addition, compartmentalizing the vault permitted it to follow the movements of the structure. Lastly, by "monopolizing" the greatest part of the thrust, the rib made possible the use of more rational solutions for the buttressing problem.

The system of the ogive vault was accompanied by several new methods of construction. In the oldest vaults the ribs are keyed, independent arches from springing to top. A fairly large supporting surface was needed to receive their springing. In the thirteenth century the solution consisted of using horizontal courses as the first course of the springing (Figures 108, 109), while the interpenetrating profiles of the ribs continued along the entire height of the weight-bearing pillar. The surfaces of the cells were curved in all directions; they were undoubtedly constructed with a centering.

The walls retained most of the characteristics of the Romanesque walls. The stones were laid in place already cut, and no resurfacing was done. The surfaces of the horizontal joints were very carefully dressed to ensure the distribution of the heavy burdens to which they were subjected. The tools used were different, however: beginning in the twelfth century the bush hammer, a tool with a serrated head, was used for stonecutting. The layers of mortar became increasingly thick. The metal clamps sometimes used were fixed by sealing them with soft lead. The Gothic architect also utilized strengthening courses; they can be seen in the sloping counterforts of Notre-Dame of Paris (a series of stone ties placed obliquely in the body of the buttress), which deflected the thrust outward, and in the cross-grain colonnettes attached to pillars. The colonnettes, which were less compressible because they were monoliths, were able to support a heavier weight than if they had been constructed in courses. They were generally taken away once the pillar had finished settling.

The Gothic buttress   The Gothic pier inherited the role of the Romanesque pillar. The stress of the vault being oblique, the weight of the springing was limited. With the appearance of supporting girders, the courses of these girders replaced the corbel; the capital was eliminated, and the ribs were continued down to the base of the pier.

Plate 62.
Vault of the church of the Hôpital-Saint-Blaise (Basses-Pyrénées),
twelfth century. *Photo Van Eyck-Rouleau, Bègles*

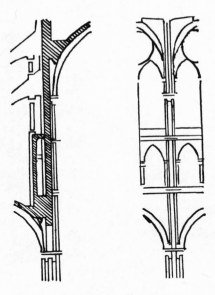

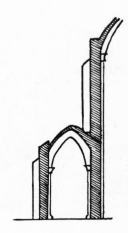

FIG. 105. Cross section and elevation of the Gothic pier (Choisy, p. 312).

FIG. 106. Buttressing by counterforts.

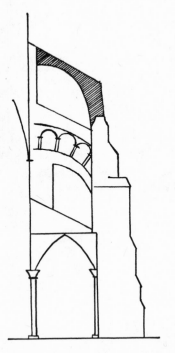

FIG. 108. Stone courses at the springing of the ribs (Choisy, p. 274).

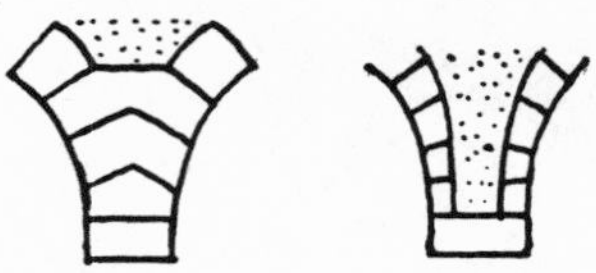

FIG. 109. Two methods used for the springing of the vaults (Choisy, p. 147).

FIG. 107. Buttressing by flying buttresses: the correction made on the buttresses of Chartres Cathedral.

The Romanesque barrel vault distributed the thrusts uniformly, and required uninterrupted abutments. The thrust of the raised groin vault resembled that of a spherical calotte, and it was prudent to have a continuous system of support in addition to the counterforts.

At first the straight counterfort of the Romanesque period was replaced by an abutment that was stepped, as if it were leaning toward the vault. The builders next returned (frequently because of lack of space) to the straight counterfort weighted at the top by a pinnacle. Thrust is an oblique force composed of a vertical thrust (weight) and an oblique thrust, and there was nothing to prevent the builders from opening a passage in the counterfort between these thrusts. They then separated the inside pier from the external portion — the true buttress.

The thrusts being concentrated, the wall between the piers and the wall ribs became nothing more than a simple space to be filled; it would later be opened and replaced by stained-glass windows. The piers were now isolated, and lost their stability, and the builders realized that the localization of the thrust was not absolute. These difficulties were resolved either by bracing the piers with a large transverse arch or by a rigid packing that filled the haunches of the vault and transmitted the thrust.

The Romanesque architects, and especially the Cluniacs, had succeeded in raising groin vaults while opening bays in the wall. After a certain period of time the buildings began to give some cause for concern, and additional methods of strengthening had to be found. Choisy notes that it was while seeking ways to shore up the Romanesque vaults that the Gothic architects discovered the flying buttress. At Vézelay, during the course of construction they had already tried to cope with the stresses by using a tying course, but had placed it too low, and a series of braces had to be placed between the haunches of the vault and the outer wall. Then they thought of using permanent stone bracing. The flying buttress thus appears as a keyed strut between the haunches of the vault and the buttress.

Timid in the beginning, hiding under roofs, utilizing the vaults of the side aisle, by the last third of the twelfth century the flying buttress had received the form that made it successful, that is, its isolation in space. The architects then sought to give it the best form possible without permitting it to lean out of line, which was a constant threat. By adopting very oblique arches, they removed a part of the vertical weight from the pier. By constructing it of two arches back to back, they made a larger opening that distributed the thrust over the full extent of the haunches of the vault. They also created flying buttresses with several flights, supported on the perpendicular line of the intermediate pier; these buttresses were invented as a corrective to a poorly placed buttress (it was quite soon discovered that the thrust was not localized exactly at the girder of the vault). A slope was sometimes added above the buttress to act both as another flying buttress and as a support for the drainage gutters.

The Gothic system was thus extremely simple. The flying buttresses eliminated the thrusts, and the piers were now merely supports. However, this support was subject to movements, and thus required a certain mobility at its base; the Gothic pier was straight, not stepped. The effects of weight were crushing and buckling (distortion by flection), with the latter being the most dangerous. Above the springing of the arcades of the side aisle, the pier consisted of a shaft resting on the column, and an upright joined to the shaft by struts. This upright was supported on a corbel, and was constructed with the help of cross-grained colonnettes; in a certain sense it was equivalent to a massive pier. To avoid buckling on the other side, the piers were braced, up to a certain level, by the blind arcades of the triforium, and above that by tying courses passing through the windows.

The flying buttress rejected the thrust onto the outermost masonry mass. The latter had a tendency to burst outward at a certain point; the builders therefore weighted these masses and made use of their slopes to support an overhanging pinnacle.

The logic of failures   It would perhaps be somewhat exaggerated to speak of the rationalism of the medieval architects, for this supposes an organization of knowledge that was certainly lacking. Undoubtedly it would be more correct to see in their work a remarkably well-organized empiricism. Work on the construction sites of the cathedrals was very slow, employing generations of architects and master builders. On these sites were transmitted the simple rules utilized by the builders, and especially the knowledge of the failures that had been encountered, and the solutions that had been found for a given problem.

The limited intellectual equipment of the builders is suggested in the catastrophes recorded in the chronicles; positive proof, obviously, is lacking. No scale drawings exist for this period; all that survives are a few elevations, especially for façades, and they are artists' drawings rather than architects' plans. A few measurements taken of the buildings prove in any case that the builders utilized extremely simple relationships for the proportions in plan and elevation. The architect drew a plan of his building (sketches can be found in the album of Villard de Honnecourt), and next to it he drew an elevation of the church. Their hesitations and failures prove chiefly that the architects were unaware of a certain number of basic features that were learned only by long tradition.

There were two preliminary problems. The problem of the foundations was on the whole solved correctly, by means of enormous substructures that in fact were sometimes out of proportion to the actual stresses. Little was known, however, about the behavior of the ground; at the beginning of the fourteenth century (1313) the vaults of the cathedral of Bourges threatened to fall because the ground had settled. A still more important factor was the choice of building material. At Bayeux, when the central tower was being constructed, the eleventh-century piers were insufficiently reinforced and the excavations were not deepened. Despite extensive repairs the building was not definitely saved until the nineteenth century. Little was known, in addition, about the compressibility of certain stones. At Beauvais, it was perhaps as much the poor quality of the stone as the lack of buttressing that led to the catastrophe of 1284; when the nave of the choir was rebuilt, the builders carefully avoided using the same calcareous stones. The cathedral of Meaux, constructed at the beginning of the thirteenth century, had to be repaired as early as 1253. In

1268 it was discovered that the materials were defective; the stone was settling. At the beginning of the fourteenth century a portion of the edifice had to be rebuilt.

Theoretical errors were made as well, especially in the buttressing of the vaults. The most typical example is that of Chartres. To resist a strong thrust from exceptionally large vaults, the architect had thrown, over the roofs of the side aisles, sturdy buttresses that rested against the wall at the point of origin of the transverse arches, and ogives reaching to the springing of the vaults (Figure 107). In the fourteenth century there was concern about the solidity of the vaults, and in 1316 Parisian experts, including Pierre de Chelles, master builder of Notre-Dame, were sent for. It was probably as a result of their visit that a third flying buttress was added higher up the wall, which it supported almost at the level of the keystone. At Amiens (second half of the thirteenth century) another flying buttress had to be added, in the fifteenth century, below the one that had already been constructed. At Évreux (1260–1310), at the end of the fifteenth century, the two abutments of the wall of the choir had to be combined to supply greater weight.

The most striking example of failure is certainly Beauvais. In 1272 the apse and a portion of the choir were finished. In 1284 several external buttresses collapsed, and a portion of the main vault fell. The catastrophe was attributed to an excessive distance between the piers. The unequal-sided quadrilateral vaults were therefore transformed into sexpartite vaults by adding an intermediate pillar. In addition, the center support of the two flights of flying buttresses was made to overhang the line of the center pillars of the ambulatory.

This is indicative of how much thought was given to solving the various problems of equilibrium. In order to economize on material and work, very simple methods had to be found. However, because of a lack of proper knowledge, "tricks" of construction rather than truly rational solutions were apparently adopted. The medieval architects achieved their results less through the creation of a theoretical body of knowledge than by a constant study of failures and successes, and by transmission from one construction site — those genuine permanent schools — to another of the results of these repeated experiments. It was at this moment that they learned to recognize the best beds of stone, and placed the flying buttress so that it had the proper curve. The medieval architect's undeniable mastery in the construction of large vaulted buildings was undoubtedly achieved through the repetition of simple plans. Not until a later age were construction methods rationalized, and only mathematical equipment made this possible. In the meantime the medieval architects were satisfied to construct buildings that might not meet the safety requirements of modern building codes. The restorers of later centuries (particularly skillful restorers appeared in the seventeenth century) often corrected the mistakes of the medieval builders.

THE ORIGINS OF A FUNCTIONAL ARCHITECTURE

It would be totally inaccurate to believe that the medieval architects and builders did not concern themselves with adapting the buildings they were commissioned to build to the use for which the structures were destined. In addition, they had before their eyes architectural specimens left them by the Romans.

Given their knowledge of various techniques, ranging from the great vaulted religious edifices to private houses which, although less complex, nevertheless posed numerous problems, the master builders possessed almost all the methods required to solve a great number of special problems. It is impossible to examine them all here, and the very brief study we shall make of some of them may perhaps serve to encourage more extensive investigation.

Public architecture   Public architecture developed from a number of already existing models. We shall not examine the varieties of monastic architecture, for which numerous and readily available studies can be found. Most of the community buildings of the monasteries were vaulted, and their conception, whether they were refectories, chapter halls, heated bathhouses, or parlors, was somewhat similar to that of the churches: they were long halls divided by rows of columns that received the springing of the vaults. Various special effects were also utilized in these buildings. In the dormitory of Maulbronn (Württemberg), for example, columns of unequal height were utilized to give the illusion that the room is larger than it actually is.

The arrangement of hospital rooms is a perfect demonstration of the fact that the architects had made a careful study of the requirements and had perfectly understood the manner in which they could be met. Hospitals were constructed at a distance from inhabited areas, generally downstream along a river. Most of them had large vaulted or paneled halls with one or two aisles. Particular attention was paid to the circulation of air: openings were left in the ceiling to form air ducts, thus permitting a change of air without causing drafts harmful to the patients. The windows were large, and were placed well above the beds. The same arrangements continued in use from the twelfth to the sixteenth centuries. The organization of the hospital services, like the techniques of building construction, were perfectly adapted to the needs of the sick.

Meeting halls were generally covered with open-timber roofs; most of them have disappeared, victims either of fire or of replacement by larger buildings. From the point of view of construction, the problems were approximately the same for a large municipal hall as for a market hall: in both cases an enormous roof was supported on stone piers. Pictures of the halls of Ypres show how elaborate these buildings could be. Ships were able to enter the water hall of Bruges through a covered canal. Many covered markets were built completely of wood; the roofs differed very little from those of such large stone buildings as, for example, cathedrals.

Warehouses looked like ordinary stone houses; the system of large vaulted halls with central pillars could be used for almost any purpose. Examples can be found in the *Grenier aux dîmes* (tithes warehouse) of Provins, and the customs house of Mainz, which dates from the beginning of the fourteenth century.

The medieval scholastic establishments were similar to the monastic buildings.

One sector of architecture, therefore, had no special techniques; the large hall with one, two, or three aisles was perfectly suitable for use as a church, refectory, dormitory, hall, meeting room, or hospital ward, and changes in detail of execution easily adapted it to these various functions.

Rural architecture   We have little information on the rural architecture of the Middle Ages. Private homes were houses with wooden frameworks, exactly like city dwellings. While urban construction began to develop beginning in the fourteenth century, the rural house remained unchanged for a long period of time, and the oldest houses still in existence were probably identical with those of the late medieval period.

It is almost impossible to supply correct information about farm buildings. They were undoubtedly constructed of wood, along the same principles as houses. The only farm buildings actually known are those of the abbeys, certain specimens of which are still extant. The granges were relatively numerous, including Meslay (Indre-et-Loire, early thirteenth century), and the Cluniac grange of Perrières (Calvados, end of the twelfth century). These were three- or five-aisled stone buildings, with open timber roofs steeply pitched to permit the passage down the center aisle of wagons laden with hay or harvests. Narrow openings in the gables ensured proper ventilation. The grange of Saultain (Nord) had parallel cart passages; that of Port-Lambert (Finistère) had a ventilating chimney.

The monastic wine cellars were generally vaulted halls with one, two, or three rows of columns.

Industrial architecture   Medieval industrial architecture has almost completely disappeared, and only a few well-studied monuments are still extant. For example, Commelle (Oise) has an early-thirteenth-century structure called by some authorities a lime kiln, by others an oven for oil. It has the form of an open fireplace — a hollow pyramid raised on four arches.

As regards mills, only those that were fortified, and whose techniques of construction consequently belonged to military rather than industrial architecture, are still in existence. No millworks have survived.

Here, again, monastic architecture succeeded in creating a building of a special type, in which all the hydraulic machines were assembled — the wheat mill, the fulling mill, the tanning mill, the pounding mill. At Fontenay the stream flowed beside the building; the wheat mill bridged the stream, and in the adjoining building were three vaulted workshops that certainly included a forge. At Royaumont, the stream crossed through the workshop along its major axis, under a high, narrow gallery. The canal was flanked by two galleries vaulted with ogive arches and buttressed by counterforts on their exterior walls. The building has been remodeled many times over the centuries, so that the arrangement of the various workshops can be distinguished only with difficulty. At Fountains Abbey in England the stream passed under the building.

These monastic edifices, together with the galleries of the saltworks of Salins, are almost the only industrial buildings of the Middle Ages that have been preserved. Miniatures and other pictorial representations of this type of installation are very rare, even impossible to find. While in certain cases the medieval architects succeeded in inventing a rational architecture, a great variety of now-vanished wooden buildings were undoubtedly the most common.

Military architecture   The early medieval military "engineers" utilized first the old Gallo-Roman fortifications (when they existed) and then hiding places during the great invasions and the period of troubles that followed. The *motte* (mount), an artificial hill shaped like a truncated cone that served as a base for the castle, and which was the origin of the keep, appeared in the tenth century; some historians attribute its origin to the Norsemen. Originally surrounded by a single fortification wall, the *mottes* were later enclosed within a series of walled areas in which villages were sometimes established. The keep, the ultimate refuge, was then placed either within the fortification wall or at its least accessible point, or in a position in which it could play a role in the defense. In this case it dominated the defenses, overlooked the entrance, was at a tangent to the wall but capable of being isolated, or commanded the passage from the lower courtyard to the castle. Most of these early constructions were built with wooden palisades. The tower, which was built on the *motte,* was also of wood, and was covered with the skins of freshly killed animals as a protection against fire; its entrance was quite high above the ground.

In countries which had quarries, the builders began at a quite early date to build in masonry; elsewhere, stone was used only for the foundation of the framework of the castle. The keep of Langeais (994) is the oldest stone keep extant. It is located in the center of a fortification wall, near an earlier *motte,* and is built on the plan of an unequal-sided quadrilateral. Flat counterforts added as an afterthought buttress the angles and center of the walls. The Gallo-Roman tradition is still very much alive in this early military construction. At Fréteval (1040), the keep is cylindrical, and is protected by a triple wall. The keep of Loches, built on a *motte* and again in the form of an unequal-sided quadrilateral, has cylindrical counterforts; its height (over 121 feet) reveals that progress had been made in construction. It was carefully faced with medium-sized ashlars. Its military interest, however, is very slight. The castle of Arques is not vaulted, which made this enormous mass of stone useless once the enemy gained a foothold inside.

The art of the military engineer appears to have actually been born at Carcassonne in the second third of the twelfth century. Three façades owed their defense solely to the science of the engineer. Here we find a rectangular area enclosed within a crenelated wall, and towers which command a considerable view over the curtain walls which they intercept. The base of the wall is oblique, to prevent sapping and to cause projectiles falling from above to ricochet. The moat is crossed by a bridge which divides in the center and has a movable bed — an early version of the drawbridge. The fire from three stories of loopholes, calculated so that the fire could be directed to the bottom of the moat, kept the enemy at bay; the only part that escaped this fire was raked by the fire of the archers in the towers. The entire medieval system of fortification is already in effect in the castle of Carcassonne. The plans of the keeps of Loches, Bressuire, and Château-Gaillard demonstrate the gradual improvement of this technique. At Ghent, the architect topped the rectangular counterforts of the wall with large bartizans acting as towers, which intercepted the ramparts at only two places.

Château-Gaillard reveals the mastery of the English engineers of the very end of the twelfth century. They utilized to perfection the contours of the land, and protected weak areas with outworks. The wall was formed of small half-circles arranged in such a manner that no area was left unbeaten by fire. The outline of the keep is reminiscent of the sixteenth- and seventeenth-century bastions with orillons.

Plate 63.
Keep of the Château-Gaillard (Les Andelys), end of the twelfth century. *Photo Roger-Viollet*

The French fortress acquired its definitive appearance in the time of Philippe Auguste (1165–1223). The castle of Dourdan (1220) is one of its first accomplishments. It is regular in shape, and has a wall intercepted by semi-cylindrical towers and a keep at one of the angles; its large cylindrical tower is separate and isolated from the wall. Here, for the first time, the keep is vaulted with ogives. The castle of Coucy (1225–1230) brought this form to an extraordinary stage of development. It is an irregularly shaped (because of the terrain) quadrilateral inscribed in a perfect geometrical figure. The keep is isolated in the center; it is 101½ feet in diameter and 177 feet high, and the walls are approximately 24½ feet thick at the base. The Constance Tower, the keep of the city of Aigues-Mortes, is of the same type. The double wall of Carcassonne corresponds to the same preoccupations, repeating the style adopted not only in the castles built in the Holy Land but also at Loches. This succession of walls was intended both to increase the obstacles in the path of attackers and to keep destructive engines, whose range was limited, at a safe distance.

In the first half of the fourteenth century there were no further innovations, and military architecture stagnated in tried and tested formulas. At the moment of the appearance of artillery (which, at least in the beginning, did not include siege machines) military architecture was thus in possession of all the basic elements of its theory. The idea was not so much to multiply obstacles (although these were frequently developed into an effective system) as to ward off the most obvious dangers. The enemy had to be kept at the greatest distance possible, whence the use of extremely high walls, when the terrain did not lend itself to simple constructions on rocky escarpments. By breaking the attack by means of a series of obstacles, the defenders avoided surprise assaults, and kept the siege instruments designed to batter the walls at a distance from the center of defense (that is, the keep). The openings for firing were stepped in such a way as to permit the surrounding terrain to be beaten for a certain depth, while the system of towers ensured that no areas were left unbeaten and completed both the defense of the foot of the fortifications and the battering of the walls. Lastly, the thickness of the walls and the series of moats made sapping operations very difficult. Undoubtedly there were also internal difficulties involved in carrying out such operations, for sappers could not very successfully guide themselves under the ground or shore up the tunnels. There are few examples of cities and fortresses being taken as a result of a collapse of the walls caused by sapping.

It now becomes understandable why the medieval military engineers attempted to batter down the walls with projectiles launched by the artillery mentioned above. But the thickness of the masonry made the success of such projects difficult. Scaling was also a highly dangerous operation.

The construction of strongholds, like that of cathedrals, reveals the existence of preconceived plans and the frequent use of geometrical figures. The plans of the castle of Coucy and the cathedral of Bourges are very revealing in this regard; moreover, this is why those fortresses whose plans were not completely dictated by the contour of the terrain are much more interesting in this regard. Not only the plans but also the direction and shape of the loopholes were designed according to geometric models rather than calculated; the extent and depth of the terrain which had to be beaten was known.

Surprise was the principal tactic used for conquering strongholds. There were few regular sieges in the Middle Ages; ruse and starvation were the major weapons. This means that the engineers who constructed these edifices were seldom consulted when it came to attacking them. The use of engineers for both defense and attack was to be one of the characteristics of the succeeding period.

MAJOR PUBLIC WORKS

The term "major public works," so familiar to our ears, is actually quite inappropriate for the Middle Ages, for it presupposes centralized states and a very advanced organization of society. In a world where political power was extremely fragmented, it was hardly possible to envisage projects on a national scale. However, collective projects that surpassed the purely local level apparently did exist.

It was for these collective works that techniques were developed whose continued existence undoubtedly permitted or at least favored the flowering of the centralized monarchies of later centuries. Some of these projects, moreover, demonstrated the need for a broader organization surpassing the traditional political and geographic limits; this was particularly the case for the major hydraulics projects, roads, and bridges. Others were certainly more limited, but they utilized techniques that were so widely practiced that their reciprocal influence already constituted a prelude to the social organization we have just mentioned; such is urban planning, which certainly profited by the universality of various types of architecture, including the great religious buildings and a few civil and military structures. The methods of execution are for the most part already known to us; building a wall for an edifice or a wharf requires the same technique in both cases, especially at a period when it was not known how to calculate the thrust of a slope. The medieval contribution to public works was the concept that the project should be viewed as a single entity, an idea that presupposes a well-developed technique of organization; even here, however, the Romans had led the way. Their lesson had not been completely forgotten as is testified to by the large number of manuscript copies of Frontinus's work on the aqueducts of Rome. The medieval West perfectly understood the importance of "environmental planning," as we would call it today. Apart from the cities, which in most cases are now quite unrecognizable, few written documents are still extant; of these few, some are nevertheless of great value to us. We have attempted, in a study in which the legal and political elements are frequently as important as the material facts, to emphasize the latter.

City planning   The urban techniques of the Middle Ages are difficult to grasp. A lack of plans, a lack of documents, and the scantiness of archaeological remains make this study still more difficult. The influence of antiquity — for the Gallo-Roman cities had not completely disappeared — and changes made by the Renaissance baffle the medieval researcher. It is nevertheless possible to pinpoint, not medieval innovations, but at least a few of the ideas that appeared at this time.

Most of the countries of western Europe had Roman cities with regular, geometrical layout and major improvements (sewers, water supply, and so on). Many of these cities were unfortified, and were therefore engulfed by the barbarian invasions. In order to save what could be saved, their area was limited so that they could be enclosed in primitive fortification walls. When commercial exchanges were resumed, "suburbs" were born that were later (between the eleventh and thirteenth centuries) included within new fortifications. When the medieval walls were square or rectangular, the plan of the city was not affected (for example, Rouen, Orléans, Strasbourg, Cologne, and Ratisbonne). Where the walls had lost this geometrical regularity, the streets were then forced to follow their contours (as at Andernach, Senlis, and Poitiers). In many cases the wall was based on a monument either ancient or of recent vintage (Bourges, Paris, Périgueux). Some elements remaining from antiquity even contained individual, independent agglomerations; such was the case of the Roman arenas at Nîmes and Arles.

The political and social structures sometimes led to the birth of cities like Angers, Rodez, and Toulouse, which actually consisted of two (three in the case of Toulouse) cities, one belonging to the count and one to the bishop. This situation created remarkable distortions in their plans.

It was in the new cities, created from the wilderness, that a spirit of organization was most in evidence. Their origin as cities that grew up around a Roman villa, feudal cities, or monastic cities, does not imply any delberate purpose. It is, however, in what have been called the "new cities" that the translation of preconceived ideas is best seen.

The radial-concentric cities and villages had a dual origin: first, their ease of defense, reminiscent of the barbarian encampments, or the attraction of a central element (the church, as at Limoges; the castle; as at Bristol; the city square, as at Bruges and Middelbourg; a fountain, as at Foix). An Eastern influence may also have played some role in this pattern. Sometimes the city grew in concentric rings along a river (Amsterdam, Oxford, Hamburg, Amiens). Regular developments were frequent in the new cities. We have definite evidence that certain individuals were placed in charge of these developments, but they do not appear to have been "city planners." In 1292 Edward the First requested twenty-four cities to elect individuals capable of apportioning lands. At Montréal (Gers) it was a notary who drew the map of the new city; he was also called to Castillonnès. However, we must not consider the builder of the new cities as a perfect geometer; he succeeded in preserving the imaginativeness that has saved many cities from the absurdity and monotony of certain modern plans, and in adapting them to technical and climatic conditions. In this regard the plan of Aigues-Mortes is one of the most remarkable. To begin with, the main square is irregular. None of the streets follow an absolutely straight line, undoubtedly for esthetic reasons but also to avoid providing a path for strong winds. The same concern is shown at Beaumont in the Périgord region, which dates from 1272. Cordes, founded in 1222, and situated on a promontory, could not adopt the rectilinear plan, and thus an "acropolis" plan was laid out, with streets following the contours of the terrain. Two exceptions to these practical ideas have been advanced, but they are, precisely, exceptions: Montpazier, founded in

1284, and Mirande, begun in 1285; their plans are rigidly geometrical. In contrast to what has been written concerning the French Midi, no distinction is made between French cities and English cities. Cities are known rather as the creations of one individual, perhaps because he utilized the services of a particular technician; thus we speak of "the cities of Alphonse of Poitiers," "the cities of Eustache de Beaumarchais."

We must be careful, then, to avoid hasty generalizations. It is true that a tendency to parallelism can be seen in the Anglo-Saxon countries, a tendency that may be due to the influence of certain Roman cities (York, Canterbury, Winchester), but which may also be spontaneous (Oxford, Salisbury). Care was even taken to leave a number of open areas, as at Winchelsea (1288). In Germany an attempt has been made to distinguish two principal zones (south and west, north and east), the first of which is supposed to have more readily adopted the regular plan; there are, for example, the geometrical plans of Freiburg-im-Breisgau (1120), Munich (1158), and Bern (1191). Radial-concentric cities are more frequent in France (Eastern influence) and in Germany (barbarian influence). Spontaneous, regular plans are characteristic of the English cities, cities acquired by France, the *bastides* (fortified "new" cities of the French Midi), and the German cities of the twelfth and thirteenth centuries.

Various other particularities of medieval "urban planning" can be mentioned. In contrast to the Roman city, the "new" medieval city generally did not have a square — space was too limited. In cases where there was an older city already in existence, the market was situated at the boundary of this older section and the "new" city. Elsewhere, it was located near an important monument, an abbey, or, in most cases, a castle. The grouping of the trades in certain streets appears to result from an Eastern influence, ethnic communities (essentially ghettos) from a Byzantine influence. The university cities did not constitute an absolute rule.

Problems of municipal administration appear to have been much more difficult to solve. The plan of the streets was, as we have seen, quite irregular. Changes of direction, irregularities of alignment, and arcade bridges undoubtedly resulted less from esthetic than from military considerations and from the necessity of protection against sun or wind. There were streets that were open to wagon traffic, others that were not; the former do not seem wide by modern standards — 26¼ feet maximum in the thirteenth century. This is because the tandem arrangement, adapted to the new system of harness, was better suited to streets of this type, and animal traffic was still very abundant, as can be seen in the manuscript of the *Vie de St. Denis*. Recessed open spaces here and there served as "garages."

The open space above the street was narrowed by overhanging upper stories, at ground level by porch roofs and openings to cellars. In certain cases laws were made (Douai, 1245; the Draconian ordinance of Ratisbonne) that were aimed at preventing excessive encroachment on the public street. At Prague in 1331, administrative authorization was required for this reason for any building bordering on a public street. The city square, when it existed, was a closed area remote from the flow of traffic, with streets opening into its corners (the Piazza della Signoria at Siena is one of the best known and best preserved examples). At Assisi in 1246 the alignment of the houses on the square was regulated. Streets were probably unpaved until the twelfth century. Philippe-Auguste ordered a start to be made on paving certain streets of Paris in 1184; London was not paved until the end of the thirteenth century, under Edward the First. In the fourteenth century paving became almost universal. The paving ordered by Philippe-Auguste consisted of square paving stones five feet long and 13½ to 15½ inches wide, with beveled edges. The stones were laid Roman-style, on a bed of cement mixed with crushed tiles, or were simply buried in a bed of fine sand. Street lighting was almost nonexistent.

The problems of water supply were equally important, and we are astonished to learn that in this regard the Middle Ages marked a regression from the standards of antiquity; methods and certain facilities were sorely lacking. Aqueducts of a certain length required a political unity and security that were very seldom present. However, some of the Roman structures were still in existence, and these were utilized and constantly maintained (the decay of the aqueduct of Segovia dates only from the nineteenth century); some of them were even imitated. St. Aldric, Bishop of Le Mans from 832 to 857, ordered the construction of an aqueduct four and a half miles long, with a vaulted reservoir; it passed under the archbishopric, and fed two fountains. The Chaillot and Arcueil aqueducts that supplied Paris had been destroyed by the Norsemen. Between the sixth and tenth centuries the monks of Saint-Laurent and Saint-Martin-des-Champs constructed the aqueducts of Pré-Saint-Gervais and Belleville. These two conduits combined supplied only 350 cubic meters of water per day; by the end of the Middle Ages their water was being distributed to the public fountains. Channels or small galleries were constructed of small stones, without mortar, laid on the impermeable layer of clay carrying the water to be harnessed; the water entered the gallery by trickling between the stones. The gallery was covered with paving stones coated with a layer of impermeable clay so as to prevent the infiltration of surface water. Shafts and manholes were dug so that cave-ins could be discovered and repaired. The water flowed into a common basin, the fountain of Pré-Saint-Gervais, whence a terra-cotta conduit approximately twenty-two miles long carried it off. Many inhabitants utilized wells. An attempt was also made to increase the number of fountains; the Saint-Lazare fountain was already in existence in the time of St. Louis, that of the Innocents or Les Halles was equally ancient.

Medieval aqueducts did exist, however, and a few splendid specimens have been preserved. At Coutances can still be seen the very beautiful ruins of a Gothic aqueduct built in 1277. The aqueduct of Sulmona in the Abruzzi dates from 1257. At Limoges there exist subterranean aqueducts from the Romanesque period, galleries cut into the tufa and wide enough to permit the passage of a man. The aqueduct of Aigoulène, supplying the city of the same name, was built between 1206 and 1216; it carried water from the Borie. The monks utilized the same technique to supply the water needed by their monasteries. Lambert, abbey of Saint-Bertin from 1095 to 1123, supplied his abbey's needs by means of a hydraulic machine. The Cistercians built numerous aqueducts — Saint-Polycarpe (Hérault), 1159; Casamari (near Rome), 1220; Obazine. The public fountains were often constructed thanks to the generosity of important personages; a num-

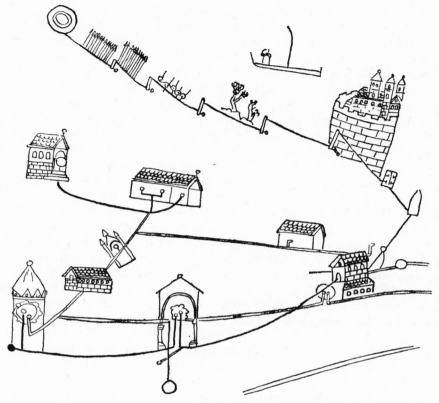

FIG. 110.   Plan of the water piping of Canterbury. Thirteenth century.

ber of them finally disappeared in the seventeenth century. Streams were harnessed and protected, and at Garde-Adhémar (Drôme) in the twelfth century, and the fountain of Houvélie at Lectoure (Gers) in the thirteenth century. The fountains of Siena and San Germignano date from the thirteenth century. Rheims had a fountain fed by an eleventh-century piping system; Provins received its fountain around 1160, Lagny around the same date. The fountain of Saint-Denis dated from the period of the Abbot Hugues (1197–1204); Italy had the fountains of Viterbo (1270) and Perugia.

Major hydraulic works   The Romans were past masters in the field of hydraulics projects, and this tradition had probably been retained in the Mediterranean countries despite the upheavals of the barbarian invasions. Aqueducts and irrigation and drainage projects had attained a high degree of perfection, and the Middle Ages would adapt them to new climatic conditions.

The work of protecting the land against submersion by rivers or the sea was practically unknown to the ancients. The gigantic project of the first dikes along the Loire was begun and completed during the reign of Louis the Pious (814–840), which undoubtedly makes it the first major public works project in the medieval West. Later modifications were completed around 1170. Along the coasts of northern Europe it was necessary to combat the invasions of the sea. The rivers and streams formed enormous alluvions that were soon conquered, drained, and cultivated, but these alluvial terrains sank and caused encroachments by the sea. It is also possible that there were tectonic collapses, which may have had repercussions throughout Europe. Thus the collapse of 839 may have simultaneously caused the diversion of the Rhine and the invasion by the sea of a large part of present-day Holland, and created the bay of Mont-Saint-Michel. The Zuider Zee was formed by the inundation of December 14, 1287, which caused much loss of life and engulfed a considerable amount of land. Here again protection was needed for the crops. In 802 Frisian law begins to speak of the seawall, but we do not know to what exactly this expression corresponds. For these early seawalls recourse was apparently had to artificial hills or dunes. The first genuine dike, called the "Golden Ring," was constructed in 1260; a few sections of it are still in existence. It was built with piles of large stones filled in with an earthen filling. The Maas (Meuse) River was dammed in 1270 to protect the country against floods from without. (This damming, which was undoubtedly poorly done, was one of the causes of the terrible catastrophe of 1421.) The lands it protected and reclaimed were drained, and numerous villages were created It is difficult, however, to determine by what method the water was pumped out.

Descending the coast, we find that similar projects were carried out in other regions. In Belgium, the dike of Count John was begun in 1289. In the twelfth century the County of Guines already had drainage canals. The inhabitants of the Baltic coast appear to have sent to Holland at an early period for Dutchmen to carry out similar projects along the Baltic; in 1103, for example, the Bishop of Bremen made use of Dutch talent in projects for controlling the water level at the embouchures of the Elbe. We find Dutchmen in 1130 in the Welster Marsch. The Norman coast also had a system of dikes by the twelfth century.

Draining of marshy lands had been done in Roman antiquity. The Middle Ages gradually relearned its principles and applied them in numerous regions. In Poitou the most solid alluvions were soon brought under cultivation. A ditch was dug around the piece of land selected, and the soil thus removed was piled up inside the enclosed area. At the foot of this ditch a second ditch was dug (the outer ditch was called the *achenal*, the pile of fill was known as the *bot*, and the inner ditch as the *contrebot*). Gradually it became the practice to coordinate the projects and to build large, continuous *achenaux*, some utilizing the slope toward the sea, others at right angles to this first network. For the former, existing streams whose beds had changed were often utilized. A small dike defended these reclaimed lands from the sea. All the slopes were shored up with piles and bundles of faggots, and trees were planted whose roots helped to maintain the stability of the *bots*. The streams were naturally dammed with movable gates or sluices that began to be mentioned in the tenth or twelfth century. These canal locks had the additional advantage of forming fishponds, and mills were added in the eleventh century. In the thirteenth century iron sluice gates and flap-valve gates (mentioned for the first time in 1249), which were closed automatically by the flood tide of the sea, were created. This description of the swamp of Poitou, the first major drainage projects of which were begun between the last third of the twelfth century and the first quarter of the thirteenth, and completed in the last quarter of the thirteenth century, gives us an idea of the techniques utilized throughout Europe in similar cases.

These projects had another advantage: they made navigation on certain streams possible. It was so when the streams of the Poitou marsh were brought under control. In 1175 the Countess of Flanders granted to the monks of Anchin the right to dig a new bed along the Scarpe waterway, on condition that the stream be navigible. In 1271, at Lille, the Deûle was made navigible and the Haubourdin Canal was created; by 1285 Spaardam in Holland had double locks. Projects were carried out between Bruges and the sea to preserve the traffic of this port.

In a desire both to eliminate floods (still frequent in modern times) and to develop irrigation, the hydraulics works of the Po River were begun, and were pursued almost without interruption from the twelfth to the sixteenth centuries. Here generations of engineers were to be trained, from the early anonymous engineers to Leonardo da Vinci in the fifteenth century. The diversion of the Ticino was begun in the twelfth century; in the thirteenth, that of the Adda at Cassano, to form the Muzza, was realized. At the beginning of the fourteenth century the diversion of the Ticino was continued as far as Milan, with the Pavia Canal, and works to control the level of the Po from Ponte Alberto to the junction of the Lambro were begun.

The Romans had been highly skilled in the techniques of irrigation projects, and numerous examples of their work can be seen both in Europe and North Africa. The invasions and disturbances that followed upon the fall of the Roman Empire were not very conducive to the maintenance and improvement of the system of canals. The Arabic invasion was the deed of nomads rather than farmers, that is, of men who took little interest in such problems. Thus it was not until the stabilization of the Arabic Empire that the Roman irrigation networks, in Spain for example, were put back into operation. The Arabs undoubtedly made additions to the system, and adapted to it the hydraulic wheels they had brought with them from western Asia. They did their work so well that it is now difficult, for lack of detailed investigation, to distinguish the various contributions of the Roman tehnician, the Arabic invader, and the medieval Christian conqueror. We should mention, however, the Arabic masonry dam of the Río Segura, 656 feet long, which raised the waters of the river upstream by twenty-five feet. Jaime I's decree of 1239 regarding the *huerta* (irrigation system) of Valencia mentions eight irrigation canals, seven of which were ceded to the inhabitants (the last was ceded in 1268), but this is simply a confirmation of the state of affairs that had existed prior to the reconquest. The dam of the Río Segura, which controls the *huerta* of Murcia, probably constructed in the tenth century, gave rise to two canals, the *acequia* of Alquibia and the *acequia* of Aljufia, the principal arteries of the network of the plain of Murcia. The *huerta* of Orihuela follows that of Murcia, and probably dates from the same period. It was also the subject of decrees made by Alfonso the Wise.

Irrigation canals are mentioned in the plain of Roussillon in the thirteenth century, the Millas Canal having been ceded in 1163. The canal regulator of the

Ille Canal, in the same area, was the product of a dam consisting of a line of large stones, boulders, and gravel, held in place by bundles of faggots, trunks of trees, and piles, and strengthened from behind by stone packing. The truth is that this was a haphazard construction that was often carried away by the violence of the current.

Strange transformations of the landscape were taking place. In the Dombes and Brenne regions it was realized in the thirteenth century that the land was losing its fertility. The monasteries of the region conceived the idea of damming the streams by means of earthen levees 6½ to 10 feet high, thus forming ponds whose waters deposited a rich alluvial mud. These ponds were periodically emptied and the land put back into cultivation. This system also created permanet swamp areas. The same system was employed in Sologne in the middle of the thirteenth and especially in the fourteenth and fifteenth centuries, thus giving this country its present-day physiognomy.

There were also failures and catastropes that could not be arrested. We have already mentioned those in Holland. We could also mention the case of the Alpine torrents that often threatened (and still threaten) cities and villages. During a flood in 1191, the Romanche River in Oisans had formed a great dam in the gorge of Livet, behind which formed a lake that engulfed Bourg-d'Oisans. Another catastrophe occurred in 1219 when this dam broke, again during a flood, carrying everything downstream for a great distance. Grenoble, too, was perpetually threatened by the floods of the Drac. Beginning in 1315, methods for diverting the bed of the river were investigated. For this purpose wooden coffers (*arca*) were used; they were filled with stones and arranged in wharfs or dikes. The city was not put out of danger until 1392.

Other cities, in contrast, made use of their rivers to supply their industries with the power they needed. Numerous diversions were created, solely with a view to turning millwheels. The city of Troyes originally had two rivers, the Seine and the Meldançon, the latter a stream of little importance. Diversions of the Seine were carried out between the eleventh and thirteenth centuries to supply drinking water and water for the moats and for power. In this way the Moline Canal was created; at the factory of Pielle it divided into three branches, the twisting river, the Planche Clément Canal that flowed back into the Meldançon, and the Pielle Canal. The Trevois Canal was created in 1174; it was the source, within the city itself, of the large and small rivers. By the twelfth century there were already eleven mills in the city or on its outskirts; in the thirteenth century, three new ones appeared. Ultimately twenty-five millwheels were turning in the city, and their number continued to increase. The utilization of the various streams that crossed Rouen reveal the same abundance of projects. At the end of the tenth century or beginning of the eleventh, there were already eight mills along the Seine. On the Robec, inside the city, there were two in the tenth century, five in the twelfth, ten in the thirteenth, and twelve at the beginning of the fourteenth; on the immediate outskirts of the city, one in the eleventh, five in the twelfth, and six in the thirteenth centuries. On the Aubette one mill appeared in the eleventh century, three in the twelfth, and six in the thirteenth. The same situation occurred in cities like Chartres and Liège.

Hydraulics projects thus appear to have been particularly important in

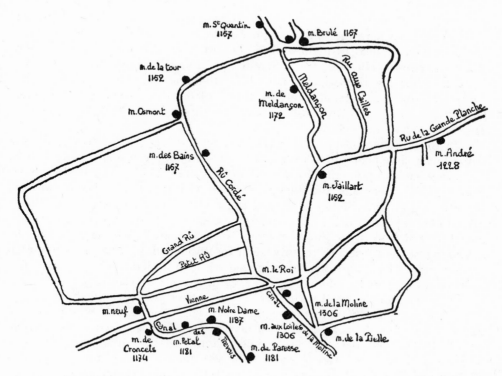

Fig. 111. Branches, canalizations, and mills at Troyes. Twelfth–fourteenth centuries.

the medieval period. We know almost nothing of the techniques used for these various projects: terracing, at a period when the wheelbarrow was perhaps not yet widely used, techniques of warping, techniques of leveling when it was a question of following the line of a slope for a certain distance. The only precise information we have concerns the large levees or barrages made of piles with a filling of stones, and the wooden coffers filled with stones that we have just mentioned. It is very evident that certain archaeological investigations would permit us to learn more of these major projects — but medieval archaeology is almost exclusively devoted to works of art. It must also be recognized that sea-walls, dikes, and irrigation canals, which have been continuously maintained and modified, have undoubtedly lost much of their original character. A particular case in point is the irrigation systems of Spain, a great part of which were improved in the eighteenth century, to the point that it is almost impossible to distinguish the respective contributions of different centuries. The only extant pictorial documents relate to the plans of the water distribution system of Canterbury, around 1167, and these supply only limited information for a very small project.

Ports   We know very little about the medieval ports, a situation that is due to the lack of detailed documents and plans. It is very probable that the Roman traditions were continued, especially in the Mediterranean, where it was not necessary to be concerned with tides and where the natural indentations in the coastline, sheltered from winds and untroubled by major difficulties with regard to their depth, are relatively numerous. The case was different in the north, where river traffic was important; for this reason ports had to be located at the mouths of, and sometimes even on, the rivers, at a certain distance from the sea and therefore more difficult of access, and continually harassed by silting caused both by the fluvial alluvions and by the ebbing of the tides.

In the Mediterranean moles were used to close the deepest coves; to construct them, large boats laden with stones were sunk between small, rocky islands. Marseille was exceptionally favored: there not even a dike was needed. Boats were rented to clear the port; they were laden with the mud to be thrown into the sea. In 1323 the commune ordered the construction of a special machine — a kind of wheel, with buckets attached to it vertically, supported by a raft. The entrance, which today is almost 328 feet wide, was artificially decreased at the beginning of the thirteenth century, reducing the opening to two passages, each 46 feet wide, which were sometimes closed off with a chain. The shipyard was located in the Plan Formiguer, on the present site of the Canebière.

Beginning in the twelfth century the consuls of Genoa decided on a series of measures for the maintenance of the port. The famous mole was already in existence by 1133, the year in which a tax was levied to lengthen it. On December 16, 1245, it was damaged by a violent storm; in 1251 it was repaired and lengthened. In 1250 Charles of Anjou, who was determined to protect the coasts of Provence from the despotism of the Genoese and the Pisans and the plunderings of the Sicilians and the Saracens, ordered the construction of a shipyard at Nice to equip and repair warships. Jacques Caïs and Guillaume Olivari, who were placed in charge of this project, were not satisfied merely to wall the area chosen (between the Ponchettes and the mouth of the Paillon): they dug a harbor, and on the south fitted out the horseshoe-shaped Port Saint-Lambert, with its two fortified breakwaters, which was reserved for coastal traffic. Thus the first harbor of the Middle Ages was created. The Port of Trani in Italy was also protected by dikes. At Naples, piles were used in preparing the foundation of the walls, dikes, and wharves (1300–1301).

In the north, as we have said, the situation was completely different. Let us first determine a few terms that are somewhat confused in the minds of the historians. The word "port" (*port*) was applied to strands, areas dug in the mud or the sand by the vessels themselves, and to beaches where boats were beached with the help of capstans — thus, primarily to areas located on coasts. The word "harbor" (*havre, hable*) designated a small bay with a narrow opening, a cove, or the enlarged mouth of a river, both improved and unimproved. In tidal waters the harbor is always empty at low tide, and the ships are beached as in ports. When applied to rivers, the word "port" referred to a dock, designated on the seacoast by the word *étaples,* from the Latin *stapula*. There was a gradual transition from one type to the other during the Middle Ages. In Norway, for example, the first maritime establishment (ninth century) was located on the right bank of the Oslo fjord; it was a simple beaching strand, utilizing a shallow cove. In the tenth century it was replaced by Tonsberg, located at the end of a long narrow fjord. At the close of the eleventh century, Oslo, situated at the end of the fjord, came into existence. Thus harbors replaced ports, moving upriver.

London was a river port linked either by coastal ships or land transport with the ports from Ipswich to the Ile of Wight (which latter formed the "Five Ports" group). At the end of the ninth century Alfred the Great constructed a port system (Ethelredshithe, later Queenshithe) which had a wharf but no basin. This was undoubtedly the first port established in western Europe.

River harbors were infinitely more difficult to establish, because of the silting problem, but the development of river traffic made the adoption of this solution an absolute necessity, in order to eliminate excessively numerous and burdensome transfers of cargo. The history of Bruges is a perfect example; it was, in any case, one of the most important river harbors. The city was located at the end of a gulf formed in the fifth century as a result of a sinking of the coast: the Zwyn, composed of two arms one of which formed the Reie, between Bruges and Damme, and the Zwyn proper, downstream from Damme. Originally the old Zwyn, the other arm, was utilized, but it had to be abandoned in the twelfth century; silting had long been a threat. Beginning in the last quarter of the twelfth century important steps were taken to cope with the problem. The Reie was equipped with dikes, and at its junction with the Zwyn the inhabitants of Bruges established around 1180 a barrage with a lock that permitted the raising of the level of the water upstream. This lock was considered a marvelous feat of technology when it was built. Unlike the ramps common at this period, which permitted only the passage of ships of limited capacity, without keels, it had a double system of gates. However, the draft could not exceed six feet, which even in the thirteenth century was already considered excessively low. Large ships could not go up to Bruges, and it was therefore necessary to establish at Damme (which acquired the name of the dike) a wharf and to transfer merchandise there. Damme thus very soon became the real port of the Zwyn. When the people of Ghent decided, around the middle of the thirteenth century, to link their city with the Zwyn, after some hesitation they decided that their canal would empty into the estuary at Damme itself. This narrow, navigable, seemingly capricious route, baptized the *Lys gantoise* and later known as the Liève, was constructed between 1262 and 1269. With the gradual increase in tonnage, all the old access routes to the inland ports were abandoned one by one, and new outer harbors were constantly being dug and developed: Mude, Monnikerede, Hoeke, Sluis, which were equipped in the second half of the thirteenth century.

It is very difficult to say how these wharves and dikes were constructed, to say nothing of the locks, of which no traces remain. It is probable that the wharves and dikes were constructed by sinking a series of piles and building them up from behind with a filling of stones and earth. This work must have been very crude, and many ports that are today famous undoubtedly began with very crude installations. In 1130 Guillaume, Duke of Aquitaine, struck by the advantages of the site of La Rochelle, built a castle there; the present basin represents the port that was encircled with fortifications in 1222, on the orders of King Henry III of England. In peacetime ships entered this harbor and had access to an arm of the sea situated to the west, outside the ramparts.

The first harbors in the north were dug by builders from the French Midi. In 1293 Philippe le Bel obtained from Genoa the services of Enrico Marchese, Huguet and Albertin Spinola, Lanfranc Tartaro, Nicolas du Parrez, and Italian carpenters under the escort of Clement de Saïn, a visitor of the Temple, to construct the *clos des galées* of Rouen on the site of the ancient parish of Saint-Martin contiguous to the bridge, in the Richebourg quarter. It was encircled by channels reinforced with a cofferdam on the Seine side; inside were ramparts with gates. In the interior a basin was dug, and covered halls were built around it. This was the first port with artificial basin, wharves, and accessory installations built in western Europe near tidal waters. The second was the *clos des galées* of Harfleur, built between 1340 and 1345, which inaugurated the era of artificial basins on the coasts of tidal waters. While a Mediterranean port always (theoretically) has deep water, an Atlantic port has low tide twice a day. It was to the advantage of the Greeks and Romans to construct moles on arcades, to avoid silting, but the case was completely different in the north: here the wall had the triple role of retaining the water as long as possible, opposing an attack coming from an open beach, and protecting the basin against alluvial deposits. For this reason the *clos des galées* at Harfleur was constructed with a solid wall. The *clos* was semicircular; its diameter did not exceed 328 feet, and it was crossed by a small stream, the Lézarde. On this stream, at the end of the fourteenth century, a lock was installed to form one of the earliest examples of a reservoir. The reservoir later became an indispensable auxiliary of river ports.

At the beginning of the Middle Ages lighthouses already had a long history. But they had been somewhat removed from their proper role in order to serve as appendages to military monuments (for example, a defense tower, as at Aigues-Mortes) or religious establishments (the Roman lighthouse at Dover had become the belltower of a church). The monuments built by the Romans, particularly the lighthouses of Dover and Boulogne, were brought back into use. The number of these structures increased when improvements in navigation permitted the development of commercial shipping. The lighthouse of Corduan was built in the eighth century at the mouth of the Gironde, where it was maintained until the sixteenth century. The lighthouse of Brindisi was constructed in 1275, following the plans, it is said, of Charles the First of Anjou. Elsewhere, other buildings were used for this purpose: a *lanterne à morts* (a monument that served as a beacon in cemeteries), as at Oléron in the thirteenth century; the façade of a church, as at Wisby in Gotland. At Aigues-Mortes the lighthouse (which is still in existence) was placed atop the Constance Tower; it is a round iron lantern in which pins held transparent sheets of horn. The great beacon of Genoa dates from the twelfth century, as is proved by the law regulating its maintenance. In 1242 it is mentioned together with another small lighthouse (*lanternino*) built at the end of the breakwater. In 1318 mention is made of the lighthouses both of Genoa and of Portofino. In 1326 oil systems were installed in the two lighthouses of Genoa — a method that had long been in existence, for the lighthouse of Bari was operating on this method by 1070. The range of these lighthouses was obviously quite limited, and they were identification signals rather than actual guides for ships coming toward the port; in most cases it was better to wait for daybreak.

Roads and bridges   Little is known about medieval roads. In many cases they must have been the successors of a more or less well-maintained Roman highway. It was very difficult for extremely fragmented countries to have a policy regarding highways, for this requires a well-organized central power. Traffic was quite limited, and depended, as we have seen, on animals; thus it did not require a network of heavy roads with solid beds. Certain difficult routes were already utilized in early times; the peaks of Mount Cenis and Mount Genèvre were being crossed regularly in the eleventh century.

In 1315 the road to Westminster was paved; this, however, was an exception. Certain portions of the roads to Santiago de Compostella, especially in northern Spain, roads that undoubtedly had changed very little since the Middle Ages, indicate that the paving was relatively limited. The famous "macadamized" roads of the Middle Ages were undoubtedly built with sand, lime, and alluvial mud, and cannot have been very sturdy. The medieval authorities were concerned primarily with maintaining their alignment and eliminating encroachments that would have blocked them.

The problem of bridges has been better studied, since it is of interest to archaeologists and because very beautiful specimens have been preserved. Extant documents unanimously proclaim the importance attached to the construction and maintenance of bridges. The Roman tradition had been well preserved, as is testified to by an eighth-century document, and the religious significance attached in many places to the building of bridges should undoubtedly be understood within this perspective.

By the eleventh and twelfth centuries we find that the bridgebuilders had become remarkable technicians. But while the bridges themselves are known to us, we are much less familiar with their construction methods, especially as regards the early medieval period. Piles were first installed, and the foundations of the abutments were rested on them. Pile driving was a difficult matter; it was necessary either to divert the water or to pump out a wooden coffering. The mortar was mixed with vinegar to increase its impermeability. After the twelfth century the abutments built on piles took the form of breakwaters that broke the current upstream and diverted floating objects, ice, and debris, and downstream eliminated eddies that would have caused the undermining of the piles; in addition, these splays facilitated the passage of laden ships.

The Gothic builders succeeded in achieving a lightness and elegance that were often lacking in the massive Roman works imitated by the Romanesque architects (for example, the bridge at Airvault). In order to give his arches strength and the correct height, the architect of the bridge of Avignon made them elliptical; to give them perfect elasticity and prevent their total destruction in case of collapse or partial crumbling, he coursed them in the Roman manner, in several closely spaced but separated arches. To avoid giving too great a foothold to the current in case of flooding, he opened large round bays above the point where the abutments become narrower at the springing of the arches. In addition, these arches acted as a bracing, which added solidity while economizing on building materials. The Pont Saint-Esprit (1265) forms an angle, like that of Avignon.

Bridges were also built of brick. The Pont des Consuls at Montauban, built between 1291 and 1335 under the direction of an engineer named Matthew of Verdun, has openings above the abutments, and passageways were opened for strategic reasons between the arches and the roadbed. The pointed equilateral arches offer greater stability and less thrust, and rise higher above the flood level. Sometimes, however, this system caused an angle in the roadbed; in such cases there were two ramps, especially in mountain bridges that spanned with a single arch a torrent that floods rapidly and carries with it large quantities of trees and stones (as Le Vigan and Mende), or when an arch for the passage of boats was created at the deepest part of the riverbed and when there was reason for leaving a wide space between the pillars. Frequently one or several arches were replaced by temporary or permanent wooden roadways. Bridges with masonry piles and wooden roadbeds also existed, as for example the Pont au Change in Paris built in 1296, after its arches had crumbled. Bridges made entirely of wood were still frequent (Basel, Lucerne). Some bridges were fortified, for example the Pont Valentré at Cahors (built in 1308), and the bridges of Orthez, Béziers, Thouars, Tournai, and many other cities.

Plate 64.
Bridge at Mende (Lozère). *Photo E. C., Paris*

CONCLUSION

THE TECHNOLOGY AND CIVILIZATION
OF THE MEDIEVAL WEST

THE INFORMATION now available does not permit us to draw a general picture of the interrelationship of technology with the other activities of the men of the Middle Ages. Too many questions would remain unanswered — some because of the lack of research, others for lack of explicit texts. At most we are able to hazard a few suggestions.

We shall therefore limit ourselves, in this chapter, to raising problems in three areas, which are more closely related specifically to the history of technology. We must first understand certain modalities of the internal development of medieval technology; that is, our problem is one of solving, not so much problems of invention (which in most cases are insoluble) as problems of origin and transmission that reveal important interrelationships. Once established, the various techniques undeniably had repercussions on the social organization, including the juridical and political aspects, and on the intellectual behavior of certain classes of society.

PROBLEMS OF ORIGIN AND TRANSMISSION

These are perhaps the most difficult problems, for in many cases the facts have not been established with all the precision desirable. Let us immediately note that the problem of invention does not exist in itself. In modern times, invention supposes a strict reasoning process that, starting from recognized principles, leads to material achievement for a precise purpose. The medieval techniques were essentially empirical; that is, the role played by human reason was extremely limited. All techniques were imperceptibly progressing; whence the absence of all feeling of material progress, an absence that gradually disappeared between the middle of the fourteenth and the middle of the fifteenth centuries.

These technological perfections and developments did, however, have specific origins — some geographical, others characteristic of certain communities. Here attributions are difficult, for the factors of the problem are still obscure.

Geographical localizations The question of geographical localizations is one of the most complex. It is dominated in the first place by the relations between the Western world and the Far East. For the past few years certain historians have been claiming a priority for the Far East that includes almost every technique, without, however, thereby having solved

557

the problem of the transmission of techniques from one end to the other of the old continent.

Some techniques were very probably born in the Far East, and there seems to be no question that they were transmitted to the West. Such is the case for the modern horse harness, which was in existence in China by the second century, and was definitely carried west by the nomads. The semantic factors outlined by A.-G. Haudricourt seem to confirm this hypothesis completely. Dates that appear to be relatively precise have also been established for the development of the Paper Route, first across central Asia and then the Near East.

Other techniques, however, leave us completely in doubt. While the wheelbarrow and spinning wheel definitely existed in China and India at periods that are not greatly at variance with their date of appearance in the West, there **is** nothing that permits us to affirm that they were transmitted from East to West. The compass is attested to in China one century before it is mentioned in Europe, but the intermediate stages are missing, and the Arabs, the only people who could have transmitted it, do not appear to have had it before Christian Europe. By the beginning of the Christian era the Chinese had, first, detonating mixtures (sulfur and saltpeter) and then powder (seventh to tenth centuries); they invented the grenade (1231) and the cannon (1259–1272). Some authors refuse for various reasons to see in powder made from saltpeter an importation from the Far East; they regard it is a Western invention of the end of the thirteenth century.

It is possible, however, that concomitances occurred and that parallel developments, sometimes separated by time lags, appeared. Definite examples of this exist: for example, the water mill was known in the Far East and in Mediterranean Europe at the same time. An influence from one region to another is thus improbable. We should not underestimate the contribution of the intermediary peoples (especially those of the Middle East), a subject of which too little study has been made; the windmill, certain peculiarities of vaults, glassmaking, and the hydraulic norias can perhaps be attributed to them.

Even if we thus accept an indigenous development for the Western techniques, or at least of most of them, this does not solve the problem of localizations in Europe itself. Our chronological points of reference are too fragmentary to permit us to draw definite conclusions. Can the use of semantics assist us? In certain cases it can, when its results are corroborated by other indications.

Thus the art of mining, although quite extensively practiced by the Romans, appears to have been reborn in the Middle Ages under the impetus of certain Germanic peoples. In regions as widely separated as Scandinavia and central Italy, the mining terms used are of Germanic origin. In a region where we would expect the continued existence of a vocabulary of Latin origin, the statutes of the mines of Massa use little but terms of Germanic origin. By the thirteenth century Swedish mining terminology was completely German. In France, too, slightly later, we find the predominance of a technical vocabulary that is of German origin. We shall see that this is corroborated by certain facts, in particular the migration of German technicians. The same situation holds true for naval technology; in France we see the use of a vocabulary that owes much to the Flemish and Scandinavian idioms.

It is obvious that such a study must not be carried to arbitrarily rigid con-clusions. For example, though the French names for a certain number of wood-working tools are of Flemish and German origin, this is not to say that they were "invented" in these countries. Exchanges of technical terms are sufficiently frequent to permit us to explain certain anomalies. The plane, the bitbrace, and the trying plane certainly are not exclusively Flemish in origin.

Localizations of the rejection of technical borrowings are more easy to establish. These are essentially geographical areas of difficult access, at a distance from the major traffic routes. Especially in eastern Europe, certain groups of people thus remained closely attached to traditional tools. The terminology of the plow in Polish, as we have emphasized, is exclusively Polish in origin, a fact that demonstrates the indigenous development of these techniques.

The technical work of the Cistercians Among the groups that contributed to the spread of technical progress, perhaps none was more effec-tive than the Cistercian Order, the branches of which covered all Europe, beginning in the twelfth century. Reacting against the exclusively spiritual tendencies of the followers of St. Benedict, St. Bernard required of his brothers the performance of manual labor to satisfy their needs. The centralization of this order, at least in the beginning, was extremely favor-able to the spread of certain techniques reputed to be the best, and it must be admitted that archaeological studies have confirmed the interest that could be revealed by exhaustive research on this subject.

It is in the architectural domain that the unity of the Cistercian work was best demonstrated. Not only were plans imposed, but even, in many cases, the techniques of construction. The Cistercians immediately adopted the ogive vault, and carried it to every corner of Europe; they probably introduced it into Germany, Italy, and England. The details of their buildings are abundant dem-onstration that the builder-monks — for they generally constructed their buildings themselves — had to learn their trade as architects and contractors in a common school. The example of the production of perforated bricks has often been men-tioned. We find them, apparently for the first time, at Foigny (Aisne) — where, incidentally, they were known as "St. Bernard's bricks." The uniformity of their production can be seen in France, Germany, and Italy.

The Rule of St. Bernard required that the Cistercian abbeys be established in regions with abundant rivers. The monks needed water both for the irrigation of their lands and for the power needed for their numerous workshops. They were very skilled in hydraulics projects, using aqueducts to carry water, as at Obazine (Corrèze), and distributing it through a system of underground pipes made of carefully cut terra-cotta tubes. The piping systems of Les Châtelliers (Deux-Sèvres), Noirlac (Cher), Bonport (Eure), Vaux-de-Cernay (Seine-et-Oise), and many others are also mentioned. Elsewhere there were small, very well-designed canals with overflow outlets and sluice gates whose levels were perfectly established, as at Royaumont (Seine-et-Oise) and Maulbronn (Württemberg).

Agriculture was one of the major occupations of the Cistercians. All the historians agree on the importance of the role they played in improving methods of cultivation and cattle breeding. It is perhaps to the Cistercians that we owe the tremendous expansion of triennial crop rotation, and to the English Cistercians the particularly important improvement in breeds of sheep capable of giving high-quality wool. The Cistercian abbeys divided their lands into a certain number of holdings called "granges." These granges, which were located in the countryside and in regions that were particularly active because they were located in zones of land clearing, in a certain sense played the role of modern model farms. Their buildings, as archaeological findings have shown, were perfectly adapted to the purpose for which they were intended. It is possible to say — and we shall see this opinion borne out by their industrial activity — that the Cistercians were perhaps the inventors of what is today called "functional architecture." The simplest monastic granges included an open shed and a long building divided into three parts by two rows of columns supporting the high, double-pitched roof, where the harvests were stored, and where the (generally vaulted) stables and cattle sheds and one or two rooms for the lay brothers were located. To one side was the dovecote. The thirteenth-century buildings are in general better conceived and constructed than those of the twelfth. This architecture impresses us by its unity of conception.

In addition to the usual crops, which they perhaps contributed less to improve than to popularize, the Cistercians also cultivated the grape extensively. Their vineyards were particularly well cared for, both from the point of view of methods of cultivation and in the choice of arrangement. We are not surprised to see that many vineyards that are famous today often originated in a Cistercian establishment. In addition to the often-mentioned Clos-Vougeot, many other examples can be found in France, Italy, Spain, Portugal, and on the banks of the Rhine or the Moselle. The monks, however, did not do the work of wine making themselves, since they could not enter the wine cellars. The construction of the latter, the most beautiful and best preserved of which is perhaps the cellar of Clairvaux at Dijon, presents the characteristics of unified conception we have already mentioned.

The monks were also obliged to produce all the industrial products necessary to their daily life. Thus the abbey of Foigny, to take only one example, had fourteen wheat mills, a fulling mill, two cable-twisting machines, three furnaces, three forges, a brewery, three winepresses, and a glassworks. In southern Champagne, around the middle of the twelfth century, we find the metallurgical industry developing considerably under the Cistercian influence; most of the forges in the forest of La Chaume, which was the great center of this area, belonged to the neighboring Cistercian abbeys. Here, again, the Cistercians definitely aided in the propagation of modern techniques, and it is not surprising that they contributed in particular to the expansion of hydraulics equipment for the forging of iron; the first iron mills mentioned in Germany, Denmark, England, and southern Italy are all Cistercian establishments.

Just as they created types of buildings for rural work, so did the Cistercians invent factories. Some of these factories contained all the major industrial equipment: wheat mills, forges, fulling mills, tanneries. Beautiful specimens can still be seen at Fontenay in Burgundy (where the river that supplies power flows along the side of the building), Royaumont (where the stream crosses the center of the factory), and Fountains Abbey (where it passes under the workshops). This type of concentration of factories is a uniquely Cistercian creation.

Plate 65.
Burgundian winepress. *Photo La Cigogne*

The Cistercian unity thus made possible to a great extent the dissemination of modern techniques throughout western Europe, beginning with construction techniques, but also including agricultural and industrial techniques. These monks were, moreover, in a good position to distinguish those technical improvements that were profitable, and to bring about their dissemination and penetration, thanks to remarkably well-organized holdings scattered throughout the rural areas.

Technical migrations   To bring up the topic of technical migrations may seem a somewhat hazardous undertaking, since the only data we have been able to collect concern the migrations of technicians. In large part, however, these human migrations were the basis for the transmission of technical progress.

Marc Bloch has already demonstrated the importance of the collective migrations for the eastward spread of the water mill. Sericulture may have been brought to southern Italy by prisoners of war; it was Flemish workers, driven from their homeland by the social and political struggle, who may have introduced perfected techniques of cloth manufacturing to Florence and Italy. While we now know that many techniques whose importation was once attributed to Crusaders returning from the East actually came from other areas, the Crusaders nevertheless were in some ways connected with the importation of certain Eastern methods. To explain the technical relationship between East and West, the example of Marco Polo has been cited. Undoubtedly he was not the only traveler to venture on these routes; it is possible that others became interested in the techniques related to the products they were handling in their business. The narrative of the travels of Marco Polo mentions only the ships with four masts and twelve sails of the Great Khan, the ships of the Persian Gulf, and the use of coal. Guillaume de Robrouk (1253–1254) was astonished to meet a Parisian goldsmith at the court of the Great Khan, and in 1292 Giovanni da Montecorvino made the acquaintance at Peking of a Lombard surgeon. Thus the exchange was not a one-way street.

It would be necessary to make a more thorough study of the migrations of workers, of which only a few examples are known to us. For instance, at the end of the thirteenth century the forges of the Dauphiné often recruited their workers from among the Savoyards and Piedmontese, who had the reputation of being skilled workers.

In any event, the more highly trained technicians frequently traveled, and the architects, the great builders of cathedrals, moved from place to place; Villard de Monnecourt, who boasted of having traveled as far as Hungary, is the most famous example, The cathedral of Seu d'Urgel was completed in 1175 by Lombard architects, in the style of their own country. The powerful figures of that age did not fail to attract to their service men of whose fame they had heard; in the thirteenth century Alfonso the Wise and Frederic II Hohenstaufen had veritable courts not only of scientists but also of technicians. At the siege of Tyre, in 1124, the Latins called upon an Armenian engineer to supervise their siege machines. Before gathering together his army, St. Louis was anxious to find men "who knew how to build engines for hurling stones, and mangonels." One Portuguese city issued a call for Italian carpenters to rebuild its fleet.

Plate 66.
"Construction of Noah's Ark." Fresco by Pietro di Puccio,
Pisa, Camposanto. *Photo Giraudon-Anderson*

Some trades even had a hierarchy indicative of different levels of knowledge. By the thirteenth century the technicians of medieval artillery had most likely banded together in a common organization. Joinville mentions the chief *mestre ingingneur* of St. Louis at Mansourah, and we know of other individuals who held similar positions during the reigns of Philippe III and Philippe le Bel.

It was perhaps the slowness of certain projects, and the permanent existence over several centuries of certain cathedral and hydraulics construction sites, that most favored the transmission of techniques (especially techniques of construction) and undoubtedly resulted in continuous progress. To compensate for a lack of public instruction, both this permanence of construction sites and the guild traditions were needed if techniques were to be maintained and perfected.

TECHNOLOGY AND SOCIAL ORGANIZATION

The technological level of a population unquestionably has an influence on its social organization. Given its complexity, it is impossible to discuss the subject in a few lines. While demographic problems undoubtedly caused a profound technical evolution in the eleventh and twelfth centuries, it is equally certain that they also affected the evolution of the social structures. In short, reciprocal influences came into play here, and it is often very difficult to determine cause-and-effect relations.

Improvements in techniques made possible a certain amount of concentration of factories. This concentration, which was essentially urban, contributed to the renaissance of the cities, but also freed the rural areas from a certain number of activities that had become part of the feudal system of obligations. These obligations, at first repurchased, gradually disappeared completely, thus contributing to the liberation of the peasant population at a period when the demographic expansion was causing it to shift either toward the cities or toward the new lands to be cleared. The expansion of metallurgical production, facilitated by more daring mining explorations and by the mechanization of part of the production, favored these land-clearing projects, just as the appearance of a greater number of tools encouraged the building of stone edifices. The improvement of agricultural techniques — we are think chiefly of the expansion of triennial crop rotation — sustained rural progress, giving to the agrarian structures the form they preserved until the eighteenth century, and encouraged that demographic expansion of which we spoke earlier.

The influence of technology was felt even in the organization of the trades. Of course, the origins of the guilds were religious and social rather than technical, but it is nonetheless true that one of their principal (if not their chief) preoccupations was the creation of a technical "code." Very probably it was not so much a case of convincing possible clients of the perfection of a product as of struggling against severe competition. Technical requirements were therefore the center of an economic problem. When pricing problems became more difficult, there was a move toward simpler, less rigid techniques. Prohibitions against methods and machines were gradually abandoned, undoubtedly because it was necessary to struggle against a rapid invasion of cotton fabrics, at a price considerably lower than that of woolen fabrics.

In some rural trades, too, local codes also developed around prohibitions, and such laws were the source of economic and social organization; witness the short mining codes, which have not yet been sufficiently studied and which would furnish a rich harvest of information, both technical and social in nature. It was necessary to organize in order to avoid cave-ins and to remove water, just as in Spain it was necessary to organize in order to establish irrigation techniques. There are technical complexes that can function only if a professional organization is organized.

The appearance of more perfected (but also more costly) methods of production also led to the formation of a special body of law that was both technical and economic in origin. Undoubtedly much discussion will continue to center on the origin of the feudal laws of obligatory use (the *banalités*) and their exact nature. The fact that conventional *banalités* existed proves in any case that this phenomenon surpassed the primitive level of a purely feudal organization. Naturally, the *banalités* made possible the construction of tools whose price was relatively high; it is worth noting that they covered far more than the three machines generally connected with them — the winepress, the iron furnace, and the mill. In addition to the famous *banalité* of the boar, which perhaps originated in a technique of improvement of breeds, we find *banalités* covering almost all the instruments that utilized hydraulic power: the forge, the fulling mill, the tanning mill, the sharpening mill, the malt mill, the beer mill. Was the *banalité* the consequence of mechanization, or the result of an expropriation of the stream by the *seigneur,* especially when that stream had become the principal source of power? It has been shown how greatly this obligatory character contributed to peasant resistance to the water mill. The *banalités* sometimes forced the peasantry to preserve very outmoded equipment; in some regions, for example, hand-driven mills continued to exist for a long time.

The *banalité* is not the only case in which the use of certain techniques led to the crystallization around them of a system of laws. Certain very costly machines also contributed to the formation of communities. In the Dauphiné, many rural communities were born or developed around a grain or hemp mill. There were also community pasturelands earmarked specifically for transhumance. In Italy, too, the communities had their tanning mill or fulling mill. The tanning mill of Chartres belonged to the community of tanners. The commune of Marseille had several machines for launching galleys; the shipbuilders rented them, and could not make use of other machines without paying a tax. More complex industrial installations required the creation of the first known medieval companies.

In any case, it was techniques that lay at the origin of the professional structures, and as proof we need only consider the extreme division of labor found in certain industries. To consider only one well-studied example, the cloth-manufacturing industry was not a centralized organization but a complex of entirely independent trades, each one of which had its separate guild organization. Labor in the salt mines of the Jura was also highly specialized, and a multitude of other examples could be cited both from among urban techniques and from the rural world. Lastly, it should be noted that the social organization in turn influenced the development of technology.

TECHNOLOGY AND THE SCIENTIFIC MENTALITY

This is yet another subject on which detailed studies are lacking. However, a few studies, all too brief, have recently shed light on the interdependence between intellectual knowledge and utilitarian techniques. If there was one area in which this interdependence could appear and increase, it was that of measurement.

Weights and measures   Medieval Europe definitely owes its general system of weights and measures to Roman antiquity. In several regions, however, Eastern systems also had a certain amount of influence. For a long time Scandinavia used the Greek and Near-Eastern system, thanks to the trade routes that crossed Russia. The establishment of colonies by Greeks and Eastern peoples in France also created various difficulties.

At a time when measuring instruments were crude, the notion of a standard unit could only be completely relative. The end of Roman unity and the power of local particularisms crystallized into a sytem differences that perhaps were originally only technical. During the time of troubles, between the fifth and twelfth centuries, weights and measures were probably left to individual initiative. A few traces of this situation are found in later documents; a contract dating from the beginning of the thirteenth century, from the Abbey of Saint-Victor de Paris, proves that the parties marked off with their seals the length of the foot that was to be used to measure a field. The unit of measure became to a certain extent the creation of a decision by the two parties. This was the only possible solution at a period when, as Wolff has so correctly noted, the feudal breakup had multiplied *ad infinitum* the areas of separate evolution, in which each city and frequently even each village possessed its own units of measurement.

Very probably it was the rebirth of Roman law that made possible the reestablishment, not of a standard unit (which was not to exist until the advent of the metric system), but of greater precision in the systems of weights and measures. Texts, and the weights themselves, become extremely common in the southern regions beginning in the second third of the thirteenth century.

The texts, which regulate rather than establish exact standards, can only be vague. To establish standards, references would have been necessary to something that did not exist, and technical methods of control would have been needed. However, the idea of a standard unit was reestablished, and all misdemeanors involving short weights and measures were punished. These standard units, moreover, became more general with the increasing clarification of the political organization. In 1222 the Count of Toulouse imposed his system of measures on Cordes when he granted the charter of commune to that city. Weights and measures were marked so that they could be identified. In the case of weights, matters were easier, for they were made of metal and were used in monetary form; the seal of the city or any other identifying mark conferred an unquestionable authenticity. The city of Toulouse appears to have been the first city to resume this ancient tradition; the earliest extant monetary weight of this city bears the date 1238. Most of the cities of southwestern France proceeded to imitate this example.

The expansion of long-distance commerce inevitably led to a simplification of systems of measure, even before this policy was attempted on a national scale. In 1317 Pézenas regulated its grain measures on those of Béziers, while it adjusted its balances and weights to those of Montpellier. Undoubtedly there were commercial and economic reasons behind these moves.

Grain measures were often made of stone during the Middle Ages. Numerous undated specimens of such measures that formerly decorated the public marketplaces are preserved at Cordes and in the museums of Toulouse, Le Puy, and Amiens. These measures generally pivoted on pins and had several cavities.

Even less is known about measuring instruments; most of them have now disappeared, and the texts are not very explicit. Roman balances were undoubtedly the most commonly used weighing instrument. Balances with two arms had also been in existence since antiquity. Measures of capacity were made either of wood or tin; how vats were measured is not known. Surface areas were measured with surveying chains and trigonometry. Certain industries and projects required fine measurements; the poem on weights and measures attributed to Priscian (fourth–fifth centuries), manuscripts in the Mappae Clavicula collection (tenth–twelfth centuries), and manuscripts of the alchemists reveal that very carefully measured quantities of substances were sometimes possible. Archimedes' discovery made it possible to determine the compositions of alloys. During the thirteenth and fourteenth centuries the critical study of the Greek words and translations of the Arabic authors introduced methods that had already been well known in earlier civilizations.

Fig. 112. Weighing with Roman balance and balance with arms. Around 1023 (Salzmann, p. 61).

The measurement of time   The centuries-old instruments and methods for measuring time continued to be used during the medieval period. Clepsydras and sundials are described elsewhere in this work. The Arabs had greatly perfected all the devices of this type, and had adapted to them those automata that had been the delight of the mechanicians of the School of Alexandria.

The birth of mechanical clocks will be discussed more fully in the second volume of this work; here we shall limit our remarks to several chronological guidelines. Texts recall the memory of extraordinary instruments whose descriptions are extremely vague. The clock of Pacificio, Deacon of Verona (ninth century), and that of Gerbert (died 1003), must have been no more than improved clepsydras. The planetary water clock of the Arab Gazari (thirteenth century) had a mechanism that it is difficult to imagine but that almost certainly involved weights and possibly even more complicated parts. Lastly, there existed astrolabes with very carefully calculated gear mechanisms. Transmissions of regular movements thus appear to have existed, but we are unable to establish in very definite fashion the steps in these discoveries.

The album of Villard de Honnecourt (second half of the thirteenth century) contains one very interesting piece of information. Villard suggested causing the statue of an angel to turn slowly in such a way that its outstretched finger would always point to the sun; this required a steady revolving motion on the part of the statue. For this he utilized a system of weights and escapement that in addition was based on the friction of ropes. The alternating movement of a wheel whose spokes struck a rope stretched by two weights formed a kind of primitive escapement.

In 1271 Robert the Englishman, in his commentary on the sphere of John of Hollywood, described a clock mechanism with weights; in contrast, other texts of this period made no mention of such clocks. The work of Alfonso the Wise (1277) speaks only of water clocks and mercury clocks. The system used in the first mechanical clocks (described in detail in the second volume) remained in use for a long time. The power mechanism was operated by weights. Two right-angle blades gave an alternating movement to a crown wheel; after this it was simply a question of regulated works. A system of lead pendulums and a small wheel regularized the functioning of the clock. In the first half of the fourteenth century the mechanical clock spread through Europe, chiefly in France, England, Germany, and Italy; between 1337 and 1344 Giovanni de Dondi, the builder of the clock of Padua, described it in a treatise (now published). Clocks played an important role in the industrial development of complex machines.

Bookkeeping   Bookkeeping is yet another system of measurement that was the subject of major modifications during this period — modifications that were to develop this technique from its primitive to its most modern forms. The earliest extant account books are simple memorandums; only credits are listed, together with the names of the witnesses to the operations.

The creation of companies led to other, more complex, obligations; the assets and the proprietorship accounts had to be shown in the books. The order was a final factor in the transformation of accounting. In the Italian books of the first half of the thirteenth century, the old method continues to appear. The fair books kept by the Ugolini between 1249 and 1263 separated debit and credit and left blank spaces between the individual accounts; the accounts receivable began to be separated. The necessity of dealing with a greater number of business affairs at greater distances led to the birth of subsidiary accounts. By the end of the thirteenth century the Sienese bankers had cashbooks and ledgers, salesbooks and books for the stockholders. The inventory was in common use by the fourteenth century. The proliferation of subsidiary ledgers shows that there were other means of checking the accuracy in addition to the verification of the results.

The books kept by the Peruzzi were divided into two parts, debit and credit. Accounts receivable were born on the day debit and credit were placed side by side; the Venetians appear to have been the inventors of the idea. The double-entry system owes its name, not to its appearance or to personal and merchandise accounts, but to the practice of making two notations in the general ledger, once to the debit of one account, and once to the credit to the other. The general ledger thus contained a complete listing of individual and company accounts. The municipal accounts of Genoa were the first example of this system (1340). In the books of Francesco Datini of Prato, who settled in Avignon around 1350, the transformation can be seen gradually taking place.

Science and technology At the beginning of the twelfth century Hugues de Saint-Victor distinguished four principal sciences; that of mechanics included wearing apparel, weaponry, navigation, agriculture, hunting, medicine, and the organization of games. A few years later, Domingo Gondisalvo formulated the fundamental distinction between the theoretical and the practical sciences; technology fitted into his system perfectly. Just as there were practical mathematics and theoretical mathematics, geometry could be used equally well by land surveyors, carpenters, or masons. Gondisalvo, however, included under doctrinal knowledge something he called the "science of engines," and which he defined as the method of conceiving and inventing methods of adjusting natural bodies by a device in conformity with a numerical calculation, so that we can make whatever use of them we desire. This knowledge of the engineer was also applied to masonry and the construction of lifting machines, musical instruments, bows, weapons, and burning mirrors.

Fig. 113. Gerbert's telescope. Thirteenth century (Singer, p. 595, Fig. 349b).

The technicians were also inclined — perhaps unconsciously at first — toward systematic solutions, just as certain schools of scientists attempted to approach closer to reality. We possess numerous examples that show that the techniques of construction were tending toward general rules it was believed could be attained only by way of geometry. One thirteenth-century architect boasted of being "a great geometer and carpenter, which is superior to being a mason." We have seen that military engineers and architects of cathedrals expressed all their problems in geometrical terms. The plans of Rheims (1250) and Strasbourg (1275) are still simple elevantions of façades, but beginning in the fourteenth century plans became infinitely more numerous and changed in nature. The plan of Lando di Pietro for the cathedral of Siena (1339) is already a scale drawing. None of these fourteenth-century plans, however, was yet a true flat projection. Suger, Abbé of Saint-Denis, was already ensuring, "by means of geometrical and arithmetical tools," that the new chevet of Saint-Denis was correctly aligned with the old nave. The plan of the mine levels of Massa and their digging is proof of the use of scientific techniques. The use of experts in some cases indicates in any event that a body of theoretical knowledge was being organized.

The construction and use of trebuchets presupposes a genuine theory of ballistics. It is not impossible that here, too, long experience with these instruments led the military "engineers" to create certain mathematical formulas. In any event, the passage from a quasi-parabolical trajectory to a flat trajectory reveals an effort of thought that far surpasses the purely empirical level.

The use of geometrical figures and mathematical data indicates that the technicians were attempting to rationalize their knowledge. The scientists, who undoubtedly had never lost sight of what could be learned from the technicians — this was clearly evident after the twelfth century — certainly followed these early attempts very attentively, while the teaching of mathematics was penetrating the universities at the same time that it was benefiting the technicians of accounting. Schools for teaching the use of the abacus were being opened almost everywhere in Italy: Genoa in 1310, Florence in 1338, Lucca in 1345; these schools popularized the use of numbers.

Notebooks and handbooks proliferated. We have already mentioned the handbooks on agriculture, including the famous treatise of Pietro de' Crescenzi, Guido da Vigevano's treatise on war machines, and the treatises on the abacus, whose number increased after the appearance of Fibonacci's treatise. We have stressed the importance of the notebook of Villard de Honnecourt, which contains instructions for building edifices; unfortunately, only a small portion of it has survived. When in 1268 the *machinator* Assaut made a written request for an audience with Alfonse of Poiters, who was preparing to go on a Crusade, what he wanted to show to the brother of St. Louis can only have been a collection of drawings like those of Villard de Honnecourt or Guido da Vigevano. This is a mentality that was to become increasingly widespread in the course of the following century, and paved the way for the great engineers and scientists of the Renaissance.

We can feel, however confusedly, that technology was coming closer to the sciences quite as much as the sciences were attempting, with difficulty, to grapple with concrete reality. All the scientists of the fourteenth century who were continuing Oxford University's efforts of the preceding century — Jean Buridan is a good example — multiplied their incursions into technology in order to clarify scientific concepts. Simultaneously, the technicians borrowed from the *universitaires* their methods, a system of logic, and methods of reasoning, in all of which the technicians were sorely lacking.

It would be useless, as G. Beaujouan notes, to try to discover whether in this period it was the practices of everyday life that inspired theoretical research, or science that made possible the appearance of new techniques. In his *Liber de ratione ponderis* (last third of the thirteenth century), Jordanus Memorarius sets out a series of propositions (some of which were completely incorrect) in which an unquestionable interest in the work of the engineer can be seen emerging. Roger Bacon and Pierre de Maricourt boast of having succeeded in practicing the most diversified crafts in order to penetrate the enigma of nature and thus to be able to establish an exact science. Bacon writes of Pierre de Maricourt:

"He knows by experience the laws of Nature, Medicine, and Alchemy, as well as the things of Heaven and Earth. . . . He has delved into the trade of the metal founders. He has learned everything concerning warfare, weaponry, and hunting. He has investigated everything which relates to agriculture, surveying, and the work of the peasants. He has even studied the procedures of the old witches, their spells, their incantations, and everything concerning magic, and also the illusions and tricks of the jugglers."

This picture of the scientists' investigations into the techniques of their times is the best possible illustration of the importance that technology was acquiring in the eyes of the men of medieval western Europe.

This importance was born of the immense, continuous effort of centuries to perfect the heritage of preceding generations, and of the remarkable results that had already been obtained by the end of the thirteenth century. It would be unjust to admire the architect who threw up the daring vaults of Europe's cathedrals and to ignore the anonymous technician whose obscurantism is generally despised. Moreover, the Church was not exclusively the brake on progress that she is often thought to be: we have seen that the Cistercian abbeys undoubtedly became centers of technological progress.

Brakes were applied, however, and first of all by a traditionalism that it was difficult to overcome and all the more difficult to uproot because contacts were still limited. The greatest progress was achieved precisely in those trades and professions where these contacts were numerous: the textile industry, which represented the great international product of exchange; religious architecture, which attracted crowds; and the military art, which pitted the powers of this world against each others. Certain gaps in mechanical knowledge, the very slow development of scientific knowledge, and the scarcity of metal played their part in maintaining in many places techniques that probably had evolved little since prehistoric times. Would it be wrong, as a measure and a weight of technical progress, to say that these primitive techniques were in many cases those that were still being practiced at the beginning of the nineteenth century?

The fish saved medieval civilization by supplying man with vitally needed proteins. Wood furnished man with his principal means of action in the material domain. Thus, for those who like extremely simplified diagrams, fish and wood remain the two poles of medieval technology.

BIBLIOGRAPHY

I. GENERAL WORKS

BLOCK, MARC, *Feudal Society*, 2 vols. (Chicago, 1961).

———, *French Rural History* (London, 1966).

BOISSONADE, P., *Life and Work in Medieval Europe* (London, 1966).

Cambridge Economic History of Europe, M. M. Postan (ed.), Vol. 1: *The Agrarian Life of the Middle Ages* (2nd ed., Cambridge, 1966).

Cambridge Medieval History, edited by Henry M. Gwatkin *et al.* (Cambridge, 1911–36).

CLAGETT, MARSHALL, *The Science of Mechanics in the Middle Ages* (Madison, Wis., 1959).

CROMBIE, A. C., *Augustine to Galileo: The History of Science*, A.D. *400–1650* (Cambridge, Mass., 1953); reprinted as *Medieval and Early Modern Science* (Garden City, N.Y., 1959).

GILLE, BERTRAND, *Esprit et civilisation techniques au Moyen Age* (Paris, 1952), "Technological Developments in Europe: 1100 to 1400," *Journal of World History*, III, 1; reprinted in Guy S. Metraux and François Crouzet (eds.), *The Evolution of Science* (New York, 1963).

HARVEY, JOHN H., *The Gothic World* (London, 1950).

HODGSON, MARGARET T., *Change and History: A Study of the Dated Distribution of Technological Innovation in England* (New York, 1952).

LATOUCHE, ROBERT, *The Birth of Western Economy: Economic Aspects of the Dark Ages* (London, 1961).

LAVEDAN, P., *Histoire d l'urbanisme* (Paris, 1926).

LOPEZ, ROBERT, *The Birth of Europe* (London, 1967).

THORNDIKE, LYNN, *History of Magic and Experimental Science* (8 vols., New York, 1923–58).

WHITE, JR., LYNN, "Technology and Invention in the Middle Ages," *Speculum*, 15 (April 1940), 141–159.

———, "Tibet, India, and Malaya as Sources of Western Medieval Technology," *American Historical Review*, 45 (1960), 515–526.

———, "What Accelerated Technological Progress in the Western Middle Ages?" A. C. Crombie (ed.), *Scientific Change* (London, 1961), Chap. 10.

———, *Medieval Technology and Social Change* (Oxford, 1962).

———, "The Medieval Roots of Modern Technology and Science," in Katherine Fischer Drew and Floyd Seaward Lear (eds.), *Perspectives in Medieval History* (Houston, Tex., 1963).

———, "The Legacy of the Middle Ages in the American Wild West," *Speculum*, 40 (April 1965), 191–202.

II. TRANSPORTATION

BOYER, MAJORIE NICE, "Medieval Pivoted Axles," *Technology and Culture*, 1 (1960), 128–138.

GREEN, CHARLES, *Sutton Hoo: The Excavation of a Royal Ship-Burial* (New York, 1963).

LANE, FREDERIC C., "The Economic Meaning of the Invention of the Compass," *American Historical Review*, 68 (April 1963), 605–617.

TAYLOR, E. G. R., *The Haven-Finding Art: A History of Navigation from Odysseus to Captain Cook* (New York, 1957).

III. FOOD AND AGRICULTURE

ANDERSON, RUSSELL H., "Technical Ancestry of Grain-Milling Devices," *Agricultural History*, 12 (July 1938), 256–270.

CURWEN, E. CECIL, and HATT, GUDMUND, *Plough and Pasture; the Early History of Farming* (New York, 1953).

FUSSELL, G. E., "The Classical Tradition in West-European Farming; the Fourteenth and Fifteenth Centuries," *Agricultural History Review*, 17 (1969), 1–8.

GLICK, THOMAS F., "Levels and Levelers: Surveying Irrigation Canals in Medieval Valencia," *Technology and Culture*, 9 (1968), 165–180.

HAUDRICOURT, A., and BRUNHES-DELAMARRE, M., L'homme et la charrue à travers le monde (Paris, 1955).

HODGETT, G. A. J., *Agrarian England in the Later Middle Ages* (London, 1966).

VAN BATH, B. H. SLICHER, *The Agrarian History of Western Europe* A.D. *500–1850* (New York, 1963).

IV. MINING AND METALLURGY

AITCHISON, LESLIE, *A History of Metals* (2 vols., London, 1960).

ALBERTUS MAGNUS, *Book of Minerals,* transl. by Dorothy Wyckoff (Oxford, 1967).

NEF, JOHN, *The Conquest of the Material World* (Chicago, 1964), Chap. 1.

RICKARD, THOMAS A., *Man and Metals* (2 vols., New York, 1932).

SCHUBERT, H. R., *History of the British Iron and Steel Industry from ca. 450* B.C. *to* A.D. *1775* (London, 1957).

V. ARMS AND WARFARE

ANDERSON, R. C., *Oared Fighting Ships from Classical Times to the Coming of Steam* (London, 1962).

BACHRACH, BERNARD S., "The Origin of Armorican Chivalry," *Technology and Culture*, 10 (1969), 166–171.

BEELER, JOHN, *Warfare in England, 1066–1189* (Ithaca, N.Y., 1966).

BLAIR, CLAUDE, *European Armour, circa 1066 to circa 1700* (New York, 1959).

FFOULKES, CHARLES, *The Armourer and His Craft* (London, 1912).

NORMAN, A. V. B., and POTTINGER, DON, *A History of War and Weapons, 449 to 1660. English Warfare from the Anglo-Saxons to Cromwell* (New York, 1966).

OAKESHOTT, R. EWART, *The Archaeology of Weapons: Arms and Armour from Prehistory to the Age of Chivalry* (New York, 1960).

VI. INDUSTRIES AND CRAFTS

CARUS-WILSON, E. M., "An Industrial Revolution of the Thirteenth Century," *Economic History Review*, 11 (1941), 39–60.

———, "The Wool Industry," in *Cambridge Economic History of Europe* (1952).

FORBES, R. J., *Short Story of the Art of Distillation* (Leiden, 1948).

FREESE, STANLEY, *Windmills and Millwrighting* (Cambridge, 1957).

GOODMAN, W. L., *The History of Woodworking Tools* (London, 1964).

HOLMYARD, A. J., *Alchemy* (Harmondsworth, England, 1957).

HUSA, VÁCLAV; PETRÁŇ, JOSEF, and SURBOTÁ, ALENA, *Traditional Crafts and Skills: Life and Work in Medieval and Renaissance Times* (London, 1967).

LENNARD, P., "Early English Fulling Mills: Additional Examples," *Economic History Review*, 2nd series, 3 (1950), 342.

STOKHUYZEN, FREDERICK, *The Dutch Windmill* (London, 1962).

THEOPILUS, *On Divers Arts,* ed. and transl. by John W. Hawthorne and Cyril Stanley Smith (Chicago, 1963).

VILLARD DE HONNECOURT, *Sketchbook,* transl. and ed. by T. Bowie (Bloomington, Ind., 1959).

WAILES, REX, *The English Windmill* (London, 1954).

VII. CONSTRUCTION

ADAMS, HENRY, *Mont St. Michel and Chartres* (Boston, 1904; paperbound, Boston, n.d.).

COLVIN, H. M. (ed.), *The History of the King's Works,* vols. I and II, *The Middle Ages* (London, 1963).

FITCHEN, JOHN, *The Construction of Gothic Cathedrals* (Oxford, 1961).

GRIVOT, DENIS, and ZARNECKI, GEORGE, *Giselbertus, Sculptor of Autun* (London, 1961).

HARVEY, JOHN, *English Medieval Architects* (London, 1954).

PANOFSKY, ERWIN, *Gothic Architecture and Scholasticism* (New York, 1957).

SALZMAN, L. F., *Building in England Down to 1540* (Oxford, 1952).

SHELBY, LONNIE R., "Medieval Masons' Tools: The Level and the Plumb Rule," *Technology and Culture,* 2 (1961), 127–130.

———, "Medieval Masons' Tools II: Compass and Square," *Technology and Culture,* 6 (1965), 236–248.

———, "The Role of the Master Mason in Medieval English Building," *Speculum,* 39 3 (July 1964), 387–403.

SIMSON, OTTO VON, *The Gothic Cathedral* (New York, 1956).

VIII. MISCELLANEOUS

BOWLES, EDMUND A., "On the Origin of the Keyboard Mechanism in the Late Middle Ages," *Technology and Culture,* 7 (1966), 152–162.

COULTON, G. A., *Medieval Village* (Cambridge, 1925).

OLIVER, H., *History of the Invention and Discovery of Spectacles* (London, 1913).

ROSEN, EDWARD, "The Invention of Eyeglasses," *Journal of the History of Medicine and Allied Sciences,* 11 (1956).

WHITE, LYNN, JR., "Eilmer of Malmesbury, an Eleventh-Century Aviator: A Case Study of Technological Innovation, Its Context and Tradition," *Technology and Culture,* 2 (1961), 97–111.

GENERAL BIBLIOGRAPHY

I. BIBLIOGRAPHIES

(A) GENERAL BIBLIOGRAPHIES

FERGUSON, EUGENE S., *Bibliography of the History of Technology* (Cambridge, Mass., 1968).

FORBES, R. J., *Bibliographia antiqua: Philosophia naturalis* (10 parts in 6 vols., Leiden, 1940–50). Two supplements have been issued: *Supplement I, 1940–1950,* and *Supplement II, 1950–1960.*

RUSSO, FRANÇOIS, *Histoire des Sciences et des Techniques* (Paris, 1954).

SARTON, GEORGE, *A Guide to the History of Science* (New York, 1952).

(B) CURRENT BIBLIOGRAPHIES

Bulletin Signaletique, Section 22, Histoire des Sciences et des Techniques (Paris: Centre de Documentation du Centre National de la Recherche Scientifique). Materials published since 1958.

"Critical Bibliography of the History of Science and Its Cultural Influences," published annually in *Isis* (see below).

"Current Bibliography in the History of Technology," compiled by Jack Goodwin, published annually in the April issue of *Technology and Culture* (see below).

II. Periodicals
(A) journals in the history of technology

Documents pour l'Histoire des Techniques, published since 1961 by the Centre de Documentation d'Histoire des Techniques (Conservatoire National des Arts et Métiers, Paris). From 1963 on, the *cahiers* (official reports) of these *Documents* appear annually as a quarterly number of the *Revue d'Histoire des Sciences et de leurs Applications.*

Industrial Archaeology: The Journal of the History of Industry and Technology, quarterly journal first published in 1964, and originally entitled *Journal of Industrial Archaelogy;* name changed in 1966, published by David and Charles, Ltd., Newton Abbot, England.

Kwartalnik Historii Nauki i Techniki (Quarterly Review of the History of Science and Technology), published since 1956 by the Polish Academy of Sciences. Most articles are in Polish, but with abstracts in Western European languages.

Newcomen Society for the History of Engineering and Technology, *Transactions.* Published annually since 1920 by the Newcomen Society (Science Museum, London).

Revue d'Histoire de la Sidérurgie, published since 1960 by the Centre de Recherches de l'Histoire de la Sidérurgie (Nancy, France).

Technikgeschichte, originally entitled (until 1932) *Beiträge zur Geschichte der Technik und Industrie.* Published from 1909 to 1941, resumed publication as a quarterly journal in 1965; published by the Verein Deutscher Ingenieure (Düsseldorf).

Techniques et Civilisations, Five volumes of this now defunct journal appeared irregularly from 1950 to 1956, as a continuation of another defunct journal (1945–46), *Métaux et Civilisations.*

Technology and Culture: The International Quarterly of the Society for the History of Technology. Published by the University of Chicago Press for the Society for the History of Technology (Case Western Reserve University, Cleveland, Ohio).

(B) journals in the history of science

These journals occasionally contain articles relating to the history of technology.

Archives Internationales d'Histoire des Sciences, published in Paris since 1947 by the Division of the History of Science of the International Union of the History and Philosophy of Science. A continuation of *Archéion,* which began publication in 1919.

British Journal for the History of Science, published semiannually since 1962 by the British Society for the History of Science.

Isis, "An International Review devoted to the History of Science and Its Cultural Influence," published since 1913 by the History of Science Society (Smithsonian Institution, Washington, D.C.).

Revue d'Histoire des Sciences et de leurs applications, published since 1947 by the Groupe français d'Histoire des Sciences (Paris).

(C) museum publications

Abhandlungen und Berichte, published since 1929 by the Deutsches Museum (Munich). Since 1950 three issues per year have been published, each issue consisting of a single article on technical history.

Blätter für Technikgeschichte, published since 1932 (annually since 1947) by the Technisches Museum für Industria und Gewerbe (Vienna).

Contributions from the Museum of History and Technology, Smithsonian Institution, Washington, D.C., published since 1959 as special numbers of the *United States National Museum Bulletin.*

Daedalus. Yearbook of the Tekniska Museet (Stockholm), published since 1931. Contents page in English, from 1959, and occasional articles in English.

III. Histories of technology
(A) multivolume works

Kranzberg, Melvin, and Pursell, Carroll W., Jr. (eds.), *Technology in Western Civili-*

zation (2 vols., New York, 1967). Vol. I, *The Emergence of Modern Industrial Society*; II, *Technology in the Twentieth Century.*

SINGER, CHARLES; HOLMYARD, E. J., HALL, A. R., and WILLIAMS, TREVOR I. (eds.), *A History of Technology* (5 vols.), London and New York, 1954–58). This encyclopedic work is the starting point for any serious investigation of the history of technology. Vol. I, *From Early Times to the Fall of Ancient Empires;* II, *The Mediterranean Civilizations and the Middle Ages;* III, *From the Renaissance to the Industrial Revolution, c. 1500– c. 1750;* IV, *The Industrial Revolution, c. 1750–c. 1850;* V, *The Late Nineteenth Century, c. 1850–c. 1900.*

UCCELLI, ARTURO, *Enciclopedia Storia delle Scienze e delle loro Applicazione* (3 vols., Milan, 1941).

(B) SINGLE-VOLUME HISTORIES

DERRY, T. K., and WILLIAMS, TREVOR I., *A Short History of Technology* (New York, 1961). Primarily a condensation of the five-volume Singer *et al.*, *History of Technology* (see above), designed as a popular reference and textbook.

ECO, U., and ZORZOLI, G. B., *The Picture History of Inventions* (New York, 1963).

FINCH, JAMES KIP, *Engineering and Western Civilization* (New York, 1951).

———, *The Story of Engineering* (paperbound, Garden City, N.Y., 1960).

FORBES, R. J., *Man the Maker* (New York, 1950).

KIRBY, RICHARD S.; WITHINGTON, SIDNEY; DARLING, ARTHUR B., and KILGOUR, FREDERICK G., *Engineering in History* (New York, 1956).

KLEMM, FRIEDRICH, *A History of Western Technology* (New York, 1959; paperbound, Cambridge, Mass., 1964). Primarily an anthology of readings connected by short editorial remarks.

LILLEY, SAMUEL, *Men, Machines and History* (rev. ed., New York, 1966). A Marxist interpretation of the history of technology.

MUMFORD, LEWIS, *Technics and Civilization* (New York, 1934; paperbound, 1963). Emphasizes cultural implications of technological developments.

(C) SPECIALIZED WORKS

AITCHISON, LESLIE, *A History of Metals* (2 vols., London, 1960).

FELDHAUS, FRANZ M., *Die Maschine im Leben der Völker. Ein Ueberblick von der Urzeit bis zur Renaissance* (Basle, 1954). Contains an excellent bibliography.

———, *Die Technik der Vorzeit, der geschichtlichen Zeit und der Naturvolker: ein Handbuch* (2nd ed., Munich, 1965).

FORBES, R. J., *Studies in Ancient Technology* (11 vols. [Leiden, 1955–]; 2nd rev. eds. of most volumes have appeared since 1964). Vol. I, Bitumen and petroleum, alchemy, water supply; II, irrigation and drainage, power, land transport and road building; III, cosmetics and perfumes, food, fermented and alcoholic beverages, salts, preservation processes, paints and pigments; IV, fibers and fabrics, washing and bleaching, fulling and felting, dyes and dyeing, spinning, weaving, sewing; V, leather and parchment, sugar and substitutes, glass; VI, heat and heating, refrigeration, light; VII, geology, mining and quarrying; VIII, metallurgy, tools, gold, silver and lead, zinc and brass; IX, copper, tin, bronze, antimony and arsenic; X, pharmacy; XI, clay, stone, brick, tiles. Vols. X and XI have not yet been published.

STRAUB, HANS, *A History of Civil Engineering* (Cambridge, Mass., 1964).

USHER, ABBOTT PAYSON, *A History of Mechanical Inventions* (rev. ed., Cambridge, Mass., 1954).

INDEX

INDEX

Numbers in *italics* refer to illustrations

A

Abacus, 570

Aborigine: dwelling of, 31; food habits of, 45, 48; hunting by, 34, 147

Achaemenid Persia, 124, 129

Acheulean period, 21, 22, 24, 26, 38–45

Acropolis, *182*

adze, 24, 35, 62, *62*, *73*, 415, 502

Aegypten und Vorderasien im Altertum (Sharf and Moortgat), 143

Agricultura nabatea, 461

Agricultural society, evolution of, 61–63, 75

Agricultural technology: agricultural societies, 84–86; Americas, 390–92, *393;* Egypt, 144–45, 147; Greece, 208–10; India, 318–19; Islam, 340–41; medieval Europe, 460–72; Mesopotamia, 114; Roman Empire, 245–46

Albert the Great, *De Vegetalibus,* 461

Alchemy: Byzantium, 384; China, 272

Alcoholic fermentation: China, 277–78; medieval Europe, 483; Mesopotamia, 129; Ro-

man Empire, 245

Alexander, Age of, 196–97

Alexandria, 213; climate, 145; glassmaking, 205; lighthouse, *191;* port, 189, *190*

Alexandria, School of, 446, 567

Algoud, 379, 380

Ali al-Mas'udi, *Golden Prairies,* 458

Ambrose, *History of the Holy War,* 458

Amphitheatre, *223, 224*

Amphora, 245

Andean peoples, technological evolution of, 388–89; agriculture, 391–92, *393;* architecture, 398, 400; ceramics, 394, 395–96; dwellings, 397, 398; food, 391, 394; masonry, 400; metallurgy, 404–05; *quipu,* 416; textiles, 409, 410–11, *411;* transport, 417, *417*

Animal skinning, 36–37, *37*

Animals, game: Byzantium, 381; medieval Europe, 478; prehistory, 38

Animals, prehistoric, 38, 39, 90

Apiculture: Americas, 394; medieval Europe, 477

Apocalypse, 442

Apollinaire, Sidoine, 370, 477

Aqueducts: Americas, 391; By-

zantium, 375–76; Greece, 196; Islam, *339,* 344; medieval Europe, 546–47, *547;* Mesopotamia, 125; Roman Empire, 225, 226, *227,* 249, 546, 547

Aratrum, 83–84, 84; in Byzantium, 380; in Greece, 208–09; in Islam, 340; in medieval Europe, 466, 467, *468, 469, 469*

Arbiculture, 246, 258

Arch: Americas, 401, *402;* Greece, 185–86, *185;* medieval Europe, 529, 530; Mesopotamia, 118; Roman Empire, 221–23, *222*

Archimedean screw, 377

Archimedes of Syracuse, 191, 194–95

Architectural plans: Egypt, 173–76, *175;* Mesopotomia, 123; Roman Empire, *218*

Architecture: Americas, 396–403; *402, 403;* Byzantium, 373, 378, 386; China, 290–91; Egypt, 144, 153–58, *155;* Greece, 182–86, *182, 184;* medieval Europe, 528–40, *533, 543, 539;* Mesopotamia, 115, 123; Roman Empire, 218–24, *222,* 258

Architecture, industrial, 539–40

Architecture, military, 541–43
Architecture, religious: Americas, 398, *400;* China, 291–92; India, 321–23, 322; medieval Europe, 528–37, *533, 534,* 539; Mesopotamia, 119
Archytas of Tarentum, 191
Aristotle, 276
Arrowhead, 44–45
Art de véneri, L' (Twici), 477
Artesian well: China, 271; Islam, 337
Assyria: building materials, 123, 124; carpets, 133; chariot, *130;* decoration, 123; ivory, 137; metals, 133; ships, 131; tools, 136; transport, *120;* weapons, 137
Astrolabe: Arabia, 356; Byzantium, 383; medieval Europe, 444–45, 568
Astronomy: Byzantium, 383; China, 263, 264, 265, 294–96
Aswan, 154, 156, 158, 170
Athens, water system, 195, 196
Atlanthropus, forerunner of man, 13; home, 51; hunting, 38–39; tools, 12, 29
Aurignacian javelins, 41, *42*
Australopithecinae: caves of, 51; diet of, 38; human traits in, 14, 38; tools of, 12–13, 29
Automata: Byzantium, 384; Greece, 193, 455; medieval Europe, 568
Ax: China, 268; Mesolithic era, 61, *62;* Neolithic age, 65, *73;* prehistoric era, 24, 35, *35*
Azilian period, 35, *44,* 56
Aztecs: agriculture, 391; gardens, 392; silos, 392

B

Babylon, 121, 125
Babylonian culture, 115; bridges, 125; building techniques, 116, 118, 119, 125; gardens, 125; metals, 135
Bacon, Roger, 310, 571
Baghdad, 347
Barbarians: breadmaking, 483; contributions to other cultures, 257, 427–28; enamelwork, 235; invasions, 544
"Basket Makers," 413
Basketry: Neolithic, 65, 71; pre-Columbian, 389, 413
Baths: Byzantium, 375; India, 333; Islam, 343, 371; Mesopotamia, 117; Rome, 228–29, *222*
Bayeux Tapestry, 438, 470, 562
Beam press: Greek, 193, 209–210, *210,* 211, *211;* Roman, 246
Bedouins, 348–49
Bellows: China, 264; Roman Empire, 253
Bellows, hydraulic, 495–96
Biface, 21–23, *22,* 26, *26,* 27, 33
Bitumen, 117
Blade tools, 24, *25,* 26, 27, 28, 30, 31, 34, 35, 136
Blast furnace, Chinese, 268, *267*
Bloch, Mark, 423, 562
Block and pulley, 244
Blum, A., 287
Bola, 40, 147
Bone tools, 28–33, *32, 33,* 44
Bookkeeping, medieval, 568–69
Books: Arabia, 358; Byzantium, 379; China, 265, 266, 289; early form of, 213; India, 327; Mesopotamia, 140; Roman Empire, 255, *256*
Boomerang, 147
Borchardt, *Die Entstchung der Pyramide,* 164
Bow: Byzantium, 383; Egypt, 147; Greece, 197–98, *197;* Islam, 368; medieval Europe, 477; prehistoric era, 44–45; Roman Empire, 251, *252*
Bowsaw, 207, *207*
Brace and bit, 242

Brass: Byzantium, 383; China, 266, 269; India, 329

Bread: Islam, 356; Mesopotamia, 126, 127; medieval Europe, 471, 483

Brhatsamhita, 329

Brick: Egypt, 153–54, 157, 158; India, 320; Islam, 341, 344; Mesoamerica, 399–400; Mesopotamia, 116; medieval Europe, 492; Rome, 220, 491

Bridges: Americas, 417; China, 265, 303–04; Islam, 351; medieval Europe, 555, 556; Mesopotamia, 125; Roman Empire, 221–22, 226

Bronze: Byzantium, 496; China, 263, 264, 267–68, *268;* Egypt, 162, 171; India, 329; medieval Europe, 497; Mesopotamia, 134; South America, 404

Bronze Age: food, 45; hides and skins, 36; tools, 35; utensils, 87

Brunhes-Delamarre, M., 423

Buddhist temples, 291, 292

Buildings, multistoried: Americas, 398, 403; Arabia, 341, 345; China, 292; Roman Empire, 221, 223–24

Buttermaking: medieval Europe, 484, 499–500; Mesopotamia, 127, *127*

Byzantine technology, 386, 429; agriculture, 380; art, 376–79; construction, 373–75; hunting and fishing, 380–81; military affairs, 382–83; navigation, 381; public works, 375–76; textiles, 379

C

Cabinetwork: Islam, 357–58; medieval Europe, 550, *521*

Cahen, Claude, 367

Cairo, *342, 343,* 344, 347

Camel: in Americas, 395; breeding and domestication of, 91–92; harness for, 98, *98;* in Islam, 352, 353

Camshaft system, medieval, 455–56

Canal: Americas, 391; China, 303; Egypt, 147, 149; medieval Europe, 549–50, *557;* Mesopotamia, 125, 126, 131; Roman Empire, 226

Cannon: China, *309,* 310; medieval Europe, 517–18, *517*

Canoe, pre-Columbian, 416–17, *416*

Canon of Medicine (Nei-ching) (Huang-ti), 263

Canticle of Canticles, 347

Canvas, use of, 254, 255

Caravan, 129, 152

Caravanserie, 343, 352

Carpetmaking: Assyria, 153; Islam, 360

Carriage: agricultural society, 94–101, 96, 97; China, 263, 264; Islam, 352–53

Castagnol, 282

Catapult: China, 310; Roman Empire, 251

Cathedrals and churches, construction of, 529–37

Caton-Thompson and Gardner, *The Desert Fayum,* 166

Cattle raising: China, 263; Egypt, 145–47; India, 319;

medieval Europe, 475–77; Neolithic era, 63, 71; Roman Empire, 246

Cave dwellings, 50–58, *52, 54, 57*

Celts: carriage of, 235; enamelware of, 235; use of horseshoe, 236

Cement, Roman, 219

Ceramics, *see* Pottery

Cereal grains: Americas, 391, 392; China, 263; Egypt, 145; India, 315; Islam, 341; medieval Europe, 470–72; Mesopotamia, 126–27; primitive culture, 66, 87

Chace dou Serf, La (de Châtillon and de Vergy), 477

Chariot: China, 95, *96,* 96–97, *97,* 263, *293,* 293–94, 300;

India, 326–27; medieval Europe, *519;* Mesopotamia, 129–31, *130;* Ur tombs, *130*

Charrue (plow), 466–69, *468*

Chartres cathedral windows, 448, 498, 508, 537

Chasses du roi Modus, Les, 477, 478

Chatelperronian culture, 19–20, 24; dwellings, 54–55; tools, 27–28; weapons, 41

Cheops, pyramid of, 160

Chia Tan, *Hai-nei hua-yi-t'u* (map), 296

Chia Sze-hsieh, *Ch'i-min yao-shu* (Arts and Sciences Concerning Agriculture, and Necessary to the People), 277

Chia Szu-hsieh, *Agriculture, or Important Facts for the Well-Being of the People (Ch'i-min yao-shu),* 281

Ch'i-min yao-shu (Arts and Sciences Concerning Agriculture, and Necessary to the People) (Chia Sze-hsieh), 277, 281

Ch'in Dynasty, 264; architecture, in, 290

China: agriculture, 280–283, *281, 286;* chemistry, 272–79, *279;* construction, 290–92; dynasties, 262–67; glass, 272–73; lacquer, 275–76; measuring devices, 292–98, *293, 295;* metallurgy, 266–70, *267, 268, 270;* military techniques, 207–10, *309;* mining, 270–72, *271;* paper, printing, 285–87, *284, 286;* pottery, 273–75, *273;* sugar, 276–77; technological development, 262–67, 310–12; textiles, 283–85, *284;* transport, 293–307, *293, 305*

Ching-të-Chën t'ai-lu (The History of the Porcelains of Kingtehchen) (Lan Pu), 275

Chisel, use of: China, *305;* Egypt, 170, 171; *Mesopotamia,* 136; primitive society, *73*

Chi hsiao-shih (Little History of Wine) (Sung Po-jen), 277

Choisy, François, 529, 531

Chou Dynasty, 263; agriculture, 280; alcoholic fermentation, 277; clocks, 292–93, *293;* mapmaking, 296, *297*

Choukoutien cave, 19, 29, 38, 52

Chronometer, *293*

Chu Yi-chung, *Pei-shan chiu-ching* (Wine Classic), 277

Chu Yü, *P'ing-chou k'u-t'an* (on artificial magnet), 294

Cicero, *Pro Rabirio Posthumo* (on glass), 233

Cistercian Order, 559–62

City as cultural center: China, 263; India, 320; Islam, 342–44, 350–51; medieval Europe, 544; Neolithic age, 69, 71; Roman Empire, 217, 544

City planning in: China, 291; Greece, 188; medieval Europe, 543–51; Mesopotamia, 121; Roman Empire, 217, 218, 240

City walls and gates in: Islam, 343; medieval Europe, 544; Rome, 225

Clark, John G. D., 87

Clay, uses of: brick, 72, 116–17; pottery, 394; utensils, 72

Clepsydra, 292–93

Climate, technological influence of: Egypt, 145; Islam, 347; medieval Europe, 425, 460; primitive society, 61

Clock, 293, *293,* 568

Clothing: India, 329–30; medieval Europe, 513–14; Mesopotamia, 132; Mousterian era, 36–40, 49–50; Neolithic age, 71; prehistory, 36–40; Roman Gaul, 254

Clothmaking, medieval technology of, 506–11

Coal, 486–87, 488–89

Codex, *see* Books

Coinage: China, 263; Greece, 211–12

Colonization, 61, 62

Colosseum, 224

Columns: Americas, 401–03; Greece, 183; India, 327–28; medieval Europe, 532, *534;* Mesopotamia, 118–19, *124,* 124–35

Communications: Americas, 416; China, 311–12; Rome, 242

Compass: Byzantium, 383, *383;* China, 293–94, 445; medieval Europe, 445

Construction: Americas, 396–

401, *397, 398, 400;* Byzantium, 373–75, *373, 386;* Egypt, 144, 153–54, 156–66; Greece, 182–86, *182, 183, 184;* India, 321–23, *322;* Islam, 344–47, *346;* medieval Europe, 525–37, *533, 534;* Mesopotamia, 116–25, *118, 124;* Roman Empire, 217–24

Cooking: China, 262; India, 320; medieval Europe, 528; Mesopotamia, 129; primitive societies, 19–20

Copper: Americas, 404, 405; Egypt, 170; Far East, 268; India, 329; medieval Europe, 485–86, 496; Mesopotamia, 133, *133;* mines, 485–86; Neolithic era, 69

Coral fishing, 363

Córdoba, 361, 364, 366

Corn, birth of, 77, *79*

Cosmetics: Byzantium, 384; China, 270; Islam, 361

Cotton: Americas, 410, 411; Egypt, 168; Far East, 283; Greece, 206; India, 329, *330, 331;* medieval Europe, 512–13; Roman Empire, 254

Crafts: Arabia, 355–56; Egypt, 357; Greece, 198; India, 327; medieval Europe, 427, 502–03, 520–21

Crane: Greece, 192–93, *192, 193;* medieval, *449,* 450

Crank, 107, 191

Crank-and-connecting-rod, 446–47

Crescenzi, Pietro, 570; *Opus ruralium commodorum,* 461

Crete, 187, 196

Cro-Magnon man, 13, 19, 30

Crops: Americas, 389, *91;* Greece, 209; India, 318–19; medieval Europe, 470–71

Crop rotation: agricultural society, 85; China, 282; medieval Europe, 463–64, 474; Roman Empire, 246, 263

Crossbow: Arabia. 368, *369;* China, 308, *309;* medieval Europe, 477, 514–15

Cupola, vaulted, 221

Cutting tools, evolution of, 21–27, *22, 23, 24, 25, 33,* 33–34

Cylindrical roller, 160

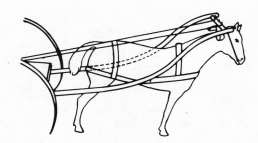

D

Damascene: Byzantium, 378, 386; Islam, 348

David, Paul, 264

da Vigevano, Guido, 422, 518, 519, 570

da Vinci, Leonardo, 549

De bello Gallico (Caesar), 242

de Clavijo, González, 310

Decoration: Egypt, 164; Greece, 203–05; India, 332; Islam, 347; Mesopotamia, 119, 123

Deffontaines, P., 433

de Honnecourt, Villard, 423, 449, 457, 502, 514, 515, 536, 568, 570

Delphi, 183

de Maricourt, Pierre, 571

de Mely, 275

Demievilel, P., *Bulletin de l'École français d'Extrême-Orient,* 292

de Millinate, Walter, *De officiis regum,* 517

De officiis regum (de Millinate), 517

Desert Fayum, The (Caton-Thompson and Gardner), 166

des Nouettes, Lefebvre, 131, 300, 435

Desroches-Noblecourt, 178

De Vegetalibus (Albert the Great), 461

Devices, lifting, in: Americas, 399; Egypt, 160, 161, *164;* Greece, 192–93; India, 323–26, *324, 325;* Islam, 337; medieval Europe, 432, *433,* 449–50, 454

De Villis capitulary, 469

Diderot, *Encyclopédie,* 448

Diet: aboriginal man, 45; Americas, 338, 390, 391; Islam, 348–49; medieval Europe, 470–71, 477, 478, 482; Neanderthal man, 49; Paleolithic man, 47

Digging stick, 82, *83,* 391, 392, *392*

Dikes: China, 303; Egypt, 149; Islam, 340; medieval Europe, 547–48, 553, 554

Diodorus, 147–49, 172

Dioscorides, 276

Distillation, 496–97, *497*

Domesday Book, 451

Domestication of animals: agricultural society, 90–92; Americas, 394; Egypt, 145–47; Islam, 352; Mesopotamia, 129

Doors: Americas, 401; Byzantium, 375; Greece, 186; medieval Europe, 521

Drainage: ancient times, 110; Greece, 196; Islam, 340; medieval Europe, 548–49; Roman Empire, 226

Drills: China, 264; Egypt, 154, *154,* 156; medieval Europe, 414, 502; Roman Empire, 242–43, *243*

Drioton and Vandier, *Les peuples de l'Orient méditerranéen,* Vol. II, *L'Égypte,* 143

Driving shaft, 446, 447

Driving wheel, 107–08

Dromedary, *see* Camel

Dwellings: Americas, 397–98, *397, 398;* Byzantium, 373, 374; China, 290; Egypt, 144, 153, *175,* 341; India, 320, 323; Islam, 341–42; medieval Europe, 525–28, 537; Mesolithic era, 56–58; Mousterian era, 50–53, *52;* Upper Paleolithic, 53–56, *54, 55*

Dyes: Americas, 411; China, 264; Egypt, 168; medieval Europe, 497–98; Mesopotamia, 140

Dynasties: Chinese, 263–67; Egyptian, 142–43

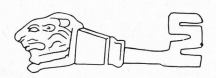

E

Economic and political system, importance of metal in, 74–75, 390

Egypt, 142–43, 174; agriculture, 144–45, *146;* animals, 145–46; architecture, 153–58, *155, 157;* cattle raising, 145; construction, 153–66; diet, 145; dynasties, 142–43; geography, 143–44; glass, paint, pottery, 169–70; graphic arts, 173; hunting and fishing, 147; irrigation, 147–49; lifting devices, 160, *161,* 164; metalworking, 170–73; monuments, 154–56, *155,* 158, *158,* 160–65, *163, 165;* roads, 152; textiles, 166–68; tools, 143, 144, 145, 154–56, *154;* transport, 158–59, *158;* woodworking, 166

Elamites, 115, 116, 124, 134

Enamelwork: Byzantium, 378; Egypt, 235; Greece, 206, 378; Mesopotamia, 139; Roman Empire, 235

Encaustic: Byzantium, 378; Egypt, 170; Greece, 208

Encyclopédie (Diderot), 448

Entstchung der Pyramide, Die (Borchardt), 104

Eskimo: clothing, 50; diet, 45, 48; dwelling, 33; metals, 74, 390; transport, 391, 416, *418*

Eupalinos of Megara, 195

Evolution, early technical, 68–75

Explosives: China, 277; Islam, 370; medieval Europe, 501

F

Faïence: Egypt, 169; Islam, 347, 364

Falconry in: Gallo-Roman era, 478; medieval Europe, 478; Mesopotamia, 128

Fallowing, 85; in Americas, 391; in China, 264, 280; medieval Europe, 462; Roman era, 246

Family in early society, 69, 71

Faraday, Michael, 6

Featherworking, 407–08

Fertilization of soil: Americas, 391; China, 280, 282; Egypt, 145; India, 318; Islam, 347; medieval Europe, 464–65; Roman Empire, 246

Fibula, 253–54, *254*

Figùrines, in Egypt, 169

Fire, in early societies, 18–19, 20, 73; in land clearance, 81, 263, 389, 391, 463

Fireplace, 527

Fishing: Byzantium, 381; Egypt, 147, *148;* India, 319; Islam, 356; medieval Europe, 478–82, *479, 480,* 484; Mesopotamia, *133*

Flake tools, 22–23, *23, 25,* 26, *26,* 27–30, 33

Flamethrowers, 382

Flax: Byzantium, 379; Egypt, 168; Far East, 283; medieval Europe, 512–13; Mesopotamia, 132

Fleta, 461, 470, 475, 482

Flood control: China, 303; Euphrates and Tigres valleys; medieval Europe, 548–49, 550; Nile Valley, 144, 145, 149

Food gathering, 45–47, *46,* 64, 77, 88

Food preservation: Americas, 392; Far East, 280; medieval Europe, 472, 482, 484; primitive society, 69, 71, 86–87

Food supplies: Egypt, 145; India, 318, 320; medieval Europe, 482–84; Mesopotamia, 127–29; primitive society, 47–48, 87

Forbes, R. J., 284

Forests, medieval: economic importance of, 472–73; exploitation of, 473; rights of usufruct in, 453, 473, 474

Forge: Chinese, 264; Gallo-Roman, *253;* Greek, *203*

Fortifications: China, 308; Islam, 351–52; medieval Europe, 540–43, *541;* Mesopotamia, 119, *119*

Framework, wood: Americas, 399; medieval, 525–27; Roman Empire, 221–22

Frescoes: Byzantine, 76; Greek, 208; Mesopotamian, 139

Froissart, 518

Frontinus, 226, 543

Fuchs, *The Mongol Atlas of China,* 298

Fuegians: clothing, 49; culture, 19; food and implements, 48; homes, 51

Fuel: China, 264, 265; Islam, 343; medieval Europe, 488; prehistory, 20; Roman Empire, 228

Furniture: Americas, 390; Byzantium, 375; China, 293; Egypt, 166; India, 327; Islam, 347; Mesopotamia, 133; Roman Empire, 255

G

Gardens: China, 282; Meso-america, 392; Mesopotamia, 125–26

Garland, John, 515

Gaul: enamelware, 235; glass-making, 489; invention of plow, 245; pottery, 231, *232*, 233, *233;* ships, 236; tinplat-ing, 253; use of drill, 242–43

Gear train, development of, 108, 191; use in China, 264; in medieval Europe, 452–53; in Roman times, 108–09

Geoponica, 380

Gilbert, William, 6

Glass, painting on, 490, *490*

Glass, stained, 499

Glass, window: Byzantium, 375, 378; Roman Empire, 221, 233

Glassblowing, 205–06

Glassmaking: Byzantium, 378; China, 272–73; Egypt, 169–70; medieval Europe, 489–70;

490; Roman Empire, 233, *233, 234,* 235, 272

Glaze: China, 273, 274; Egypt, 169

Gnomon, 294–95, *295*

Gold: Americas, 404–07, *406;* Byzantium, 378; Egypt, 171–73, *173;* Islam, 362; Roman Empire, 253

Golden Prairies (al-Mas 'udi), 458

Goldworking: Americas, 404–07, *406;* Egypt, 172–73; Mesopotamia, 134

Grafting of trees, 282

Grains, *see* Cereal grains, Wheat

Grand Canal of China, 264, 265, 303

Granet, M., 311

Grape culture, 466

Graver, 30–31, *32,* 35, *35,* 170

Great Pyramids (Egypt), 144, 160, *161, 162, 163,* 164,

164

Great Wall of China, 264, 266, 290

Greece, 181–82, 213; agricul-ture, 208–11, *204, 209, 210;* architecture, 182–86, *182, 183, 184,* 185; coinage, 211–12; dwellings, 186; fishing, 381; glass and enamelwork, 205–06; hydraulic power, 193–96, *194, 195;* measuring devices, 211; mechanisms, 191–93, *192, 193;* military machines, 196–98, *197, 198;* navy, 189, *190;* painting, 208; papyrus and parchment, 212–13; pottery, 200–05, *201, 203;* textiles, 206, *207;* wood-working, 207

Greek fire, 370, 382, *382,* 501

Groslier, G., 306

Grosseteste, Robert, *Rules,* 461

Grousset, R., 274

H

Hai-nei hua-yi-t'u (Map of China and the Barbarian Countries Within the Seas) (Chia Tan), 296

Hammer, hydraulic, 456–57

Han Dynasty, 262; agriculture, 280–81; cavalry, 308; glass, 273; harness, *300;* ink, 285, 287; looms, 295; neo-Con-fucianism, 311; pottery, 273; technological stimulus, 310; water clock, 293

Hand-mill: agricultural socie-ties, 103, 104; Greece, 249; India, 341; medieval Europe, *447,* 461

Harada, Antonius, *Contemptus Mundi,* 285

Harness: Byzantium, 380, *380;* China, 264, 300, *300, 301;* Egypt, 340; Greece, 188; medieval Europe, 435–36; Roman Empire, 236, *236, 237*

Harness, breast, 98–99, 435; "Byzantine," 97, *98;* dog and

reindeer, 95, *95,* 96; *douga,* 99–100, *100;* draft-animal, 93, 94–95, *96,* 97, 98, *98,* 300, 435–36; horse collar, 98, 100, 131, 300; human trac-tion, 95; millstone, 102; Nea-politan, *99;* shaft, 300; shoul-der, 435; types of, 92–101

Harpoon: Egypt, 147; Eskimo, 42; Magdelian era, 44, *44;* Mesopotamia, *136;* medieval Europe, 480; prehistory, 31

Harrowing, 85, 281, 470

Harvester, 245

Harvesting: agricultural societies, 85–86; Americas, 393; Egypt, 146; medieval Europe, 465, *465;* Mesolithic age, 65

Hatchet, 31, *32, 35*

Health care, 333–34

Hearth: Mouserian, 52–53; Paleolithic, 48, *48;* prehistorical, 19–20

Heating: Greece, 186; medieval Europe, 527; Mesopotamia, 117; Roman Empire, 228, 527

Hellenistic era: achievements in, 213; glass and pottery in, 205; influence on Rome, 216, 219; lifting devices in, 249; scientific contributions of, 256–57

Hemp: Far East, 283; in medieval history, 512–13

Hermann the Lame, 444

Hero of Alexandria, 191–92, 193, 455

Herodotus, 126, 147–49, 168

Herrad of Landsberg, 450, 454; *Hortus deliciarum,* 432

Hesiod, 84, 85

Hides, preparation and use of, 500–01, 513

Hipparchus of Nicaea, 191

History of the Berbers (Ibn Khaldun), 337

History of the Holy Grail, 432

History of the Holy War (Ambrose), 458

History of the Sui Dynasty, The, 276

History of the T'ang Dynasty, The, 277

Hittites: ironworking, 136; jewelry, 136, sandals, 133; use of stone and wood, 124

Hoe, copper and flint, *73;* use in Americas, 389, 392; China, 268; primitive areas, 82; wood, 82, *83*

Hogitaro, Inada, *The Sword Book in Honchō Gunkikō,* 269

Hooker, A. H., 171

Horse: 394; ascendancy of, 101; first appearance of, 88; Greece, 188–89; Islam, 352; Mesopotamia, 129–30; use as draft animal, 102

Horseshoe: Byzantium, 380; China, 300, 435; Gallo-Roman period, 236; Islam, 353; medieval Europe, 435

Horticulture in China, 280, 282–83

Hortus deliciarum (Herrad of Landsberg), 432

Housebondrie, 477

Hrabanus Maurus, 432, 487, 490

Hsien-ch'ing ngou-ki, 267

Hsin-yi-hsiang-fa-yao (Su Sung), 296

Hsu Kuang-sh'i, *Nung-cheng ch'iuan-shu,* 282

Hsui-shih-lu (Huang Ch'êng), 276

Huan K'uan, *Yen-t'ieh lüen* (Speech on Salt and Iron), 271

Huang Ch'êng, *Hsui-shih-lu,* 276

Huang-ti, *Canon of Medicine (Nei-ching),* 263

Hubert of Barcelona, 444

Hunting: Acheulian era, 38–45; Byzantium, 380–81, *386;* Egypt, *147;* India, 319; medieval Europe, 477–78, *479;* Mesopotamia, 128, *128;* protopastoral period, 63

Hunters, prehistoric, 38–45

Husbandry (de Henley), 461

Huygens, Christiaan, 5

Hydraulic tools, medieval, 455, 456, *457*

Hydraulics: ancient times, 106, 107, 109–10; China, 303; Egypt, 147–49; Greece, 192–94, *194,* 195; Islam, 337; medieval Europe, *452, 457,* 547–57; Roman Empire, 249–50

Hypocaust, medieval, 527; Roman, 228, *228,* 229

I

ibn-Batuta, 353, 363

ibn-Khaldun, 344, 347; *History of the Berbers,* 337

Idrisi, 363

Inca culture: agriculture, 391–92, *393;* building construction, 399–403; metalworking, 404–05; pottery, 394, 395; spinning, 411; weaving, *411*

Incendiary machines: China, 308–09; Islam, 370

India: agriculture, 318–19; buildings, 320–24; *322;* esthetic crafts, 331–32; food preparation, 320; hunting and fishing, 315; information materials, 315, 316–17; irrigation, 323–26, *323, 324;* metal-lurgy, 328; physical and mental discipline, 332–34; textiles, 329–31, *330*

Indus Valley: cities, 320; heating, 228; irrigation, 323; sewage, 196

Influences orientales et l'éveil économique de l'Occident (Lopez), 354

J

K

Karnak, Temple of, 165, *165*
Kashmir, agriculture in, 320
Keng-cheh-t'u (Album of Agriculture and Weaving), 285
Khanikoff, *Memoire sur la partie méridionale de l'Asie central,* 337

Kiln: Americas, 396, China, 273; Greece, 199, 200, *200;* medieval Europe, 492; Roman times, 231–33
Kitchen, medieval, 528
Kite, 307
Knife, bone and flint, *26,* 27, *37*
Korea, typography, 288
Koukoules, P., 381
Ku-chin, 289
Ku Hung, *Pao-p'u-chü,* 279
Ku Tsung-shih, *Pen-ts'ao yen-yi,* 294

L

Labor, division of, in family group, 69–70; among hunters and food-gatherers, 45–46
Labor force: China, 303; Egypt, 356, 357; Greece, 181; Islam, 340, 352, 357, 371; Neolithic period, 68; Roman Empire, 258
Lacquer, technology of making, 275–76
Lake cities, 68
Lan Pu, *Ching-të-Chën t'ai-lu,* 275
Land clearance, in agricultural societies, 81–82; in medieval Europe, 463
Language differentials, 58–59, 311–12, 423
Lantern gear, 108, 446, 453
Laplanders, clothing of, 50; pastoralism of, 63
La Roerre, Guilleux, 307
Lathe, 448, *448, 449;* Greek, 207; medieval, 455
Lauer, *Le problème des pyramides d'Égypte,* 164
Laufer, B., 307

Lead, in China, 270–71
Lead mining, medieval techniques of, 485–86
Lemerle, P., 373
Leo VI, 383
Leroi-Gourhan, André, 6, 84, 85, 87
Lever: Egypt, 160, 164; Greece and Rome, 249; medieval Europe, 446; primitive societies, 105
Levey, Martin, 139
Li, H. L., 283
Li Shih-chén, *Materia medica,* 271
Li yu, *Hsien-ch'ing ngou-ki,* 267
Libraries of China, 289, 298
Libro de ratione ponderis (Memorarius), 571
Lifting devices, *see* Devices, lifting
Lighthouses: Byzantium, 383; Egypt, 191; medieval Europe, 544–55; Roman Empire, 242
Lighting: Byzantium, 375;

Egypt, 144; India, 327; Islam, 349, *350;* medieval Europe, 483, 500; Mesopotamia, 117; Paleolithic era, 56
Lion-Goldschmidt, D., 265, 274
Liutprand, 384
Livestock: in Americas, 394; economic importance of, in agricultural societies, 88; in medieval Europe, 475–77; in Mesolithic era, 63
Locks, ship: Byzantium, 375, *375;* Greece, 186–87, *186;* Islam, 347–48, *348;* Roman Empire, 243
Loom: Americas, 411, *411, 412, 413;* China, 283–85; Egypt, 168, *168;* Greece, 206, 207; India, 330; medieval Europe, 508–09, 512; Mesopotamia, 132, *132;* Persia, 360
Lopez, Robert, *Les influences orientales et l'éveil économique de l'Occident,* 354
Lucas, A., 169
Luen Heng (Wang-Chung), 294
Luttrell Psalter, 435, 465, 478

588

M

Machiavelli, Niccolò, 312
Magdalenian era, technology of: clothing, 50; dwellings, 53, 56; fishing, 41; food, 48; harpoon, 44, *44*; javelin, 41–43, *42, 43, 44;* lamps, 48, 56; preparation of hides and skins, 36; spearheads, *42,* 44; tools, 27–28, 31, 33, 34, 36; utensils, 49
Maglemosian era, technology of: dwellings, 56; hatchet, 35; hunting, 31, 44; tools, *62;* utensils, 49; woodworking, 33
Magnetic needle, use of, 294, 445
Magnetism, knowledge of, 293–94
Magnifying glass, 233
Maize, 389, 391, 483
Mangonel, 516–17
Manual ability and technological progress, 13–15
Mapmaking, 296–98, *297*
Marçais, William, 356; *L'Islamisme et la vie urbaine,* 342
Marcellinus Ammianus, 251
Marcus the Greek, 373
Markets, 341, 342–43, 360, 371
Masonry: Americas, 400; Egypt, 156, 157; Greece, 182–83; medieval Europe, 523–24; Roman Empire, 219, *221*
Maspero, H., 264
Massignon, 360
Materia medica (Li Shih-chen), 271
Mathematics: Byzantium, 373; Mesopotamia, 140; medieval Europe, 569–70
Mayan culture, technology in: agriculture, 392; architecture, *399,* 401, 403, *403;* mosaic, 407–08; pottery, 496; roads, 416; stonecutting, 414
Measuring devices: Byzantium, 383; China, 293–98, *295;* Greece, 212; Islam, 361–62; medieval Europe, 566–68, *567;* Roman Empire, 247; "Roman scale," 247, 362, *362*
Mechanica, 191
Medicine, veterinary, 475
Mémoire sur la partie méridionale de l'Asie central (Khanikoff), 337
Memorarius, Jordanus, *Libro de ratione ponderis,* 571
Mercier, Maurice, 370, 382
Mesoamerica, technological evolution in, 388–90; technol-

ogy of agriculture, 391; architecture, 398–99, *400,* 401–03, *403;* basketry, 413; ceramics, 394–95, 396; construction, 399–400, *400,* 401–03; dwellings, 397; featherwork, 407; metallurgy, 404–07, *406;* mosaics, 407–08; paper, 408; stoneworking, 413; tools, 415; transport, 417–18; weaving, 410, *412*
Mesolithic culture, 60–63, 75; technology in agriculture, 65–66; hunting, 44; tools, 24, 34, 35; utensils, 49
Mesopotamia, 114–16, *114;* technology of agriculture, *126,* 135–37, *136;* architecture, 118–25, *118, 119, 121, 123, 124;* chemical products, 139–40; clothing, 132–33; construction, 116–18; dwellings, 116, 117; food, 127–28, *127, 128;* furniture and music, 133; metallurgy, 133–37; pottery, 137–38, *137;* transport, *120,* 129–31, *130, 131, 132;* water supplies, 125; weaponry, *134,* 136–37, *136;* writing, 138–39, *138*
Mexico, *see* Mesoamerica
Microlith, 44
Middle Ages: agriculture, 460–74, *467, 468, 469, 470;* architecture and construction, 528–43, *531, 533, 534, 541;* cattle raising, 475–77; chemical processes, 497–501; city planning, 543–47; dwellings, 525–28; food, 482–84; glass and pottery, 488–97, *490;* hunting and fishing, 477–82, *479, 480;* hydraulics, 547–51; joinery, 520–22, *521;* machines, 446–60, *447, 448, 449, 452, 455, 457, 459;* metals, 493–96, *494;* mining, 484–87; textiles, 505–14, *507;* tools, 501–03, *504;* transport, 431–45, *439, 440, 441;* weaponry, 515–18, *515, 517, 519*
Middle Kingdom (Egypt): bronze, 171; looms, 168; stoneworking, 156; walls, 154
Military machines: Byzantium, 383; China, 308–10; Greece, 196–98, *197, 198;* Islam, 368–70, *369;* Roman Empire, 251
Mill, development of, 101–05, *102, 103, 104, 105*

Mill, hydraulic: China, *281;* development of, *106,* 107–10, *108;* floating, 454; Greece, 210–11, *210;* Islam, 341; medieval Europe, 451–56, *452, 453;* tide, 454; Roman times, 249–51
Mill wheel, 107–08, 451–53, *545–55;* harness for, 102
Ming Dynasty, technological evolution in, 265–66; architecture, 291, 292; books, 285; bronze, 267; fortifications, 308; lacquer, 292; porcelain, 274–75; typography, 288
Mining: China, 271–72, *271;* Egypt, 172; Greece, 198–99, *199;* Islam, 362–63; medieval Europe, 484–88; Mesopotamia, 137
Mirror-making: India, 323; Rome, 233
Mirrors, bronze, in China, 267; in India, *328*
Mo-fa chi-yao (Shen Chi-sun), 288
Mogul period, monuments in, 321, 323
Monastery, influence of, on architecture, 539–40; *see also Cistercian Order*
Monastery of the Great Bell, 266, 267
Monetary system: China, 264; Greece, 211–12
Mongol Atlas of China, The (Fuchs), 298
Mongoloid in Americas, 388, 389
Mongols: astronomical instruments, 296; mapmaking, 298
Montaigne, 289
Montandon, 307
Montet, Pierre, 168
Mosaics: Byzantine, 376, 377; pre-Columbian, 407; Roman, 229–31, *230*
Mosques, 342–43, 344, 349
Mousterian people: animal skinning, 37, *37;* clothing, 49–50; dwellings, 51–52; hunting, 39; tools, 36; weapons, 33, 34
Mumification, 178–79
Mung-ch'i pi-t'an (Essays Written at the Villa "Torrent of the Dream") (Shen Kua), 294
Musical instruments: Byzantium, 384, 386; India, 331–32; Mesopotamia, 133
Mustard Seed Garden, 282

N

Natural History, **XVII** (Pliny the Elder), 245

Navigation: Byzantium, 381, 386; China, 304, 307; Greece, 189–91; Islam, 354, 356; Middle Ages, 431, 436, 444–45

Neanderthal man: appearance of, 49; implements of, 30, *32, 33, 35;* use of fire, 19

Near East: agricultural and pastoral life, 65–66; domestication of animals, 90; farming, 72; food, 77, 87; invention of mill, 104

Needham, Joseph, 264, 269, 271, 304, 310, 312

Neo-Confucianism, 310–11

Neolithic culture: agrarian economy, 65–66, *65*

New Kingdom (Egypt): cereal products, 145; copper, use of, 170, 171; horse, appearance of, 145; monuments, 164; mumification, 179

Ngao Tung Yang lien ti t'u (large-scale map), 298

Nile River Valley: agriculture, 145; building materials, 144, 153; economic importance, 144, 147; flooding, 145; papyrus and paper, 176; shipping, 354

Nilometer, 149–52, *150*

Nineveh: aqueduct, 125; gold-working, 134; painting, 139; palace, 118, *118, 123*

Nomadism, in agricultural economy, 66; Arabian, 356, 429; in Americas, 388

Noria, use of, in agrarian society, 105, 106, 107; China, 281; Egypt, 337, *338, 339; 455;* Roman Empire, 249

Nung-cheng ch'iuan-shu (agricultural treatise) (Hsu Kuang-ch'i), 282

Nung-sang chi-yao (agricultural treatise) (Wang Chen), 282

Nung-shu (agricultural treatise) (Wang Chen), 282

O

Oar, use of, Byzantium, 381; China, 306–07; Greece, 189; Mesopotamia, 131; Roman Empire, 239, *239;* Scandinavia, 437, 438

Obelisk, Egyptian: construction of, 154, *155,* 158, *158;* raising, 160, *161*

Old Kingdom (Egypt): agriculture, 144–45; construction, 144, 153; metalworking, 170; pyramids, 162–64, *163*

Optical telegraph, 383

Opus ruralium commodorum (Crescenzi), 461

Orientalische Steinbücher und Persische Fayencetechnik, 366

P

U

V

Vases, attic, 202
Vault: Americas, 401; Byzantium, 373; Egypt, 157, *157;* Greece, 185–86, *185;* medieval Europe, 521–22, *522,* 529–36, *533, 534;* Roman Empire, 221–22

Vegetius, 251
Vehicles, shaft in: agricultural societies, 94–100; China, *299, 300;* medieval Europe, 432–35, *434*
Vernier, 171
Vie de Saint Denis, 436, 545

Vincent of Beauvais, 490
Viollet-le-Duc, 449, 515
Vita sancti Arnulfi, 449
Vitruvius, 191–92, 249, 251; *De architectura,* 238
Volney, 368
Volta, Alessandro, 6

W

Walls: Americas, 400, *400;* Egypt, 153–54, 156; Greece, 183; India, 323; Islam, 347; medieval Europe, 532; Roman Empire, 220
Wang Chen, *Nung-sang chi-yao,* 282; *Nung-shu,* 282
Wang Cheng, *Yüan-si ch'i-ch'i t'u-shuo-lu-tsui,* 267
Wang-Chung, *Luen Heng,* 294
Wang Shao, *T'ang-shuang-p'u,* 277
Water clock: China, 263, 292–93; Islam, 340, 360; medieval Europe, 568
Water closet: Byzantium, 375; Islam, 347; Mesopotamia, 117
Water system: Americas, 399; China, 303; Egypt, 152; Greece, 195–96; India, 323; Islam, 344; medieval Europe,

528, 546–47; Mesopotamia, 125
Waterways, European, 436, 446
Weaponry: Byzantium, 282–83, *382;* China, 307–10; Greece, 196–97, *197, 198;* Islam, 366–71, 367, 369; medieval Europe, 514–19, *515, 517, 519;* Roman Empire, 251
Weaving: Americas, 410–13; Byzantium, 379–80; Egypt, 166; Greece, 206, *207;* India, *330,* 331; medieval Europe, 508–13, *509;* Mesopotamia, 132
Wen Yü, 272
Wen-yüan-ku, 289
Wharf: medieval, 553; Roman, 240
Wheat, *see* Agriculture *and* Cereal grains
Wheel, development of, as driv-

ing power, 106–11, *106, 108, 109, 110*
Wheelbarrow: Chinese, 300; medieval, 432, *433*
Wilson, E. H., 282
Windlass: China, 271; Greece, 209, 249; Roman Empire, 249
Windmill: agricultural societies, 110, *111;* China, 272; Iran, 458; medieval Europe, 458–59, *459*
Woodworking: Americas, 414–16; Egypt, 166; Islam, 357–58; medieval Europe, 502, 520–23; prehistoric cultures, 33–35
Writing tools: Byzantium, 379; China, 263–65; Egypt, 289; Mesopotamia, 115, 138
Wu-ching chïng-yao (Principles of Military Science) (Tseng Chüng-liang), 279, 294

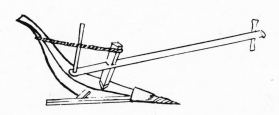

X

Xylography, 264, 265, 287–88, 289

Y

Z

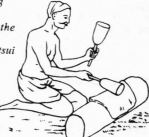